Multifunctional Mesoporous Inorganic Solids

NATO ASI Series

Advanced Science Institutes Series

A Series presenting the results of activities sponsored by the NATO Science Committee, which aims at the dissemination of advanced scientific and technological knowledge, with a view to strengthening links between scientific communities.

The Series is published by an international board of publishers in conjunction with the NATO Scientific Affairs Division

A Life Sciences **B Physics**	Plenum Publishing Corporation London and New York
C Mathematical and Physical Sciences **D Behavioural and Social Sciences** **E Applied Sciences**	Kluwer Academic Publishers Dordrecht, Boston and London
F Computer and Systems Sciences **G Ecological Sciences** **H Cell Biology** **I Global Environmental Change**	Springer-Verlag Berlin, Heidelberg, New York, London, Paris and Tokyo

NATO-PCO-DATA BASE

The electronic index to the NATO ASI Series provides full bibliographical references (with keywords and/or abstracts) to more than 30000 contributions from international scientists published in all sections of the NATO ASI Series.
Access to the NATO-PCO-DATA BASE is possible in two ways:

– via online FILE 128 (NATO-PCO-DATA BASE) hosted by ESRIN, Via Galileo Galilei, I-00044 Frascati, Italy.

– via CD-ROM "NATO-PCO-DATA BASE" with user-friendly retrieval software in English, French and German (© WTV GmbH and DATAWARE Technologies Inc. 1989).

The CD-ROM can be ordered through any member of the Board of Publishers or through NATO-PCO, Overijse, Belgium.

Series C: Mathematical and Physical Sciences - Vol. 400

Multifunctional Mesoporous Inorganic Solids

edited by

César A.C. Sequeira

Instituto Superior Técnico,
Universidade Técnica de Lisboa,
Lisboa, Portugal

and

Michael J. Hudson

Department of Chemistry,
University of Reading,
Reading, U.K.

Kluwer Academic Publishers

Dordrecht / Boston / London

Published in cooperation with NATO Scientific Affairs Division

Proceedings of the NATO Advanced Study Institute on
Multifunctional Mesoporous Inorganic Solids
Sintra, Portugal
April 5–17, 1992

A C.I.P. Catalogue record for this book is available from the Library of Congress.

ISBN 0-7923-2289-4

Published by Kluwer Academic Publishers,
P.O. Box 17, 3300 AA Dordrecht, The Netherlands.

Kluwer Academic Publishers incorporates the publishing programmes of
D. Reidel, Martinus Nijhoff, Dr W. Junk and MTP Press.

Sold and distributed in the U.S.A. and Canada
by Kluwer Academic Publishers,
101 Philip Drive, Norwell, MA 02061, U.S.A.

In all other countries, sold and distributed
by Kluwer Academic Publishers Group,
P.O. Box 322, 3300 AH Dordrecht, The Netherlands.

Printed on acid-free paper

Printed in the Netherlands

PREFACE

1. Introduction.

There is much interest in the general subject of porous inorganic materials with respect to their use as sorbents or catalysts. Such inorganic solids may be microporous, mesoporous or macroporous according to the sizes of the pores within the solid. Often there is a range of pore sizes within any given solid and so there is special interest in the synthesis, characterisation and application of porous inorganic solids with well defined pores. Pores of diameter larger than 50 nm are generally termed *macropores*. Those with diameters of less than 2 nm are *micropores* and pores of intermediate size are called *mesopores*. Solids, which contain only mesopores, are correctly called *mesoporous* but very often there is a combination of different types of porosities within one given solid. The synthesis, characterisation and application of microporous solids is much more advanced than is the case with mesoporous substances. Moreover, the synthesis of crystalline mesoporous materials is one clear goal for the future but which has not been attained so far. Consequently, it is of interest to examine the current state of our knowledge of microporous materials and to examine how this may apply to mesoporous materials. Both catalytic and sorption processes could benefit from studies of mesoporous solids because the mesopores could permit diffusion of larger reactants or products than is the case in microporous materials.

2. Pore Size Distribution

One source of interest in the pore-size distribution for inorganic solids stems from the need to predict the effective diffusivity in a porous solid in relation to the ease of access of reactant molecules to the interior of a sorbent. Another source stems from the need to relate sorptive or catalytic properties to pore-sizes and surface areas. The first part of this book deals with the theory and practice of some aspects of the measurements of pore-size distributions. Many sorbents or catalysts such as inorganic oxides have bimodal pore-size distribution curves. The fine pores are frequently in the range of 1 to 10 nm. That is to say the oxides have micropores and mesopores, which account for most of the surface area. The large pores are macropores typically in the range of 10^2 to 10^3 nm. The mixed porous nature of such solids is sometimes useful as the macropores allow mass transfer into the reaction sites within the micro- and mesopores. The fine pores are considered to be within the particles whereas the larger pores are the residual spaces between the particles. The fine pores may be developed within the particles by calcination and/or reduction procedures. However, there is always clear need to attempt to improve the synthesis and methods of characterisation of mesoporous materials.

Clearly, mesopores may be formed when the particles are abutting one another but it is important to consider mesoporous materials in which the pores are within structural parts of the particles themselves. In the case of zeolites, which are microporous, there are well defined pores which may be prepared by methods such as the hydrothermal methods discussed later. It is of interest to examine as to whether these and other methods could be extended to mesoporous materials. In layered solids such as oxides or hydrogen phosphates, the layers may be separated by means of pillars of either an organic or inorganic nature. There is interest in establishing whether the distances between the layers and between the pillars themselves could be above 2 nm so that the materials could be mesoporous rather than microporous. Sol-gel methods, which are discussed later, have lead to the synthesis of porous materials with a range of pore sizes and there is now a great interest in such methods of synthesis.

3. Porous Crystalline Materials

Zeolites are, by definition, microporous but it is important to understand the mechanisms whereby such crystalline porous materials are formed so that it will be possible to suggest methods by which mesoporous materials are made. There are two alternative views regarding the mechanisms of controlled synthesis. One is the conventional "secondary building unit" model and the other involves the extended structures model. Many related structures based on the same or similar sheets, columns and large cages, cocrystallise or intergrow with one another. These data, based on examining the products of synthesis rather than the precursor gel structures indicate that the process of building the structures comprises two steps. The first involves the formation of an intermediate extended structure and the second step is the clipping or joining of them together in specific ways. The gathering body of high resolution microscopy data indicates ordered and disordered intergrowths and overgrowths of related structures. Consequently, there is interest in seeing whether these and other ideas can be extended to the synthesis of mesoporous materials.

4. Pillared Layered Solids

In pillared layered solids (PLS), there is a pillar between the layers host formed by an organic or inorganic guest molecule or ion. At the moment, these PLS are principally microporous materials but there are clear indications that they can have microporosity combined with mesoporosity and, as reported in this book, there may be pillared layered solids which are solely mesoporous.

5. Sol-gel Methods

Recently, there have been important developments in the synthesis of oxides by the sol-gel route and these new materials promise to be of high purity with well defined pore sizes. Sol-gel chemistry is based on the polymerisation of molecular precursors such as metal oxides of the type $M(OR)_n$. Hydrolysis and condensation of these alkoxides lead to the formation of oxo-polymers which are then transformed into an oxide network. Sol-gel chemistry offers the possibility of designing molecular precursors and to control the polymerisation process so that, in principle, materials of the required porosities and surface areas can be prepared. Thus molecular design and engineering of metal alkoxides could permit the synthesis of mesoporous solids. Monodispersed colloids or long chain polymers can be synthesised via the chemical modification of metal alkoxides by nucleophilic reagents. Organic and inorganic polymerisation reactions can be performed during the sol-gel synthesis leading to the formation of new materials made of an organic-inorganic polymeric network.

Thus, it can be seen that it was timely to have a NATO Advanced Study Institute on the subject of Multifunctional Mesoporous Inorganic Materials (MMIS) in which specialists in the different areas were brought together.

Michael Hudson

ACKNOWLEDGEMENTS

It is a pleasure to acknowledge with gratitude the financial support for the Course on Multifunctional Mesoporous Inorganic Solids by NATO, the Luso-American Foundation, the Portuguese Air Force, the Bank Pinto & Sotto Mayor, and the Tourism Office in Sintra.

We also wish to thank our colleagues in the field for many helpful suggestions on the organization of the Course. In addition, we would like to thank all the lecturers, particularly to my co-director Dr. Michael J. Hudson, the discussion groups chairmen and the participants for making the program so stimulating and successful.

Finally it is necessary to thank all of those people involved in the local organization of the Institute. In particular, a great deal of the administrative load was carried by my wife, Maria Elisa, who helped also with the production of this volume.

César Sequeira

North Atlantic Treaty Organization
ADVANCED STUDY INSTITUTE
on
MULTIFUNCTIONAL MESOPOROUS INORGANIC SOLIDS
Hotel Tivoli Sintra, Sintra, Portugal, April 5-17, 1992

DIRECTORS

Prof. César A C Sequeira
Lab. de Electroquímica
Instituto Superior Técnico
Av. Rovisco Pais
1096 Lisboa Codex
PORTUGAL

Dr. Michael J. Hudson
Dept. of Chemistry
University of Reading
Whiteknights
P.O. Box 224
Reading RG6 2AD, U.K.

SCIENTIFIC COMMITTEE

Prof. Giulio Alberti
Dipartimento di Chimica
Università di Perugia
06100 Perugia
ITALY

Prof. Abraham Clearfield
Texas A&M University
Dept. of Chemistry
College Station
Texas 77843-3255, U.S.A.

Prof. N.H.J. Gangas
Straton hi-tec Ltd
108, Kifissias Ave
GR-115 26 Athens
GREECE

Prof. J.D. Mackenzie
Univ. of California, LA
Dept. of Mat. Sci. & Eng.
405 Hilgard Avenue
LA California 90024-1595
U.S.A.

Prof. Jacques Rozière
URA CNRS 79
Place Eugène Bataillon
34060 Montpellier Cedex
FRANCE

Prof. Shoji Yamanaka
Dept. of Applied Chem.
Hiroshima University
Higashi-Hiroshima 724
JAPAN

North Atlantic Treaty Organization
ADVANCED STUDY INSTITUTE
on
MULTIFUNCTIONAL MESOPOROUS INORGANIC SOLIDS
Hotel Tivoli Sintra, Sintra, Portugal, April 5-17, 1992

LECTURERS

CYPRUS

Prof. Charis R. Theocharis
Dept. of Natural Science
University of Cyprus
P.O. Box 537
Nicosia

FRANCE

Dr. Deborah J. Jones
Univ. des Sci. et Techn. du Languedoc
U.A. 79 CNRS
Place Eugéne Bataillon
34060 Montpellier Cedex

Prof. J. Livage
Univ. Pierre et Marie Curie
URA CNRS 1466
4, place Jussieu (T54-E5)
75252 Paris Cedex 05

Prof. Jean Rouquerol
CNRS - CTM
26 rue du 141éme RIA
13003 Marseille

Prof. Jacques Rozière
Univ. Sci. Techn. Languedoc
URA CNRS 79
Place Eugène Bataillon
34060 Montpellier Cedex

GERMANY

Prof. P. Behrens
Universitat Konstanz
Fakultat fur Chemie
Postfach 55 60
D-7750 Konstanz 1

GREECE

Dr. Dimitris Petridis
"Demokritos" National Centre for Scientific Research
Institute of Materials Science
153 10 Ag. Paraskevi Attikis
P.O.B. 60228

ITALY

Prof. Giulio Alberti
Dipartimento di Chimica
Lab. di Chimica Inorganica
Universitá di Perugia
Via Elce di Sotto, 8
06100 Perugia

JAPAN

Prof. Shoji Yamanaka
Dept. of Applied Chemistry
Faculty of Engineering
Hiroshima University
Higashi-Hiroshima 724

PORTUGAL

Prof. C. A C. Sequeira
Instituto Superior Técnico
Av. Rovisco Pais
1096 Lisboa Codex

SPAIN

Prof. Enrique Rodriguez-Castellón
Departamento de Química Inorgánica
Facultad de Ciencias
Universidad de Málaga
29071 Málaga

UNITED KINGDOM

Prof. Richard M. Barrer
Department of Chemistry
Imperial, College of Science and Technology
South Kensington
London SW7 2AY

Dr. Robert Brown
Catalysis Research Unit
Leeds Polytechnic
Leeds

Dr. M. J. Hudson
University of Reading
Dept. of Chemistry
Whiteknights
P.O. Box 224
Reading Berkshire RG6 2AD

Dr. W. Jones
Department of Chemistry
University of Cambridge
Lensfield Road
Cambridge CB2 1EW

Dr. Ashley North
Physics Laboratory
The University
Canterbury
Kent CT2 7NR

Prof. N. Quirke
Applied Physics Branch
BP Research
Chertsey Road
Sunbury on Thames
Middlesex TW16 7LN

UNITED STATES

Prof. Abraham Clearfield
Texas A & M University
Department of Chemistry
College Station
Texas 77843 -3255

Prof. Thomas J. Pinnavaia
Michigan State University
Center for Fundamental Materials Research
435 Chemistry Building
East Lansing
Michigan 48824

Dr. D.E.W. Vaughan
Exxon Research and Engineering Company
Clinton Township
Route 22 East
Annandale
New Jersey 08801

North Atlantic Treaty Organization
ADVANCED STUDY INSTITUTE
on
MULTIFUNCTIONAL MESOPOROUS INORGANIC SOLIDS
Hotel Tivoli Sintra, Sintra, Portugal, April 5-17, 1992

PARTICIPANT ROSTER

CANADA

Prof. J. B. Moffat
Dept. of Chemistry
University of Waterloo
Waterloo, Ontario
Canada N2L 3GL

FRANCE

Dr. Faiza Bergaya
CNRS - CRMD
1B rue de la Férollerie
45071 Orleans Cedex 2

Mr. Thierry Cassagneau
Lab. des Agrégats Moléculaires et Matériaux Inorganiques
Université Montpellier 2
34095 Montpellier Cedex 5

Dr. Laurent Michot
URA 235 du CNRS, BP 40
54501 Vandoeuvre Cedex

Dr. John Ramsay
CNRS, Institut de Recherche Sur la Catalyse
Villeurbanne Cedex 69626

Dr. Helene Suquet
Univ. Pierre et Marie Curie
URA 1106 - CNRS
Tour 54-55 - 2è étage - 4 place Jussieu
75252 Paris Cedex 05

Miss Virginie Vendange
ONERA - OM
ONERA BP 72
92322 Chatillon Cedex

GERMANY

Prof. A. Neimark
Institut fur Anorgan. Chemie
u. Analytische Chemie
Universitat Mainz
Huxelrebenweg 30
Mainz 42
D-6500

GREECE

Dr. P. Falaras
Inst. of Physical Chemistry
Inorganic Chemistry Lab.
National Centre for Scientific Research "Demokritos"
153 10 AG. Paraskevi Attikis
POB 60228

Mr. Fotis Katsaros
Inst. of Physical Chemistry
Inorganic Chemistry Lab.
National Centre for Scientific Research "Demokritos"
153 10 AG. Paraskevi Attikis
POB 60228

Dr. J. Kovanis
Inst. of Physical Chemistry
Inorganic Chemistry Lab.
National Centre for Scientific Research "Demokritos"
153 10 AG. Paraskevi Attikis
POB 60228

Mr. Nicholas Pantelis
Democritus Nuclear Research Center
7 Tmolou Street
16233 Athens

Prof. Philip Pomonis
Dept. of Chemistry
University of Ioannina
45332 Ioannina

Mr. C. Scordilis
Univ. of Ioannina
Dept. of Chemistry
Section of Industrial and Food Chemistry
453 32 Ioannina

Mr. Stelios P. Skaribas
Univ. of Ioannina
Dept. of Chemistry
Section of Industrial and Food Chemistry
453 32 Ioannina

INDIA

Dr. S. Khaja Masthan
Indian Institute of Chemical Technology
Hyderabad 500 007

ITALY

Prof. Rosario Alello
Università della Calabria
Dipartimento di Chimica
87030 Arcavacata di Rende (CS)

Dr. Carla Ferragina
National Council of Research
Via Salaria km 29300/C.N.R.
00016 Monterotondo (Roma)

Dr. Fabio Marmottini
Dept. of Chemistry
Univ. of Perugia
Via Elce di Sotto, 8
06100 Perugia

Dr. G. Moretti
Centro CNR "SACSO"
Dip. di Chimica "La Sapienza"
00185 Roma

Dr. Riccardo Vivani
Perugia University
Chemistry Dept.
Via Elce di Sotto, 8
06100 Perugia

Dr. Piergiorgio Zappelli
Eniricherche
Via E. Ramarini, 32
00015 Monterotondo, Roma

KUWAIT

Dr. Khalid Habib
Materials Application Department
Kuwait Institute For Scientific Research
P.O. Box 24885
13109 Safat

NETHERLANDS

Dr. J. M. Zueter
University of Twente
P.O. Box 217
7500 AE Enschede

PORTUGAL

Eng. Luís P. S. Araújo
Instituto Superior Técnico
Av. Rovisco Pais
1096 Lisboa Codex

Eng. Paulo S. D. Brito
Instituto Superior Técnico
Av. Rovisco Pais
1096 Lisboa Codex

Dra. Maria Odete Caldeira
Instituto Tecnológico para
a Europa Comunitária
Av. Almirante Reis, 178-r/c
1000 Lisboa

Dra. Ana C. P. Carrasco
Instituto Tecnológico para
a Europa Comunitária
Av. Almirante Reis, 178- r/c
1000 Lisboa

Prof. Anselmo M. Elias
Faculdade de Ciências
Universidade de Lisboa
Departamento de Química
Campo Grande, C1 - 5ª
1700 Lisboa

Enga. Maria Augusta G. Martins
Lab. Central EDP
Rua Cidade de Goa, 4
2685 Sacavém

Dra. Maria Manuela Nunes
Dept. de Química
Faculdade de Ciências
Universidade de Lisboa
Campo Grande, C1 - 5ª
1700 Lisboa

Dr. Adriano M. G. Pacheco
Lab. de Electroquímica
Departamento de Química
Instituto Superior Técnico
Av. Rovisco Pais
1096 Lisboa Codex

Enga. Isabel S. Paiva
Instituto Tecnológico para
a Europa Comunitária
Av. Almirante Reis, 178-r/c
1000 Lisboa

Enga. Maria Joäo Plancha
LNETI
Dept. de Tecnologia de Materiais
Travessa dos Lameiros ao Paço do Lumiar
1699 Lisboa Codex

Dr. Luiz F. F. Rodrigues
Instituto Tecnológico para
a Europa Comunitária
Av. Almirante Reis, 178-r/c
1000 Lisboa

RUSSIA

Prof. V. M. Gryaznov
A.V. Topchiev Institute of Petrochemical Synthesis
6 Miklukho-Maklaya
117198 Moscow

Dr. N. Ye. Mokrousov
Institute of Microelectronics
Universitetskaya, 21
150007 Yaroslavl

Dr. K. L. Padukov
Center of SHS, 164
Leninsky prospect 4,
Moscow 117936

SPAIN

Dr. Mª Dolores Alba Carranza
Univ. of Seville
Dpto Q. Inorgánica
Apdo 874
Sevilla 41012

Dr. Enrique Jaimez Falagan
Oviedo University
Dept. Inorganic Chemistry
Julian Claveria Sin
Oviedo
Asturias 33008

Dr. Vicente Fornes
CSIC/UPV (Universidad Politécnica)
Camino de Vera S/N (ITQ)
Valencia

Dr. Pedro Maireles Torres
Dpto. Química Inorgânica, Crist y Miner.
Universidad de Málaga
Campus de Teatinos
29071 Málaga

Dr. Amparo Mifsud
CSIC/UPV (Universidad Politécnica)
Camino de Vera s/n (ITQ)
46071 Valencia

Dr. Sonia Murcia
CSIC/UPV (Universidad Politécnica)
Camino de Vera s/n (ITQ)
46071 Valencia

TURKEY

Dr. Aysel Atimtay
Middle East Technical Univ.
Faculty of Eng.
Dept. of Environmental Eng.
Inono Bulvari 06531
Ankara

Dr. Gamze Gul Avci
Marmara Research Centre
P.O. Box 21
41470 Gebze-Kocaeli

Dr. Suna Balci
Dept. of Chem. Eng.
Middle East Tech. Univ.
06531 Ankara

Dr. Adil Denizli
Hacettepe University
Chemical Eng. Dept.
06532, Beytepe
Ankara

UNITED KINGDOM

Dr. Christopher Breen
Sheffield City Polytechnic
School of Science
Pond Street
Sheffield S1 1WB

Mr. Xu-Liang Cao
Environment Science Division
Lancaster University
Lancaster LA1 4YQ

Mr. Simon Carino
University of Reading
Chemistry Department
Reading, Berksheire RG6 2AH

Miss Michelle A. Gardner
The University of Kent at Canterbury
Physics Laboratory
Kent CT2 7NR

Miss Winnie Kagunya
University of Cambridge
Dept. of Chemistry
Cambridge CB2 1EW

Mr. Robert Mokaya
University of Cambridge
Chemistry dept.
Cambridge CB2 1EW

Miss Morana Vucelic
University of Cambridge
Dept. of Chemistry
Cambridge CB2 1EW

Mr. A. Workman
University of Reading
Dept. of Chemistry
Reading, Berkshire RG6 2AD

UNITED STATES

Mr. Tarek M. Abdel-Fattah
Northeastern University
Chem. Dept.,
Boston MA 02115

Dr. Joseph Herbst
Mobil R&D Corp.,
Paulsboro
NJ 08066

Prof. Marwan Houalla
University of Pittsburgh, PA
4742 Centre Ave #702
Pittsburgh
PA 15213

Prof. Thomas E. Mallouk
Texas A & M University
Dept. of Chemistry
College Station
TX 77840-3255

Dr. Paul Sylvester
Texas A&M University
Dept. of Chemistry
College Station
TX 77840-3255

CONTENTS

I. THE MEASUREMENT OF POROSITIES

II. POROUS CRYSTALLINE MATERIALS

III. PILLARED LAYERED SOLIDS

V. OTHER METHODS OF CHARACTERISING INORGANIC MATERIALS

VI. SOME APPLICATIONS OF MULTIFUNCTIONAL MESOPOROUS INORGANIC MATERIALS

SECTION ONE

THE MEASUREMENT OF POROSITIES.

THE MEASUREMENT OF MESOPOROSITY

Charis R. Theocharis,
Department of Natural Science, University of Cyprus, P.O. Box 537, Nicosia, Cyprus, formerly at, Porous Solids Group, Brunel University, Uxbridge, Middlesex, UB8 3PH, UK

1. ABSTRACT

This paper discusses two ways of assessing mesoporosity, namely mercury intrusion porosimetry, and from isothermal nitrogen adsorption data. Each technique will be discussed, and the way information obtained will be described. The limitations of each technique, and the range of pore size distributions for which each one is most valid will be discussed. The interpretation of adsorption data will be discussed briefly, together with the various types of hysteresis, and their interpretation. Low pressure hysteresis, and the non-closure of the hysteresis loop will be presented.

2. INTRODUCTION

The ability of porous solids to adsorb large quantities of gaseous or other extraneous matter into their apparent bulk, with or without obvious swelling, has long been known and used[1]. For example, charcoal can adsorb air (gas) and is used medically for that reason, but it is also utilised by the organic chemist in the laboratory as a decolourising agent. A similar material is used in protective equipment used by firemen to protect them from noxious volatiles.

Every chemistry undergraduate has come across a solid called "molecular sieve 4A", which if one is lucky can be used to keep solvents dry. That is an example of a zeolite, a class of porous material of considerable interest, which cannot only adsorb water, but also other polar or -with some persuasion- non-polar molecules, and exchange certain of its ions for others in the environment. This last ability has been used as long ago as the 3rd Century AD by a dishonest merchant in Alexandria to convert worthless stones into counterfeit emeralds[2].

Zeolites and other porous solids have, however, probably their most important application in the field of heterogeneous catalysis[3], where the presence of porosity imparts the solid with a series of desirable properties, such as delayed sintering or selectivity.

The study of solids by adsorption can be said to have been first put on a firm theoretical basis by Langmuir[4], who assumed that gas molecules adsorbed on a surface in such a way as to completely cover it with a layer of one molecular thickness, the so-called monolayer capacity of the solid -when expressed per unit weight-. Brunauer Emmett and Teller[5], then suggested what amounted to an extension of that theory, by assuming that adsorption could go beyond the monolayer stage, to result in multilayer adsorption. This is the BET theory[5], which with its extension the BDDT theory[6] is still used in interpreting adsorption data to yield surface areas, at least in the case of finely divided non-porous solids.

There are two broad types of interaction between the adsorbed molecules -the adsorbate- and the solid -adsorbent-, one physical or of the van der Waals type giving rise

C.A.C. Sequeira and M.J. Hudson (eds.), Multifunctional Mesoporous Inorganic Solids, 3–18.

to physical adsorption or physisorption, and the other chemical, giving rise to chemisorption. For completeness, it should be noted that the adsorbate in the gaseous state is called the adsorptive.

Zeolites are examples of a so-called microporous solid. A pore is defined as being an area of the surface, access to which is through a restricted opening. Porous solids, more strictly speaking pores, can be classified into the following three categories according to the width of the pore[1]:

Micropores of width less than 2.0 nm
Mesopores of width between 2.0 and 50.0 nm
Macropores with widths more than 50 nm

Of interest to the present ASI is the second of these.

The reason for this classification, is because ingress, egress and adsorption in a pore can be different from the same process on an open surface, and because these can be influenced by the width of the opening. As it shall be seen later, the shape of the opening, but also that of the body of the pore, is of importance.

Adsorption isotherms are classified by the IUPAC according to the so-called BDDT scheme (Figure 1). This strictly speaking only refers to types I to V, but a sixth one, VI, has been added. Type VI isotherms are rare. Some isotherms have hysteresis loops, the shape of which imparts information about the porous structure of the solid. These are classified in four types, H1 to H4 (Figure 2). The presence of hysteresis generally implies that capillary condensation has taken place.

3. THE BET THEORY

The quantity of gas adsorbed per unit weight of a solid, can be expressed as:

$$n=f(p,T,gas,solid)$$

If the experiment is carried out under isothermal conditions, the expression becomes

$$n=f(p)_{T,gas,solid}$$

and below the critical temperature of the gas,

$$n=f(p/p^{o})_{T,gas,solid}$$

where p is the pressure, and p^o is the saturation vapour pressure for the adsorptive, at the temperature of the experiment.

In general, the interaction between a solid and an adsorbate is due to various types of forces, and can be expressed as follows:

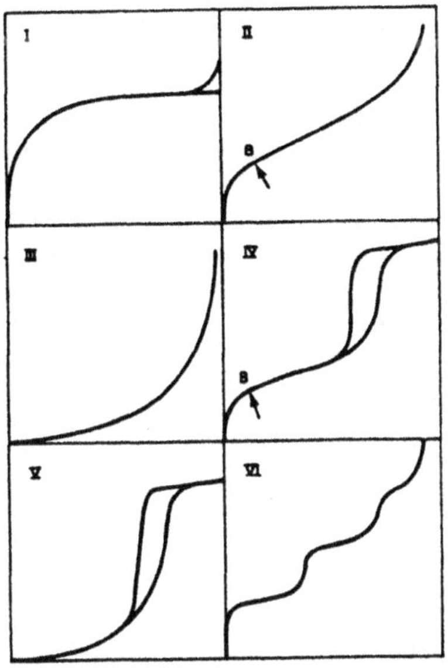

Figure 1 The IUPAC classification of adsorption isotherms

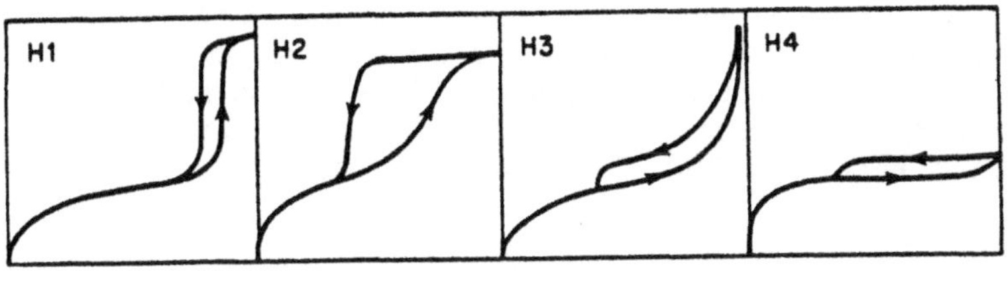

Figure 2 The classification of hysteresis loops

$$\Phi_Z=\Phi_D+\Phi_P+\Phi_{F\mu}+\Phi_{FQ}+\Phi_R$$

where Φ_z is the overall interaction energy for a molecule at distance z from the surface, and Φ_D, Φ_P, $\Phi_{F\mu}$, Φ_{FQ}, and Φ_R are the interaction energies due to dispersion forces, the electric field caused by the polarity of the surface, the electric dipole of the molecule, the interaction of the quadrupole of the molecule with the surface field gradient, and to short term repulsion forces. Of these, Φ_D and Φ_R always obtain and give rise to non-specific interaction, whereas the others are present if the requisite property of the gaseous molecule or the surface is present, and give rise to specific interaction.

The most widely used expression for the adsorption isotherm, as mentioned before, is the BET equation

$$\frac{n}{n_m}=\frac{c(p/p^o)}{(1-p/p^o)(1+(c-1)p/p^o)}$$

It is normally used in its linear form,

$$\frac{p/p^o}{n(1-p/p^o)}=\frac{1}{n_m c}+\frac{c-1}{n_m c}\frac{p}{p^o}$$

where c is a function of the energy of interaction between adsorbent and adsorptive, expressed as

$$c=e^{\frac{q_1-q_L}{RT}}$$

In the BET expression, n is the number of molecules of adsorbate adsorbed per unit weight of adsorbent and n_m is the monolayer capacity. In the expression for c, q_1 is the heat of interaction for the first layer, and q_L is the latent heat of condensation for the adsorbate. This last is taken to be the interaction energy between adsorbate and adsorbent for all layers above the first. From the linear form of the BET equation, by plotting

$$\frac{p/p^o}{n(1-p/p^o)}\ V_s p^o$$

we obtain the slope s and intercept i, which are equal to

$$s=\frac{c-1}{n_m c}$$

$$i=\frac{1}{n_m c}$$

respectively, thus giving

$$n_m = \frac{1}{s+i}$$

and

$$c = \frac{s}{i} + 1$$

4. THE TYPE IV ISOTHERM AND HYSTERESIS

The characteristic of the type IV isotherm is that at high partial pressures beyond a further point there is no more uptake of gas with increasing pressure, and that on desorption the isotherm follows a different path to that of adsorption down to a partial pressure for nitrogen of, in general 0.45. Below that partial pressure, types II and IV isotherms are similar in shape (Figure 3)[1]. According to the BET theory, as partial pressure is increased from 0, adsorption is taking place at discrete adsorption sites, until a monolayer is first formed. That point at which the monolayer is formed corresponds to the first change in slope of the isotherm (point B) beyond which multilayer adsorption takes place. Type II isotherms are characteristic of non-porous or macroporous solids. Thus, whatever gives rise to hysteresis in type IV isotherms is a process which commences after a number of layers of adsorbed molecules have formed on the surface. This process is capillary condensation: in pores of diameters between 2 and 50 nm (mesopores), the walls are near enough that once the pressure has increased to a given value and thus adsorption has progressed beyond a certain number of layers, then condensation of liquid adsorbate takes place, and a meniscus is formed. When the mesopores have filled up, then no more adsorption can take place (flattening of isotherm at high partial pressures). On desorption, evaporation is taking place from the meniscus in the filled pore, leading to hysteresis.

The relationship between partial pressure and the radius of the meniscus is given by the so-called Kelvin equation:

$$\ln(p/p^{o}) = -\frac{2\gamma V_L}{RT}\frac{1}{r_m}$$

where γ is the surface tension of the meniscus, V_L is the liquid molar volume, and r_m the meniscus radius. The Kelvin equation is a special case of the Young-Laplace equation

$$p^{\beta} - p^{\alpha} = \gamma\left(\frac{1}{r_1} + \frac{1}{r_2}\right)$$

which relates the radii of a meniscus (allowing for a non-spherical meniscus), with the difference in pressure of the two phases that the meniscus separates. Given that when the meniscus forms there is already an adsorbed layer on the walls of the pore, the core radius as opposed to the pore radius should be considered, the difference between the two being the thickness of the adsorbed layer. The relationship between core radius r^k and the meniscus radius is:

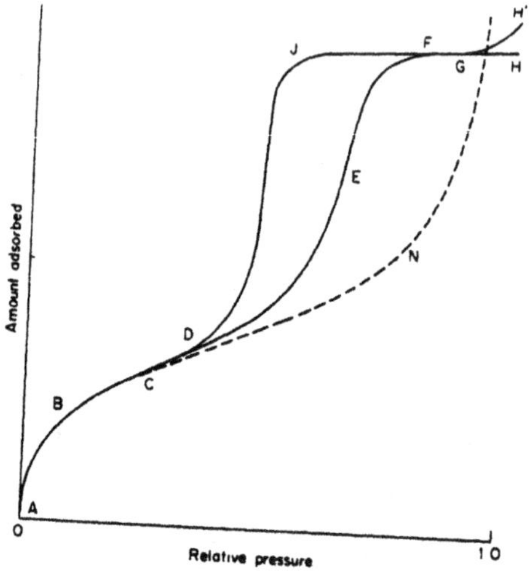

Figure 3 Comparison of types II and IV isotherms

Figure 4 Cylindrical pore, closed one end

Figure 5 Cylindrical pore, open both ends

$$r^k = r_m \cos\theta$$

In this relationship, θ is the contact angle between the meniscus and the walls of the pore, which cannot be observed, and thus is taken to be equal to 0, making the meniscus in a cylindrical pore semispherical.

Hysteresis can be explained in the following fashion. K is defined to be

$$\frac{\gamma V_L}{RT} = K$$

allowing the Kelvin equation to be rewritten thus:

$$p/p^o = e^{-\frac{2K}{r_m}}$$

If the pore in question is cylindrical in shape (see Figure 4) and is closed at one end, then as the partial pressure is increased, a meniscus will form at point B, at pressure[1]:

$$p/p^o = e^{-\frac{2K}{r^k}}$$

During the desorption cycle, evaporation will start at point A from a hemispherical meniscus. Since while the pore is filling there is no change in meniscus radius, evaporation will take place at the same partial pressure as condensation, and thus no hysteresis is observed. But, if the pore is open at both ends (Figure 5), then the meniscus is cylindrical during filling[1], but becomes hemispherical when the pore is full, and is from this meniscus that evaporation takes place. Therefore, the relationship between meniscus radius and core radius during the filling process is

$$r_m = 2r^k$$

and condensation occurs when the partial pressure is

$$p/p^o = e^{-\frac{2K}{r_m}} = e^{-\frac{K}{r^k}}$$

Evaporation takes place at a partial pressure given by the same expression as for the cylinder closed at one end. Thus, evaporation occurs at a different partial pressure to that at which the same pore had filled up, thus resulting in hysteresis.

In an ink-bottle shaped pore[1] (see Figure 6) a meniscus forms at point B at the bottom of the pore, at a partial pressure equal to:

$$(p/p^o)^r = e^{-\frac{2K}{r_w}}$$

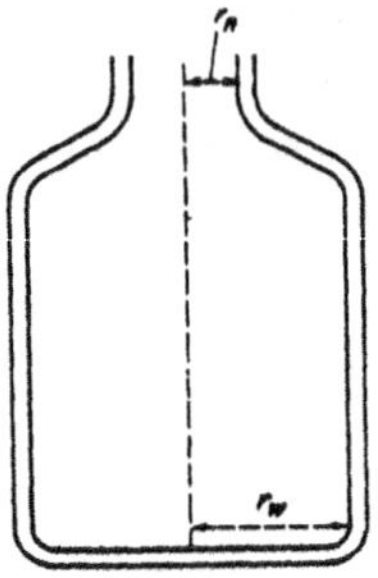

Figure 6 Ink-bottle shaped pore

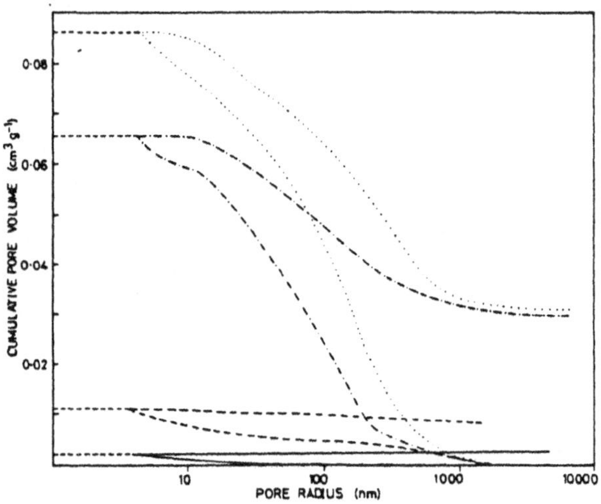

Figure 7 Examples of mercury porosimetry curves for cement after various heat treatments. Extrusion as well as intrusion curves are shown

At the neck, a meniscus forms at

$$(p/p^{o})^{II} = e^{-\frac{K}{r_n}}$$

If

$$\frac{r_w}{r_n} < 2$$

then

$$(p/p^{o})^{I} < (p/p^{o})^{II}$$

Condensation starts at point B and continues to fill the whole pore, including the neck, at a partial pressure:

$$p/p^{o} = e^{-\frac{2K}{r_w}}$$

On desorption, evaporation will start at the neck at a pressure:

$$(p/p^{o})^{III} = e^{-\frac{2K}{r_n}}$$

and will continue, until the body empties, since

$$(p/p^{o})^{III} < (p/p^{o})^{I}$$

But, if

$$\frac{r_w}{r_n} > 2$$

then

$$(p/p^{o})^{I} = e^{-\frac{K}{r_n}} < (p/p^{o})^{II} = e^{-\frac{2K}{r_w}}$$

In this case, condensation will take place in the neck, but not in the body, until the pressure rises to partial pressure I. Evaporation takes place as for the previous example. The core therefore empties at a partial pressure

$$(p/p^{o})^{III} = e^{-\frac{2K}{r_n}}$$

and therefore, hysteresis occurs.

Given the relationship between meniscus radius and the partial pressure at which the pore is filled or emptied with liquid, a pore size distribution (psd) can be calculated from type IV isotherms, using the Kelvin equation. The calculation is carried by dividing the isotherm into steps of partial pressures, or of pore radii, r^p, starting from the high pressure end. The amount of gas given up at step i, is expressed in terms of liquid volume, δv_i. The contribution of the thickness of the adsorbed film, δv_f has to be estimated. Thus, the core volume δv^k associated with mean core radius r^k can be obtained.

$$\delta v_i - \delta v_f = \delta v^k$$

In a subsequent step, δv^k has to be converted to δv^p, the pore volume, and thus obtain r^p, the pore radius. The methods available for carrying out these calculations, in general make use of a model for a pore shape. Furthermore, knowledge is needed of t, the statistical thickness of the adsorbed film[1]. n/n_m is taken to be the number of layers in the adsorbed film, t is thus equal to:

$$t = \frac{n}{n_m}\sigma$$

where σ is the thickness of a single layer, which for nitrogen is taken as 0.352 nm. Thus,

$$\frac{n}{n_m} = \frac{t}{\sigma}$$

In the expression above, values for t are obtained from standard isotherms. The idea of a standard isotherm has been discussed by several authors, including the criteria for selecting an appropriate solid, which has to be non-porous. A widely used standard isotherm is that on fused silica[7]. Whereas for the calculations relating to micropore filling which takes place at low partial pressures and without the presence of an adsorbed multilayer on the surface, mesopore filling takes place in the presence of such layers. Since for layers beyond the first the energy of interaction is taken to be equal to the heat of liquefaction of the adsorptive, one standard is sufficient, in spite of chemical differences to the analytical solid. Several methods are available for the estimation of pore size distribution from adsorption data. The pore size distribution is obtained from plots of

$$\frac{\delta v^p}{\delta r^p} \; V_s \bar{r}^p$$

The range of pore radii for which the Kelvin equation is valid, is between 1.7 and 50 nm, both limits being approximate. The lower limit is related to the stability of a meniscus at low pressures, and the upper limit to the fact that a 50 nm radius corresponds to a partial pressure of 0.95, thus making it impossible to determine pore diameters wider than those.

5. MERCURY POROSIMETRY

Mercury porosimetry is a widely used technique for pore size distribution in the mesopore to macropore range. Mercury is a non-wetting liquid, which means that the contact angle, θ is more than 90° (the contact angle for mercury is generally taken to be 140°). As a result, excess pressure Δp is required to be applied to the system, in order to force mercury into the pores. The wider pores, or voids, are filled first, and as the applied pressure is increased, mercury can penetrate into narrower and narrower pores. This means that porosimetry is the opposite to nitrogen adsorption measurements, where the narrower pores are filled before the wider ones. Under the conditions for porosimetry, the Young-Laplace equation can be rewritten as

$$p^{Hg}-p^{g}=-\gamma\left(\frac{1}{r_1}+\frac{1}{r_2}\right)$$

γ Is the surface tension of liquid mercury and is taken to be equal to 480 mN m^{-2}. p^{Hg} is the pressure in the liquid mercury, and p^{g} is the vapour pressure of mercury over the sample. As with gas sorption, the assumption is made that the meniscus is hemispherical, and thus:

$$r_1=r_2=r^{p}\cos\theta$$

Then by putting

$$p^{Hg}-p^{g}=\Delta p$$

we obtain the Washburn equation, which takes the form:

$$r^{p}=-\frac{2\gamma\cos\theta}{\Delta p}$$

with Δp in practice being equal to the pressure required to introduce mercury into pores of radius r^{p}. The range in which the Washburn equation is valid, is a function of the force needed to insert the mercury into the pores, and mechanical limitations. The limits are:

$$35\text{Å}<r^{p}<7.5\mu m$$

Measurements are taken by measuring the volume of mercury v(Hg) introduced into interstices in the sample at applied pressure p (p^{Hg} - p^{g}). The value of v_i(Hg) at pressure p_i is a measure of the total volume of pores with radius equal to, or larger than, r_{ip}, the so-called cumulative pore volume. The pore size distribution is obtained by plotting

$$\frac{dv^{p}}{dr^{p}}V_s r^{p}$$

Almost always, the mercury intrusion curve does not follow the same path as that for extrusion, giving rise to hysteresis. This hysteresis is often due to the collapse of the structure thus trapping mercury which cannot escape. Alternatively, the insertion of mercury into pores may lead to an opening of the pore diameter, or reveal access to areas of the solid hitherto inaccessible. Another source of hysteresis may be a change in the contact angle between mercury and the walls of the pores; this may be different during extrusion than during intrusion. But, hysteresis may also be possible in the absence of structural change. To obtain a pore size distribution from mercury porosimetry results, we need to make an assumption about the shape of the pores. For example for ink-bottle shaped pore, mercury fills the neck at pressure

$$p_n = -\frac{2\gamma\cos\theta}{r_n}$$

but cannot leave the body of the pore until the pressure has fallen to

$$p_w = -\frac{2\gamma\cos\theta}{r_w}$$

Therefore, analysis of the intrusion curve gives a value for, r_n and of the extrusion curve, of r_w. An example of mercury porosimetry curves is given in Figure 7.

6. LOW PRESSURE HYSTERESIS

Porosity in most solids is the result of aggregation or agglomeration of particles leaving behind voids. During adsorption into these areas, especially if the pores are slit-shaped, swelling may take place by the particles being pushed further apart within the aggregate[8]. As a result, upon desorption hysteresis occurs which may persist to very low partial pressures, since the surface from which desorption has taken place is not the same as that on which adsorption occurred. For an example of such an isotherm, see Figure 8.

For some samples, especially where microporosity is also present, there is no flattening of the isotherm at high partial pressures, (Figure 9) in spite of the presence of a hysteresis loop. This is usually attributed to the presence of ill-defined areas of porosity, such as those that may arise from the way particles pack together[9].

7. EXPERIMENTAL METHODS

A variety of experimental setups are available for the determination of adsorption isotherms, which can be readily purchased from a manufacturer, although the existence of a good glass blowing workshop in a laboratory, would allow the building of an apparatus in situ. The methods are broadly speaking of two kinds[10], one the so-called volumetric, and the second the gravimetric method.

In both methods, the sample is enclosed in a vacuum system which is kept at an appropriate constant temperature (in the case of nitrogen, 77K). The pressure is first measured, normally the initial pressure is in the region of 10^{-5} to 10^{-4} Torr, and then small amounts of adsorptive are introduced to it. The system is allowed to equilibrate and then the pressure is measured again. The amount of adsorbate taken up is measured either from the weight increase of the sample (gravimetric method) or from the pressure difference, the volumetric method. In the volumetric method, the adsorptive is introduced at a known pressure in a known volume of space, at a known temperature, thus allowing a calculation of the amount introduced. The pressure drop after adsorption has taken place, and making allowances for volume changes, gives the amount adsorbed. Figure 10 shows a volumetric experimental setup, and Figure 11, a gravimetric one.

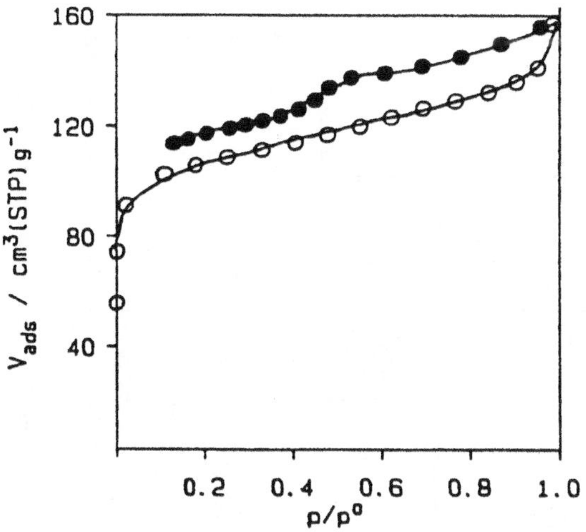

Figure 8 Nitrogen adsorption isotherms for ALPO-5, an aluminophosphate molecular sieve, exhibiting low pressure hysteresis

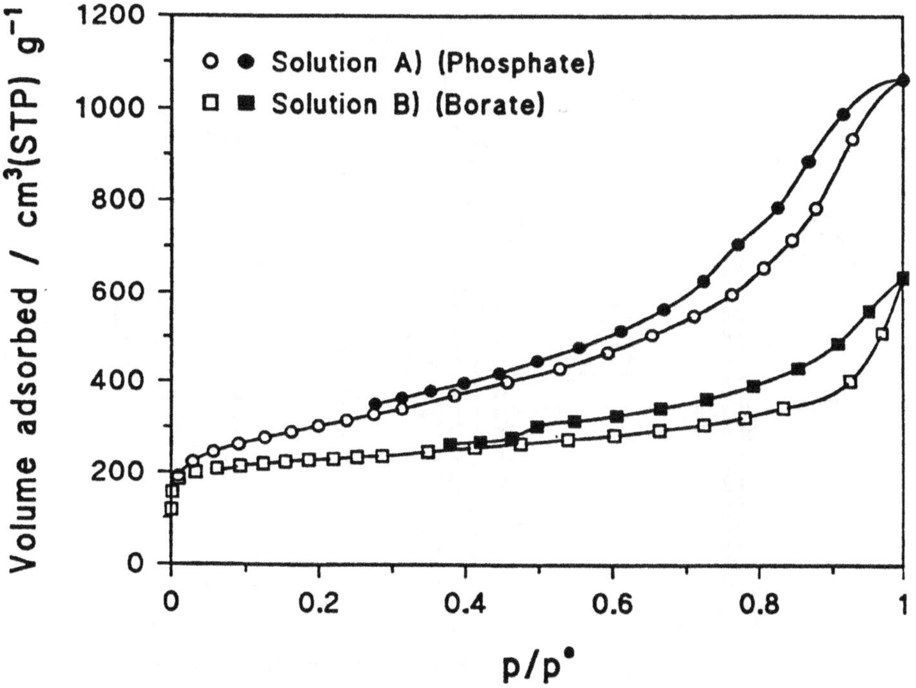

Figure 9 Nitrogen adsorption isotherms for steam-activated chars pre-treated with phosphate or borate solutions

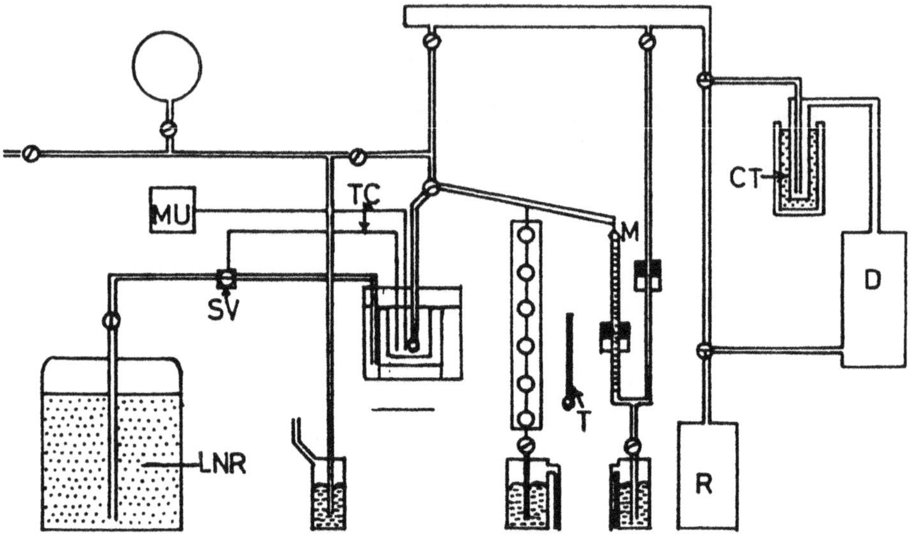

Figure 10 Schematic diagram of a volumetric adsorption rig

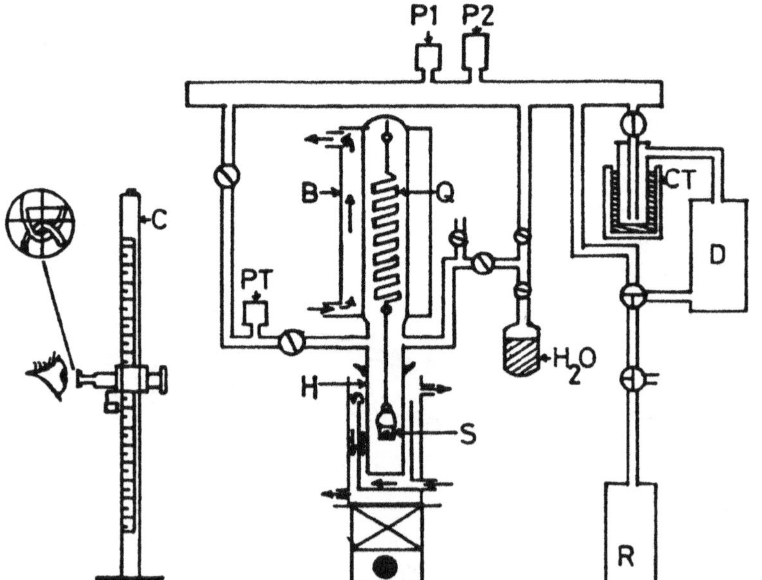

Figure 11 Schematic diagram of a gravimetric adsorption rig

It is always necessary prior to sorption measurements to clean the surface (and pores) by outgassing the sample to the starting pressure. The length of the outgassing step and the temperature at which it will take place are crucial parameters, as the conditions have to be adequate to remove all molecules adsorbed on the surface, without however altering it. For example, heating may decompose the sample, or alter its porosity, or on the other hand fail to make all pores accessible to the adsorptive.

8. CONCLUDING REMARKS

The determination of the surface textural characteristics of porous solids is a very important analytical tool which can be applied to a wide variety of solids. The choice of adsorptive and of outgassing conditions is crucial and can influence greatly the usefulness of the measurements. Also, the choice of pore model to be used in the calculations can be important. All the models used, whether of cylindrical, slit-shaped or ink-bottle-shaped pores are extreme, idealised, cases which may not necessarily portray the reality in a satisfactory way. Figure 12 depicts one such instance, in the case of a slit-shaped pore. Work is needed, and is currently under way in a variety of Institutions, to develop modelling techniques for achieving better methods of interpretation. Even the classification of an isotherm as of one type or another, can be problematical as it may lie in between various types. In all, but especially such cases, the adsorption data should be complemented with other methods, as well as the use of more than one adsorptive.

9. ACKNOWLEDGEMENTS

The author thanks the Department of Chemistry at Brunel University, and colleagues and research students there, for the provision of examples and data used in this paper, and for many useful discussions. The SERC, Exxon Chemical Ltd, Rhone-Poulenc, ICI Plc, and the European Community (SCIENCE programme) are thanked for their financial support. Last, but by no means least, the author gladly acknowledges a debt of gratitude to his teachers and mentors, Professors Sir John Meurig Thomas and Kenneth S.W. Sing, for introducing him to the field of porous solids.

10. REFERENCES

(1) S.J. Gregg and K.S.W. Sing, "Adsorption, Surface Analysis and Porosity", 2nd Edition, Academic Press, London, (1982)
(2) P.L. Llewellyn and C.R. Theocharis, J. Chem. Tech. Biotechnol., 52, 473, (1991)
(3) J.M. Thomas and C.R. Theocharis, in (Ed. R. Scheffold) "Modern Synthetic Methods", 5, 249, (1989)
(4) I. Langmuir, J. Am. Chem. Soc., **38**, 2221, (1916)
(5) S. Brunauer, P.H. Emmett, and E. Teller, J. Am. Chem. Soc., **60**, 309, (1938)
(6) S. Brunauer, L.S. Deming, W.S. Deming, and E. Teller, J. Am. Chem. Soc., **62**, 1723, (1940)
(7) J.D. Carruthers, P.A. Cutting, R.E. Day, M.R. Harris, S.A. Mitchell, and K.S.W. Sing, Chem. and Ind., 177, (1968)
(8) C.R. Theocharis and M.R. Gelsthorpe, in (Eds. K.K. Unger, J. Rouquerol, K.S.W. Sing and H. Kral), "Studies in Surface Science and Catalysis", **39**, (Characterization of Porous

Solids), Elsevier Science Publishers, Amsterdam, 541, (1988)
(9) J.B. Tomlinson, J.J. Freeman, and C.R. Theocharis, Carbon, In the Press
(10) M. Yates, PhD Thesis, Brunel University (1991)

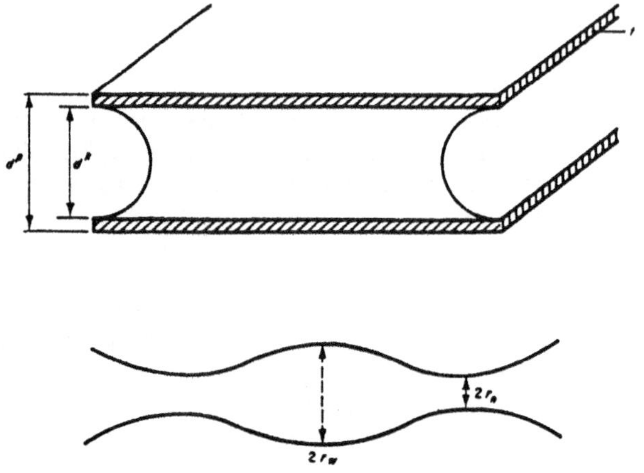

Figure 12 Idealised slit-shaped pore, and a more realistic model, exhibiting a variation in the layer to layer separation

ABOUT SURFACE AREA AND POROSITY MEASUREMENTS IN PILLARED CLAYS

F. BERGAYA, L. GATINEAU and H. VAN DAMME
Centre de Recherche sur la Matière Divisée
CNRS and Université d'Orléans
45071 Orléans cedex 02
France

ABSTRACT. We discuss the problems raised by chemical heterogeneity and geometrical confinement in pillared clays for the measure of their surface area and their porosity by physical adsorption.

1. Introduction

Pillared layered materials are (micro)texturally complex and geometrically constrained materials. The textural complexity stems from the distribution of interlayer distances, from the deformation (bending) of the host layers and from their entanglement. The (micro)texture is controlled by the wet preparation chemistry and, to a large extent, by the drying method. The geometrical constraints arise from the fact that the pillared galleries are (so far) microporous or just mesoporous spaces in which the small distance between walls is of the order of one or a few molecular diameters and is therefore expected to perturb the arrangement of the pore-filling molecules.

In addition to (micro)textural complexity and geometrical constraints, one has also to consider a possible chemical heterogeneity. Indeed, the pillaring species (those which control the gallery height) are often not the only species which occupy the gallery space. Additional simple or oligomeric ions and "precipitated" hydroxide-like species may also encumber the void space and, in particular, the galleries.

As far as surface area and micro- or mesoporosity determinations by physical adsorption are concerned, this overall complexity introduces interpretation difficulties. It is the purpose of this short paper to consider two aspects of this problem in pillared clays. In the first section we will illustrate, on the basis of results obtained with mixed Al-Fe pillared laponites, the influence of the gallery height and the chemical heterogeneity. In the second section, we will describe a model macroscopic experiment which gives some indications on the expected consequences of geometrical confinement.

C.A.C. Sequeira and M.J. Hudson (eds.),Multifunctional Mesoporous Inorganic Solids, 19–26.

2. Surface area, porosity and chemical heterogeneity

We reported previously (Bergaya and Barrault, 1990; Bergaya et al., 1992; Barrault et al., 1992) the preparation of mixed Al-Fe pillared laponite by an "in situ" method, in which both the hydroxide (NaOH) and the metal ion (Al and Fe) solutions were simultaneously added to a laponite suspension. Several Al/Fe ratios were studied, leading to samples which were noted as Al_{100}, $Al_{90}Fe_{10}$, $Al_{80}Fe_{20}$, $Al_{70}Fe_{30}$, $Al_{50}Fe_{50}$ and Fe_{100}, respectively, where the subsript denotes the % molar fraction of each ion in the solution, the total concentration of M(Al,Fe) being constant.

XRD examination of the samples showed that the laponite pillared in all cases. For the Al-rich solutions (Al_{100}, $Al_{90}Fe_{10}$ and $Al_{80}Fe_{20}$), the 001 distance is about 1.8 nm, which corresponds to the distance obtained with "Al_{13}" pillars. Samples prepared from Fe-richer solutions show a significant broadening of the 001 peak, indicating a loss of order along the c axis. The 001 distance is about 2.2 nm, larger than in the Al_{13}-pillared clays.

As far as the nature of the pillars is concerned, we were able to show from the converging results of three methods (chemical analysis, H_2-TPR and ^{27}Al MAS NMR) that in the Al_{100}, $Al_{90}Fe_{10}$, $Al_{80}Fe_{20}$ samples, there are Al_{13} and $Al_{13-x}Fe_x$ pillars, together with non-pillaring Fe-containing species, probably in oxyhydroxide form. The substituted iron in $Al_{13-x}Fe_x$ pillars is not reduced by hydrogen at 500°C.

On the other hand in the more iron-rich $Al_{50}Fe_{50}$ and Fe_{100} samples, no Al_{13} pillars were detected. The selectivity of the clay is such that no aluminium is retained. Two kind of Fe-containing species can be detected, one of them being the pillaring species, the other being non-pillaring oxyhydroxide. Both species can be distinguished by TPR since the pillar Fe is reduced at higher temperature than Fe outside the pillars. The $Al_{70}Fe_{30}$ samples exhibit an intermediate behaviour.

As shown in Table 1 and fig.1, the BET surface area measured by nitrogen adsorption increases smoothly with iron content as long as there are $Al_{13-x}Fe_x$ pillars (Al_{100}, $Al_{90}Fe_{10}$, $Al_{80}Fe_{20}$, and $Al_{70}Fe_{30}$). Surprisingly, the microporosity determined from t-plot calculations, decreases as the surface area increases (Table 1) and vanishes in the $Al_{70}Fe_{30}$ sample, which has the largest surface area. This regime is followed by another regime ($Al_{50}Fe_{50}$ and Fe_{100}) where the surface area decreases as the iron content of the material increases.

The previous behaviour illustrates the difficulty of interpreting surface area and microporosity measurements. The main factors to take into account are the increase of the 001 distance from 1.8 to 2.2 nm on one hand, and the increasing concentration of non-pillaring oxyhydroxide species on the other hand. The first factor is expected to allow for more molecules to be accomodated in the galleries (hence increasing the BET surface area), whereas the second one might either reduce the number area if the oxyhydroxide species encumber the galleries, but it may also increase the surface area if the oxyhydroxide colloids are located outside the galleries. The fact that the surface area increases whereas the microporosity decreases in the first regime suggests that, in this regime, they are probably present in both situations. The

TABLE 1 . Surface area values obtained by nitrogen adsorption and microporosity determined by t-plot calculations for Na and different pillared laponites

	BET Surface m^2/g	Microporosity mm^3/g
Lap Na	316	42.7
Lap Al_{100}	369	32.9
Lap Al_{90} Fe_{10}	376	26.4
Lap Al_{80} Fe_{20}	418	18.6
Lap Al_{70} Fe_{30}	472	0
Lap Al_{50} Fe_{50}	410	
Lap Fe_{100}	370	

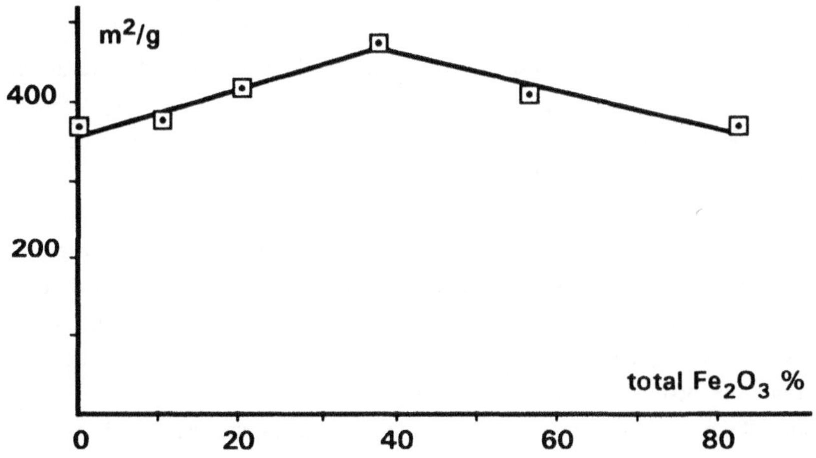

Fig.1 Evolution of the nitrogen BET surface area of mixed Al-Fe pillared laponites as a function of the total Fe2O3 content (weight percent), with respect to the weight of laponite sheets.

oxyhydroxide species outside the galleries might be responsible for an increase of the orientational disorder (with deformations) of the layered microtexture, thereby increasing the surface area whereas, concomitantly, they might clog the galleries, thereby diminishing the microporosity. Further increase of their concentration (beyond $Al_{70}Fe_{30}$) may finally lead to an overcrowding of the (meso)porous space outside the pillared stacks and to a decrease of the surface area.

3. Influence of geometrical confinement

What is measured by physical adsorption or capillary condensation is a *number of molecules*, which is further converted via an adequate model (BET, Kelvin...) into an area, a pore size or a volume, assuming some well-defined arrangement of these molecules, usually close to liquid-phase packing. Thus, the only physically meaningful relationship to understand is that between the dimensions and the shape of a "box" and the number of molecules which can be adsorbed on its walls or condensed in its void space.

As a model system for pillared materials, let us consider the packing of spheres in a narrow gap. "Narrow" means a gap which is of the order of a few sphere diameters, i.e. a gap where the molecules are never far from the walls. The question to be addressed is the influence of the walls on the packing of the molecules. More specifically, one might wish to know the evolution of the density of the confined fluid as the distance between the wall, Δ, is increased. Obviously, one knows that the density of the confined fluid has to tend towards that of the bulk fluid as Δ becomes very large, but what about regimes where Δ is only a few molecular diameters? Direct experimental results are difficult to obtain, but a simple macroscopic simulation study shows clearly the trends.

The experience may be performed with two pieces of glass of 0.5mx0.5m, separated by three pieces of wood of thickness Δ. This produces a slit-shaped pore (fig.2) in which glass beads of diameter $2a$ can be poured. The same experiment may be performed with a few fixed cylinders in the cell simulating pillars in the case of pillared smectites (fig.3). After gentle vibration, the saturated set-up is weighted and, knowing the weight of one glass bead, the total weight converted into the total volume occupied by the glass beads.

Fig.4 shows the evolution of the volume *fraction* occupied by the beads, $1\text{-}\Phi$, as a function of the dimensionless distance between the walls, $\Delta/2a$. The curve, which is the result of a number of measurements performed with several sphere diameters, shows two remarkable features. The first is that the density of the confined "fluid" is always lower than that of the bulk "fluid". Clearly, the presence of walls prevents the packing from reaching its optimum density which, for an infinite random close packing of spheres, is close to 0.65 (the porosity is of the order of 35%). Not surprisingly, this is precisely the asymptotical value measured at large

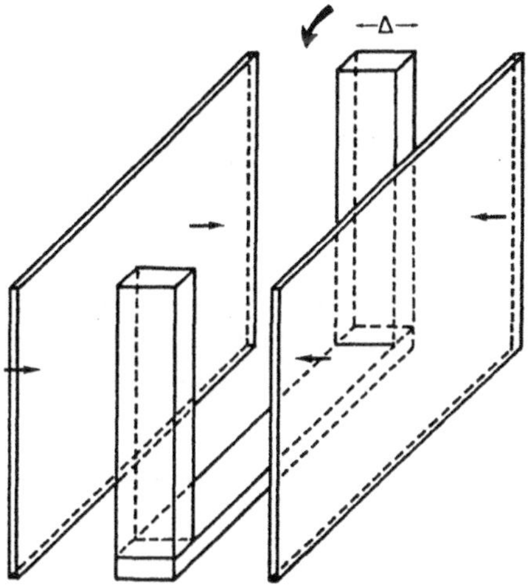

Fig.2 Schematic drawing of the experimental macroscopic slit-shaped cell used for studying the packing of hard spheres in a narrow gap.

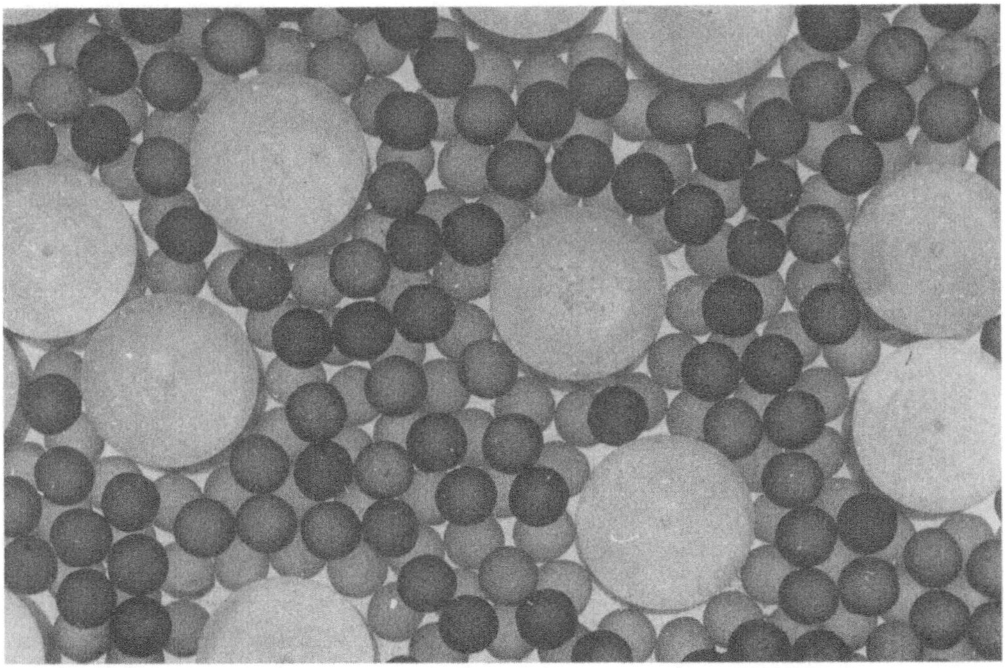

Fig.3 Top view of glass spheres filling a cell like in fig.2, with cylindrical obstacles simulating pillars in a pillared smectite.

$\Delta/2a$, when the gap is of the order of 10 "molecular" diameters or more. In other words, the steric hindrance due to the presence of walls vanishes at distances of the order of 10 molecular diameters.

The second remarkable feature of fig.4 is that the evolution of 1-Φ is not smooth. It shows several well defined minima and maxima, with a periodicity of the order of 0.95 $\Delta/2a$. The amplitude of the min-max difference decreases as the confinement effect diminishes. This is nothing else than the development of the three-dimensional disorder in the system, starting from a single monolayer with (at least local) two-dimensional order. In small gaps, the build-up of a new monolayer requires a minimum free volume. As long as this threshold is not reached, the average density decreases.

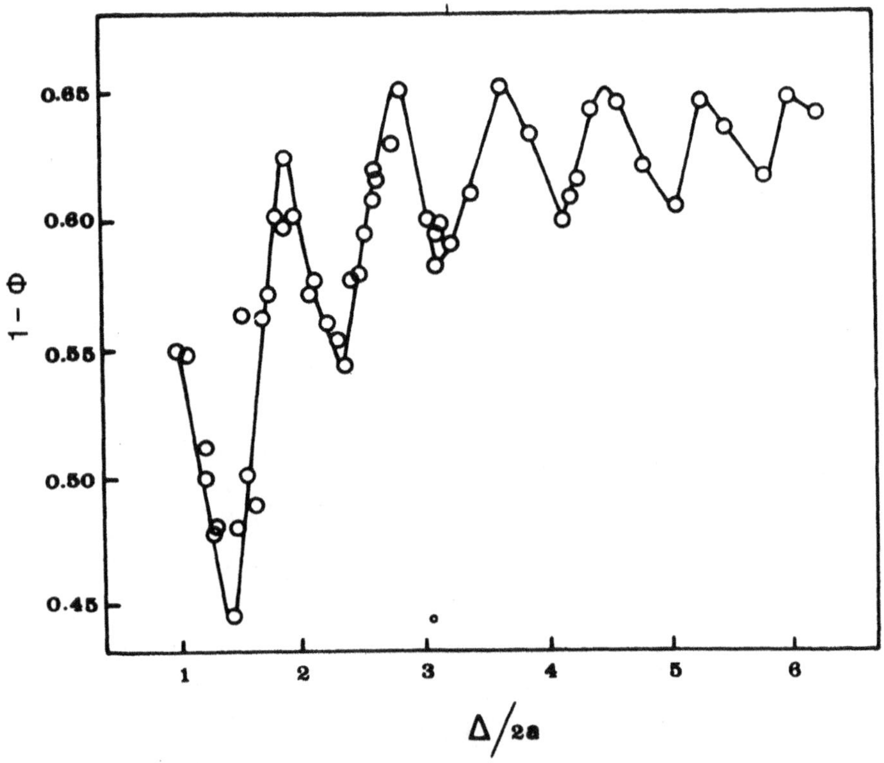

Fig.4 The structure of a packing of hard spheres between two walls is strongly dependent on the distance Δ between the walls. The density (Φ is the porosity) oscillates with a period which is slightly smaller than the sphere diameter 2a. The influence of confinement vanishes at about ten diameters

As far as practical consequences are concerned, what fig.4 tells us is that surface area or microporous volume measurements in pillared materials should take into account a d_{001}-dependent correction factor. More precisely, the volume of the gallery, V_{gal}, may be expressed as:

$$V_{gal} = V.N_{mol}(1/0.65)\beta$$

where V is the van der Waals volume of one molecule, N_{mol} is the number of molecules in close packing in the gallery and β (<1) is the correction factor. In the worst situation (the density minimum between one and two monolayers in fig.4), β may be as large as $0.65/0.45=1.44$. In the case of an Al_{13}-pillared smectite and nitrogen molecules, this situation would be observed for d_{001} around 1.56 nm. It is at this spacing that the pore volume of the gallery (gallery height close to 0.6 nm) would suffer from the largest underestimation if one does not take into account the correction factor.

It is more difficult (and, in fact, of little physical significance) to calculate the wall surface area within the pillared galleries from V_{gal}. What may be useful though is to calculate the external surface area of the pillared stacks. Provided the distance between stacks is larger than ~10 nitrogen diameters (about 4 nm), the BET method may be applied to the corrected adsorption isotherm obtained by substracting the intragallery molecules from the total isotherm.

4. Conclusions

Surface area and porosity measurements in pillared clays are hampered by a number of factors. The chemical heterogeneity arising from the simultaneous presence of pillars, exchangeable ions and precipitated species can lead to a complex behaviour of the surface area and microporosity, as examplified by Al-Fe pillared laponite.

On the other hand, the molecular confinement in the galleries is a major source of underestimation of the gallery pore volume. It makes the surface area determination difficult and of little physical significance. Calculation of the gallery porosity from adsorption data should take into account a correction factor which is a damped oscillatory function of gallery height, Δ, as shown by a simple macroscopic model measurement with hard spheres. More realistic Monte Carlo simulations with Lenard Jones molecules are under progress.

References

BERGAYA F. and BARRAULT J. (1990), Mixed Al-Fe Pillared Laponites: Preparation, Characterization and their Catalytic Properties, in Pillared Layered Structures, I.V. MITCHELL Ed., Elsevier Applied Science, 167-184

BERGAYA F., HASSOUN N., BARRAULT J. and GATINEAU L. (1992), Pillaring of Synthetic Hectorites by Mixed ($Al_{13-x}Fe_x$) Pillars, submitted to Clay Minerals

BARRAULT J., GATINEAU L., HASSOUN N. and BERGAYA F. (1992), The Selective Syngas Conversion over Mixed Al-Fe Pillared Laponites, accepted and to be published in Energy and Fuels

NEW THEORETICAL APPROACHES TO CHARACTERIZATION OF MESOPOROUS SOLIDS

A.V.NEIMARK
Inst. Phys. Chem., Academy of Sciences,
117915, Moscow, Russia
Inst. Anorg. Anal. Chem., J.-G.-Universitat,
6500 Mainz, Germany

ABSTRACT. New trends in characterization of mesoporous solids are analysed. The attention is paid to the fractal analysis of surface irregularities and to the percolation methods of calculating the pore size distribution using adsorption measurement data.

1. Introduction.

According to the IUPAC classification (Gregg and Sing, 1982), mesoporous materials are those containing the pores of sizes from several nanometers to several hundreds nanometers. These pores are called mesopores, smaller pores are known as micropores and larger pores are macropores. Mesoporous materials include porous glasses, silica gels and other oxides, carbon materials and so on. Mesoporous solids are widely used in modern technology and scientific research as adsorbents, catalysts, membranes, filters, etc. Their technological properties depend in many respects on the geometrical parameters of porous structure: porosity, specific surface area, pore size distribution. Therefore, the problem of characterising mesoporous solids is of essential value.

The analysis of gas adsorption isotherms is one of the main methods of exploring the mesoporous structure. At the moment the experimental measurement of adsorption isotherms is a routine procedure. Firms such as "Micromeritics", "Carlo Erba Instruments" and the others produce automated instruments for adsorption measurements, principally for the adsorption of nitrogen at 77K. The main problems are those of the interpretation of experimental data and of the calculation of pore structure parameters. Perhaps it is one of a few examples in modern research when the reliability of the finite results determines not by the experimental possibilities but the insufficiently advanced theory.

The conventional methods of calculating the geometrical parameters of porous solids use models which are too simplified for the adsorption process and for the pore structure. For example, while determining the specific surface area, one supposes that the adsorption by porous solid, at least up to the beginning of capillary condensation, is quite similar to the adsorption by a non porous solid of the same origin with the same value of surface area, i.e. by a flat surface. While calculating the pore size distribution, the pore space is represented by a system of unrelated pores of the simplest shape, principally by a system of cylindrical capillaries of different radii. These models - of a flat surface and of a system of unrelated pores - are rather distant from reality.

C.A.C. Sequeira and M.J. Hudson (eds.), Multifunctional Mesoporous Inorganic Solids, 27–35.

Over the last ten years, the development of new theoretical methods for the characterization of porous solids went into two directions reflecting the main tendencies of the theoretical thought development. These two directions are based on the application of the fractal geometry and the percolation theory. Moreover, the fractal approach is the development of the methods of determining the surface area of porous materials, and the percolation approach is the development of the methods of determining the pore size distribution.

2. Determining the surface fractal dimension from adsorption data.

After the pioneering papers of B.Mandelbrot, D.Avnir, P.Pfeifer et al. the fractal dimension has been introduced as a new structure characteristic of a porous solid which must be considered together with porosity, specific surface area and the other usual characteristics . Numerous published data testify to the fractal properties of different mesoporous materials: carbons, coals, clays, natural minerals, aerogels, ets.

Many examples have been presented at recent conferences, i.e. "Characterization of Porous Solids" (Unger et al. 1988; Rodrigues-Reinoso et al. 1991), "Fundamentals of Adsorption" (Mersmann 1990) and in a very informative book edited by Avnir (1989), etc.

A set of alternative procedures has been proposed for determining the fractal dimension experimentally. Among these there are several methods based on adsorption experiments.

2.1. DEFINITION OF FRACTAL DIMENSION

There are several theoretical definitions of fractal objects and fractal surfaces (Mandelbrot 1982). They are distinguished by the level of the strictness of the mathematical formulations. However, from the viewpoint of such objects as real disperse materials, the strictness of the mathematical formulations is not the main factor. The main problem is that the theoretical definition of the fractal dimension has to be linked with the conditions and with the possibilities of real experiments.

The basis of all experimental methods for its determination is the need for a particular method of measuring some metric parameters. While carrying out such measurements it is important to take into account not only the size of the object explored but also the size of a gauge or yardstick used for the measurement. In other words, it is important to take into account the magnification capability of a given instrument. Increasing the magnification leads to an increase in the amount of irregularities visible for the given object.

We are interested in measuring surface area and it is natural to use as gauges spheres of different radii. As a rule the solid phase has very rough indented surface with pores of different radii. If we use spheres whose radius is larger than the maximum radius of the pores we can only measure the external surface, as such gauges do not penetrate the pores. By decreasing the radius of the spheres we provide the opportunity for exploring the internal surface of the pores and, hence, the smaller the radius employed the increase in surface area obtained. For each measurement, the corresponding surface area obtained by spheres of a given radius is determined from the following proportionality:

$$S(a) \sim a^2 N(a) \quad \text{at} \quad a \to 0 \tag{1}$$

i.e. the surface area $S(a)$ is proportional to the radius squared or the cross-section of the sphere multiplied by the number of spheres $N(a)$ necessary for compact monolayer coverage, or monolayer capacity.

For a two-dimensional surface, the monolayer capacity is in inverse ratio to the cross-section of a gauge at the limit of the radius, a , approaching zero:

$$N(a) \sim a^{-2} \qquad \text{at} \quad a \to 0 \tag{2}$$

The resultant surface area, S_0 is the limit of equation (1) as the radius, a, approaches zero:

$$S(a) \to S_0 \qquad \text{at} \quad a \to 0 \tag{3}$$

For an ideal fractal surface, the monolayer capacity $N(a)$ is an inverse ratio to the radius raised to a certain non-integer power which is equal to the fractal dimension D:

$$N(a) \sim a^{-D} \qquad \text{at} \quad a \to 0 \tag{4}$$

Relation (4) is one of the possible definitions of a fractal dimension. This definition is not very strict from a mathematical point of view and there are some mathematical objects for which this definition leads to a contradiction. But for our purposes, i.e. those of the physical chemistry of disperse systems, this definition is the most convenient for a fractal dimension. It should be pointed out that it is possible to use not only spheres as gauges, i.e. cubes or other particles with a definite form may be employed. By using gauges of different sizes we obtain relation (4) with the same fractal dimension independent of the form of the probe particle. In all cases the surface area $S(a)$ is proportional to the radius, a, raised to a power equal to $2-D$

$$S(a) \sim a^{2-D} \qquad \text{at} \quad a \to 0 \tag{5}$$

When a approaches zero, the surface area $S(a)$ goes to infinity when the fractal dimension $D>2$ and goes to zero when the fractal dimension $D<2$. We are interested in the case of the surface of mesoporous solids for which the fractal dimension can vary between the limits 2 and 3. A value equal to 2 corresponds to a smooth surface and non-integer values greater than 2 correspond to surfaces with a fractal roughness.

In the general case, for real porous media relation (4) applies only over certain scale intervals. Thus, for natural objects, inner a_{min} and upper a_{max} limits of the fractal region exist. When the size of the yardstick lies between these values, the object shows fractal properties according to relation (4). These limits depend on the texture of the given material; hence a_{min} is always greater than the size of the molecules which form the substratum of the solid, and a_{max} is less than the linear size of the particle or body being explored.

2.2. METHOD OF MOLECULAR TILING

We are now in a position to discuss adsorption methods as a means of determining the surface fractal dimension. From the theoretical point of view, the most irreproachable method is that of molecular covering or molecular tiling introduced by Pfeifer and Avnir (1983). The basis of this method involves a comparison of the monolayer capacities of different adsorbates. If for a set of adsorbates with different size, the monolayer capacity, N_m, which is a function of the molecular radius, a, satisfies the equation

$$Nm = N_{m,0}(a/a_0)^{-D} \tag{6}$$

then the exponent D is the surface fractal dimension of a given adsorbent. Here $N_{m,0}$ and a_0 are the monolayer capacity and molecular radius, respectively, of a given adsorbate, i.e. nitrogen. Such a definition corresponds strongly to the initial definition (4), with other molecules playing the role of yardsticks.

Two experimental methods are available for the determination of monolayer capacity: (a) adsorption from the gaseous phase, where N_m is determined by the BET method and (b) adsorption from the liquid phase, where N_m is determined by Langmuir's method.

The method of molecular tiling has been used by various authors for different adsorbents. However, at least four problems need to be taken into account: (a) the monolayer capacity can only be determined with some error; (b) the radius of the adsorbed molecule has to be calculated from the density of the liquid adsorbate and this introduces an additional source of errors; (c) the effects of specific molecular interactions are neglected; (d) the possib2le scale ranges for adsorbed molecules are very narrow.

The latter problem is the main problem encountered. Generally the inner limit a_{min} is about 0.2 *nm* for nitrogen while the upper limit a_{max} is always less than 1 *nm*. So the range of scales available in this method is that of molecular sizes not of mesopore sizes.

2.3. THERMODYNAMIC METHOD

A new thermodynamic method has been proposed for calculating the surface fractal dimension from the adsorption or desorption isotherm in the region of capillary condensation (Neimark, 1990a, 1990b, 1991a). This method does not rely on any model of the porous structure or of the adsorption process and does not use any physicochemical parameters, but only the experimental data. It is based on the following suppositions:

(a) It is assumed that the role of a gauge of varying radius can be played by the average radius of curvature of the meniscus at the interface between condensed adsorbate and gas. This radius a_k increases as the relative pressure $X=P/P_s$ increases in accordance with the Kelvin equation:

$$a_k = 2\sigma v/RT(-\ln X) \qquad (7)$$

Here σ and v are the surface tension and molar volume, respectively, of the liquid adsorbate, R is the universal gas constant and T is the absolute temperature. This value is called the Kelvin radius. In equilibrium adsorption experiments, the radius of curvature, a_k, varies from about 1 to 100-200 *nm*. This is quite a wide scale range and corresponds to the whole range of mesoporous structure.

(b) Under particular experimental conditions, i.e. at a given relative pressure, the surface area of the adsorbed liquid-like film, can be calculated using the integral thermodynamic relationship:

$$S(X) = (RT/\sigma v)\int_{N}^{N_s}(-\ln X)\,dN \qquad (8)$$

where N_s is the adsorption when $X\to 1$. This relationship, which is known in literature as the Kiselev equation, is a balance equation involving the work of adsorption and the work involved in the formation of the interface. The physical meaning of this equation is that account is only taken of capillary forces and it is assumed that the surface of the adsorbate is one of constant curvature, a_k.

(c) On the basis of equation (8) it is possible to interpret the surface area of the adsorbed film as that of the adsorbent which would be measured by spheres with radius a_k. For the fractal surface, this surface area $S(a_k)$ is proportional to the radius of curvature raised to a power equal to $2-D$ in conformity with the basic relationship (5), i.e.

$$S(a_k) \sim a_k^{2-D} \tag{9}$$

To obtain the value D from this proportion it is necessary to calculate the surface area S and the radius of curvature, a_k, over a set of relative pressures and to provide a regression in the logarithmic coordinates.

The fractal properties can be obtained only in a certain range of scales, where the proportionality (5) is valid. For a current point of the isotherm $N(X)$ in this region of fractality the equations (7)-(9) lead to the following linear correlation

$$\ln\int_{N}^{N_s}(-\ln X)\,dN = const + (D-2)\ln(-\ln X) \tag{10}$$

This correlation contain only the experimental adsorption isotherm $N(X)$. Therefore, the unknown value of surface fractal dimension and the upper and inner limits of the region of the fractality can be obtained without using any additional parameters except the experimental data.

The thermodynamic method was used for calculating the surface fractal dimension of active carbons (Neimark, 1990b) and can be recommended for the fractal analysis of other polydisperse mesoporous materials.

3. Percolation methods for calculating the pore size distribution

The presence of adsorption hysteresis is the special feature of all adsorbents with a mesopore structure. The adsorption and desorption isotherms differ appreciably from one another and form a closed hysteresis loop. Experimental adsorption and desorption isotherms in the hysteresis region provide information for calculating the pore size distribution. Traditional methods for such calculations are based on the assumption of an unrelated system of pores of simple form, as a rule, cylindrical capillaries. The calculations are based on either the adsorption or the desorption isotherm, ignoring the existence of hysteresis in the adsorption process. This leads to two different pore size distributions. The question of which of these is to be preferred has been the subject of unending discussion. The models of unrelated pores fail to describe the capillary hysteresis phenomena in mesoporous materials mainly because of the essential validity of the effect of pore intersections (Everett, 1979; Gregg and Sing, 1982). Only using the network models and the methods of percolation theory makes it possible to get a reliable picture of capillary hysteresis phenomena and to calculate the pore size distribution based on the both branches of the hysteresis loop. The pioneering papers in this field were those of Wall and Brown (1981); Neimark (1981); Mason (1982); Kheifets and Neimark (1982). Their development has led to the new percolation methods which are valid, in the first place, for such mesoporous solids as porous glasses, silica gels and other xerogels.

3.1. NETWORK MODELS

In real materials the pores are connected to one another and form a three-dimensional network. The interconnection of pores accounts for the cooperative character of adsorption phenomena. During capillary condensation the effect of the initiation of the condensation in the wide pores appears after condensation in the narrow pores adjacent to them (Neimark, 1981). The delay in desorption from the wider pores is stipulated by its blocking by the narrower ones (Everett, 1979). These cooperative effects cannot be allowed for by a model of unrelated pores: the requirements for filling or emptying of a given pore depend not only on its own characteristics but on the characteristics of adjacent pores as well.

There is no common pore structure model which can be used in all the cases. The variety of porous materials requires a necessity of the employment of different structure models. The shape of the hysteresis loop reflects on peculiarities of the pore structure. According to the IUPAC classification four main types of hysteresis loops can be distinguished: H12, H2, H3 and H4 (Gregg and Sing, 1982). The network effects are essential for the mesoporous solids with hysteresis loops of type H1 and H2.

Two different network models can be taken into account. The first, used for adsorbents exhibiting a hysteresis loop of type H1, is the network of channels. The second, used for the adsorbents exhibiting a hysteresis loop of type H2, is the network of cavities and constrictions. These models are particular examples of a more common model of a network of cavities and channels. In the first case we assume the pore space consists of intersecting channels of different sizes and the main pore volume is concentrated in the network bonds (Fig.1a). In the second case we assume the main pore volume is concentrated in the network sites imitating the pore cavities connected by more narrow pore constrictions (Fig.1b)

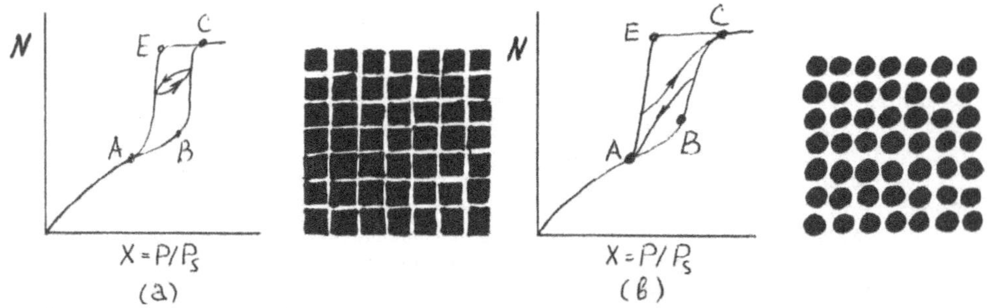

Fig.1. Hysteresis loops of type H1 (a) and of type H2 (b) and corresponding pore space models: network of channels (a) and network of cavities and constrictions (b).

By means of the percolation theory it was shown that the difference in the properties of the hysteresis in the adsorbents characterized by the loops of type H1 and type H2 can be explained in the frameworks of these models (Neimark,1991). The particular attention has to be spared to the scanning isotherms. The behaviour of isotherms scanning loop of type H1 is quite different from the behaviour of isotherms scanning loop of type H2. In the first case the scanning isotherms closed loops inside the main loop (Fig.1a). In the second case the scanning isotherms of desorption, starting from the main adsorption branch, finish in the point A of the beginning of hysteresis. On their turn the scanning isotherms of adsorption, starting from the main desorption branch, finish in the point C of the end of hysteresis. Analogous behaviour is typical for isotherms scanning internal hysteresis loops as well. This difference was explained by the peculiarities of porous structure, which are taken into account in the network models mentioned above.

3.2. PERCOLATION METHODS

The special methods for calculating the pore size distribution in mesoporous solids having loops of type H1 and H2 have been suggested (Neimark, 1983; 1986; 1991b). Their principal innovation is that they employ experimental information simulneously obtained from both the adsorption and the desorption branches of the isotherm. These methods are used in new versatile software for the characterization of porous solids (Neimark et al. 1990).

The problem of desorption from the network of pores along the main desorption branch is the classical problem of bond percolation. At a given value in the network two types of bonds (channels in the network of channels or pore constrictions in the network of cavities and constrictions) are distinguished: overcritical - of size $r>r_-(X)$ and undercritical - of size $r>r_+(X)$. The critical pore size of unrelated pores in which the desorption is observed at relative pressure X. This equivalent size of a single pore can be introduced by means of well-known thermodynamic equations of two phase equilibrium. The simplest one is the Kelvin equation (7) for a cylindrical pore:

$$r_-(X) = 2\sigma v/RT(-\ln X) \tag{11}$$

In principle, the difference between a network of pores and a system of unrelated pores is as follows. At the process of pressure reduction down to a given value the desorption occurs not from the all undercritical pores (as would be in a system of unrelated channels) but only from those pores which are forming connected system of undercritical pores looking on the external surface of the sample. The point E of the transition from the gently sloping section of desorption isotherm to the steep section corresponds to the percolation transition - to the forming of connected system of overcritical pores and the beginning of desorption from the sample's volume. In the point E the portion, q, of overcritical bonds in the network is equal to the percolation threshold q_c.

The percolation threshold is the minimum portion of marked bonds (in our problem, overcritical ones) forming an infinite connected system or an infinite cluster. Its value q_c depends on the coordination number, Z, of the three-dimensional network by the following approximate formulae

$$q_c \approx 1.5/Z \tag{12}$$

In the percolation theory the distribution in the network of two different types of bonds is characterized by the two functions: the connectivity function, $Q_c(q)$, is the portion of the bonds contained in the infinite cluster and the percolation probability, $Q_p(q)$, is the portion of sites contained in the infinite cluster. Below the percolation threshold the infinite cluster does not exist and consequently,

$$Q_c(q) = Q_p(q) \equiv 0 \qquad \text{at} \quad q<q_c \tag{13}$$

For the dependences $Q_c(q)$ and $Q_p(q)$ at $q>q_c$, the reliable interpolation formulae have been proposed to be used at calculating the pore size distribution (Neimark, 1986b).

These functions determine the portion, $Q_-(q)$, of pores in the network free of capillary condensate during desorption at a given value of X, namely, in the network of channels the portion of empty pores (channels), $Q_-(q)$, is equal to the connectivity function, $Q_c(q)$,

$$Q_{-}(X) = Q_c(q) \tag{14a}$$

and in the network of cavities and constrictions the portion of empty pores (cavities), $Q_{-}(X)$, is equal to the percolation probability, $Q_p(q)$,

$$Q_{-}(X) = Q_p(q) \tag{14b}$$

It should be taken into account that the portion q of the overcritical pores is nothing else but the integral pore size distribution function $q_{>}(r)$ - the portion of pores of size larger than r:

$$q(X) = q_{>}(r_{-}(X)) = \int_{r_{-}(X)}^{r_{max}} \Phi(r)\,dr \tag{15}$$

Here $\Phi(r)$ - the numerical differential distribution function for the channels in the network of channels and for the constrictions in the network of cavities and constrictions. Hence, for a given pore network model, the right part of the equation (13) is the known theoretical function of the pore size distribution, which can be used for calculating the portion of empty pores, $Q_{-}(X)$.

On the other hand, this value can be estimated from the experimental data. The assumption, that in the network the equivalent sizes and other geometrical characteristics of pores are not correlated, produced the following equation between the adsorption $N_{+}(X)$ and the desorption $N_{-}(X)$ isotherm at $X > X_B$ (B is the point of the beginning of irreversible capillary condensation, see fig.1):

$$N_{-}(X) = N_c - Q_{-}(X)\ (N_c - N_{+}(X)) \tag{16}$$

Equation (16) allows us to calculate the portion, $Q_{-}(X)$, of free pores at desorption for both models. On the basis of the comparison of this value obtained from experimental data with the theoretical value (14) we can determine the portion, $q_{>}(r_{-}(X))$, of overcritical pores. So the integral pore size distribution function is equal to

$$q_{>}(r) = Q_{c/p}^{(-1)}[[N_c - N_{-}(X_{-}(r))]/[N_c - N_{+}(X_{-}(r))]] \tag{17}$$

where $Q_{c/p}^{(-1)}$ is the function inverse to the connectivity function, $Q_c(q)$, for the network of channels and the function inverse to the percolation probability, $Q_p(q)$, for the network of cavities and constrictions, respectively, and $X_{-}(r)$ - the relative pressure corresponding to the desorption. The limited volume of this paper forces us from a single pore of size p to pay attention only to the basic theoretical ideals. Some examples are presented in our poster contribution (Neimark et al., 1992).

4. Conclusion.

We have discussed two new trends in the characterization of mesoporous solids using adsorption measurements which are based on modern ideas of fractal geometry and percolation theory. Fractal and percolation approaches have provided many positive results. Among them: (a) widening the knowledge about the geometry of mesoporous solids and its role in adsorption and related subjects; (b) introduction of fractal dimencion as a new chacteristec of porous solids; (c) development of new theoretical methods for calculating the fractal dimension; (d) establishment of fractal properties in various real systems such as aerogels, active carbons, coals, fine particles, products of diffusion limited aggregation, etc.; (e) development of network models for mesoporous solids; (f) quantitative description of the process of capillary condensation and desorption in mesoporous materials characterized by the adsorption hysteresis loop of type H1 and H2 by IUPAC classification; (g) development of the percolation methods for calculating the pore size distribution in mesoporous materials such as in xerogels, porous glasses, powder compacts, etc.

Acknowledgment: Supported by the Alexander von Humboldt Foundation.

5. References

Avnir D. Ed. (1989) *Fractal Approach to Heterogeneous Chemistry.* Wiley, New York.

Everett D.H. (1979) 'Network effects at capillary condensation' in S.J.Gregg et al. Eds. Characterization of porous solids. Soc. Chem. Ing. , London, p.229.

Gregg S.J., Sing K.S.W. (1982) *Adsorption, Surface Area and Porosity.* Academic Press, London.

Kheifets L.I., Neimark A.V. (1982) *Multiphase Processes in Porous Media*, Khimia, Moscow.

Mandelbrot B. (1982) *Fractal geometry of nature.* Freeman, San-Francisco.

Mason G. (1982) 'The effect of pore space connectivity on the hysteresis of capillary condensation in adsorption - desorption isotherms', *J. Coll Int. Sci.*, **88**, 36.

Mersmann A. Ed. (1990) *Proc. Third Int. Conf. Fundamentals of Adsorption*. Engineering Foundation, New York.

Neimark A.V., (1981) 'The theory of capillary phenomena in porous media and its application', *Sov. Chem. Ind.*, **11**, 656.

Neimark A.V. (1983) 'Development of a theory of capillary condensation and desorption in mesoporous adsorbents on the basis of lattice models of porous structure', *Rep. Acad. Sci. USSR (Phys. Chem.)*. **273**, 867.

Neimark A.V. (1986a) 'A percolation method for calculating the pore size distribution in materials of intermediate porosity based on the adsorption and desorption isotherms in the hysteresis region', *Russian J.Phys. Chem.* **60**, 1045.

Neimark A.V. (1986b) 'Calculation of percolation system characters', *Sov. Phys. - Tech. Phys.*, **56**, 2235.

Neimark A.V. (1990a) 'Determination of surface fractal dimensionality from adsorption experimental data', *Russian J. Phys. Chem.* **64**, 987.

Neimark A.V. (1990b) 'Thermodynamic method of calculating surface fractal dimension', *J. Exp. Theor. Phys. Lett.* **51**, 607.

Neimark A.V., Rabinovich A.B., Kheifets L.I. (1990) 'New versatile software for characterization of porous solids', *Preprints of COPS-II.* Poster 31. Alicante.

Neimark A.V. (1991a) 'Calculating surface fractal dimension of adsorbents', *Ads. Sci.& Tech.* 7, 210

Neimark A.V. (1991b) 'Percolation theory of capillary hysteresis phenomena and its application for characterization of porous solids', in F.Rodrigues-Reinoso et al. (Eds) *Characterization of Porous Solids II.* Elsevier, Amsterdam, 67.

Neimark A.V., Unger K.K., Krebs K.-F., Eisenbeiss F. (1992) 'Characterization of mesoporous silica using percolation method', *Poster contribution. NATO ASI on Multifunctional Mesoporous Inorganic Solids*. Sintra. Portugal.

Pfeifer P., Avnir D. (1983) 'Chemistry in noninteger dimensions between two and three', *J. Chem. Phys.* **79**, 3558.

Rodrigues-Reinoso F.,Rouquerol J., Sing K.S.W., Unger K.K. (Eds) (1991) *Characterization of Porous Solids-II.* Elsevier, Amsterdam.

NITROGEN ADSORPTION ON ZIRCONIUM BIS MONOHYDROGENPHOSPHATE WITH α-TYPE STRUCTURE.

F. Marmottini
Department of Chemistry, University of Perugia, Via Elce di Sotto, 8, 06123 Perugia, Italy.

ABSTRACT. Different samples of layered α-zirconium bis (monohydrogen phosphate) monohydrate, prepared by HF method, were outgased from room temperature to 300°C and their specific surface area determined by nitrogen adsorption. The specific surface area increases from 0.64 to 0.77 m^2/g as the outgasing temperature increases from r.t. to 200°C. A small decrease is observed at 300°C. Adsorption and desorption isotherms showed that this layered compound possesses only a negligible porosity. Thus the α-zirconium phosphate prepared according to the HF method can be used as a non porous reference material for the study of the porosity of the other M(IV) phosphates or/and phosphonates which have similar surface chemical properties. An empirical equation, interpolating adsorption data, is proposed and compared with standard isotherms reported in literature.

1. INTRODUCTION.

Nitrogen adsorption measurements are presently widely applied for the determination of specific surface area and pore size distribution of powder materials. The nitrogen adsorption isotherms on the α-zirconium bis (monohydrogen phosphate), $Zr(HPO_4)_2$, have been reported in literature since 1973; however, those investigations were essentially carried out for determining the variations of the surface area of the material due to its preparation method [1] or ion exchange processes [2].

In the last two decades α-zirconium hydrogenphosphate and, generally, α-layered M(IV) phosphonates, α-M(IV)$(RPO_3)_2$, where M(IV) is Zr, Ti, Hf, etc. and R is -OH, -H or organic radical, have raised an increasing interest for their ion exchange, intercalation, catalytic and proton conduction properties as well as for the possibility to design their

C.A.C. Sequeira and M.J. Hudson (eds.),Multifunctional Mesoporous Inorganic Solids, 37–48.

composition and structure by partial or full substitution of R groups with different chemical groups [4,5]. Their general formula can be now written as $M(IV)(RPO_3)_x(R'PO_3)_{2-x}\cdot nH_2O$ where $0 \leq x \leq 2$. The R and/or R' groups point into the interlayer region. Therefore porous materials can be obtained by a suitable choice of R and R' groups with different size or by joining adjacent layers with diphosphonic groups. Furthermore, it was recently reported that an appreciable porosity can be created in the interlayer region of M(IV) layered phosphonates [6,7] by intercalation of suitable organic or inorganic molecules which act as pillars to get compounds analogous to pillared clays [8]. A more detailed study of the nitrogen adsorption curves on α-zirconium phosphate seemed to be of interest since this compound could be used as a non-porous reference material for the surface and pore size distribution calculations of related porous materials. The IUPAC recommendation [3] indicates that the chemical nature of the material is the best criterion for selecting the most adequate standard adsorption isotherm.

The structure of α-zirconium bis (monohydrogen phosphate) monohydrate is formed by the packing of layers held together by Van der Walls forces [9]. This compound can be prepared according to two different main methods: the first consists in the formation of an amorphous precipitate that is subsequently crystallized by refluxing in phosphoric acid solutions [10], the second one in the precipitation of crystalline compound as a consequence of slow release of zirconium ions from their fluorocomplexes in the presence of phosphoric acid [11]. The material employed in this work was synthesized following the latter method because it gives, as confirmed by previous papers [1,12], materials with crystallinity degree higher than the former one.

2. EXPERIMENTAL.

2.1. SAMPLES AND THEIR PREPARATION FOR THE ANALYSIS.

A large batch of zirconium bis (monohydrogen phosphate) with α-layer structure was synthesized according to the Alberti and Torracca method [11]. $ZrOCl_2\cdot 8H_2O$ was dissolved in hydrofluoric acid solution, phosphoric acid and water were added to this solution until to have $[H_3PO_4]$ = 5.2M, $[F^-]/[Zr(IV)]$ = 8 and $[H_3PO_4]/[Zr(IV)]$ = 40. Due to the presence of Zr fluorocomplexes no precipitate formation occurred. The solution was put on water bath at 80 °C to reduce the concentration of hydrofluoric acid. During the precipitation the level of solution was kept constant. After five days the white precipitate was filtered, washed until pH=4 and, finally, stored over P_4O_{10}.

The compound was characterized by X-ray diffractometry, chemical analysis and thermogravimetry. The molar ratio Zr/P resulted 1 to 2 and the interlayer distance was 7.56 Å.
About 10g of this material were subsequently outgased at room temperature (ZP0), at 100°C (ZP1), at 200°C (ZP2) and at 300°C (ZP3). After each outgasing procedure the samples were used for the nitrogen adsorption measurements at 77K.
The experimental data reported in this work were collected by carrying out several measurements on different samples. The comparison of the various series of data is used to determine the level of correlation for the interpolating curves and to obtain an estimate for the reproducibility of adsorption data.

2.2. EXPERIMENTAL APPARATUS AND ANALYTIC METHODS.

The adsorption and desorption measurements were made automatically by a SORPTOMATIC 1800 (Carlo Erba Strumentazione) supplied with an outgasing unity.
The X-ray diffraction (XRD) patterns were taken with a Philiphs Diffractometer PW1710 using Ni filtered CuKα radiation.
Thermogravimetric and differential thermal analysis were performed by a Stanton ST780 thermoanalyser.
Phosphates analysis was made both by ion chromatography (DIONEX chromatograph) and colorimetric determinations (Beckman colorimeter) [13]. The zirconium content was obtained by gravimetry, after precipitation with cupferron [14].

3. RESULTS AND DISCUSSION.

The XRD patterns and the weight loss of the samples, after each outgassing step, were recorded. No change of XRD patterns and a small weight loss (0.03%) were observed after outgassing at room temperature for 24 hrs. Samples outgased at 100, 200 and 300°C showed, as expected, a weight loss of 6.03-6.10%, due to loss of the crystallization water, and a change of interlayer distance from 7.56 to 7.41 Å, value typical of anhydrous α-zirconium phosphate [15]. Figure 1 shows the XRD patterns of the samples ZP0 and ZP1.
Figure 2 shows the nitrogen adsorption isotherms obtained in the four experiments. All the isotherms can be classified as type II according to Brunauer, Deming, Deming e Teller [16]. The adsorption and desorption isotherms of ZP1 sample are shown in figure 3. The comparison between them shows that the hysteresis is practically absent; thus no evident mesoporosity is present in the solid. This behaviour is also exhibited by the other samples investigated.
Table 1 reports the values of surface area calculated by the

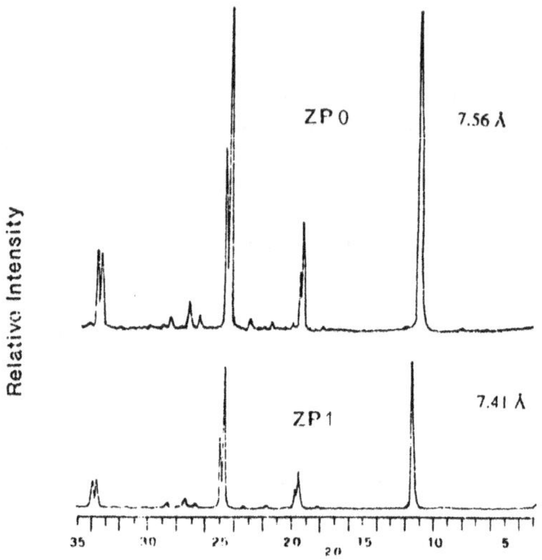

FIGURE 1:X-ray diffractograms of samples ZP0 and ZP1.

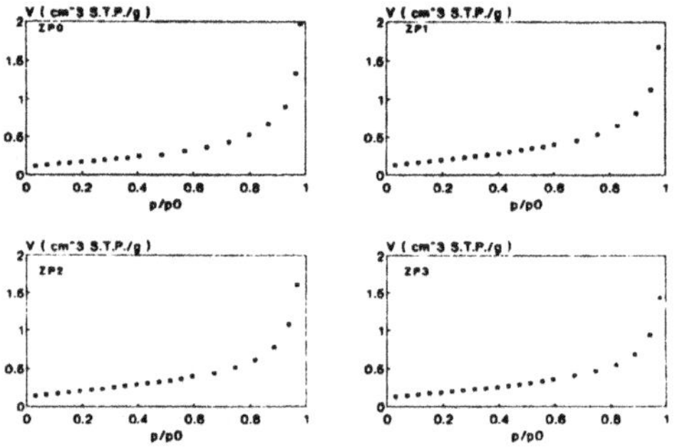

FIGURE 2: Nitrogen adsorption isotherms of indicated samples.

B.E.T. theory (V_m is the volume of adsorbed nitrogen necessary to form a monolayer of molecules adsorbed on the solid surface).
It can be observed that there is an increase of 0.12 m^2/g in the specific surface area going from ZP0 to ZP1. This increase is only in part imputable to weight loss (the calculated increase should be about 0.04 m^2/g) and may be ascribed to a loss of crystallinity degree due to dehydratation process. It is known [2] that the ion exchange processes are accompanied by cleavage of layer lattice structure. The water deintercalation is also a process which involve the interlayer region. Thus it is possible to have a decrease of the crystallinity degree of material during this

TABLE 1. Results obtained from B.E.T. application to nitrogen adsorption data as function of outgasing temperature.

Sample	Temperature (°C)	Surface (m^2/g)	C	V_m (cm^3S.T.P./g)
ZP0	room	0.64	45	0.146
ZP1	100	0.76	32	0.174
ZP2	200	0.77	46	0.177
ZP3	300	0.69	73	0.158

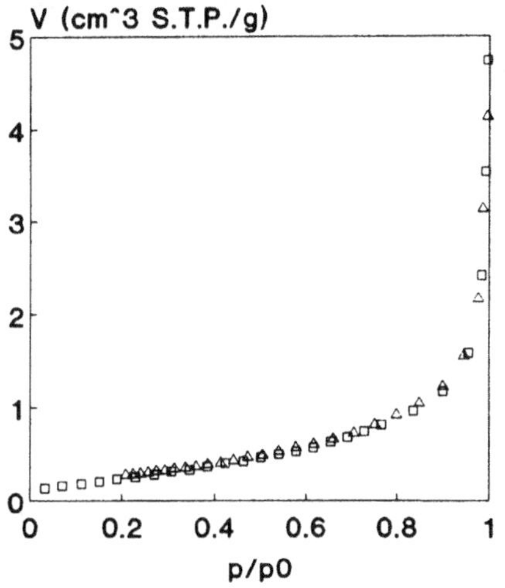

FIGURE 3: Nitrogen adsorption (□) and desorption (△) isotherms of ZP1.

process but the effect is less pronounced than that observed for the ion exchange process. The small decrease of α-$Zr(HPO_4)_2$ specific surface observed after outgasing at 300°C can be related to the condensation of P-OH groups belonging to adjacent surfaces of different grains with formation of polycrystalline aggregates. The hypothesis that some condensation of hydrogenphosphate groups can occur at the temperature of 300°C was also made in previous papers [17].
Other authors [1] report specific surface area values higher than those listed in table 1; the discrepancy can be ascribed to the different crystallinity degree arising from the different preparative methods.
The C values of the four samples obtained from the B.E.T. theory are comprised between 32 and 73. Results of measurements on different samples indicate that the values

listed in table 1 are reproducible for the various thermal treatments. Furthermore, it may be observed that the values of surface area and of C of the samples outgased in different conditions are comprised in a narrow range. This is consistent with the fact that the various samples have similar chemical composition and that, in the investigated temperature range, thermogravimetric and differential thermal analysis do not show other effect than the crystallization water loss. Thus a thermal treatment in this temperature range has not relevant influence on the surface adsorption properties.
Table 2 reports the experimental values of V/V_m, referred to the various p/p_0, for the four samples (p is the equilibrium pressure, p_0 is the saturation pressure, V is gaseous nitrogen volume, at S.T.P., adsorbed for gram of solid).

TABLE 2. Nitrogen adsorption data obtained for the indicated samples.

ZP0		ZP1		ZP2		ZP3	
p/p_0	V/V_m	p/p_0	V/V_m	p/p_0	V/V_m	p/p_0	V/V_m
0.03	0.77	0.03	0.71	0.03	0.76	0.03	0.78
0.07	0.87	0.07	0.79	0.07	0.86	0.07	0.86
0.11	0.96	0.11	0.91	0.11	0.95	0.11	0.95
0.16	1.05	0.15	0.99	0.15	1.04	0.15	1.06
0.20	1.13	0.19	1.08	0.19	1.12	0.19	1.12
0.24	1.21	0.24	1.20	0.23	1.20	0.23	1.23
0.28	1.33	0.28	1.27	0.27	1.29	0.27	1.29
0.32	1.41	0.32	1.38	0.31	1.41	0.31	1.41
0.36	1.50	0.36	1.51	0.35	1.48	0.35	1.48
0.40	1.64	0.40	1.59	0.39	1.59	0.39	1.58
0.49	1.79	0.44	1.72	0.43	1.68	0.43	1.66
0.57	2.13	0.48	1.84	0.47	1.77	0.47	1.80
0.65	2.47	0.52	2.01	0.51	1.87	0.51	1.89
0.72	2.89	0.56	2.12	0.55	2.01	0.55	2.03
0.80	3.58	0.60	2.29	0.59	2.20	0.59	2.21
0.87	4.55	0.68	2.60	0.67	2.44	0.67	2.51
0.93	6.13	0.76	3.07	0..74	2.89	0.74	2.88
		0.83	3.74	0.82	3.45	0.82	3.45
		0.89	4.69	0.88	4.36	0.89	4.31
		0.95	6.46	0.94	6.04	0.94	5.92

The ratio V/V_m can be considered as a measure of the average number of adsorbed layers on the surface of solid. For the study of porosity, Halsey proposed an empirical equation (here reported in the Wheeler modified form), that gives the ratio t/σ as a function of p_0/p (where t is the thickness of adsorbed nitrogen layers at p/p_0 and σ is the thickness of nitrogen monolayer); namely:

$$t/\sigma = (5/\ln(p_0/p))^{1/3} \qquad (1)$$

In this work it was assumed that the V/V_m as a function of p_0/p may have the same form of equation 1; i.e.:

$$V/V_m = (a/\ln(p_0/p))^{1/b} \qquad (2)$$

and the parameters a and b were optimized in order to obtain the best fit of experimental data of table 2. The interpolation has been made with the least square method by writing equation 2 in the following logarithmic form:

$$\ln(V/V_m) = (1/b)\ \ln a - (1/b)\ \ln(\ln(p_0/p)) \qquad (2bis)$$

Thus, if the empirical equation 2 is able to model the experimental data of table 2, the plot of $\ln(V/V_m)$ against $\ln(\ln(p_0/p))$ will be linear and the slope, s, and the intercept, i, of the straight line will give a and b according to the following equations:

$$s = -(1/b) \qquad (3)$$

$$i = (1/b)\cdot\ln a \qquad (4)$$

that is:

$$b = -(1/s) \qquad (5)$$

$$a = e^{-(i/s)} \qquad (6)$$

Figure 4 shows the plots $\ln(V/V_m)$ as function of $\ln(\ln(p_0/p))$ for the four cases examined in this work. From an analysis of these plots it can be seen that the region of linearity extends up to p/p_0 values of about 0.90 (corresponding to $\ln(\ln(p_0/p) = -2.25)$.
The a and b values obtained for $p/p_0<0.90$ are listed in table 3 together with the standard deviation of experimental V/V_m values.

TABLE 3. a and b parameters of equation 2 obtained by application of equation 5 and 6 to plots of figure 5.

	ZP0	ZP1	ZP2	ZP3
a	2.07	1.98	2.12	2.16
b	1.75	1.74	1.87	1.89
σ	0.03	0.14	0.06	0.08

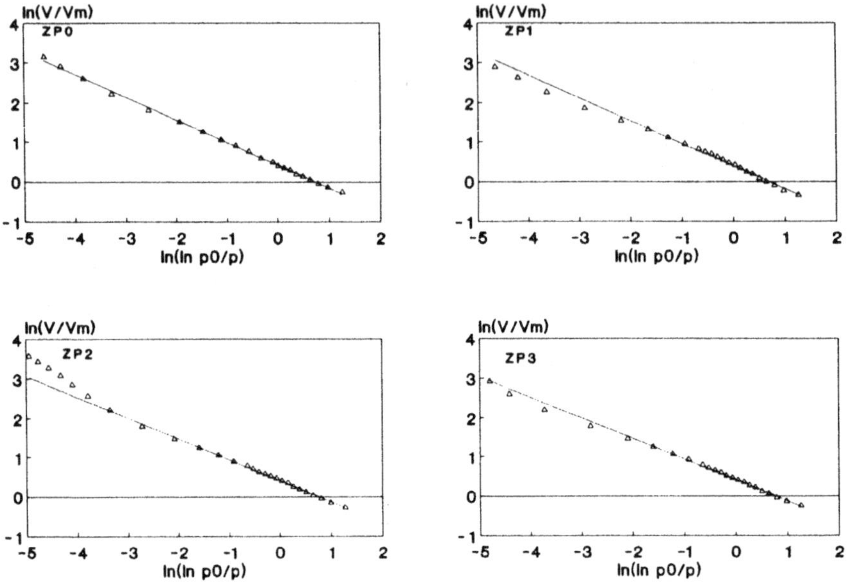

FIGURE 4: $\ln(V/V_m)$ against $\ln(\ln(p_0/p))$ plots of indicated samples

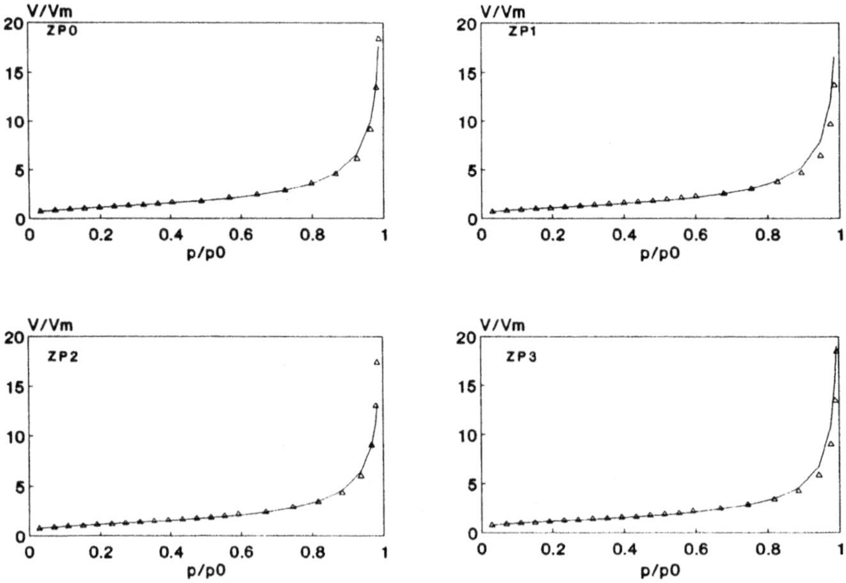

FIGURE 5: V/V_m plots of indicated samples. The continuous lines are the interpolating functions.

The interpolating curves are plotted in figure 5 and compared with the experimental data. It may be seen that the agreement is quite good.
It was of interest to check if all the experimental data, listed in table 2, could be interpolated by a single equation with acceptable standard deviation. The simultaneous interpolation of all the four series of data leads to the following equation:

$$V/V_m = \left(2.08/\ln(p_0/p) \right)^{1/1.81} \qquad (7)$$

for ($p/p_0 < 0.90$)

Figure 6 shows the plot of this equation (continuous line) together with the adsorption data listed in table 2. The standard deviation found is 0.11.
Finally, figure 7 shows the comparison between the plot of equation 7 (curve a) and those of other standard isotherms reported in literature, namely equation 1 (curve b), Lecloux and Pirard data (for C parameter values between 40 and 100) [18] (curve c) and De Boer et al. equation [19] (curve d). Within given intervals of p/p_0 the agreement between the equation proposed in this work with the others is good. The best one is obtained in the case of curve c, for p/p_0 values less than 0.3, and curve d, for p/p_0 values less than 0.6.

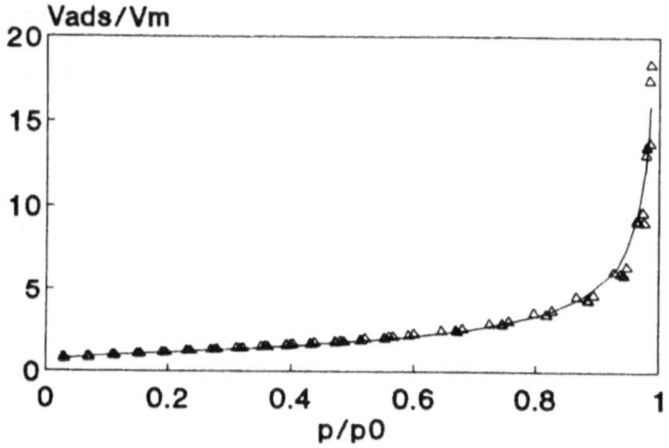

FIGURE 6: V/V_m plot obtained by interpolation of all data of table 2.

4. CONCLUSIONS.

The specific surface area measured in this work is lower than that reported in ref. [1]. The discrepancy can be ascribed to the different preparation conditions that in the present work allow to obtain more crystalline samples which

do not show an evident porosity.
It has also been found that the dehydratation process causes an increase in specific surface area probably due to some cleavage and/or breaking of the microcrystals.
The nitrogen adsorption isotherms for zirconium hydrogen phosphate with layered structure of α-type, treated from room temperature to 300°C, have been examined. It has been found that the outgasing temperature do not sensibly affect the adsorption isotherms, but the difference between them are significant. The experimental data have been interpolated and the obtained equation has been compared with standard isotherms proposed by other authors.
We are presently using the equations proposed in the present work with good results to calculate the surface and porosity characteristics of various zirconium phosphates and phosphonates.

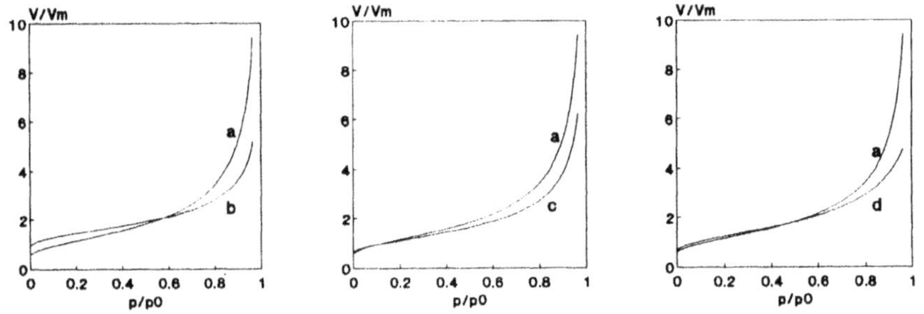

FIGURE 7: Comparison of the equation interpolating the experimental data (curve a) with Halsey's equation (curve b), curve obtained by interpolation of the Lecloux and Pirard data (curve c) and De Boer et al. equation (curve d).

ACKNOWLEDGEMENT. The author thanks Prof. Alberti and Prof. Costantino for the encouragement and many useful discussions and M.U.R.S.T. 40% for financial support.

REFERENCES.

1. Horsley, S.E. and Nowell, D.V. (1973) "The Preparation and Characterisation of Crystalline α-Zirconium Phosphate", J. Appl. Chem. Biotechnol. 23, 215-224.
2. Horsley, S.E. and Nowell, D.V. (1974) "The Particle Characteristics of Cation Exhanged α-Zirconium Phosphate" J. Colloid and Interf. Sci. 49, 394-402.

3. Sing, K.S.W., Everett, D.H., Haul, R.A.W., Moscou, L., Pierotti, R.A., Rouquerol, J. and Siemieniewska T. (1985) "Reporting physisorption data for gas/solid systems with special reference to the determination of surface area and porosity (Recommendations 1984)" Pure Appl. Chem. 57, 603-619.
4. Clearfield, A. (1982) "Zirconium phosphates" in A. Clearfield (ed), Inorganic Ion Exchange Materials, CRC Press, Boca Raton (FL) pp. 1-74.
5. Alberti, G. (1978) "Synthesis crystalline structure and ion exchange properties of insoluble salts of tetravalent metals and their salt forms" Acc. Chem. Res. 11, 163-170.
6. Clearfield, A. (1987) "New approaches to the design of materials via preparative inorganic chemistry" in D.L. Cocke and A. Clearfield (eds.) Design of New Materials, Plenum Press, New York, pp. 121-134.
7. Alberti, G. and Costantino, U. (1991) "Intercalation of zirconium phosphates and phosphonates", in J.L. Atwood, J.E.D. Davies and D.D. MacNicol (eds.), Inclusion Compounds, Oxford University Press, 5, pp. 136-176.
8. Pinnavaia, T.J. and Kim, H. (1992) "Molecular engineering of lamellar solids. Principles derived from the pillaring of smectite clays" in E.G. Deronane, F. Lemos, C. Naccache and F.R. Ribeiro (eds.), Zeolite Microporous Solids: Synthesis, Structure and Reactivity." , Kluwer Academic Publisher, pp. 79-90.
9. Clearfield, A. and Smith, G.D. (1969) "The crystallography and structure of zirconium bis (monohydrogenphosphate) monohydrate" J. Inorg. Chem. 8, 431-436.
10. Clearfield, A. and Stynes, J.A. (1964) "The preparation of crystalline zirconium phosphate and some observation on its ion exchange behaviour" J. Inorg. Nucl. Chem. 26, 117-129.
11. Alberti, G. and Torracca, E. (1968) "Crystalline insoluble salts of polybasic metals. II. Synthesis of crystalline zirconium or titanium phosphate by direct precipitation" J. Inorg. Nucl. Chem. 30, 317-318.
12. Allulli, S., Massucci, M.A. and Tomassini, N. (1979) "The standard entalphy of formation of anhydrous, monohydrated and dihydrated zirconium bis(monohydrogen phosphate)" J. Chem. Thermodyn. 11, 613-618.
13. Bernhart, D.N. and Wreath, A.R. (1955) "Colorimetric determination of phosphorus by a modified phosphomolibdate method" Anal. Chem. 27, 440-441.
14. Elving, P.J. and Olson, E.C. (1955) "Gravimetric and titrimetric determination of titanium, zirconium and hafnium with cupferron" Anal. Chem. 27, 1817-1820.
15. La Ginestra, A., Ferragina, C., Massucci, M.A., Patrono, P., Di Rocco, R. and Tomlinson, A.A.G., (1983) "Thermal,

redox, and catalytic characterisation of synthetic inorganic ion exchangers" Gazz. Chim. Ital. 113, 357-365.
16. Brunauer, S., Deming, L.S., Deming, W.E. and Teller, E. (1940) "A theory of the van der Waals adsorption of gases" J. Amer. Chem. Soc. 62, 1723-1732.
17. Alberti, G., Casciola, M., Costantino, U. and Radi, R. (1979) "The mechanism of proton transport in powder compounds of acid salts with α-layered structure" Gazz. Chim. Ital. 109, 421-426.
18. Lecloux, A. and Pirard, J.P. (1979) "The importance of standard isotherms in the analysis of adsorption isotherms for determining the porous texture of solids" J. Colloid Interf. Sci. 70, 265-281.
19. De Boer, J.H., Lippens, B.C., Linsen, B.G., Broekhoff, J.C.P., Heuvel, A. van den and Osinga, T.J. (1966) "t-Curve of multimolecular nitrogen adsorption" J. Colloid Interf. Sci. 21, 405-414.

STUDY OF THE ACCESSIBLE MICROPORE VOLUME IN PILLARED MONTMORILLONITES

M.D.ALBA, R.ALVERO, M.A.CASTRO and J.M.TRILLO
Dept. Inorganic Chemistry. University of Sevilla
P.O.Box 874
41080 Sevilla
Spain

ABSTRACT. The evolution upon heating of the interlamellar accessible volume of two alumina pillared montmorillonites has been studied. The samples under examination have been selected with quite different charge deficit distributions between the tetrahedral and octahedral sheets to test the influence of their different reported pillar-lattice link nature on their microporosity. A comparison between theoretical accessibility values and experimental ones using two approaches has been stated by means of XRD data and nitrogen adsorption isotherms on the samples heated at different temperatures. The experimental accessible volumes obtained from t-plots and a subtraction method are in good agreement, and much lower than those calculated from crystallographic data. Identical results have been obtained for their evolution upon heating. From this, the accepted conclusion that alumina pillars are stable at temperatures higher than 500°C could give a misleading general picture of the pore structure evolution with calcination.

1. Introduction

Pillared clays have appeared in the search for large-pored materials to substitute zeolite-based catalysts. Improvement of their thermal stability is the main aim of most researchers in this field.

Within the studies on the alumina pillar-sheet interaction, a division has been established regarding the deficit charge distribution between octahedral and tetrahedral layers. So, whereas temperatures higher than 500°C are necessary to anchor the olygomer to the lattice in the case of montmorillonites without tetrahedral substitution (Tennakoon et al., 1986), this transformation is attained for samples with this kind of isomorphous substitutions at temperatures lower than 400°C (Plee et al., 1985). These results have led to different proposed pillar-lattice bonding models.

C.A.C. Sequeira and M.J. Hudson (eds.),Multifunctional Mesoporous Inorganic Solids, 49–53.

Evolution of the interlamellar accessible volume upon thermal treatment of pillared smectites has essentially been analyzed by basal spacing and BET area measurements, correlations between both parameters being frequently reported (Sterte, 1991a). It should be noted (Sing et al., 1985) that the BET method does not include in its formulation the micropore filling and, therefore, is not appropriate to characterize this type of solids where pores narrower than 20 Å are induced by the pillaring process, even though it is usual to report BET areas and pore-size distribution data on pillared clays (Takahama et al., 1992). Consequently, it is advisable to apply proper procedures to obtain a more real picture of the thermal stability of pillared smectites with respect to accessible volume.

The aim of the present work is to evaluate the accessible volume of two pillared montmorillonites selected with quite different charge distributions, by using two different experimental approaches: the *t*-method (Lippens et al, 1965) and a subtractive one taking sodium montmorillonites as references . A comparison between them and the one theoretically calculated has also been performed. Along with the standard pillaring agent (Keggin cation), a similar solution containing lanthanum was employed to test its promoter effect as reported in the literature (Sterte, 1991b).

2. Materials and Methods

The montmorillonites used were a Wyoming sample designated SWy-1, with a charge deficit per unit cell of 0.60 corresponding predominantly to isomorphous substitutions in the octahedral sheet (Jaynes et al., 1987) and a Trancos sample from Gádor (Spain) with a charge deficit value of 0.87 distributed as 0.36 and 0.51 between tetrahedral and octahedral layers (Trillo et al., 1990). The corresponding sodium-saturated samples, Na-Mt, were used as starting materials in the pillaring process, as well as reference ones.

The montmorillonites pillared with alumina, as well as those containing lanthanum and alumina in a molar ratio 1:25, Al-CLM and Al(La)-CLM, were prepared according to the method described by Singh et al. (1988). Air calcination treatments were performed at 300, 500 and 700°C for 6 hours on all the samples, prepared in order to determine their adsorptive properties upon heating.

X-ray powder diffraction diagrams were obtained with a Siemens Kristalloflex D-500 instrument, using Cu $K\alpha$ radiation, 36 kV and 26 mA.

Nitrogen adsorption isotherms were carried out with a Micromeritics Flowsorb II 2300 instrument, at 77.35 K, after "in situ" outgassing of the samples at 250°C for 4 hours. To apply the *t*-method correctly, a reference sample had to be selected, fulfilling the requirements of being non-porous and having a surface and chemical structure similar to those under study. A Hillsboro pyrophyllite ground for ten minutes (Sánchez et al., 1992) was shown to be an appropriate reference sample.

3. Results and Discussion

To measure the accessible volume of the two samples studied, we have used two experimental approaches. The first is the IUPAC-recommended t-plot, using an adequate reference non-porous solid (see figure 1 for selected results). The second consists of the elimination from the nitrogen adsorption isotherm of the multilayer contribution by subtracting a reference isotherm, the result thus providing the microvolume contribution to the total experimental adsorption.

The requirements of the reference sample to be used correctly with the second method are: absence or negligible amount of micropores, and chemical/surface similarity to the studied sample, thus implying similar external surface areas, measurable from S_t values. The sodium-saturated Trancos and Wyoming montmorillonites were chosen as references, and each pillared sample, calcined at a different temperature, was compared to the corresponding sodium one. This procedure gave rise to results from which a set is included in figure 2, where the suitability of the subtractive method is shown. Type I isotherms are obtained from which the micropore volumes were estimated.

The results obtained are included in tables 1 and 2, corresponding to Trancos and Wyoming montmorillonites, respectively. S_{BET} values are not shown due to their lack of meaning for this kind of microporous samples. Instead, S_t are presented, the similarity among the set of samples at the same calcination temperature being remarkable in all the cases. This fact let us validate the subtractive method employed and the micropore volumes obtained (V^b_{mic}). Comparison of these values with those from t-plots (V^a_{mic}) leads to a good agreement which can be considered as a consequence of both the accurate application of the method and the reliability of data.

Accessible theoretical volumes are presented in tables as V^c_{mic}. They were calculated by subtracting the volume occupied by the pillars from the total volume generated per gram in the pillaring process in each case. For the first value, the amount of aluminium incorporated to the sample, known by chemical analysis, was used, considering that the whole aluminium is presented as Al_{13} cations and employing crystallographic data reported by Plee et al. (1985). For the second one, basal spacing d(001) and density data were used.

By comparison of the experimental and theoretical values, the latter are higher in all cases. An observation of tables 1 and 2 reflects a decay of the whole V_{mic} values when the calcination temperature is increased, even though the theoretical ones, from the basal spacing, are less sensitive to this variation. This lower sensitivity could be against the wide employment of the c parameter as a single figure, without paying attention to either the intensity or the shape of the corresponding band in the diffractogram. Thus, the general conclusion from d(001) data that alumina pillars are stable at temperatures higher than 500°C could give a misleading picture about the pore structure evolution with calcination.

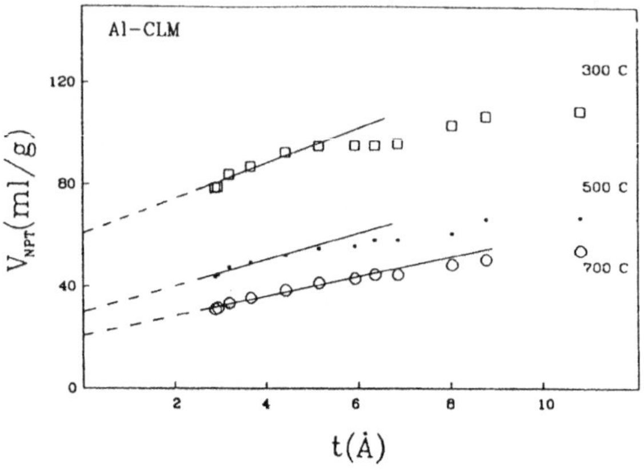

Figure 1. *t-plots of Trancos Al-CLM calcined at different temperatures.*

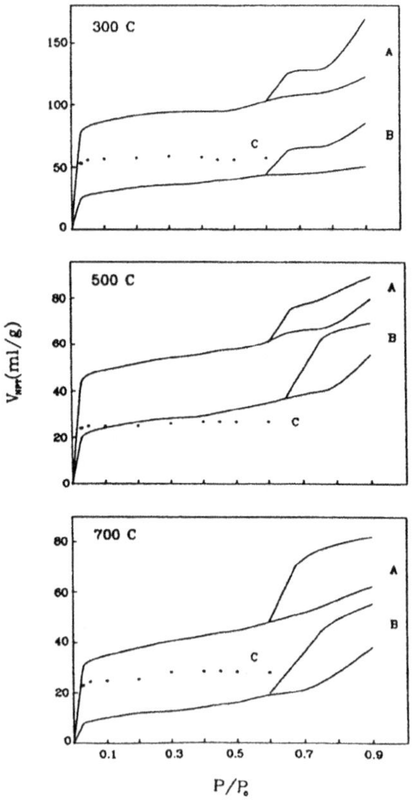

Figure 2. *Nitrogen adsorption isotherms for Trancos Al-CLM* (A) *and Na-Mt* (B) *calcined at the temperatures indicated. Curve* C *corresponds to the difference between* A *and* B.

Table 1
Trancos Montmorillonite

SAMPLE	Heating (C)	St (m2)[d]	V mic[a] (ml)[d]	V mic[b] (ml)[d]	d(001) (A)	V mic[c] (ml)[d]
Na-Mt	300	94	0.012		12.6	
Na-Mt	500	87	0.006		9.9	
Na-Mt	700	43	0.000		9.9	
Al-CLM	300	129	0.066	0.065	18.2	0.22
Al-CLM	500	79	0.046	0.038	17.2	0.19
Al-CLM	700	69	0.029	0.036	b.r.	
Al(La)-CLM	300	114	0.066	0.060	18.2	0.22
Al(La)-CLM	500	82	0.065	0.056	16.2	0.16
Al(La)-CLM	700	77	0.030	0.041	b.r.	

Table 2
Wyoming Montmorillonite

SAMPLE	Heating (C)	St (m2)[d]	V mic[a] (ml)[d]	V mic[b] (ml)[d]	d(001) (A)	V mic[c] (ml)[d]
Na-Mt	300	31	0.003		9.8	
Na-Mt	500	28	0.004		9.6	
Na-Mt	700	21	0.000		9.6	
Al-CLM	300	72	0.119	0.112	18.3	0.24
Al-CLM	500	24	0.040	0.036	17.9	0.23
Al-CLM	700	17	0.023	0.021	16.8	0.20
Al(La)-CLM	300	41	0.101	0.100	17.7	0.22
Al(La)-CLM	500	56	0.065	0.071	17.3	0.21
Al(La)-CLM	700	23	0.053	0.052	17.0	0.20

[a] values obtained from "t" plots
[b] values obtained by subtraction of reference sodium sample
[c] values obtained using theoretical calculations
[d] surface and volume values expressed per gram

SECTION 2

POROUS CRYSTALLINE MATERIALS

HYDROTHERMAL SYNTHESIS AND POROUS CRYSTALS

R. M. BARRER
Chemistry Department
Imperial College
London SW7 2AY
United Kingdom

ABSTRACT. Procedures for making minerals include hydrothermal growth, pyrolysis, vapour phase deposition and pneumatolysis. In synthesis of porous crystals the hydrothermal method is standard, although the pyrolytic procedure yields water-swelling layered fluorsilicates and microporous felspathoid and zeolite frameworks in which salts replace zeolitic water. Pneumatolysis can also be important, not for synthesis but for zeolite modification, as in preparing USY from zeolite Y. Some physicochemical aspects of, and recent trends in, hydrothermal synthesis are considered. The many types of crystal which are, or can be made, permanently porous include 3-dimensional 4-, 6, 4- and 6-connected nets as host frameworks and, inter alia, layer structures such as clay minerals, hydrotalcites and certain phosphates, titanates and phosphonates. Even some organic crystals can display permanent microporosity. Molecule sieving and lock-and-key sorption are exemplified for several less familiar microporous crystals.

1. Introduction

There has been a remarkable growth in the science and technology of porous crystals in the last 50 years. It is the purpose of this paper to comment on mineral synthesis, especially upon trends in hydrothermal chemistry, and upon the range and character of some currently available porous crystals, both silicate and non-silicate.

2. Mineral Synthesis

Early researches on the synthesis of minerals were performed by earth scientists wishing to learn about reactions and transformations in the lithosphere. In Nature these processes can be vast in extent and in the times and pressures involved, whereas in the laboratory both amounts and times are necessarily limited. Table 1 [1] gives procedures which have been used, exemplified by synthesis of important silicate groups. Of the four methods indicated in column 1 of the table hydrothermal synthesis and pyrolysis are notably versatile in silicate chemistry. In vulcanism all four processes can have major roles. For synthesising porous crystals, however, it is the hydrothermal method which has been outstandingly successful.

3. Hydrothermal Synthesis

3.1. STABILISATION OF HOST BY GUEST

As Table 1 indicates, in hydrothermal synthesis crystals form from an aqueous often multi-

C.A.C. Sequeira and M.J. Hudson (eds.), Multifunctional Mesoporous Inorganic Solids, 57–71.

component magma. Zeolites in particular grow readily from such mixtures at high pH (§ 2.3.). The alkalinity increases the solubility of the amphoteric oxides involved in the solvent, water. Water plays a second essential stabilising role as a guest species filling the pores and channels in the host frameworks. The stabilisation has a thermodynamic basis which was examined in 1988 [2].

TABLE 1. Methods of crystal growing exemplified by several important types of silicate minerals [1]

Growth Conditions	Examples *	Variants
Vapour only Pressure, P usually controlled. High Temperature, T.	Cristobalite	Decomposition of vapour at hot wire
Pneumatolytic Reactive gas plus solids, T high, P often low.	Felspars Quartz	Reactive gas may be H_2O, CO_2, HF, etc
Hydrothermal Aqueous solutions plus solids. T medium high. Often autogenous P.	Clay minerals Zeolites Felspathoids Felspars Crystalline silicas	Isothermal Temperature gradient Cooling
Pyrolytic High T. P may be very small to very great. Solids or melts or both.	Fluor-smectites Fluor-vermiculites Fluor-micas Felspathoids Felspars Crystalline silicas	Flux-melts Crystal pulling from melt Sintering Hot pressing Diamond forming process

* References are given in [1].

The crystal plus its zeolitic water represents a type of two component solution which, like any other, obeys the Gibbs-Duhem equation. The chemical potential of the water-bearing host component (μ_H) is thus lowered relative to that of the water-free host (μ_H^θ). Integration of the Gibbs-Duhem relation gives for the stabilisation free energy of the host, $\Delta \mu_H$,

$$\Delta\mu_H = \mu_H - \mu_H^\theta = V_H P - RT\nu \int_0^a (\theta/a)\,da = \mu_1 + \mu_2 \qquad (1)$$

The activity, a, of the guest component is zero when the host is guest-free. θ denotes the fractional saturation of the host by the guest when the activity of the guest is a; and ν is the ratio of moles of guest to moles of host when θ is unity. V_H is the molar volume of the host which in rigid zeolite frameworks is nearly independent of the uptake of guest. P is the total hydrostatic pressure, often only the vapour pressure of the water. If the vapour pressure, or the relative pressure, is used as a measure of the activity the integral can be found graphically from the sorption isotherm of the water plotted as θ/a against pressure. For silicas the natural choice of the mole of host is SiO_2, and by analogy a corresponding choice of mole for the zeolite is $M_ZAl_ZSi_{1-Z}O_2$ where $0<z<0.5$ and M is one equivalent of cations.

Most sorption isotherms in zeolites are of Type 1 in Brunauer's classification [3]. Langmuir's isotherm is the simplest theoretical isotherm of this type and was used to estimate the integral $\Delta\mu_2 = -RTv\int_0^a(\theta/a)da$ in the most highly and the least highly hydrated zeolites (respectively faujasite and bikitaite). Some of the results are given in Table 2 [2] for different values of the Langmuir equilibrium constant $K = \theta/(1-\theta)x$ where x is the relative pressure of the water vapour. It is seen that

TABLE 2. $\Delta\mu_2$ in kJ mol^{-1} at x = 1 for different values of K, using the Langmuir model [2]

K	θ at x = 1	$-\Delta\mu_2$ at 100°C for x = 1 v = 1.25 (faujasite)	 v = 0.33 (bikitaite)
1	0.500	2.687	0.717
2	0.66˙	4.259	1.136
3	0.750	5.374	1.433
5	0.833˙	6.946	1.853
10	0.90909	9.295	2.479
15	0.9375	10.75	2.867
25	0.96154	12.63	3.368
50	0.98039	15.24	4.064
100	0.99010	17.89	4.771
500	0.99800	24.10	6.427
1000	0.99900	26.78	7.141

the more highly hydrated the zeolite the larger is $-\Delta\mu_2$ at x = 1, and that $-\Delta\mu_2$ can be substantial. $\Delta\mu_1 = V_Hp$ was also calculated for a range of zeolites and temperatures at autogenous pressures. In the range of temperatures suitable for zeolite synthesis its value was much less than $-\Delta\mu_2$ at x = 1, so that $\Delta\mu_H = (\Delta\mu_1 + \Delta\mu_2)$ is dominated by $\Delta\mu_2$.

Several other conclusions in [2] were as follows:

(i) Stabilisation of zeolites by zeolitic water at x = 1 is larger the more rectangular the isotherm and therefore the lower the synthesis temperature and/or the more aluminous (and so the more hydrophilic) the zeolite.

(ii) Because equ. 1 with appropriate V_H, v, θ, a and P applies to any guest molecule, stabilisation of the guest-bearing zeolite framework relative to the guest-free structure is a completely general phenomenon. Where the host framework is neutral and hydrophobic (as in the porosils) the best stabilising guests will be organic species rather than water, as shown by the contours of the respective isotherms.

(iii) Stabilisation of the host by the guest is not the same as a template action which favours nucleation of a particular zeolite. Thus, water is a universal stabiliser of zeolites, but it can hardly be a universal template. There is, however, the possibility that a specific organic molecule may function both as a template which initiates nucleation of a particular zeolite and also as a space filling component which stabilises it.

3.2. NUCLEATION

Events at the molecular level which occur during nucleation of zeolites still remain unclear despite various suggestions. However a formal general account of nucleation is possible. Willard Gibbs considered phase separation in terms of two kinds of fluctuation [4]. The first of these is minute in volume but large in extent of departure from the mean in density or composition. It may lead to binodal phase separation. The second type of fluctuation is large in volume but minute in departure from the mean density or composition. It may lead to spinodal phase separation. Hydrothermal nucleation is of the binodal type. What factors oppose the immediate appearance of viable nuclei growing spontaneously? If a particular fluctuation comprises of j lattice-forming units its free energy formation, Δg_j, can be considered in terms of three components:

$$\Delta g_j = \Delta g_\delta + \Delta g_s + \Delta G(j/N_A) \qquad (2)$$

In equ. 2, Δg_δ and Δg_s denote respectively interfacial and strain free energies and ΔG is the free energy of formation of that amount of bulk zeolite containing one Avogadro number, N_A, of lattice-forming units. For zeolite formation to be possible, since Δg_δ and Δg_s are positive, ΔG must be negative. Δg_s and $\Delta Gj/N_A$ are each expected to be proportional to the volume of the fluctuation ($\Delta g_s = C j$; $\Delta G\, j/N_A = -Aj$) and Δg_δ is proportional to the surface area ($\Delta g_\delta = Bj^{2/3}$). Thus

$$\Delta g_j = Bj^{2/3} + Cj - Aj \qquad (3)$$

where A, B and C are coefficients ($A = -\Delta G/N_A$). In a fluctuation involving gel or solution we expect C to be small or zero (little or no strain free energy and A>C). As j increases we expect the positive term $Bj^{2/3}$ to increase less rapidly than the negative term $-(A-C)\,j$. If, as expected for small j, the positive term exceeds the negative term a plot of Δg_j, against j will at first increase from zero to positive values (Fig 1). The fluctuation is thus unstable and more likely to disappear than to grow. However equ. 3 has a maximum given by $d\Delta g_j / dj = 0$ and so from equ. 3

$$j_m = 8B^3/27(A-C)^3 \quad ; \quad \Delta g_{jm} = j_m\,(A-C)/2 \qquad (4)$$

where subscript m denotes "at the maximum". Any increase in j beyond the maximum results in a decrease in free energy and so the fluctuation will thereafter tend to grow spontaneously.

In favourable synthesis conditions and often after aging, a current of germ nuclei can be expected to begin to cross the saddle point of the curve of Δg_j vs. j and so to become viable. As such nuclei grow, their surface area increases with time, and they therefore consume nutrients more and more successfully in competition with germ nuclei. This tends to decrease the chance that germ nuclei become viable, and so to decrease the current of germ nuclei across the saddle point of Fig. 1 as time increases. Kinetic analysis for growth of Na-X showed that the rate at which viable nuclei formed went through a maximum and then decayed toward zero in agreement with the above expectation [5]. The linear growth rate of individual crystals was measured and found to be constant until near-exhaustion of the chemical nutrients [5,6]. The above observations account for the sigmoid form of curves of percent crystallisation vs. t which is almost universally observed [6]. The sharper the maximum in the rate of formation of viable nuclei versus time, the narrower should be the half-width of the curve, and the more uniform the crystal size in the final yield. Uniformity in size of zeolite Y crystals was reported if gelatin was present in the crystallising aqueous gel [7]. This may limit sedimentation and so maintain a more uniform environment for growing crystals. The greater the gelatin content the narrower was the crystal size distribution and the smaller the average particle diameter.

3.3 THE ROLE OF pH AND THE FLUORIDE PATHWAY

pH is an important variable in hydrothermal chemistry. In the highest pH range alkali-rich **tectosilicates** (zeolites, felspars and felspathoids) crystallise readily. **Micas**, which are likewise rich in alkali, can also form at high pH at high temperatures. **Smectites**, which are considerably less rich in alkali, form at intermediate more mildly alkaline pH values. **Kandites**, which ideally contain no alkali, form under neutral to slightly acid conditions.

The pH range for growing **microporous $AlPO_4$'s** (AlPO's) is much lower than for zeolites and tends to be on the acid side. A criterion for a good pH range for hydrothermal synthesis is one in which a precursor gel forms readily from the reactant solutions when mixed. With AlPO's this is on the acid side of neutral. Indeed, aluminium phosphate gels and microporous AlPO's dissolve in alkali.

Mineralogists recognized early not only that F can replace OH in some minerals but also that F^- like OH^- was a good mineraliser. This observation has recently been re-discovered or re-emphasised. The hydrothermal "fluoride pathway" enables zeolites to grow at pH near neutral and so much below those required when OH^- alone is the mineraliser. Kessler [7] considers that:-

(i) Because of the lower pH supersaturation is less than in high pH systems and that this results in fewer meta-stable phases.

(ii) Because growth rates are slower the number of defects is thereby reduced.

Framework topologies of zeolites, and also of some AlPO's, made by the fluoride pathway include [8,9,10]

Topology	T atoms	Topology	T atoms
MFI	Si	NUO	Si + Al
	Si + Al, B or Ga	CJS-I [9]	Si + Al
	Si + Fe, Ti or Ge	AFI	Al + P
FER	Si		Si + Al + P
	Si + Al		B + Al + P
	Si + Fe		Li + Al + P
TON	Si	AEL	Al + P
MTT	Si	Cloverite [10]	Ga + P
MTN	Si		

3.4 ORGANIC BASES IN HYDROTHERMAL SYNTHESIS

A paper of Barrer and Denny in 1961 [11], based on a Ph.D. programme started in 1956, established the use of ammonium hydroxide, alkylammonium hydroxides and especially quaternary ammonium bases as alternatives to inorganic bases in zeolite synthesis. It was suggested and shown that the zeolites produced (which included faujasite and zeolite A) were more siliceous than their counterparts Na - X and Na - A. This method of making zeolites soon became of major importance and has been immensely successful in making novel zeolites and zeolites very rich in silica. It is used exclusively for synthesising microporous AlPO's and related compounds and has culminated, inter alia, in the preparation of the range of porous crystalline silicas (porosils) given in Table 3. Clathrasils are porosils in which the windows are not wide enough to admit other guest molecules or to allow

TABLE 3. Microporous crystalline silicas (porosils) sub-divided into clathrasils and zeosils

Clathrasils	SiO_2 per 1000 Å^3	Zeosils	SiO_2 per 1000Å^3
Octadecasil (AST)	16.7	SSZ-24 (AFI)	17.5
Si-sodalite (SOD)	17.4	Decadodecasil 3R (DDR)	17.6
Dodecasil 1H (DOH)	18.5	Silicalite 1 (MFI)	17.8
Dodecasil 3C (MTN)	18.6	Silicalite 2 (MEL)	17.9
Melanophlogite (MEP)	19.0	Si-ZSM-50 (EUO)	18.2
Nonasil (NON)	19.2	Si-ZSM-12 (MTW)	18.5
		Si-ferrierite (FER)	19.3
		Si-ZSM-22 (TON)	19.7
		Si-ZSM-48	19.9
		Si-ZSM-23 (MTT)	20.0

escape of the guest imprisoned during synthesis. The zeosils, on the other hand, have windows wide enough to admit guest molecules of the right shape and size just as do the zeolites. SSZ-24 with the $AlPO_4$-5 structure is the most open, having a 1-dimensional channel system of free diameter ~ 7.3 Å, circumscribed by 12-rings. The letters in brackets (eg AST) are the codes giving the framework topologies [12]. The number of SiO_2 units per 1000 Å^3 is a measure of total microporosity. Comparison of octadecasil and SSZ-24 shows that the former has a greater total microporosity. However, accessibility of micropores is determined by the window sizes, not the total microporosity.

3.5. SYNTHESIS WHERE WATER IS REPLACED BY ORGANIC LIQUIDS

Zeolites and AlPO's have recently been grown successfully in systems where organic liquids have replaced all, or nearly all, the water, but under conditions otherwise analogous to those used in the hydrothermal procedure. Preparations by this route include the following 3-D, 4-connected nets:-

$AlPO_4$-5 (AFI)	[13]	ZSM-5, silicalite 1, (MFI)	[16,17]
$AlPO_4$-11 (AEL)	[13]	ZSM-35 (FER)	[16]
$AlPO_4$-21	[13]	ZSM-39 (MTN)	[17]
Si-sodalite (SOD)	[14]	ZSM-48	[17]
Mordenite (MOR)	[15]		

Neutral organic species can play the role of water (§3.1.) as space-fillers and stabilisers (eg glycol in Si-sodalite). Sometimes large organic cations alone serve as both space-filler and charge neutraliser, as in the hydrothermal synthesis of Si-rich sodalite with one $N(CH_3)^+_4$ ion filling each sodalite cage [18].

4. Pyrolysis

4.1. POROUS 3-D FRAMEWORKS MADE PYROLYTICALLY

Microporous host frameworks are also made by pyrolysis, melt-fluxed or sintered (Table 1). Here, non-volatile salts replace zeolitic water as the space-fillers and stabilisers. Table 4 [19] gives examples of the guests which have been included.

TABLE 4. Examples of salt-bearing tectosilicates

Mineral Type	Examples	Salts present in natural or synthetic species
Scapolites	Marialite Wernerite Meionite	$NaCl$, Na_2SO_4, Na_2CO_3
Sodalite-nosean group	Sodalite Nosean Hauyne Ultramarine Helvite Danalite	$NaCl$, $NaBr$, $NaOH$, Na_2SO_4, $CaSO_4$, Na_2CO_3, Na_2SO_3, $Na_2S_2O_3$ $NaNO_3$, $NaNO_2$, Na_2S_x, Na_3PO_4, Na_3AsO_4, Na_3VO_4, Na_2SeO_4, Na_2CrO_4, Na_2WO_4, Na_2MoO_4, $NaMnO_4$, $NaClO_3$, $NaClO_4$, $NaBrO_3$, $NaIO_3$, NaN_3, H_2O, $Na_2C_2O_4$
Cancrinite group	Cancrinite Davynite Vischnerite Microsommaite	$NaCl$, Na_2SO_4, $CaCO_3$, $NaNO_3$, Na_2SeO_4, Na_2MoO_4, Na_3VO_4, $NaMnO_4$, $Na_2S_2O_3$, Na_2S, H_2O
Cancrinite-related	Liottite Afghanite Franzinite	$(Na,K,Ca)_{1\text{-}2}(CO_3,SO_4,Cl,OH)_{1\text{-}2}$
Zeolite frameworks	Zeolite ZK-5 K-F (edingtonite) Phase A (merlinoite)	$BaCl_2$, $BaBr_2$, H_2O KCl, KBr, KI, H_2O $BaCl_2$

Not all the examples in Table 4 are made by pyrolysis because certain of the salts are not stable to heating (e.g.'s. $NaClO_3$, $NaClO_4$, $NaBrO_3$, $NaMnO_4$ or NaN_3). However, the hydrothermal method is equally facile in the synthesis of salt-bearing zeolites and felspathoids, at temperatures which may be below 100°C [20, 21]. Thermally unstable salts are then readily incorporated.

In hydrothermal synthesis of salt-bearing zeolites or felspathoids, there is competition between the salt, water and alkali for the intracrystalline cavities. With sodalite and cancrinite, one can go continuously from the frameworks containing only water as guest to these same frameworks containing only the salts as guests [21]. The salt uptake isotherms had contours of Type 1 [3], resembling the isotherm of Langmuir, while the water uptake declined as that of salt increased. Salt inclusion isotherms were modelled quantitatively assuming one salt, two NaOH or four water molecules are required to fill a sodalite cage. Analysis shows this to be correct, at least for salt and water. The isotherm equations for each of the three components in their mixtures were obtained by detailed balancing with sodalite cages as single (salts), double (NaOH) or fourfold (H_2O) sorption sites. With one disposable coefficient per intercalated component calculated isotherms were similar in shape to those observed [21].

We have also quantitatively investigated controlled and safe release of oxygen from oxysalts of chlorine and of nitrogen from NaN_3, all packaged in sodalite [22]. Metallic sodium entrained from its vapour in outgassed sodalite hydrate yielded blue to black phases in which Na was protected from all but a very slow attack by water giving a controlled release of H_2 [23]. Stein et al [24] have suggested the use of Ag-exchanged oxalato-sodalite for the reversible storage of optical data.

4.2. PYROLYTIC SYNTHESIS OF SWELLABLE LAYER SILICATES

Of equal interest to pyrolytic and / or hydrothermal synthesis of porous salt-bearing tectosilicate frameworks is that of layer silicates by sintering, or melt-fluxed (Table 1). In this way water-swelling fluor-smectites, fluor-vermiculites and fluor-micas are readily made [25]. These materials are among those which serve for the preparation of pillared clays and permanently microporous organo-clay molecular sieves (Table 5 and § 6.4).

5. Pneumatolysis and Zeolite Modification

Pneumatolysis, though not significant in the synthesis of zeolites, may serve to modify them. Pneumatolytic conditions exist with water vapour at high temperature and low pressure in deep-bed dealumination of zeolites by heating their NH_4^+- exchanged forms. The formation of USY catalyst from zeolite Y is an important example.

6. Diversity among Porous Crystals

Two striking features of porous crystals are their structural variety and, for a given framework topology, their compositional diversity. These will be illustrated in turn.

6.1. STRUCTURAL DIVERSITY OF POROUS FRAMEWORKS

Table 5 exemplifies some types of crystal which either have permanent porosity or from which permanently porous sorbents can be made. Among the 3-D, 4-connected nets about 80 framework topologies have been made and structurally determined in detail. This number does not include nets which have been synthesised but which are still of unknown or only partially known structures. Among 3-D, 4-connected nets so far prepared, the largest free dimensions of windows giving access to cavities and channels are found among the porous phosphates:

Crystal	Window	Free Dimensions (Å)
$GaPO_4$ (cloverite [10]) :	20-TO_4 ring	~ 6.0 x 13.2
VPI-5	18-TO_4 ring	~ 12
$AlPO_4$-8	14-TO_4 ring	~ 8.7 x 7.9

In addition there are various structures with 12-TO_4 ring windows having free diameters between 7 and 8Å (eg. faujasite, mordenite, LTL, AST, mazzite, zeolite β).

In wide, straight 1-D channel systems, long axial flights of molecules are possible but the chance of capture in the periodic energy wells (sorption sites) along the channels must be high. Among intracrystalline cavities, the largest supercage so far is that in cloverite with a body diagonal of 29-30Å [10]. This raises the question whether other crystals with mesoporosity may be synthesised.

In this connection, Iijima [26] has recently reported the synthesis of hollow tubular crystals of carbon with typical internal diameters ~20 to ~50Å. Wada et al [27] reported a hollow spherical form of aluminosilicate, a kind of bubble, the wall a monolayer thick and with external diameters 80 to 100Å. This material had a cation exchange capacity of ~230 meq / 100g and a water content of 32%. In silicates where YO_6 octahedra (Y = Al, Mg, Ni, Co) are attached to a layer of SiO_4 tetrahedra, misfit may produce bending of the double sheets. This curvature in the 1:1 layer can result in hollow tubes,

TABLE 5. Examples of porous crystals, or those from which permanently porous sorbents can be made

A. Permanently porous Frameworks	
(a) 3-D, 4-connected nets based on tetrahedra TO_4	Zeolites Porosils Some phosphates and arsenates
(b) 3-D, 4, 6-connected nets based on linked tetrahedral and octahedral units	Some molybdenum phosphates Some titanium silicates
(c) 3-D, 6-connected nets based on linked octahedral units	Some cyanometallates
(d) Other types	Some aluminoborates Cs-graphite Dianin's compound (a chroman) Schardinger dextrins

B. Layer structures which can be made permanently porous

Smectites, vermiculites, fluor-micas
Hydrotalcites
Trititanate
Some layered phosphates and phosphonates

like a rolled-up carpet. The curvature depends in part on the size of Y atoms in octahedral co-ordination, as illustrated in Table 6, due to Leibau [28]. The inside, or free, diameter of the tubes in chrysotile is about 50Å, again in the mesopore range, but access to the interior of such tubes may be prevented by detrital material or made slow because of the fibrous habit which can result in long needles with entry only at the two ends. Nevertheless the principle involved in these rolled-up

TABLE 6. Some chrysotile-type tubular structures [28]

Y in YO_6	1	2	3	4	Name	Formula
Al^{3+}	0.53	5.05	650	35	Metahalloysite	$Al_2[Si_2O_5](OH)_4$
Ni^{2+}	0.70	5.40	154	9	Pecoraite	$Ni_3[Si_2O_5](OH)_4$
Mg^{2+}	0.72	5.39	156	11	Chrysotile	$Mg_3[Si_2O_5](OH)_4$
Co^{2+}	0.735	5.50	101	4	Synthetic	$Co_3[Si_2O_5](OH)_4$

1 = radius of cation Y in Å
2 = lattice period in Å within the hexagonal or pseudo-hexagonal layer of the hydroxide of the cation.
3 = mean external diameter of the tubes in Å
4 = average number of 1:1 layers in the wall of the tube

carpet structures merits attention by chemists. If hollow cylindrical detritus-free tubes of controlled lengths and different internal diameters can be made, they will have potential as sorbents and carriers of catalysts such as enzymes.

Imogolite [29] is another example of a hollow tubular aluminosilicate, the external and internal diameters of a tube being ~23 to 27Å and ~10Å, respectively. It occurs naturally and is readily synthesised. The outer surface consists of AlOH groups and the inner surface of SiOH groups.

6.2. DIVERSITY IN CHEMICAL COMPOSITION

For a given framework topology one may have ranges in chemical composition leading to many variants. Among clay minerals, zeolites and felspathoids this may be done by direct ion exchanges of the type

$$Na^+ \leftrightarrow K^+ \;;\; 2Na^+ \leftrightarrow Ca^{2+} \;;\; 3Na^+ \leftrightarrow La^{3+}$$
$$Na^+ \leftrightarrow NH_4^+ \;;\; Na^+ \leftrightarrow (CH_3)_4N^+, \quad \text{and the like.}$$

Secondly, in a given framework, appropriate syntheses can vary the tetrahedral (T) atoms. These may, for example, be Zn, Be, Al, Ga, B, Fe, Si, Ge, P or As in a variety of combinations. Replacements of one T atom by another may be complete or only partial, and they may be of a simple kind, such as

$$Al \leftrightarrow Ga \;;\; Si \leftrightarrow Ge \;;\; P \leftrightarrow As \;;$$

or of a less simple kind, still involving only T atoms:

2Al ↔ Be, Si (sodalite and tugtupite);
2Si ↔ Al, P (some AlPO's and SiAPO's);

or they may involve T atoms and extra-framework cations as exemplified by

Si ↔ Na, Al ; Na, Si ↔ Ca, Al ; Na, Al ↔ Mn, Be (sodalite and helvine);
2(Na,Al) ↔ Be, 2Be (analcime and Be-analcime).

Finally, one may replace extra-framework but intracrystalline salts as components of porous frameworks, as has been illustrated in some detail in Table 4. One may exemplify these abundant compositional variations using the sodalite topology (Table 7) as example.

A much respected rule relating to 3-D, 4-connected aluminosilicate nets is that of Lowenstein [35] which states that Al-O-Al bonds do not occur in such nets. The mineral bicchulite, with Si/Al =1/2 departs substantially from this rule, and there are various aluminates in Table 7 which are thermally very stable and in which the sodalite framework contains only Al-O-Al bonds. However, within the composition range Si/Al ≥ 1 MASNMR studies of Al-Si ordering have established the Al-O-Al avoidance rule of Lowenstein as a very good guide.

6.3. SOME NON-ZEOLITE MOLECULAR SIEVES

Table 5 includes some less familiar types of porous crystal. The two microporous aluminoborates of unknown structure [36,37] could be forerunners of others of the same family. After hydrothermal

TABLE 7. Variations in composition among phases of sodalite topology

Composition	Comment	Ref
$[Si_{12}O_{24}]$2 glycol	Silica gel in glycol at 150°C	[14]
$[Al_6P_6O_{24}]$	AlPO-20	[30]
Li_6 $[Be_6P_6O_{24}]$ 2 Li Cl	Hydrothermally	[30a]
Li_6 $[Be_6As_6O_{24}]2Li(Cl,Br)$	Hydrothermally	[30a]
$\{N(CH_3)_4\}_2[Al_2Si_{10}O_{24}]$	Silica gel in $N(CH_3)_4OH$	[18]
$Na_6[Al_6Si_6O_{24}]2NaCl$	Natural sodalite	
$Ca_4[Al_8Si_4O_{24}]4Ca(OH)_2$	Natural bicchulite	[31]
$Ca_6[Al_{12}O_{24}]2CaWO_4$	$CaCO_3$, $Al(OH)_3$, $CaWO_4$ at 1350°C	[32]
$Ca_6[Al_{12}O_{24}]2CaO$	Aluminate 4CaO. $3Al_2O_3$. $3H_2O$ at 650°C	[33]
$Sr_6[Al_{12}O_{24}]2SrWO_4$	Pyrolysis	[34]
$Sr_6[Al_{12}O_{24}]2SrMoO_4$	Pyrolysis	[34]

synthesis and outgassing they sorbed gases zeolitically. In water-free Ca Al_2 B_{19} O_{33} $4.2H_2O$, for example, the N_2 isotherm at 77K was very rectangular and reached ~12% by weight. Free dimensions of windows of 5-6Å were reported.

Porous 3-D, 6-connected cyanometallates include $K_2Zn_3[Fe(CN)_6]_2xH_2O$, $Zn_3[Co(CN)_6]_2$ and $Zn[Fe(CN)_5NO]$. The first of these contains ellipsoidal cavities, each able to accommodate ~8.5 H_2O molecules and with window dimensions ~3.9 x 5.2Å. The outgassed material readily sorbed CO, N_2, C_2H_4 and C_2 to C_4 paraffins [38]. The smaller windows in $Zn[Fe(CN)_5NO]$ admit molecules such as CO_2 [39]. $Zn_3[Co(CN)_6]_2$ has windows with free dimensions ~5.6 x 8.6Å. It readily sorbed n-C_6H_{14} and 3-methyl-pentane but excluded 2,2-dimethyl butane [39].

Table 5 also refers to Dianin's compound, which has been termed an "organic zeolite" [40] because the microporous structure of its crystals persists in presence or absence of guest molecules.
It is the chroman

which is the structural unit in the hydrogen bonded host structure. In the framework are columns of elongated cavities each with a slight waist, which creates notable selectivity in uptake of guests. Paraffins with methyl or ethyl side groups near the middle of the backbone are excluded because of the waist in the hour-glass shaped cavity. This lock-and-key specificity was elegantly demonstrated by Goldup & Smith [41] as shown below:

Molecules which fit and are sorbed	**Molecules which do not fit and are excluded**
n-C_6, n-C_7	n-C_8 (too long for cavity)
2-Me-C_6; 2-Me-C_5	3-Me-C_6; 3-Me-C_5
2,5-di-Me-C_6	2,2-di-Me-C_4; 2,3-di-Me-C_5
	2,2,3-tri-Me-C_4; 3-Et-C_5
(Me = methyl; Et = ethyl)	2,4-di-Me-C_6; 2,2,4-tri-Me-C_5

We discovered that rates of uptake of guest molecules were greatly accelerated by agitation of host crystals of Dianin's compound in presence of the guests with small steel shot. This behaviour was also shown to be characteristic of clathration, as demonstrated for water, phenol and other clathrate-forming host species [42,43,40].

The α-,ß- & γ Schardinger dextrins of Table 5 are permanently hollow molecules formed respectively by chemical bonding of 6, 7 and 8 glucose residues [44]. They have the shapes of sections of gently sloping hollow cones with free diameters of ~5, 6.7 and 8.5Å. Although much studied in complex formation in solution, evaluation of the outgassed crystals as selective sorbents of gases and vapours appears to be lacking.

6.4 EXPANDABLE LAYERED CRYSTALS

Table 5 includes some hydrothermally prepared types of swellable layered crystals which are of increasing interest. These will be discussed by others in this ASI and therefore will here be referred to only briefly. The way in which such crystals may be rendered permanently microporous was suggested and demonstrated by Barrer & McLeod in 1955, using smectites [45]. The principle involved is shown in Fig. 2. [46]. Small charge-balancing cations are replaced by larger, somewhat globular, ions which may in addition carry multiple charges. The large cations prop the smectite sheets apart permanently to give an intersheet free distance, d_1, while leaving adequate free distances, d_3, between any adjacent pair of intersheet cations. Molecules of the right shape and size can then permeate the galleries of the resultant interlayer regions. While d_1 could increase further during imbibition d_3 is unlikely to alter since the positive charges of the intersheet cations will tend to create a 2-D regular intersheet net of these ions.

Following the 1955 paper many molecular sieves and selective sorbents, derived from expanded clay minerals, were made and examined in my laboratory, some results of which were recently summarised [47,48]. Molecule sieving using these permanently microporous clay minerals can be as sensitive to the shapes and sizes of molecules as with zeolites.

Recently much attention has been given to pillaring clay minerals by exchange of interlayer cations with oligomeric oxycations carrying multiple charges, thereby improving the thermal stability of the calcined products (piilared interlayered clays of PILCS). The vertical free distances, d_1, after calcination are given for some of these materials in Table 8. The vertical free distances and monolayer equivalent areas are usually large but variable, suggesting that differing reactions may occur on exchange with oligomeric cations followed by calcining. Isotherms usually indicate from their contours that there is meso- as well as micro-porosity. Little is known about the magnitude or regularity of the free distance, d_3, between adjacent pillars or about the charge, if any, remaining on the pillars after calcination.

Layered double oxides of hydrotalcite type can also yield permanently microporous sorbents according to the same principle as with clay minerals, illustrated in Fig.2. In hydrotalcites, however, the sheets carry nett positive charge and the charge neutralising interlayer ions are therefore anions. The anion exchange capacity is high (410 to 240 meq / 100g). Accordingly, to develop good permanent interlayer porosity, anionic props with multiple charges are best able to give adequately large lateral free distances, d_3, between adjacent anions in the interlayer galleries. The nearest parallels to smectites (exchange capacities 60 to 100 meq / 100g) with interlayer props such as $N(CH_3)_4^+$ or $[Co(en)_3]^{3+}$ (en = ethylenediamine) are hydrotalcites with interlayer $Fe(CN)_6^{3-}$ or $Co(CN)_6^{3-}$

TABLE 8. Interlayer free distances d_1 in Å and monolayer equivalent areas (MEA) in m^2g^{-1} for some PILCS

Pillar based on:	Pillaring via:	After calcination d_1	MEA	Ref
Al	Oligomeric cation	8.4-9.1	~400	[49]
Al	" "	6.8	>300	[50]
(Si,Al)	" "	6.0-10.2	85-450	[51]
(Si,Al)	" "	6.6-8.8	340-500	[51]
Si	$[Si(acetonylacetate)_3]^+$	3.2(max)	40-190	[52]
Si	Silsesquioxane	6.8-10.4	140-400	[53]
Zr	Oligomeric cation	8.6	260	[49]
Zr	" "	12.4	290	[49]
Zr	" "	12.4-16.2	250	[49]
Ti	" "	15.6	300	[54]
Ti	" "	19.4	330	[54]
Cr	" "	11.6	430	[55]
Fe	$[Fe_3(O.CO\,CH_3)_7OH]^+$	7.3	280	[56]

Examples of reported vertical free distances, d_1, between the double oxide layers are :

Interlayer Anion	Free distance (Å)
$Cr_2O_7^{2-}$	5.2 [57]
$Fe(CN)_6^{4-}$, $Fe(CN)_6^{3-}$	6.1 [57,58]
$Co(CN)_6^{3-}$	6 to 6.5 [58]

The $Co(CN)_6^{3-}$-hydrotalcite sorbed n-hexane, 2-methylpentane, cyclohexane and methylcyclohexane [58].

The first pillared layered phosphate with a high specific surface area and a narrow pore size distribution was reported by Olivera-Pastor et al in 1989 [59]. It was made from α-$Sn(HPO_4)_2H_2O$ in several steps, finally leading to a pillared structure thought to contain the Keggin cation. After calcination at 400°C the specific surface was reported to be 228 m^2g^{-1} as compared with 11 m^2g^{-1} in the parent α -$Sn(HPO_4)_2.H_2O$.

7. Concluding Remarks

Most, but not all, of the types of microporous framework referred to in Table 5 are made by the versatile hydrothermal method. Many zeolites and also some clay minerals will form at temperatures below 100°C so that the distinction between hydrothermal and ordinary wet way synthesis can be relatively arbitrary. The examples in Table 5, with their rich variety both topologically and in terms of compositional variations within a given topology, illustrate a rapidly growing area in chemical science leading, inter alia, to many actual or potential applications. Finally, it is seen in §6.1 that porous crystalline phases exist which bridge the gap between micro- and meso-pores.

References

1. Barrer, R.M. (1982) Hydrothermal Chemistry of Zeolites, Academic Press, Chapter 2, Table 1.
2. Barrer, R.M. (1988) in Zeolite Synthesis, A.C.S. Symposium Series 398 Eds M.L.Occelli and H.E.Robson, Chapter 2.
3. Brumauer, S. (1948) The Adsorption of Gases & Vapours, Oxford University Press, p.150.
4. Gibbs, J.W. (1948) Collected Works, Yale University Press, Vol.1, pp.105-115.
5. Zdhanov, S.P. and Samuelevich, N.N. (1980) Proc. 5th Internat. Conference on Zeolites, Ed. L.V.C.Rees, Heyden, p.75.
6. Kacirek, H. and Lechert, H. (1976) J. Phys. Chem., 80, 1291.
7. Kessler, H. (1989) in Recent Advances in Zeolite Science, eds. J.Klinowski & P.J. Barrie, Elsvier, p.30.
8. Ref.7, p.22.
9. Zhao Daqing and Pang Wenqin. (1990) J. Chem. Soc. Chem. Comm., p.884.
10. Estermann, M., McCusker, L.B., Baerlocher, C., Merrouche, A. and Kessler, H. (1991) Nature, 352, 320.
11. Barrer, R.M., and Denny, P.J. (1961) J. Chem. Soc. p.971.
12. Meier, W.M., and Olson, D.H. (1987) Atlas of Zeolite Structure Types, Butterworths.
13. Huo Qisheng and Xu Ruren. (1990) J. Chem. Soc. Chem. Comm. p.783.
14. Bibby, D.M., and Dale, M.P., 1985, Nature, 317, 159.
15. Xu Wenyang et al. (1987) Chinese Pat. 88 100 228.3.
16. Xu Wenyang, Li Jianqan, Li Wenyuang, Chang Huiming and Liang Bingchang. (1985) Zeolites, 9, 486.
17. Huo Qisheng, Feng Shouhua and Xu Ruren. (1988) J. Chem. Soc. Chem. Comm., p.1486.
18. Baerlocher, C., and Meier, W.M. (1969) Helv. Chim. Acta, 52, 1853.
19. Barrer, R.M. (1982) Hydrothermal Chemistry of Zeolites, Academic Press, Chapter 7.
20. Barrer, R.M., Cole,J.F., and Sticher, H. (1968) J. Chem. Soc. A, p.2475.
21. Barrer, R.M., and Cole, J.F. (1970) J. Chem. Soc. A, p.1516.
22. Barrer, R.M., Daniels, E.A., and Madigan, G.A. (1976) J. Chem. Soc. Dalton, p.1805.
23. Barrer, R.M. and Cole, J.F. (1986) J. Phys. Chem. of Solids, 29, 1755.
24. Stein, A., Ozin, G.A., and Stucky, G.D. (1990) Nippon Shashin Gakkaishi, 53, 322.
25. Barrer, R.M., and Jones, D.L. (1970) J. Chem. Soc. A, p.1531.
26. Iijima, S. (1991) Nature, 354, 56.
27. Wada, K., Wilson, M., Kakuto, Y. and Wada, S. - I. (1988) Clays and Clay Minerals, 36, 11.
28. Liebau, F. (1985) Structural Chemistry of Silicates, Springer-Verlag, p.225.
29. Farmer, V.C., Adams, M.J., Fraser, A.R., and Palmieri, F. (1983) Clay Minerals, 18, 459.
30. Wilson, S.T., Lok, B.M., Messina, C.A., Cannan, T.R. and Flanigen, E.M. (1982) J. Amer. Chem. Soc., 104, 1146.

30a. Gier, T.E., Harrison, W.T.A. and Stucky, G.D. (1991) Angew. Chem. 103, 1191.

31. Sahl, K., and Chatterjee, N.D. (1977) Z. Kristallogr., 146, 35.
32. Depmeier, W. (1979) J. Appl. Cryst., 12, 623.
33. Ponomarev, V.I., Kheiker, D.M. and Belov, N.V. (1971) Soviet Physics - Crystallogr., 15, 799 (English translation).
34. Depmeier, W., and Buehrer, W. (1991) Acta Crystallogr. Sect. B: Structural Sci., B47, 197.
35. Lowenstein, W. (1954) Amer. Mineralog., 39, 92.
36. Wang, J., Feng, S. and Xu, R. (1990) Gaodeng Xuexiao Huaxue Xuebao, 11, 449.
37. Wang, J., Feng, S. and Xu, R. (1989) J. Chem. Soc. Chem. Comm. p.265.

38. Cartraud, P., Cointot, A. and Renaud, A. (1981) J. Chem. Soc. Faraday Trans. I, 77, 1561.
39. Boxhoorn, G., Moolhuysen, J., Coolegem, J.G.F. and van Santen, R.A. (1985) J. Chem. Soc, Chem. Comm., p.1305.
40. Barrer, R.M. and Shansen, V.H. (1976) J. Chem. Soc. Chem. Comm. p.333.
41. Goldup, A. and Smith, G.W. (1971) Separation Sci., 6, 791.
42. Barrer, R.M. and Edge, J.A.V. (1967) Separation Sci., 2, 145.
43. Allison, S.A. and Barrer, R.M. (1968) Trans. Faraday Soc., 64, 549 and 557.
44. Saenger, W. (1984) in Inclusion Compounds Vol II. Editors Atwood, J.L., Davies, J.E.D. and McNicol, D.D., Academic Press, p.231.
45. Barrer, R.M. and McLeod, D.M. (1955) Trans. Faraday Soc., 51, 1290.
46. Barrer, R.M. 1986 J. Inclusion Phenomena, 4, 109.
47. Barrer, R.M. 1989, Clays & Clay Minerals, 37, 385.
48. Barrer, R.M. 1989, Pure & Applied Chem., 61, 1903.
49. Vaughan, D.E.W. and Lussier, R.J. (1980) Proc. 5th Internat. Conf. on Zeolites, Editor L.V.C. Rees, Heyden, p.94.
50. Schultz, A., Stone, W.E.E., Poncelet, G., and Fripiat, J.J. (1987) Clays and Clay Minerals, 35, 251.
51. Sterte, J., and Shabtai, J. (1987) Clays & Clay Minerals, 35, 429.
52. Endo, T., Mortland, M.M., and Pinnavaia, T.J. (1980) Clays & Clay Minerals, 28, 105.
53. Lewis, R.M., Ott, K.C., and van Santen, R.A. (1985) U.S.P. 4, 510, 257.
54. Sterte, J. (1986) Clays & Clay Minerals, 34, 658.
55. Pinnavaia, T.J., Tsou, M.S., and Landau, S.D. (1985) J. Amer. Chem. Soc, 107, 4783.
56. Yamanaka, S., Doi, T., Sako, S., and Hattori, M. (1984) Mat. Res. Bull., 19, 161.
57. Chibeve, K., and Jones, W. (1989) J. Chem. Soc. Chem. Comm., 926.
58. Suzuki, E., Idemura, S., and Ono,Y. (1989) Clays & Clay Minerals, 37, 173.
59. Olivera-Pastor, P., Jimenez-Lopez, A., Maireless-Torres, P., Rodriguez-Castellon, E., Tomlinson, A.A.G. and Alagna, L. (1989) J. Chem. Soc. Chem. Comm., p.926.

PORES IN TETRAHEDRAL FRAMEWORKS

P. Behrens
Faculty of Chemistry
University of Konstanz
P.O. Box 5560
D-7750 Konstanz
Germany

ABSTRACT. Porous crystalline tetrahedral frameworks, *i.e.* zeolites and related compounds, are the most important class of *micro*porous substances due to their uniform pore size, the periodic arrangement of pores and to the fact that the framework structure and chemistry can be controlled rather deliberately. Numerous attempts to extend the favourable characteristics of these compounds to the *meso*porous regime have failed so far. This article deals with the factors influencing the size and the shape of the pores in zeolite-type compounds. The primary factor is the kind of linking of $[TO_{4/2}]$ tetrahedra determining mainly the size and shape of the windows and cavities. This topological aspect of the structure of tetrahedral frameworks is discussed in the first part of the article. On that basis, topological and other concepts to extend the apertures of periodic structures with well-defined uniform pores into the mesoporous range are discussed in the second part. The third section of this article focusses on "framework mechanics". It has often been realized that the pore sizes of zeolite-type compounds as determined from absorption and diffusion experiments differ from those estimated on the basis of the crystal structure. It will be shown that the conformation of the framework - as determined by its dynamic response on temperature and pressure changes and by its adaption to the spatial and coordinative demands of the guest species - strongly influences the "real" size and shape of the pores and thus their effective apertures.

1. Introduction

Zeolites and related porous compounds are three-dimensional frameworks that topologically belong to the class of tetrahedral networks. These are built from atoms or arrangements of atoms, that possess four free valences pointing in the directions of the corners of a tetrahedron. The presence of directional valences in tetrahedral network structures implies a substantial amount of covalent bonding in these compounds. Typical other examples for this class of substances are the elements of Group IV (hexagonal and cubic diamond, Si, Ge, grey Sn), isostructural III-V (GaP), II-VI (ZnS), I-VII (AgI) compounds as well as water and its clathrate compounds, where $[OH_{4/2}]$ is the tetrahedral building unit (TBU) (Fig. 1). Together with the dense phases of the SiO_2, $AlPO_4$ and similar systems, zeolite-like compounds constitute the family of oxidic (4,2)-bonded framework structures, which typically contain $[AlO_{4/2}]$, $[SiO_{4/2}]$,

C.A.C. Sequeira and M.J. Hudson (eds.), Multifunctional Mesoporous Inorganic Solids, 73–97.

and $[PO_{4/2}]$ as TBUs (Fig. 1). Other elements like Be, Fe, Co, Zn, Ga, Ge, As ... have been shown to be able to substitute as four-bonded elements in oxidic (4,2)-frameworks, and sulfide and selenide, but also (new and more exciting) nitride appears as the two-connected bridging element in the recently discovered compound $Zn_7[P_{12}N_{24}]Cl_2$ (SCHNICK and LÜCKE 1992) with sodalite structure.

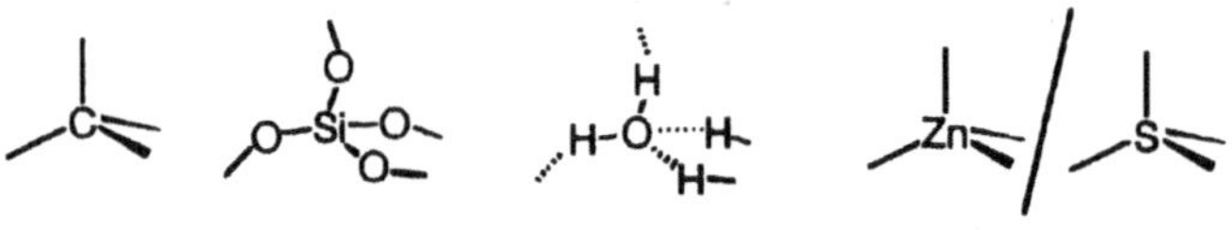

Figure 1. Tetrahedral building units for simple inorganic solids with the topologies of the cubic and the hexagonal diamond structure.

Oxidic tetrahedral framework structures provide access to micro- and mesoporous materials in several different ways:

- as microporous materials with an ordered structure, *i.e.* zeolites and related solids;
- as mesoporous products of their (partial) destruction by various treatments, and
- as meso- and microporous gels.

Following MEIER (1986), the general formula for oxidic tetrahedral framework structures may be written as

$$A_a\ A'_{a'}\ B_b\ [T_t T'_{t'} \cdots\ O_{2(t+t'+\ldots)-\epsilon}\ (OH)_{2\epsilon}]\ X_x \cdot nH_2O \cdot M_m,$$

where A and B are inorganic cations (that usually can be exchanged) and organic cations (which can usually be burnt off), resp. T are the tetrahedral atoms of the four-connected threedimensional network, which in the case of interrupted frameworks, contains 2ϵ terminal OH groups. The framework components are designated by square brackets. Besides the cations, the voids of the network may contain anions X, water molecules and neutral molecules M. As an example, the formula of zeolite A powder, currently engaged in water softening in a washing machine, may be written as

$$Na_a\ Mg_{a'}\ Ca_{a''}\ [Al_{12}Si_{12}O_{48}] \cdot n\ H_2O.$$

Note that the term "zeolites" (*sensu stricto*) is used only for compounds with an aluminosilicate framework.

An as-synthesized zeolite ZSM-5 prepared using the tetrapropylammonium cation as template is described as (VAN KONINGSVELD *et al.* 1987):

$$Na_{0.04}\ [N(C_3H_7)_4]_{0.5}\ [Al_{0.04}Si_{11.96}O_{24}]\ (OH)_{0.5} \cdot nH_2O;$$

Note that according to this formulation the OH groups are not part of the framework, but are non-framework guest anions. However, OH groups can appear as terminal, non-connecting units in "interrupted frameworks". These may be crystalline as in the case of cloverite, a new gallophosphate possessing cage-like voids with an inner diameter of about 30 Å (ESTERMANN *et al.* 1991):

(quinuclidinium)$_{192}$ [$Ga_{768}P_{768}O_{2976}(OH)_{192}$] F_{192},

These cage-like voids make cloverite the first zeolite-type material containing voids in the mesoporous range. However, the access to these pores is partly blocked by the terminal OH groups protruding into the large 20-ring window of the cages.

An example for a partly crystalline interrupted framework is the sodium silicate hydrate magadiite (SCHOLZEN *et al.* 1991). The mesopores formed during the dealumination of zeolite Y (typical diameter 10 nm) are supposed to be coated with terminal OH groups.

This article starts with the description of basic concepts of the topology of tetrahedral framework structures; special emphasis is put on the topological requirements, that have been spotted as essential for the construction of solids with low framework density and large pores. Topological and other concepts guiding the way to the extension of zeolite-like compounds to the regime of mesopores are summarized in several slogans in the second part. Different measures exist for the porosity of tetrahedral framework structures: One is the ring-size or diameter of the largest window of a structure type, measured as the number of T atoms surrounding the micropore or as its free diameter (in Å); another parameter is the diameter or the volume of the largest cage-like void. "Framework density" as defined by BRUNNER and MEIER (1989) is measured as the number of T atoms per cubic nanometer.

The third part focusses on the "framework mechanics" encountered in real zeolite-type frameworks. The investigation of framework mechanics provides an insight on the "real" size and shape of the pores as a function of temperature, pressure and of the spatial and coordinative requirements of the non-framework components. As the O–T–O angles in the tetrahedra exhibit only small variations from the ideal value of 109.54°, framework mechanics deal mainly with the T–O–T angle around the oxygens which bridge these tetrahedra. This angle is rather flexible, ranging from about 125° to 180°. Recent progress (BAUR 1992) in our understanding of the relations between the values of the different T–O–T angles of a framework give a deeper insight on the dynamic conformation of tetrahedral framework structures and yield possible explanations for the fact that pore sizes determined experimentally by adsorption and diffusion experiments differ from those estimated on the basis of the kinetic diameter of the guest species and the knowledge of the host structure.

2. Tetrahedral frameworks: Topological concepts

As the valence theory of chemistry defines the number and the directions of the connections, an atom or a group of atoms may make, the mathematical concept of topology is used to investigate the different ways, how the "connectedness" of the atoms can be accommodated in periodically repeating, *i.e.* crystalline, patterns in three-dimensional space, thereby obeying the specific laws of this space. The topologic approach to the structure of solids, most impressingly presented by WELLS (1977, 1985), treats the connectivity and the environments of the building units, but ignores the numerical values of bond-lengths and angles.

2.1 TOPOLOGY OF SIMPLE TETRAHEDRAL FRAMEWORKS

Despite of the fact that there exist infinitely many tetrahedral networks and no systematic procedure to derive all of them is available, the topology of tetrahedral framework structures is at an well-elaborated state. This is of course due to the interest in zeolites and related materials, that has driven the exploration of "open" structure types (which are of less importance for other framework types) in order to be able to catalogize these structure types and to predict new ones. It is not my intention to give a full account of the topology of tetrahedral networks in this paper. This has been done by SMITH (1988). Some new developments are covered by the papers of KOKOTAILO *et al.* (1989), BRUNNER and MEIER (1989), AKPORIAYE (1989, 1992) and VAN KONINGSFELD (1991, 1992). A "Consortium for Theoretical Frameworks" has been founded (SMITH 1989). Clearly, the infinite number of different possible frameworks can never be synthesized, and criteria have to be eleborated that allow to distinguish between "possible" and "realizable" structures (GRAMLICH-MEIER AND MEIER 1982, SMITH 1989, BRUNNER 1990). Only then, topological concepts can be used to predict new structure types, also ones containing larger pores than those that have been observed in microporous solids up to now.

The two simplest ways to connect tetrahedral building units are represented by the structures of cubic and hexagonal diamond (Figs. 2a and b). Simple structures that can easily be derived are those of the zinc sulphide modifications zincblende and wurtzite (by substituting half of the carbon atoms by Zn, the other half by S), the dense SiO_2 modifications cristobalite and tridymite (by insertion of oxygen into the connections between tetrahedral atoms) and two structures of ice, namely Ice-I_c and Ice-I_h, where oxygen takes the function of the T atom and hydrogen bonds act as bridging elements. Another rather simple tetrahedral network is represented by the sodalite structure (Fig. 2c). Sodalite frameworks contain pores and are realized with a large variety of combinations of chemical elements. Sodalites *sensu stricto* possess an aluminosilicate framework with the formula $[Al_6Si_6O_{24}]^{6-}$ for the unit cell of the ideal cubic structure. The "end members" $[Al_{12}O_{24}]^{12-}$ and $[Si_{12}O_{24}]^0$ are known as well as many isostructural T substitutions including the elements Be, Mg, P, Fe, Co, Zn, Ga, Ge, As. The substitution of the anionic framework component O by N was recently shown to be possible in the compound $Zn_7[P_{12}N_{24}]Cl_2$ (SCHNICK and LÜCKE 1992). A sodalite framework built mainly from water molecules appears in the clathrate hydrates of the type $[O_{10}F_2H_{24}][PF_6]_2$ and $[N(CH_3)_4]_2[O_{12}H_{23}]$, where the PF_6^- anion and the tetramethylammonium cation are the guest species, resp.

The sodalite framework (structure code SOD, the structure codes are those used in the "Zeolite Atlas" from MEIER and OLSON (1987)) is unique among the known porous tetrahedral networks as it can be built by simple translation from only one type of cage. The sodalite cage (Fig. 2d), described as $[4^66^8]$ polyhedron, has the shape of a truncated octahedron (and not of a cuboctahedron, as is often wrongly stated), which is one of the five Fedorov polyhedra that can fill space completely by simple translation without the need for using different orientations or other polyhedral building units. This gives sodalites a special character regarding many structural features. It may thus serve as a model case for structural studies of zeolite-type compounds.

The special distribution of the four- and six-ring planes of the sodalite cage allows its use as a polyhedral building unit itself. The structure of the industrially important zeolites X/Y (FAU topology, Fig. 2e) can be built using four of the eight six-ring planes as a TBU, arranging them according to the topology of the diamond structure and connecting them via double six-ring (D6R) units. From the topological point of view, it is most satisfying that the "sodalite analog" of the hexagonal diamond structure (EMT topology, Fig. 2f), predicted long ago by BRECK (1974), could recently be synthesized by the use of the crown ether 18-crown-6 as a template (DELPRATO *et al.* 1990). Furthermore, it is possible to conduct the synthesis to controlled intergrowths of the cubic and the hexagonal form by using a mixture of 15-crown-5 (templating for FAU if used alone) and 18-crown-6 as templates (DELPRATO *et al.* 1990, ANDERSSON *et al.* 1991, DOUGNIER *et al.* 1992), reminding of the polytypism of, *e.g.*, ZnS. Zeolites X/Y were for a long time the microporous solids with the largest cage (the supercage with a diameter of 11.8 Å) and largest windows (twelve-ring, 7.4 Å). The EMT topology provides a one-dimensional channel system of super-cages which are linked *via* twelve-ring windows to oblate cages (6.9 × 7.4 Å). In total, both topologies provide a three-dimensional channel system.

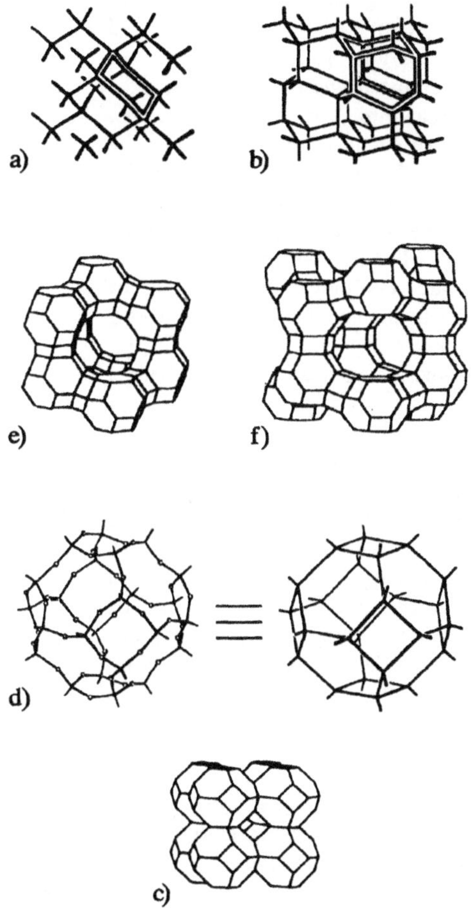

Figure 2. The topologies of the two simplest tetrahedral networks, that of cubic (a) and hexagonal (b) diamond are the basis for the topologies of the zeolites with structure type FAU (e) and EMT (f), where sodalite cages (d) are used as tetrahedral building units. Sodalite cages themselves can be used to give the space-filling arrangement of the structure type SOD (c). Usually, the oxygen bridges of tetrahedral frameworks are not drawn for clarity (d).

2.2 TOPOLOGY OF MICROPOROUS SOLIDS

2.2.1 *One-dimensional channel structures*. Microporous solids with one-dimensional channel systems, *i.e.* channels running parallel each to other and not being interconnected by other channels with at least eight-ring diameter, possess only limited practical applications, as the parallel running channels are easily blocked by defects occurring in the crystal structure or by side products of catalytic reactions. For some applications, however, the strict alignment of channels of one sort may be advantageous, for example to align molecules which exhibit directional-dependent non-linear optical effects (CARO *et al.* 1992). Typical examples are the framework topologies of the aluminophosphates AlPO-5 (AFI, Fig. 3a) and VPI-5 (VFI, Fig. 3b). Figure 3 shows the topological arrrangement of atoms in layers perpendicular to the channel direction. VPI-5 possesses circular 18-ring pores with a free diameter of about 13 Å, the largest pores observed for a uni-dimensional channel system (DAVIS *et al.* 1988).

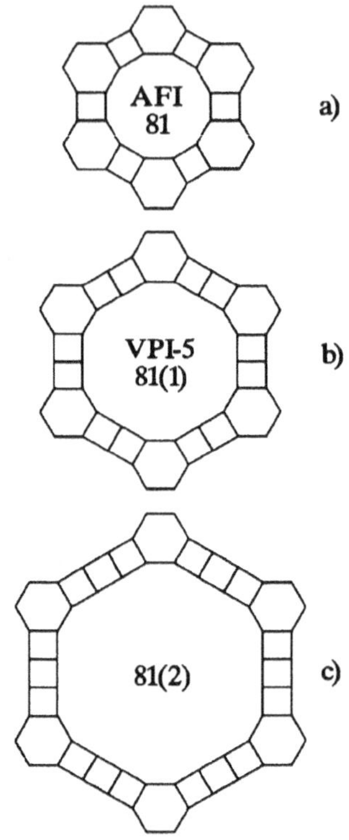

Figure 3. Framework topologies of the one-dimensional channel structures AFI/net 81 (a), VFI/net 81(1) (b) and net 81(2) (c, compare section 3.1).

2.2.2 *Three-dimensional channel structures*. Most of the microporous solids with practical applications possess two- or three-dimensional channel systems for the reasons discussed above. The FAU topology shown in Fig. 2 in its low-alumina aluminosilicate variant is used in large amounts as a catalyst in hydrocarbon processing. Interesting applications might be envisaged for zeolites with EMT topologies with their more complicated (and more versatile ? See for example DOUGNIER *et al.* 1992) arrangement of channels and cages. However, in contrast to the zeolite X/Y variants, hexagonal faujasite can only be synthesized in pure form using expensive crown-ethers.

Here, I will consider a class of microporous solids with three-dimensional channel systems which is topologically very simple and very interesting from the structural point-of-view. The structures are derived from the most simple topology for corner-sharing octahedra, that is the ReO_3 structure. There, the octahedral building unit (OBU) is a $[ReO_{6/2}]$ octahedron (Fig. 4a). Even the relatively small void provided by linking oxygen octahedra can be filled, resulting in the structures of perovskites and tungsten bronzes.

The concept of "building" porous structures, especially those with large cages and channels, by linking OBUs according to the ReO_3 topology is in accord with BRUNNER's (1990) considerations of zeolite-type materials. He finds that in these materials the points of highest symmetry remain unoccupied, *i.e.* they tend to "repel" atoms, thus allowing the construction of large voids. Using different kinds of OBUs, placing them at the corners of a cube and connecting them gives (topologically) highly symmetric frameworks with usually large cages at (1/2, 1/2, 1/2), *i.e.* the center of the cube. These cages are interconnected by windows centered at (1/2, 1/2, 0), *i.e.* framing the centers of the faces of the cube.

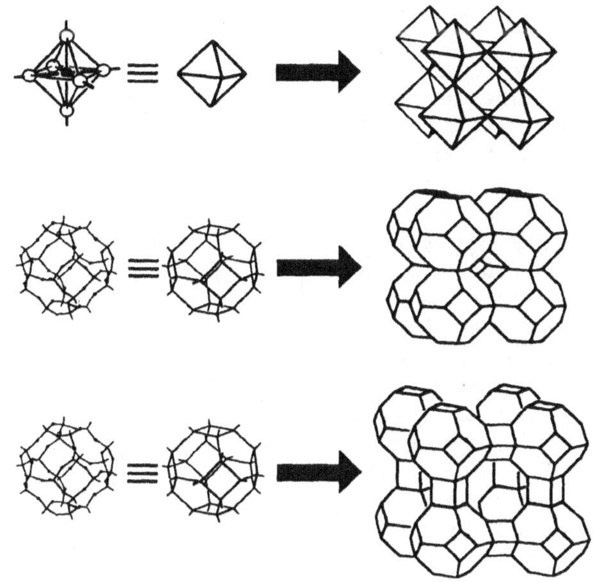

Figure 4. The octahedral framework topology of ReO_3 (a), according to which sodalite cages can be used to construct the tetrahedral framework structures of the SOD- (b) and the LTA- (c) type.

The principle is shown in Fig. 4b for the topology of SOD and in Figure 4c for that of LTA. In both cases, sodalite cages act as OBUs using their octahedral arrangement of four-ring windows. Sodalite cages are connected directly in the case of SOD topology, whereby a further sodalite cage is formed at the center of the cubic unit cell, or *via* a double four-ring (D4R) unit in the case of LTA topology. In zeolite A with LTA topology (the most-produced zeolite due to its ion-exchange applications in water-softening) a larger cage with a diameter of 11.4 Å is formed. These α-cages are interconnected by circular eight-ring windows with a diameter of 4.1 Å. Besides in aluminosilicatic zeolite A, the LTA topology is also found as gallophosphate crystallizing from a fluoride medium. Fluoride ions are incorporated in the double four-ring (D4R) units (SIMMEN *et al.* 1991).

Other OBUs leading to existing framework structures if linked according to the ReO_3 prescription are shown in Fig. 5. Chabazite, a network formed from connecting double D6R units *via* six four-rings (CHA, Fig. 5a), is known as an aluminosilicate zeolite occurring in nature and can be synthesized as aluminophosphate, silicoaluminophosphate and as different metallo-aluminophosphates. The AST topology possesses only six-ring windows as largest entrances; however, it contains large cages ($4^6 6^{12}$ polyhedra) which are composed to give a space-filling arrangement together with D4R units (Fig. 5b). It is realized as aluminophosphate ($AlPO_4$-16, BENNETT and KIRCHNER 1991) and as a clathrasil with a pure silica framework: octadecasile, where again the D4Rs contain F^- ions (CAULLET *et al.* 1991).

In the RHO topology, α cages are connected *via* double eight-rings (Fig. 5c, see also Fig. 9a, top), leading to two independent three-dimensional eight-ring channel systems, which are connected only via six-ring windows. RHO is known as aluminosilicate and occurs in nature as the beryllophosphate pahasapaite. A berylloarsenate variant has also been prepared (GIER and STUCKY 1991). α cages are also the main building element of the zeolite-type material with the largest cages found yet (diameter $\approx$ 30 Å): the gallophosphate cloverite (CLO, Fig. 5d). Here, they are interconnected via so-called rpa cages known from the structure of $AlPO_4$-22 (the "front" rpa cage is not shown in Fig. 5d for clarity). All the four-membered rings of these two buiding units are parts of double four-rings, half of which are not fully connected. The coordination spheres of the corresponding Ga and P atoms are completed by terminal hydroxyl groups (not shown in Fig. 5d) protruding into the large windows of the "super-super" cages, thus leading to an interrupted framework and giving these windows a clover-leaf like shape (ESTERMANN *et al.* 1991). The unique pore shape and the presence of a large number of terminal hydroxyl groups may give cloverite exceptional catalytic properties (DAVIS 1991, MERROUCHE *et al.* 1992). Again, the D4R units have incorporated F^- ions during synthesis.

"ReO_3-type" linkages are also found in the structures of microporous solids with a mixed tetrahedral/octahedral framework. The OBU is shown in Fig. 5e (note that the presentation used here is different from the other pictures of Fig. 5: the oxygen coordination polyhedra are shown): four cations Me, disposed tetrahedrally, are each coordinated by an oxygen octahedron. The four octahedra share edges. These $[(Me_4O_4)O_{12/2}]$ clusters occupy the corners of a cube and are interlinked by tetrahedral $[TO_{4/2}]$ groups. This structure type exhibits a wide variety of

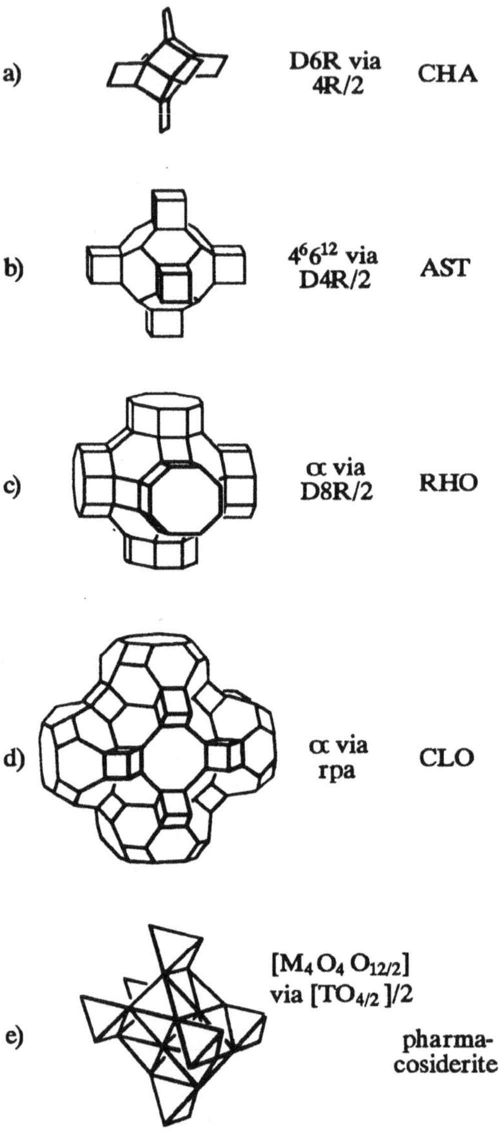

Figure 5. Octahedral building units for the construction of the topologies CHA (a), AST (b), RHO (c), CLO (d) and of zeolite-type compounds with the pharmacosiderite structure (e).

Me/T combinations in the framework with mixed coordination: It is found in nature as the minerals pharmacosiderite (Me: Fe, T: As) or aluminopharmacosiderite (Me: Al, T: As). A class of "germanate zeolites" with a $[Ge_7O_{16}]^{4-}$ framework (Me and T: Ge) also has this structure (WITTMANN (1966), ROBERTS and FITCH (1991)) as well as some molybdenum phosphates with a $[Mo_4P_3O_{16}]^{3-}]$ framework (KING *et al.* 1991) and some titanosilicates (Me: Ti, T: Si; CHAPMAN and ROE 1990). These compounds can be considered as microporous, since they show the de- and rehydration behaviour typical of these materials. For the germanate and the titanosilicate material, the possibility of ion exchange has also been proven.

2.2.3 *Topological concepts betrayed !* The processes of unraveling the structures of the two zeolite-type record materials regarding pore and cage size, namely VPI-5 (VFI) and cloverite (CLO), show clearly the advantages and the limitations of the topological approach to the structural analysis of microporous tetrahedral framework structures. In the case of VPI-5, DAVIS *et al.* (1988) could rely on SMITH and DYTRYCH's (1984) prediction of a topology named "net 81(1)" by these authors. Details of the structure, however, were not compatible with the criteria BRUNNER (1990) had developed to decide between "permissible" and "realizable" framework structures. Consequently, first structure refinements were not satisfactory, as stated by MCCUSKER *et al.* (1991). Then, ^{31}P MAS NMR spectroscopy (GROBET *et al.* 1989) indicated the presence of three different P atoms instead of the two crystallographically distinct positions possible in the assumed space group $P6_3cm$. ^{27}Al MAS NMR spectroscopy (GROBET *et al.* 1989) provided evidence for the presence of both tetrahedral and octahedral Al. With this information, namely that VPI-5 possesses a mixed tetrahedral/octahedral framework, at hand and on the basis of high-resolution powder x-ray diffraction data measured at a synchrotron on highly crystalline material, MCCUSKER *et al.* (1991) succeeded in refining the arrangement of the framework atoms and that of the water molecules, which form a hydrogen-bonded triple helix inside the channels. To my opinion, this is one of the most beautiful examples of the structure of a microporous solid and of its host-guest interactions.

In the case of cloverite (ESTERMANN *et al.* 1991) it was the "interruption" of the gallophosphate framework by terminal OH groups, a situation which is not included in the topology of tetrahedral frameworks, which made powder diffraction data resistent to a structure solution *via* the common approach (MCCUSKER 1991) of topological concepts, adsorption characteristics and model building. The structure was solved only when single crystals became available.

3. Old and new ways to periodically arranged, uniform large micropores

Naturally, one of the most important goals in the synthesis of microporous solids has been the controlled synthesis of materials containing ever larger voids in periodic, *i.e.*, crystalline arrangement, and any material establishing a new record is sure to become the "zeolite of the year" (regardless of its chemical composition, which probably will not be this of an aluminosilicate anymore). It cannot be the aim of this paper to treat this subject to full extent. Several excellent reviews have done this (see, for example, BARRER 1986, MEIER 1986, VAUGHAN 1989) and many of the ingenious ideas put forward there remain to be realized. The slogans showing possible ways to larger micropores which are given in the following collection, are of course by no means ideas of a single individual, especially not of the present author, but are strongly influenced by literature and by personal communications.

The following topics should be considered with regard to the realization of large micropores and open networks:

3.1 USE TOPOLOGICAL CONCEPTS

After the treatment of the topology of tetrahedral framework structures given above, there are some obvious simple (and maybe partly simple-minded) approaches to the theoretical construction of zeolite-type compounds with any desired size of windows and cages: The mere extension of the "spacing units" between the "corner-stones" of well-known topologies leads to structures with -in principle - infinitely large pores. For example, addition of further four-rings in between the six-rings in the series of aluminophosphates shown in Fig. 3 leads to pores with ever larger diameters. This series was discovered by SMITH and DYTRYCH (1984), when only the first member, $AlPO_4$-5 with the net described as "81", was known (Fig. 3a). Net 81(1) (Fig. 3b) was established as VPI-5 in 1988 (DAVIS *et al.*), whereas net 81(2) (Fig. 3c) and further extensions 81(n) still await their synthesis. A similar approach extends the "spacers" between the OBUs shown in Fig. 4b,c) and Fig. 5. A "super-zeolite A" would have triple four-rings bridging the sodalite units, a "super-cloverite" would for example possess additional double eight-rings between the rpa-cages. The pore-size of the pharmacosiderite structure could be extended by using linkages with longer chains of tetrahedra. Not all of these examples will withstand closer inspection, because in some cases unfavourable distortions from tetrahedral coordination geometry are necessary to close all links between the tetrahedra. However, as shown by the example of many AlPOs, for instance VPI-5, extension of the coordination number of strained T atoms to five or six can release the strain.

The structural elements characteristic of open or large-pore tetrahedral networks (at least some of them) have been identified by BRUNNER (1979, 1981, 1990) and MEIER (1986) and BRUNNER and MEIER (1989), the most noteworthy being the conclusion that the average ring size in framework structures is practically constant and so, in order to have some large rings (and consequently large pores), many small rings should be present in the structure. Recently, some examples for tetrahedral frameworks containing three-ring arrangements have been found. These include the beryllosilicate mineral lovdarite (MERLINO 1981, 1990) and the related synthetic zincosilicate VPI-7 (ANNEN *et al.* 1991), both of which possess building units composed of two three-rings arranged in a *spiro* configuration, and by the aluminosilicate ZSM-18 (LAWTON and ROHRBAUGH 1990). However, these three-ring containing structures do not yet exhibit especially large pores. Joined four-rings seem to be a more favourable way to obtain large-pore structures, as evidenced by the topologies of the two record materials VPI-5 and cloverite which both contain many four-rings arranged in D4R cubes. It is tempting to speculate that the presence of F^- ions in the D4R units of many of the materials synthesized in a fluoride medium (see above) indicates a templating role of this ion; stabilizing Ga-F interactions have been recognized by x-ray structural analysis and ^{71}Ga MAS NMR (SIMMEN *et al.* 1991, ESTERMANN *et al.* 1991, MERRROUCHE *et al.* 1992). Are there templates stabilizing the *spiro* five-ring (probably not) or double three-ring (perhaps) arrangements ? Note that the D4R stabilization also seems to work in pure silicate systems as exemplified by octadecasile (CAULLET *et al.* 1991); therefore, CLO-like topologies might be accessible as zeosiles, *i.e.* pure silica frameworks, too (DAVIS 1991).

3.2 SUBSTITUTE T ATOMS

The fact that the maximum window and cage sizes have been stagnating for a long time in the aluminosilicate system and that only the use of different oxidic systems provided the possibility to synthesize the aluminophosphate VPI-5 and the gallophosphate cloverite underlines the necessity to test other T,T',T''... combinations. The investigation of metal aluminophosphates

(MeAPOs) and metal aluminosilicophosphates (MeAPSOs) have contributed a number of new framework structure types, that are not known as (alumino)silicates. It has also been realized that formation of three-rings with unfavourably small T–T–T angles is simplified by the availability of two sorts of T atoms with different T–O bond lengths.

Incorporating metal atoms into the framework allows the introduction of specific reactivities as compared to the case where these are present as guest cations. Many of these new T atom combinations will leave the frame of purely tetrahedral frameworks, as exemplified by VPI-5 and other AlPOs as well as by titanium silicalite-1 (BEHRENS *et al.* 1991), structures which contain five- and/or six-fold coordinated metal atoms. Titanium silicalite is a nice example for the new kind of reactivity introduced by framework incorporation of a metal (ROFFIA *et al.* 1990, TATSUMI *et al.* 1990, HUYBRECHTS *et al.* 1990).

3.3 USE MORE "SPECIALIZED" TEMPLATES

The role of templates during the synthesis of microporous solids is not yet fully understood. They may in some cases act as mere void-fillers, only preventing the crystallization of dense tetrahedral frameworks. This view is supported by the fact that the same microporous structure can be stabilized by different templates. Consider for example the MFI topology, represented by the aluminosilicate ZSM-5, which is probably the most important zeolite found by using organic template cations, in this case tetrapropylammonium ions. Not only does the use of other $[NR_4]^+$ cations (di- and tripropylammonium, GUTH *et al.* 1986) lead to ZSM-5, but also the "template-free" synthesis from purely inorganic solutions is possible (JACOBS and MARTENS 1987, TISSLER *et al.* 1989). On the other hand, a strong structure-directing influence is apparent in many other cases where the structure of the framework seems to be "wrapped" around the guest molecule enclathrated during synthesis. This is true, for example, for the case of ZSM-18 (LAWTON and ROHRBAUGH 1990). A recent basic study of the influence of the template on the structure of the compound formed concentrated on the simple system SiO_2–template–H_2O, where some of the additional variables complicating zeolite synthesis (Si/Al ratio, presence of additional anions and metal cations) are absent. GIES and MARLER (1992) showed that in the pure silica system, strong structure-directing effects occur and that the size and the shape of the template molecule are the two most relevant parameters determining the framework structure formed, whereas its chemistry is of less importance. For example, different five-membered cyclic organic compounds (cyclopentane, tetrahydrofurane, pyrrole, thiophene) all direct the crystallization of the same structure type, namely dodecasil 3C (structure code MTN).

A valuable new approach to analyze the kind of interactions that give rise to templating effects in the synthesis of zeolites and related compounds is the investigation of the low-temperature crystallization products of typical reaction gels and solutions, that – if treated at temperatures between 80 °C and 200 °C – lead to microporous solids. It has recently been shown (WIEBCKE 1991) that the corresponding products obtained by crystallization between room temperature and 0 °C possess structures with mixed tetrahedral frameworks of $[SiO_{4/2}]$ (or $[AlO_{4/2}]$) and $[OH_{4/2}]$: (alumino)silicatic D4R (or D3R) units (with covalent bonding) and water molecules (with hydrogen bonding) build open three-dimensional frameworks which often exhibit the simple topologies known from zeolitic structures. The terminal oxygens of the double ring units are involved in covalent bonding to the core of the double ring and in hydrogen bonding in the water framework. The template molecules (tetraalkylammonium cations) occupy the cages of these open frameworks. Whereas the structural relationship of the first member of this group of compounds to the LTA topology was immediately realized (BISSERT and LIEBAUS (1987) work on a tetrabutylammonium silicate hydrate), some similar compounds dwelt in the literature as alkylammonium silicate hydrates until their character as mixed ("heterogeneous") frameworks was discerned by the reinterpretations of their structures by WIEBCKE (1991). As a further example I note that the silicate hydrate with the formula $Na[N(CH_3)_4]_7$ $[Si_8O_{20}]\cdot 54H_2O$ crystallizes in a structure with AST topology (WIEBCKE und KOLLER 1992). These crystalline mixed silica/water frameworks with entrapped organic molecules ("templates") provide a link between the chemistry of the reaction gel or solution and the structure of the solid formed at high temperatures and thus shed some new light on the "template effect" observed in zeolite synthesis (WIEBCKE 1991).

In spite of our incomplete knowledge of the mechanism of template effects, the use of other structure-directing agents than those which have commonly been employed (*e.g.* tetraalkylammonium cations) continues to yield new framework topologies. Consider for example the use of crown ethers directing topologies FAU and EMT (DELPRATO *et al.* 1990); or the use of quinuclidine which gave the first examples of the AST topology in $AlPO_4$ (BENNETT and KIRCHNER 1991) and SiO_2 systems (CAULLET *et al.* 1991) and also the CLO topology in the $GaPO_4$ system (ESTERMANN *et al.* 1991), all products having been described in 1991. Irregular template shapes can lead to irregularly shaped voids, most impressingly shown by the structure of ZSM-18 (LAWTON and ROHRBAUGH 1990). New templates might include cryptates ("three-dimensional" crown ethers) or specially functionalized oligo-/polymers.

3.4 TRY DIFFERENT FRAMEWORK COMPOSITIONS

The formation of crystalline porous solids is by no means restricted to the usual oxidic systems. The following small (and incomplete) selection names just a few other compounds which contain micropores. Linkage *via* cyanide ions is observed in $(Me_3Sn)_4Fe(CN)_6$ (ELLER *et al.* 1989), $Cd(CN)_2\cdot 1/3(C_6H_{12}N_4)$ (ABRAHAMS *et al.* 1991), and $N(CH_3)_4[CuZn(CN)_4]$ (HOSKINS and ROBSON 1991) and other compounds discussed by BARRER (1986), BOXHOORN *et al.* (1985) and IWAMOTO (1991). Other anions replacing the oxide ion of the traditional frameworks are sulfide (BEDARD *et al.* 1989) and recently nitride (SCHNICK and LÜCKE 1992) in the above-mentioned P-N-sodalite. Mixed tetrahedral/octahedral frameworks are exemplified by substances with the pharmacosiderite structure (see section 2.2.2) and other molybdenum phosphates (HAUSHALTER *et al.* 1989). Octahedral microporous frameworks (discussed by VAUGHAN 1989) are found among Ti and Mn oxide minerals. Non-zeolitic crystalline ion exchangers have been reviewed by CLEARFIELD (1988).

3.5 ESTABLISH NEW SYNTHETIC ROUTES

All the above-mentioned approaches to new microporous solids with hopefully better properties rely on the same synthetic method, namely the (hydrothermal) crystallization from a solution or gel (mostly aqueous) containing the framework components together with a mobilizer (if necessary) and a structure-directing agent, which acts at least as a void-filler to prevent the formation of dense frameworks. Although syntheses in non-aqueous systems have been tried and very short reaction times at high temperatures (crystallization of AlPO-5 in some minutes at 600 °C, HUO and XU 1992) have recently been used, the basic chemistry involved is always very similar. Can we expect nature to build crystalline structures with ever larger pores and ever more beautiful frameworks by using always the same chemistry ?

Consider again the approach described in the beginning of section 3.1, namely extension of pore sizes by enlarging the spacing units between the "corner-stones" of known structure types. This kind of "paper chemistry" of course does not state the conditions under which the corresponding structures can be prepared and it is doubtful, whether conventational zeolite syntheses will lead to success. One can however think of another synthetic approach, schematically depicted in Fig. 6. Whereas the uncontrolled reaction between a multifunctional "corner-stone" element A and a bifunctional spacer B will in most cases lead to a non-crystalline arrangement with small surface area and broad pore-size distribution (see below), the stepwise process shown in Fig. 6 may lead to a crystalline porous network, with regular pores of any desired pore size. This approach may be designated as "three-dimensional pillaring". It reminds of the highly elaborated syntheses in biochemistry, where large proteine or nucleic acid molecules are generated by a multi-step sequence of repeating, near-100%-yield reactions, including specific blocking and deblocking as well as connecting reactions, running preferably on a polymer surface to avoid problems of solubility. Once the corresponding methods for linking inorganic building units of type A and B to build three-dimensional frameworks have been worked out, the process could be automatized like the biochemical polymerisations are today.

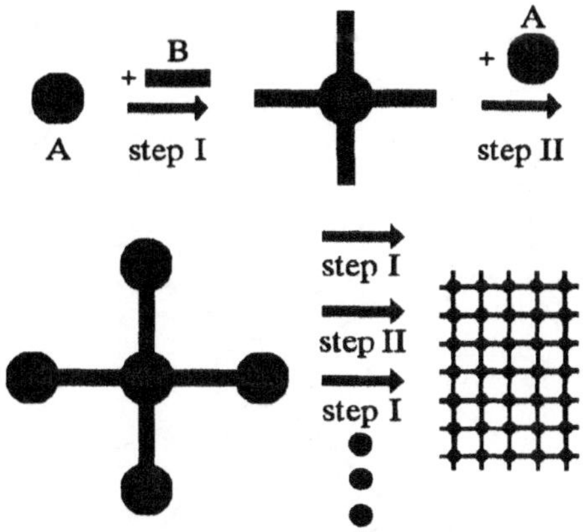

Figure 6. "Three-dimensional pillaring": a hypothetical new route to micro- and mesoporous solids (explanation see text).

A first step in this direction is provided by recent work from the group of HOEBBEL (HOEBBEL *et al.* 1990, 1991), making use of the reaction between functionalized silica D4R units. Although the reactions were conducted as uncontrolled one-step experiments, they provide examples for typical building units that could be used in "three-dimensional pillaring". In one of their experiments, HOEBBEL *et al.* reacted vinyl-functionalized D4R units $[(CH_2=CH_2)(CH_3)_2Si]_8Si_8O_{20}$ with dimethylsilylated D4R units $[H(CH_3)_2Si]_8Si_8O_{20}$ (two "A" units according to Fig. 6), giving an amorphous polymer the structure of which is schematically depicted in Fig. 7. The specific surface of this substance is still small. Maybe, the directed blocking of some of the connectivities provided by the functional groups of the D4Rs and the use of stiffer linking units would give these materials some microporous character.

4. Framework mechanics and effective pore sizes

It has been realized for a long time that the pore sizes estimated on the basis of the known crystal structure of a microporous solid differ in many cases from those determined experimentally by absorption and diffusion methods. Using reasonable sizes for the framework oxygen atoms that surround the window and kinetic diameters of molecules that have been determined independently, one usually finds that molecules which are larger than the estimated window aperture by about 10 to 20 %, are still able to pass that window. This is especially true for higher temperatures.

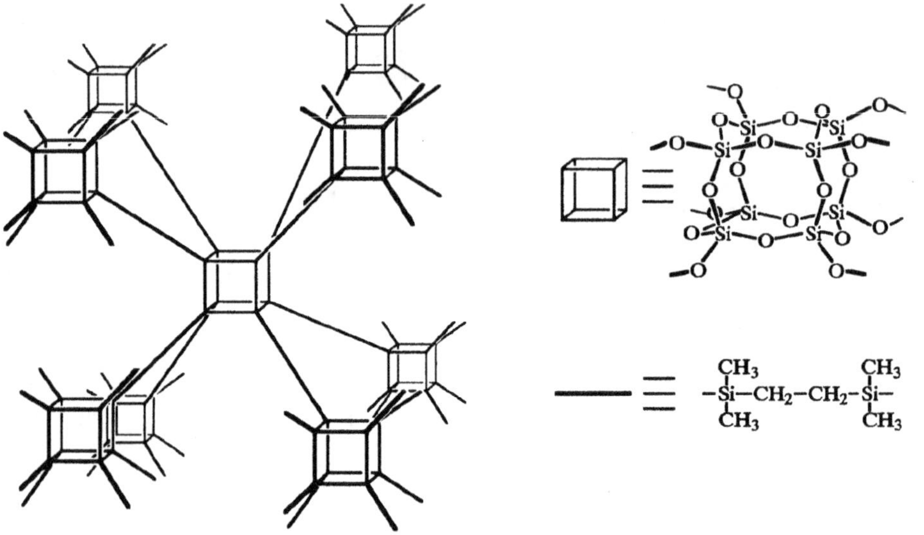

Figure 7: Organo-silicate polymer obtained by an uncontrolled "three-dimensional pillaring" reaction (see text).

The reason for this peculiar behaviour can be traced back to the fact that the estimated pore diameters are based on the determination of a static crystal structure. The passage of a guest species through a window of a porous solid is, however, a highly dynamical process with strong interactions between the moving guest and the host reacting upon its spatial requirements. A complete description of the size and the shape of the pores has to include, therefore, the possible changes of these properties with the dynamical movement of the framework atoms, *i.e.* it requires a dynamic conformational analysis of the framework. The above-mentioned fact that the pore-sizes apparently seem to increase with increasing temperature was related to the excitation of so-called "breathing modes", *i.e.* lattice vibrations that "open" and "close" the window apertures. However, the concept related to the somewhat nebulous term "breathing modes" was never explicitely worked out and obvious questions as:

- "Why do the framework tetrahedra join into a concerted "breathing" action rather than to vibrate independently (which would in average lead to smaller effective pore sizes) ?"
- "Do all the different windows of a microporous framework open and close simultaneously (an unfavourable process due to the volume work necessary) or are some closing as others are opening (a process where the volume of the microporous stays constant) ?"
- "Why do some microporous structures exhibit stronger 'breathing modes' than others ?"

remained unanswered until recently.

The structural element responsible for the peculiar changes in oxidic framework structures is the T–O–T link. Whereas only small distortions are found to occur in the primary building unit (the $[TO_{4/2}]$ tetrahedron), the T–O–T angle may adopt values from about 125° to 180°, with a preference for angles between 135° and 160°. The T–O–T linkages may thus be regarded as "hinges" connecting the rigid tetrahedra (BAUR 1992).

We will study the action of these hinges first for the model case of the sodalite structure, showing how this framework – which possesses the most simple topology among the tetrahedral framework structures - reacts upon temperature, pressure and upon the spatial requirements of the non-framework components. Already PAULING realized in 1930 during his structural analysis of the mineral sodalite $Na_8[Al_6Si_6O_{24}]Cl_2$ that the rotation of the tetrahedra around the T–O–T hinges causes the existence of a "fully expanded" and "partly collapsed" framework structures, the degree of collapse being related to the size of the non-framework components. Framework collapse leads directly to a decrease in the cubic lattice constant a_0 of sodalites. A measure that is independent of the T–O bond length and thus independent of the framework composition is the so-called "tilt angle" φ (see Fig. 8), defined according to TAYLOR (1972); the fully expanded state is characterized by $\varphi = 0°$, whereas partly collapsed frameworks possess non-zero values of φ; φ is related in a simple way to the T–O–T angle (HASSAN and GRUNDY 1984).

That the tilt mechanism is responsible for the adaption of the sodalite framework to the size of the non-framework components was verified by many room temperature crystal structure refinements of compounds with different cage cations and cage anions. Extreme examples for aluminosilicate sodalite structures determined at room temperature are provided in Fig. 8f and 8g where the framework arrangement of the strongly collapsed lithium chloride sodalite $Li_8[Al_6Si_6O_{24}]Cl_2$ (WELLER and WONG 1989) and of the nearly fully expanded sodium perchlorate sodalite $Na_8[Al_6Si_6O_{24}](ClO_4)_2$ (VEIT *et al.* 1990) are shown. The fully expanded state is realized in the case of a sodalite with a pure silica framework containing ethylene glycol in its cages, $(C_2H_6O_2)_2[Si_{12}O_{24}]$ (Fig.8a, RICHARDSON *et al.* 1988).

The tilt mechanism was also made responsible for the extraordinary thermal expansion behaviour of sodalites by TAYLOR (1968, 1972) and by TAYLOR and HENDERSON (1978). Only recently, however, we have been able to show that it is really tilting that governs the framework mechanics during thermal expansion of sodalites (Fig. 8d and 8e): In anhydrous silver sodalite $Ag_6[Al_6Si_6O_{24}]$, φ decreases from 16.3° at 298 K (Fig. 8d) to 7.8° at 623 K (Fig. 8e), as evidenced by Rietveld refinements of x-ray powder diffraction data (BEHRENS *et al.* 1992a). At 673 K, near to the fully expanded state with $\varphi = 0°$, a phase transition takes place because a further unfolding of the framework by reducing φ is not possible anymore (BEHRENS *et al.* 1992a). That also the decrease of volume upon applying pressure on a sodalite structure proceeds mainly via tilting has been shown by FÜTTERER *et al.* (1992), who performed pressure-dependent neutron powder diffraction measurements and Rietveld refinements of another variant of silica sodalite, namely that containing trioxane in its cages: $(C_3H_6O_3)[Si_6O_{12}]$. Note that in this silica sodalite the structure at ambient conditions is characterized by a value of φ of 8.1° (as verified also by single-crystal structure analysis, BEHRENS *et al.* 1992b), *i.e.* in contrast to its ethylene glycol analogue it is not fully expanded (Fig. 8c). Pressurizing to 0.58 GPa causes an increase of φ to 14.3° (Fig. 8b, FÜTTERER *et al.* 1992).

Another example for a framework that can exist in an expanded and in partly collapsed states is that of zeolite Rho. Fig. 9 compares the fully expanded state of the aluminosilicate framework of dehydrated zeolite Rho containing protons as counter-ions, $H_{12}[Al_{12}Si_{12}O_{48}]$ (MCCUSKER 1984), to the most strongly collapsed framework arrangement in a zeolite containing Ca^{2+} and D^+ as counter ions (CORBIN *et al.* 1990). Note the strong elliptical distortion of the eight-rings. The lattice constant of H-Rho is 8.1 % larger than that of Ca,D-Rho (BAUR 1992), corresponding about to the difference between the smallest and largest room temperature lattice constants of aluminosilicate sodalites. There are three crystallographically distinct oxygen atoms in the framework of partly collapsed zeolites Rho, corresponding to three different hinges (in the fully expanded state, there are only two different oxygen atoms); O2 and O3 surround the eight-ring windows (O2 in the case of the fully expanded Rho framework). Their T–O–T angles change only by about 10° to 20° between the fully expanded state of H–Rho and the strongly collapsed state of Ca,D-Rho (Table 1 lists the corresponding data). Nevertheless, a strong distortion of the eight-ring window is observed in Ca,D-Rho, even stronger than the distortion of the six-rings in sodalite structures (compare Fig. 8a to 8f). Both, SOD and RHO, may be classified as framework topologies which possess a tendency to collapse (BAUR 1992).

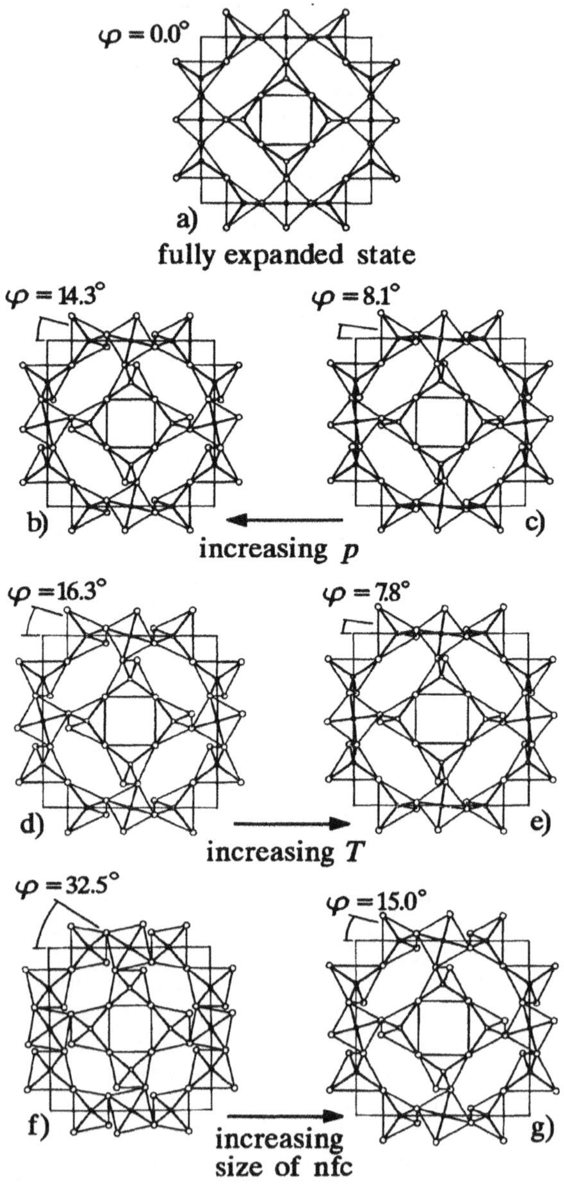

Figure 8. The sodalite framework reacting upon pressure, temperature and upon the size of the non-framework components (nfc) via changing the tilt angle φ: a) the fully expanded framework of ethylene glycol silica sodalite $(C_2H_6O_2)_2[Si_{12}O_{24}]$: $\varphi = 0°$ (RICHARDSON *et al.* 1988); b) and c) the action of pressure on the framework of trioxane silica sodalite $(C_3H_6O_3)_2[Si_{12}O_{24}]$: φ increasing from 8.1° (c) at ambient pressure to 14.3° (b) at 0.58 GPa (FÜTTERER *et al.* 1992); d) and e) the action of temperature on the framework of anhydrous silver sodalite $Ag_6[Al_6Si_6O_{24}]$: φ decreasing from 16.3° (d) at 298 K to 7.8° (e) at 623 K (BEHRENS *et al.* 1992a); f) and g) influence of the size of the non-framework components on the framework of aluminosilicate sodalites: φ = 32.5° (f) in the case of $Li_8[Al_6Si_6O_{24}]Cl_2$ (WELLER and WONG 1989) and φ = 15.0° (g) in the case of $Na_8[Al_6Si_6O_{24}](ClO_4)_2$ (VEIT 1990). Small and large circles represent T and oxygen atoms, resp.

Table 1. Values of T–O–T angles, cubic lattice constants a_0 and volumes V of the framework structures of sodalites (SOD) and of zeolites of topology LTA and RHO (see the text for further explanations and references). Remarks: a) composition not explicitely given; b) compositions, lattice constants and volumes given for zeolites A refer to the pseudocubic unit cell; c) see Fig. 3 of BAUR (1992).

topology	compound	see Figure	T–O–T angle (°) O1	O2	O3	a_0 (Å)	V (Å^3)
SOD	$Ag_6[Al_6Si_6O_{24}]$ at 623 K	8e	157	–	–	9.142	764
	$Li_8[Al_6Si_6O_{24}]Cl_2$	8f	124.5	–	–	8.444	602
RHO	$H_{12}Al_{12}Si_{36}O_{96}$	9a	143.5	154.8	143.5	14.982	3363
	Ca,D-aluminosilicate[a]	9b	133	132	130	13.965	2723
LTA[b]	$K_{12}[Al_{12}Si_{12}O_{48}]$	–[c]	128.5	178.4	153.7	12.31	1865
	$Rb_{9.8}K_{0.2}Na_{0.4}Ba_{0.6}$ $[Al_{11.4}Si_{12.6}O_{48}]$	10a	131.8	172.4	150.3	12.282	1853
	$H_{3.25}Ag_{8.5}$ $[Al_{11.1}Si_{12.9}O_{48}]$	10b	150.1	157.7	143.5	12.275	1850
	$Li_{12}[Al_{12}Si_{12}O_{48}]$	–[c]	171.6	140.4	133.4	11.96	1710

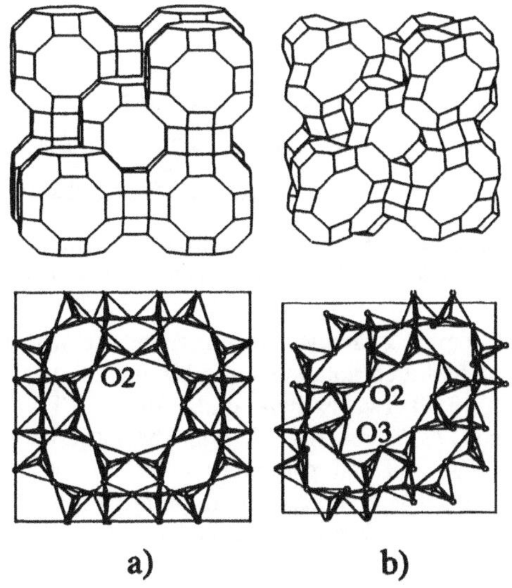

Figure 9. Two views of the of the fully expanded (a) and of the most strongly collapsed (b) zeolite Rho framework; Fig. a) was derived from the crystallographic data of $H_{12}[Al_{12}Si_{12}O_{48}]$ (MCCUSKER 1984), Fig. b) from those of an aluminosilicatic Rho zeolite containing Ca^{2+} and D^+ as counter ions (CORBIN *et al.* 1990). Note the strong elliptical distortion of the eight-ring windows, due to the corotation of the framework tetrahedra.

The fact that there exists a different group of zeolitic frameworks, the members of which, although exhibiting large variations in the T–O–T angles, show only small variations in the lattice constants, was discerned recently by BAUR (1992). A typical example is the framework of LTA topology, which possesses three crystallographically distinct framework oxygen atoms. The values of the T–O–T angles around O1 and O2, those oxygens which surround the eight-ring windows, cover the full range from 128° to nearly 180° and thus exhibit a much stronger variation than the T–O–T angles in SOD and RHO frameworks. Nevertheless, the variation in the lattice constant and in the unit cell volume amounts only to about one half of the corresponding changes in the collapsible frameworks of SOD and RHO. Accordingly, the LTA topology, although extremely flexible with regard to the T–O–T angle, is termed a non-collapsible framework by BAUR (1992). This author ascribes the fact, that flexible frameworks may be either collapsible or non-collapsible, to the special arrangements of hinges in these two classes of frameworks. In collapsible frameworks, like RHO and SOD, the hinges rotate in the same direction, thus effectively reducing the volume of the structure (termed "corotation" by BAUR). In non-collapsible frameworks, the hinges rotate in opposite directions ("antirotation"). Fig. 10 gives a simplified mechanic model of the both types of frameworks: The arrangement in Fig. 1a is unstable (collapsible) because all props move in the same direction upon loading. On the contrary, the arrangement of Fig. 10b is stable (non-collapsible), because rotation of one class of parallel props is hindered by anti-rotation of the other props.

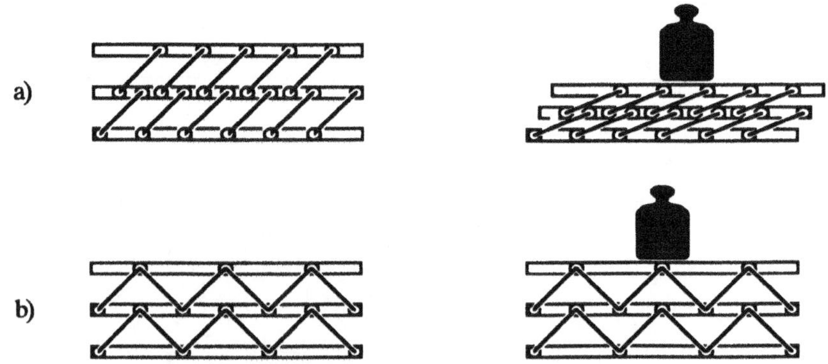

Figure 10. Mechanic model for a framework with corotating hinges that collapses upon loading (a) and for a framework with antirotating hinges that is stable (b).

Chemical examples demonstrating the properties of a non-collapsible framework are provided by dehydrated potassium and lithium zeolites A (PLUTH and SMITH 1983a; SEFF and SHOEMAKER 1969), two extreme cases for the LTA topology (Table 1.): The T–O2–T angle changes from 178.4° for K-LTA with the largest unit cell constant to 140.4° for Li-LTA with the smallest unit cell parameter. On the contrary, the T–O1–T angle opens up from 128.5° to 171.6°, thus compensating the effect of the closing of T–O2–T hinges, leading to only a small difference (2.9 %) in unit cell constants. This example is shown in Fig. 3 of BAUR's work (1992). Even more important for the chemistry of non-collapsible frameworks is the existence of a region where the framework is "soft" (BAUR 1992). For zeolite A, this situation occurs at lattice constants of about 12.3 Å, where very different framework conformations which possess the same volume can be realized. It is explained by using Fig. 11 which shows the frameworks of Rb- and Ag-exchanged zeolites A (PLUTH and SMITH 1983b; GELLENS *et al.* 1983); these compounds possess nearly equal lattice constants a_0 (see Table 1). Rb-LTA possesses angles T–O1–T and T–O2–T of 131.8° and 172.4° and an eight-ring window with four oxygens drawn to its center, giving this window nearly the shape of a square. On the contrary, for Ag-LTA, T–O1–T and T–O2–T are nearly equal (150.1° and 157.7°), giving a circular aperture. Changes between these two conformational states are very easily performed due to the fact that the volume remains constant. BAUR (1992) takes this feature as an indication of a considerable softness of the framework around lattice constants of 12.3 Å: It may be assumed that guest species passing through the eight-ring window of a zeolite A can easily deform the window according to their own spatial requirements and thus can move through pores, which would be too small if they were static. Corresponding volume changes in collapsible frameworks are more difficult to perform since these are associated with volume changes.

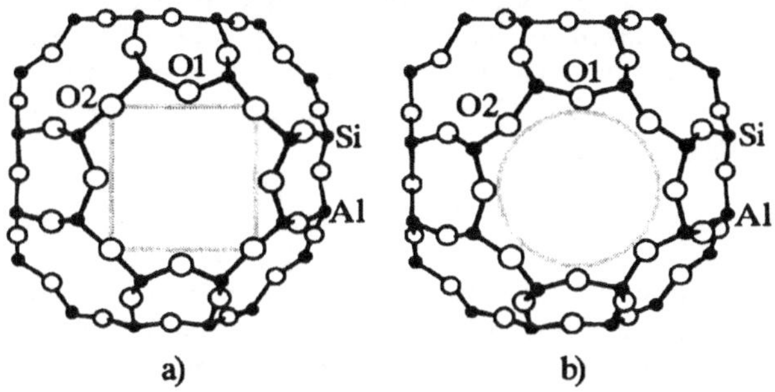

Figure 11. Framework structures of Rb exchanged (a) and Ag silver exchanged zeolite A (b), which possess nearly identical lattice constants, but different framework conformations, giving the eight-ring window approximately the shape of a square (a) or a circle (b).

The question whether a framework is collapsible or non-collapsible in the terms of BAUR (1992) is a question of the interconnection of the hinges and thus a question of topology. Another example of a non-collapsible framework is the FAU topology, and it is tempting to speculate that it is the "softness" of LTA and FAU topologies that gives them their special properties allowing their use as catalysts. It will be of importance to classify known and predicted tetrahedral framework topologies according to their tendency to collapse. This is especially true if mesoporous tetrahedral frameworks are to be constructed. Due to the large pore size aimed at, the walls of the framework surrounding them will be structurally less stable than in microporous solids. Then, collapsible hinge connectivities might lead to irreversible losses in pore size, whereas non-collapsible topologies will stiffen the framework ("backbone") of the porous solid.

The investigation of the collaps tendency of tetrahedral framework structures is related to an actual research area in mathematics, namely the investigation of the rigidity of polyhedra (DEWDNEY, 1991). The theories developed in this field should be tested for possible applications to three-dimensional infinite framework structures.

Similar mechanics influencing the size and the shape of the pores probably are operative in other porous solids, too, for example in pillared layered compounds; they are, however, much more difficult to unravel due to the strong structural disorder present in these compounds. The introduction of the "layer rigidity parameter" gives at least a static description of the deformation of the layers and thus of the size and the shape of the pores.

5. Conclusions

It has been shown that topological considerations are of prime importance in the characterization of the pore size and shape in microporous solids with a tetrahedral framework as well as in yet hypothetical extensions of this type of framework to the mesoporous range. The connectivity of the primary $[TO_{4/2}]$ building units not only determines the coarse value of the pore size as measured in terms of "*n*-ring windows", but also the delicate conformational dynamics of the frameworks, the framework mechanics. Suggestions how to increase the pore size of microporous frameworks with their favourable properties as sharp pore size distribution, high stability and periodic structure to the mesoporous range have been given. It will be interesting to see whether (and which of) these can be realized and how mesoporous frameworks can compete with other mesoporous inorganic solids.

Acknowledgement

I like to thank W. Depmeier, G. Engelhardt, J. Felsche and M. Wiebcke for many fruitful discussions. St. Witte helped me a lot with the drawings. S. Aßmann, W. Depmeier, J. Felsche, K. Fütterer, P.B. Kempa and M. Wiebcke gave their permission to use material from our cooperative work (BEHRENS *et al.* 1992a, 1992b; FÜTTERER 1992) prior to publication. This work was supported by the Bundesminister für Forschung und Technologie.

References

ABRAHAMS, B.F., HOSKINS, B.F., LIU, J. and ROBSON, R. (1991) "The Archetype for a New Class of Simple Extended 3D Honeycomb Frameworks. The Synthesis and X-ray Crystal Structures of $Cd(CN)_{5/3}(OH)_{1/3}\cdot 1/3(C_6H_{12}N_4)$, $Cd(CN)_2\cdot 1/3(C_6H_{12}N_4)$, and $Cd(CN)_2\cdot 2/3H_2O\cdot tBuOH$ ($C_6H_{12}N_4$ = Hexamethylenetetramine) Revealing Two Topologically Equivalent but Geometrically Different Frameworks", J. Am. Chem. Soc., **113**, 3045-3051.

AKPORIAYE, D.E. (1989) "Structural relationships of zeolite frameworks: '5-ring zeolites'", Z. Krist., **188**, 103-120.

AKPORIAYE, D.E. (1992) "Structural relations of zeolite frameworks: ABC-6 and other frameworks derived from hexagonal 4.6.12 and $(4.6.8)_1(4.8.12)_1$ 3-connected nets", Zeolites, **12**, 197-201.

ANDERSON, M.W., PACHIS, K.S., PREBIN, F., CARR, S.W., TERASAKI, O., OHSUNA, T. and ALFREDSSON, V. (1991) "Intergrowths of Cubic and Hexagonal Polytypes of Faujasitic Zeolites", J. Chem. Soc., Chem. Commun., 1660-1664.

ANNEN, M.J., DAVIS, M.E., HIGGINS, J.B. and SCHLENKER, J.L. "VPI-7: The First Zincosilicate Molecular Sieve Containing Three-Membered T-Atom Rings", J. Chem. Soc., Chem. Commun., 1175-1176.

BARRER, R.M. (1986) "Porous crystals: a perspective", in Y. Murakami, A. Iijima and J.W. Ward (eds.), New Developments in Zeolite Science and Technology, Studies in Surface Science and Catalysis, vol. 28, Elsevier, pp. 3-11.

BAUR, W.H. (1992) "Self-Limiting Distortion by Antirotating Hinges Is The Principle of Flexible but Noncollapsible Frameworks", J. Solid State Chem. **97**, 234-247.

BEDARD, R.L., WILSON, S.T., VAIL, L.D., BENNETT, J.M. and FLANIGEN, E.M., "The Next Generation: Synthesis, Characterization, and Structure of Metal Sulfide-Based Microporous Solids", in P.A. Jacobs and R.A. van Santen (eds.), Zeolites: Facts, Figures, Future, Studies in Surface Science and Catalysis, vol. 49, Elsevier, Amsterdam, pp. 375-387.

BEHRENS, P., FELSCHE, J., VETTER, S., SCHULZ-EKLOFF, G., JAEGER, N.I. and NIEMANN, W. (1991) "A XANES and EXAFS Investigation of Titanium Silicalite", J. Chem. Soc., Chem. Commun., 678-680.

BEHRENS, P., KEMPA, P.B., ASSMANN, S., WIEBCKE, M. and FELSCHE, J. (1992a) "The Structure of Anhydrous Silver Sodalite $Ag_6[Al_6Si_6O_{24}]$ at 298 K, 623 K and 723 K from Rietveld Refinements of Powder X-ray Diffraction Data: Mechanism of Thermal Expansion and Phase Transition at 673 K", submitted.

BEHRENS, P., FELSCHE, J., VAN DE GOOR, G. and WIEBCKE, M. (1992b) "Single-Crystal Structure Refinement and Thermal Behaviour of Trioxane Silica Sodalite $(C_3H_6O_3)_2[Si_{12}O_{24}]$", in preparation.

BENNETT, J.M. and KIRCHNER, R.M. (1991) "The structure of as-synthesized $AlPO_4$-16", Zeolites, **11**, 502-506.

BISSERT, G. and LIEBAU, F. (1987) "Crystal Structure of $[N(n\text{-}C_4H_9)_4]H_7[Si_8O_{20}]\cdot 5.33\ H_2O$ - a zeolite A-like double-ring silicate with protonated water clusters $[H_{41}O_{16}]^{9+}$", Z. Krist., **179**, 357-371.

BOXHOORN, G., MOOLHUYSEN, J., COOLEGEM, J.G.F. and VAN SANTEN, R.A. (1985) "Cyanometallates: an Underestimated Class of Molecular Sieves", J. Chem. Soc., Chem. Commun., 1305-1307.

BRECK, D.W. (1974) "Zeolite Molecular Sieves", Wiley, New York.

BRUNNER, G.O. (1979) "The Properties of Coordination Sequences and Conclusions Regarding the Lowest Possible Density of Zeolites", J. Solid State Chem., **29**, 41-45.

BRUNNER, G.O. (1981) "Parameters for framework and space-filling polyhedra", Z. Krist., **156**, 295-303.

BRUNNER, G.O. and MEIER, W.M. (1989) "Framework density distribution of zeolite-type tetrahedral nets", Nature, **337**, 146-147.

BRUNNER, G.O. (1990) "Criteria for the evaluation of hypothetical zeolite frameworks", Zeolites, **10**, 612-616.

CARO, J., FINGER, G., KORNATOWSKI, J., RICHTER-MENDAU, J., WERNER, L. and ZIBROWIUS, B. (1992), Adv. Mater., **4**, 273-276.

CAULLET, P., GUTH, J.L., HAZM, J., LAMBLIN, J.M. and GIES, H. (1991) "Synthesis, characterization and crystal structure of the new clathrasil phase octadecasil", Eur. J. Solid State Inorg. Chem., **28**, 345-361.

CHAPMAN, D.M. and ROE, A.L. (1990): "Synthesis, characterization and crystal chemistry of microporous titanium-silicate materials", Zeolites, **10**, 730-737.

CLEARFIELD, A. (1988) "Role of Ion Exchange in Solid-State Chemistry", Chem. Rev., **88**, 12-148.

CORBIN, D.R., ABRAMS, L., JONES, G.A., EDDY, M.M., HARRISON, W.T.A., STUCKY, G.D. and COX, D.E. (1990) "Flexibility of the Zeolite RHO Framework. In Situ X-ray and Neutron Powder Structural Characterization of Divalent Cation-Exchanged Zeolite RHO", J. Am. Chem. Soc., **112**, 4821-4830.

DAVIS, M.E., SALDARRIAGA, C., MONTES, C., GARCES, J. and CROWDER, C. (1988) "A molecular sieve with eighteen-membered rings", Nature, **331**, 698-699.
DAVIS, M.E. (1991) "Grand openings for cloverite", Nature, **352**, 281-282.
DELPRATO, F., DELMOTTE, L., GUTH, J.L. and HUVE, L. (1990) "Synthesis of new silica-rich cubic and hexagonal faujasites using crown-ether-based supramolecular templates", Zeolites, **10**, 546.
DEWDNEY, A.K. (1991) "The theory of rigidity, or how to brace yourself against unlikely accidents", Scient. Amer., May 1991, p. 89.; "Mathematische Unterhaltungen", Spektrum der Wissenschaft, März 1992, pp. 12-15.
DOUGNIER, F., PATARIN, J., GUTH, J.L. and ANGLEROT, D. (1992) "Synthesis, Characterization, and catalytic properties of silica-rich faujasite-type zeolite (FAU) and its hexagonal analog (EMT) prepared by using crown-ethers as templates", Zeolites, **12**, 160-166.
ELLER, S., BRANDT, P., BRIMAH, A.K., SCHWARZ, P. and FISCHER, R.D., "Effiziente Ionenaustauscher und Rezeptoren für organische und metallorganische Kationen", Angew. Chem., **101**, 1274-1275.
ESTERMANN, M., MCCUSKER, L.B., BAERLOCHER, C., MERROUCHE, A. and KESSLER, H. (1991) "A synthetic gallophosphate with a 20-tetrahedral-atom pore opening", Nature, **352**, 320-322.
FÜTTERER, K., DEPMEIER, W., ALTORFER, F., BEHRENS, P. and FELSCHE, J. (1992) "The Compression Mechanism in Trioxane Silica Sodalite $(C_3H_6O_3)_2$ $[Si_{12}O_{24}]$", submitted to Z. Krist.
GELLENS, L.R., SMITH, J.V. and PLUTH, J.J. (1983) "Crystal Structure of Vacuum-Dehydrated Silver Hydrogen Zeolite A", J. Am. Chem. Soc., **105**, 51-55.
GIER, T.E. and STUCKY, G.D. (1991) "Low-temperature synthesis of hydrated zinco(beryllo)-phosphate and arsenate molecular sieves", Nature, **349**, 508-510.
GIES, H. and MARLER, B. (1992) "The structure-controlling role of organic templates for the synthesis of porosils in the system SiO_2/template/H_2O", Zeolites, **12**, 42-49.
GRAMLICH-MEYER, R. and MEYER, W.M. (1982) "Constituent Units and Framework Conformations in Zeolite Networks", J. Solid State Chem., **44**, 41-49.
GROBET, P.J., MARTENS, J.A., BALAKRISHNAN, I., MERTENS, M. and JACOBS, P.A. (1989) "The Very Large Pore Molecular Sieve VPI-5: An Aluminophosphate Hydrate !", Appl. Catal., **56**, L21-L27.
GUTH, J.L., KESSLER, H. and WEY, R. (1986) "New Route to Pentasil-Type Zeolites Using a Non Alkaline Medium in the Presence of Fluoride Ions", in Y. Murakami, A. Iijima and J.W. Ward, New Developments in Zeolite Science and Technology, Studies in Surface Science and Catalysis, Volume 28, Elsevier, Amsterdam, pp. 121-128.
HASSAN, I. and GRUNDY, H.D. (1984) "The Crystal Structures of Sodalite-Group Minerals", Acta Cryst., **B40**, 6-13.
HAUSHALTER, R.C., STROHMAIER, K.G. and LAI, F.W. (1989) "Structure of a Three-Dimensional, Microporous Molybdenum Phosphate with Large Cavities", Science, **246**, 1289-1291.
HOEBBEL, D., PITSCH, I., HEIDEMANN, D., JANCKE, H. and HILLER, W. (1991) "Über die Reaktion der käfigartigen vinylsilylierten Doppelvierringkieselsäure $[(CH_2=CH_2)(CH_3)_2Si]_8Si_8O_{20}$ mit HSi-haltigen Verbindungen und die Herstellung eines neuen Polymers", Z. Anorg. Allg. Chem., **583**, 133-144.
HOEBBEL, D., PITSCH, I. and HEIDEMANN, D. (1991) "Herstellung und ^{29}Si-NMR-Untersuchung von Polymeren mit definierten Kieselsäurebaueinheiten", Z. Anorg. Allg. Chem., **592**, 207-216.

HOSKINS, B.F. and ROBSON, R. (1990) "Design and Construction of a New Class of Scaffolding-like Materials Comprising Infinite Polymeric Frameworks of 3D-Linked Molecular Rods...", J. Am. Chem. Soc., **112**, 1546-1554.

HUO, Q. and XU, R. (1992) "A New Route for the Synthesis of Molecular Sieves: Crystallization of AlPO-5 at High Temperature", J. Chem. Soc., Chem. Commun., 168-169.

HUYBRECHTS, D.R.C., DE BRUYCKER, L. and JACOBS, P.A. (1990) "Oxyfunctionalization of alkanes with hydrogen peroxide on titanium silicalite", Nature, **345**, 240-242.

IWAMOTO, T. (1991) "Clay-like and Zeolite-like Structures Built of Polymeric Cyanocadmates", in T. Inui, S. Namba and T. Tatsumi, "Chemistry of Microporous Crystals", Studies in Surface Science and Catalysis, vol. 60, Elsevier, Amsterdam, pp. 3-10.

JACOBS, P.A. and MARTENS, J.A. (1987) "Synthesis of High-Silica Aluminosilicate Zeolites", Studies in Surface Science and Catalysis, Volume 33, pp. 134-143.

KING, H.E., MUNDI, L.A., STROHMAIER, K.G. and HAUSHALTER, R.C. (1991) "A Synchrotron Single Crystal X-Ray Structure Determination of $(NH_4)_3Mo_4P_3O_{16}$: A Microporous Molybdenum Phosphate with $Mo_4O_4^{6+}$ Cubes", J. Solid State Chem., **92**, 154-158.

KOKOTAILO, G.T., FYFE, C.A., GIES, H. and COX, D.E. (1989) "Hypothetical Two and Three Dimensional Channel Zeolite Framework Structures", in P.A. Jacobs and R.A. van Santen (eds.), Zeolites: Facts, Figures, Future, Studies in Surface Science and Catalysis, vol. 49, Elsevier, Amsterdam, pp. 715-729.

VAN KONINGSFELD, H., VAN BEKKUM, H. and JANSEN, J.C. (1987) "On the Location and Disorder of the Tetrapropylammonium (TPA) Ion in Zeolite ZSM–5 with Improved Framework Accuracy", Acta Cryst., **B43**, 127-132.

VAN KONINGSFELD, H. (1991) "Structural Subunits in Silicate and Phosphate Structures", in H. van Bekkum, E.M. Flanigen and J.C. Jansen (eds.), Introduction to Zeolite Science and Practice", Studies in Surface Science and Catalysis, vol. 58, Elsevier, pp. 35-76.

VAN KONINGSFELD, H. (1992) "Structural relationsships and building units in the family of 5-ring zeolites", Zeolites, **12**, 114-120.

LAWTON, S.L. and ROHRBAUGH, W.J. (1990) "The Framework Topology of ZSM-18, a Novel Zeolite Containing Rings of Three (Si,Al)-O Species", Science, **247**, 1319-1322.

MCCUSKER, L.B. (1984) "Crystal Structures of the ammonium and hydrogen forms of zeolite rho", Zeolites, **4**, 51-55.

MCCUSKER, L.B. (1991) "Zeolite Crystallography", Acta Cryst, **A47**, 297-313.

MCCUSKER, L.B., BAERLOCHER, Ch., JAHN, E. and BÜLOW, M. (1991) "The Triple Helix Inside the Large-Pore Aluminophosphate Molecular Sieve VPI-5", Zeolites, **11**, 308-313.

MEIER, W.M. (1986) "Zeolites and Zeolite-like Materials", in Y. Murakami, A. Iijima and J.W. Ward (eds.), New Developments in Zeolite Science and Technology, Studies in Surface Science and Catalysis, vol. 28, Elsevier, pp. 13-22.

MEIER, W.M. and OLSON, D.H. (1987) "Atlas of Zeolite Structure Types, IZA Structure Commission, Butterworths, London.

MERLINO, S. (1981) "Lovdarite: three-membered rings in a framework silicate", Acta Cryst., **A37** suppl., C-189.

MERLINO, S. (1990) "Lovdarite, $K_4Na_{12}(Be_8Si_{28}O_{72})\cdot 18\ H_2O$, a zeolite-like mineral: structural features and OD character", Eur. J. Mineral., **2**, 809-817.

MERROUCHE, A., PATARIN, J., KESSLER, H., SOULARD, M., DELMOTTE, L., GUTH, J.L. and JOLY, J.F. (1992) "Synthesis and characterization of cloverite: a novel gallophosphate molecular sieve with three-dimensional 20-membered ring channels", Zeolites, **12**, 226-232.

PAULING, L. (1930), Z. Krist. **74**, 213-225.

PLUTH, J.J. and SMITH, J.V. (1983a) "Crystal Structure of Dehydrated Potassium-Exchanged Zeolite A. Absence of Supposed Zero-Coordinated Potassium. Refinement of Si,Al-Ordered Superstructure", J. Phys. Chem., **83**, 741-749.

PLUTH, J.J. and SMITH, J.V. (1983b) "Crystal Structure of Dehydrated Rb-Exchanged Zeolite A. Absence of Zero-Coordinated Rubidium. Preferential Ion Exchange of Barium Impurity", J. Am. Chem. Soc., **105**, 2621-2624.

RICHARDSON, J.W., SMITH, J.V. and PLUTH, J.J. (1989) "Theoretical Nets with 18-Ring Channels: Enumeration, Geometrical Modelling and Neutron Diffraction Study of $AlPO_4$-54", J. Phys. Chem., **93**, 8212-8219.

RICHARDSON, J.W., PLUTH, J.J., SMITH, J.V., DYTRYCH, W.J. and BIBBY, D.M. (1988) "Conformation of Ethylene Glycol and Phase Change in Silica Sodalite", J. Phys. Chem., **92** (243-247).

ROBERTS, M.A. and FITCH, A.N. (1991) "The Crystal Structure of $Ag_4Ge_7O_{16}\cdot 6D_2O$ and $Na_3(ND_4)Ge_7O_{16}$ Refined from High Resolution Synchrotron Radiation and Neutron Powder Diffraction Data", J. Phys. Chem. Sol., **52**, 1209-1218.

ROFFIA, P, LEOFANTI, G., CESENA, A., MANTEGAZZA, M., PADOVAN, M., PETRINI, G., TONTI, S. and GERVASUTTI, P. (1990) "Cyclohexanone Ammoximation: A Break Through in the 6-Caprolactam Production Process", in G. Centi and F. Trifiro (eds.), New Developments in Selective Oxidation, Elsevier, Amsterdam, pp. 43-52.

SCHNICK, W. and LÜCKE, J. (1992) "$Zn_7[P_{12}N_{24}]Cl_2$ – ein Sodalith mit einem Phosphor-Stickstoff-Grundgerüst", Angew. Chem., **104**, 208-209.

SCHOLZEN, G., BENEKE, K. and LAGALY, G. (1991) "Diversity of Magadiite", Z. Anorg. Allg. Chem., **597**, 183-196.

SEFF, K. and SHOEMAKER, D.P. (1969) "The Structures of Zeolite Sorption Complexes. I. The Structures of Dehydrated Zeolite 5A and its Iodine Sorption Complex", Acta Cryst., **22**, 162-170.

SIMMEN, A., BAERLOCHER, Ch., MERROUCHE, A., PATARIN, J. and KESSLER, H. (1991) "Synthesis and Rietfeld refinement of a Fluoride-containing Gallophosphate with the LTA Structure type", in H.G. Karge and J. Völter (eds.), Third German Workshop on Zeolite Chemistry, Berlin, P26.

SMITH, J.V. (1988) "Topochemistry of Zeolites and Related Materials. 1. Topology and Geometry", Chem. Rev., **88**, 149-182.

SMITH, J.V. (1989) "Towards a Comprehensive Mathematical Theory for the Topology and Geometry of Microporous Materials", in P.A. Jacobs and R.A. van Santen (eds.), Zeolites: Facts, Figures, Future, Studies in Surface Science and Catalysis, vol. 49, Elsevier, Amsterdam, pp. 29-47.

SMITH, J.V. and DYTRYCH, W.J. (1984) "Nets with channels of unlimited parameter", Nature, **309**, 607-608.

TATSUMI, T., NAKAMURA, M., NEGISHI, S. and TOMINAGA, H. (1990) "Shape-selective Oxidation of Alkanes with H_2O_2 catalysed by Titanosilicate", J. Chem. Soc., Chem. Commun., 476-477.

TAYLOR, D. (1968) "The thermal expansion of the sodalite group of minerals", Mineral. Mag., **36**, 761-769.

TAYLOR, D. (1972) "The thermal expansion behaviour of the framework silicates", Mineral. Mag., **38**, 593-604.

TAYLOR, D. and HENDERSON, C.M.B. (1978) "A computer model for the cubic sodalite structure", Phys. Chem. Minerals, **2**, 325-336.

TISSLER, A., POLANEK, P., GIRRBACH, U., MÜLLER, U. and UNGER, K.K. (1989) "Control of catalytic properties of ZSM-5 made by fast and template-free synthesis", in H.G. Karge and J. Weitkamp, Zeolites as Catalysts, Sorbents and Detergent Builders, Studies in Surface Science and Catalysis, vol. 46, pp. 399-408.

VAUGHAN, D.E.W. (1989) "Zeolites and other microporous materials", in P.A. Jacobs and R.A. van Santen (eds.), Zeolites: Facts, Figures, Future, Studies in Surface Science and Catalysis, vol. 49, Elsevier, Amsterdam, pp. 95-116.

VEIT, Th., BUHL, J.-Ch. and HOFFMANN, W. (1990) "Hydrothermal Synthesis, Characterization and Structure-Refinement of Chlorate- and Perchlorate-Sodalite", Ext. Abstr. 2nd German Workshop on Zeolite Chemistry, Hamburg; Acta Cryst. **A46**, C-254.

WELLER, M.T. and WONG, G. (1989) "Characterization of novel sodalites by neutron diffraction and solid-state NMR", Solid State Ionics, **32/33**, 430-435.

WELLS, A.F. (1977) Three-dimensional nets and polyhedra, Wiley, New York.

WELLS, A.F. (1984) Structural Inorganic Chemistry, Clarendon Press, Oxford.

WIEBCKE, M. (1991) "Structural Links between Zeolite-Type and Clathrate Hydrate-Type Materials", J. Chem. Soc., Chem. Commun., 1507-1508.

WIEBCKE, M. and KOLLER, H. (1992) "Single-Crystal X-ray Diffraction and Variable-Temperature MAS NMR Study on the Heterogeneous Network Clathrate $Na[N(CH_3)_4]_7[Si_8O_{20}]\cdot 54H_2O$", Acta Cryst., **B48**, in the press.

WITTMANN, A. (1966) "Beiträge zur Strukturchemie der Germanate", Fortschr. Miner., **43**, 230-272.

Realistic Simulation of Mesoporous Materials by Random Packing Arrays of Hard Spheres

N. H. PANTELIS

Institute of Physical Chemistry
"Democritos" National Research Center
Ag. Paraskevi, Athens, Greece

ABSTRACT. A model based on random packing arrays of hard spheres is proposed to describe the structure of mesoporous materials. The equilibrium positions of the spheres are obtained by a 'vibration' technique and the network of irregular tetrahedra that represent the porous structure are identified by a 'breaking' procedure. The three dimensional network thus obtained, with the appropriate boundary conditions imposed, can be used for the simulation of gas adsorption and condensation through the material. Equilibrium sorption and desorption isotherms are then obtained, for any selected material.

1. Introduction

Electron-microscopic studies of many porous materials have revealed two main types of structures: a corpuscular one, where individual pores represent interstices between particles and spongy systems, where individual particles cannot be distinguished. Mesoporous adsorbent materials consisting of aggregates of spherical or spheroidal particles can be realistically described by random packing arrays of hard spheres. Such arrays have proved to be more realistic representations of the pore structures of these materials than the regular packing of hard spheres in contact, widely used in the past [1]-[3]. To achieve randomness of the packing array of hard spheres a realistic method of 'shaking' the spheres must be chosen. Such a method, called 'vibration' is shown to produce the desired randomness. This technique works under two basic principles: a satisfactory periodic boundary condition imposed on the edges of the 'box' that contains the spheres and an appropriately high number of iteration cycles to ensure for the final randomness. When equilibrium is achieved a random packing array of spheres is obtained with a defined pore radial distribution, depending on the porosity of the material chosen. All spheres are considered to be of equal size for this simulation.

This geometrical representation is shown to be a simple one because only the sphere center coordinates are required to define the total pore space. To define individual pores a geometrical model must be chosen. Irregular tetrahedra are commonly used for this purpose because of their geometrical simplicity. A method of 'breaking' the array of hard spheres, as obtained by the vibration technique, must be subsequently used. A method proposed by Collins [4] is adopted here. This method is shown to produce a uniquely defined array of irregular tetrahedra representing the porous structure of the material considered. Once this is done the porous structure can be geometrically defined and the simulation of gas

C.A.C. Sequeira and M.J. Hudson (eds.), Multifunctional Mesoporous Inorganic Solids, 99–105.

adsorption and condensation through the porous medium be performed.

2. Description of the Model

2.1 THE VIBRATION TECHNIQUE

A computer algorithm is developed to simulate the vibration process. Each sphere is moved by small displacements in space (dx,dy,dz) randomly produced in the range 0 to 1/10 of its diameter. Sphere positions are checked against the periodic boundary conditions imposed at the edges of the 'box' (spheres moving out of an edge reappear from the opposite side) and for validity of the position itself (spheres that penetrate each other are not accepted). Only one chance is given to each sphere to move during each cycle. Every a certain number of cycles a histogram of the relative distances of the spheres is drawn. Equilibrium is considered to be reached when the shape of consecutive histograms does not change appreciably. This usually requires more than 500 cycles. For a system consisting of 100 spheres each sphere is moved at an average of 10^5 times. Simulations with 1000 or even 2000 spheres can be performed in reasonable computer time.

2.2 THE BREAKING ALGORITHM

A computer algorithm is developed to 'break' the random packing array of hard spheres, obtained by the previous method, into a geometrically defined array of irregular tetrahedra. The breaking procedure adopted in this work follows a scheme proposed by Collins [4]. This scheme works as follows: the array of relative distances (the distances between each possible pair of spheres) is sorted in ascending order. Connections of each pair of spheres (following the sorted array) are made, starting with the closest pair and proceeding with the next one in the array. During the procedure the following rule is taken into account: each line drawn cannot intersect internally any existing triangle formed by previously connected spheres. That is, any three spheres already connected in the array form an impenetrable, to any candidate neighbors, triangular shield.

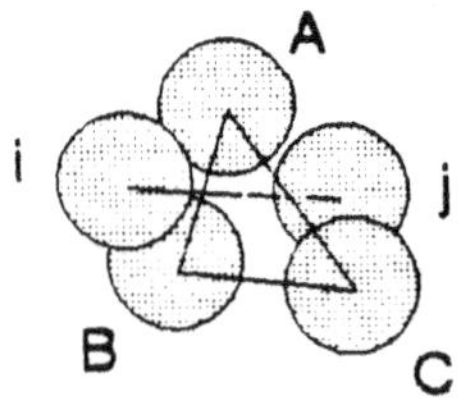

Figure 1: The Collins Procedure for joining two spheres i-j

This is shown in Figure 1: the line i-j, connecting the sphere i with the sphere j is discarded (not drawn) because it intersects the triangle ABC, formed by the previously connected spheres A, B and C. If the intersection point S lied externally to the triangle ABC the line i-j would be accepted. This algorithm is shown to produce a uniquely defined array of irregular tetrahedra, used to define the nearest 'neighbors' of each sphere (for the concept of nearest neighbors see reference [7]) and represent the pore structure of the material. The number of nearest neighbors, for a material of porosity 0.37, is about 12 to 13. Experimental values for random packing arrays are not very different. Mason [6] gives 13.66 as the average number of spheres neighboring a central one, while Bernal and Finney [7] place this figure to 14.28.

Once the geometry of the system is defined the pore size distribution can be calculated and the interconnections between pores can be determined. It is assumed that each tetrahedrical pore in the array communicates with the other pores through four triangular 'windows', the size of which can be calculated. The network thus obtained can be used for the simulation of the gas sorption and condensation.

2.3 SIMULATION OF THE GAS SORPTION

The approximate amount (V_m) of the gas sorbed on a spherical particle of radius R is calculated by equation 1:

$$V_m = \frac{4\pi}{3}[(R+t)^3 - R^3]\frac{s}{4\pi} - \sum_i V_{hi} \tag{1}$$

where the summation is taken from i=1 to 6. V_{hi} is the part within a tetrahedron of the volume of a 'cap' between two of the spherical particles which define the vertices of the tetrahedron. Its value for an edge length a_i connecting the centers of the two spheres is given by equation 2:

$$V_{hi} = \frac{\lambda_i}{2}\left[\frac{2}{3}(R+t)^3 - \frac{a_i}{2}(R+t)^2 + \frac{a_i^3}{24}\right] \tag{2}$$

The thickness t of the sorbed layer is a function of the sphere curvature (R) and can be determined from equation 2 [8], if t_0 (thickness of adsorbed layer on flat surface at a given pressure) is known.

$$\frac{t_o}{R} = \frac{t}{R} + \frac{t}{R^2} + \frac{t^3}{3R^3} \tag{3}$$

The adsorbed layer thickness t_0 (in nm) is calculated, as a function of the relative pressure, by the Halsey equation:

$$t_o = A\left(\frac{-5}{\ln(p/p_o)}\right)^{1/3} \tag{4}$$

For nitrogen at 77.4 K the constant A = 0.354.

The total volume of condensate V_c is calculated as

$$V_c = V_p + V_f \tag{5}$$

where V_p is the volume of the pendular rings formed between the neighboring spheres and V_f is the volume of the filled cavities, i.e. the space between spheres left by the adsorbed layer and the pendular rings. The geometric calculations of these volumes follow the approach of Kanellopoulos et.al. [5].

3. Discussion of Results

Two different mesoporous materials, of porosity 0.37 and 0.44, were simulated with this model. The sphere diameter for both is equal to 100 Angstrom, a typical size for Alumina pellet membranes. The distribution of neighbor spheres, in a random packing array of 500 spheres, is shown in Figures 3 and 4, respectively.

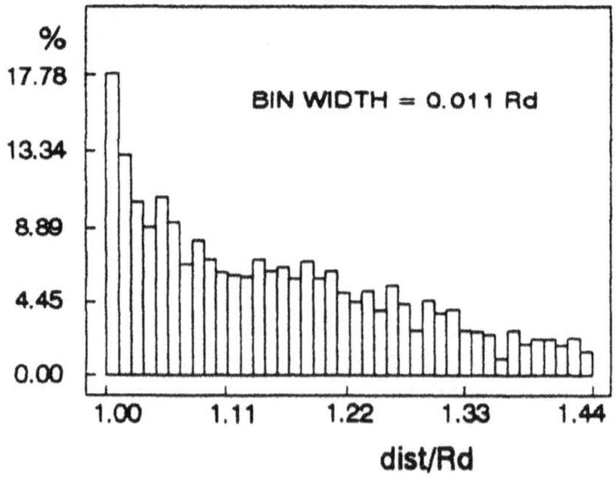

Figure 2: Distribution of sphere neighbors (porosity 0.37)

These histograms were drawn in the interval 1.0 to 1.44 sphere diameters, with each bar having a width of 0.036 sphere diameters. The distribution of the first material (porosity 0.37) shows that the average number of neighbors in the interval 1 to 1.44 diameters, is about 12.5. This value is not very different from those reported by Mason [6] and Bernal and Finney [7] (13.66 and 14.28, respectively). The second material (of porosity 0.44) gives a lower value (10.8), as expected for a more loosely packed array.

The equilibrium sorption-desorption isotherms, using N_2 at 77.4 K, are plotted in Figures 5 and 6, for the two materials chosen. The total volume of sorbate, in cm^3/g, is plotted against the relative pressure, p/p_0. The curves, compared with experimental ones, show good agreement. It can be noticed that as the porosity of the material increased (from 0.37 to 0.44) the desorption

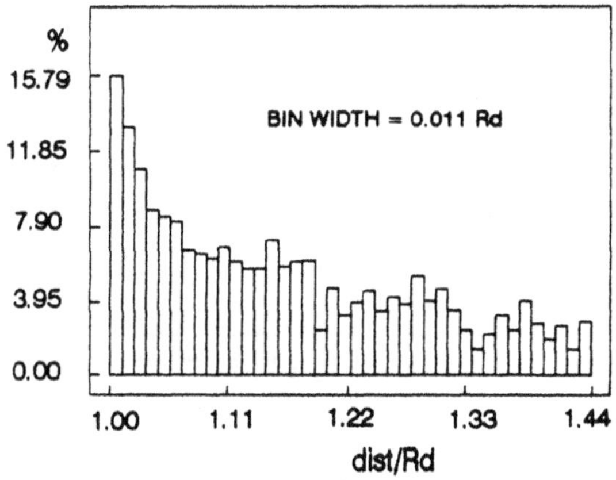

Figure 3: Distribution of sphere neighbors (porosity 0.44)

hysteresis narrowed. This is an expected result as in the less packed material the pore cavities can be emptied more easily during the desorption cycle.

4. Conclusion

A simple geometrical model for mesoporous materials is proposed, based on the tetrahedrical representation of the porous structure. The model, built using the vibration technique, for obtaining the equilibrium positions, and Collins' algorithm, for breaking the system into an array of irregular tetrahedra, used to simulate the porous spaces, seems to produce reasonably accurate sorption-desorption isotherms of N2 at 77.4 K. The prediction of equilibrium isotherms for mesoporous systems could be a crucial piece of information in designing commercially important materials, such as membranes and catalysts.

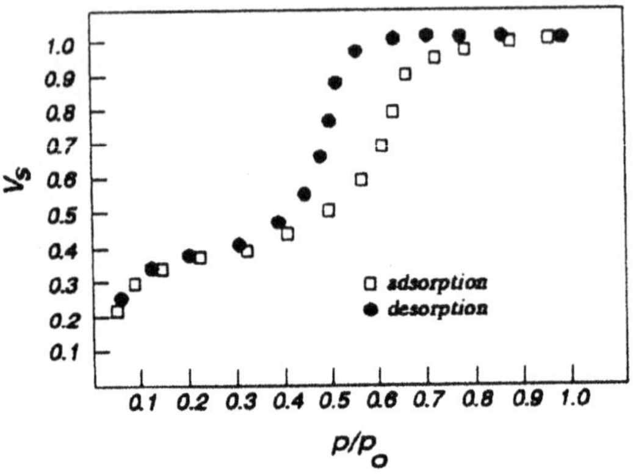

Figure 4: Sorption-desorption isotherms for N_2 at 77.4 K (porosity 0.37)

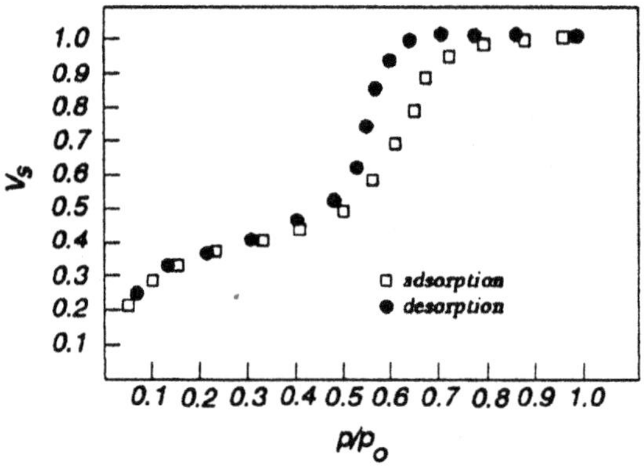

Figure 5: Sorption-desorption isotherms for N_2 at 77.4 K (porosity 0.44)

List of Symbols

p/p_0	relative pressure (p_0 = saturation pressure)
Rd	sphere diameter
R	sphere radius
t	thickness of the adsorbed layer on a curved surface
t_o	thickness of the adsorbed layer on a flat surface
V_s	total volume of sorbate, cm^3/g
V_c	volume of condensate, cm^3/g
V_f	the space between spheres left by the adsorbed layer and pendular rings
V_{hi}	part within a tetrahedron of the volume of a "cap"
V_m	total volume of the adsorbed layer within a tetrahedron
V_p	volume of pendular rings formed between the neighboring spheres

References

1. Aristov, B.G., Karnaukov, A.P., and Kiselev, A.V., *Russ. J. Phys. Chem.,* 1159 (1962)
2. Dollimore, D., and Heal, G.R., *J. Collod Interface Sci.* **42**, 233 (1973)
3. Wade, W.H., J. Phys. Chem. 69, 322 (1964)
4. Collins, R., Proc. Phys. Soc., 86 (1965)
5. Kanellopoulos, N. K., Petrou, J. K., and Petropoulos, J.H., *J. Colloid Interface Sci.,* **96** (1983)
6. Mason, G., *J. Colloid Interface Sci.,* **35** (1971)
7. Bernal, J.D., and Finney, J.L., *Discussions Faraday Soc.* **43**, (1967)
8. Dollimore, D., and Heal, G. R., *J. Colloid Interface Sci.* **33**, 508 (1970)

SYNTHESIS AND CHARACTERIZATION OF VANADIUM OXIDE LAYERED COMPOUND

N.ULAGAPPAN and V.KRISHNASAMY
Department of Chemistry
College of Engineering
Anna University, Madras - 600 025
India

ABSTRACT. A new layered aluminium-vanadium oxide compound has been hydrothermally synthesized using Benzyltrimethyl ammonium chloride salt. The compound crystallized in tetragonal system with unit cell parameters a = 6.08A° and b = 3.899A° and with interlayer spacing of 8.824A°. The compound was further characterized by TGA, chemical analysis and IR.
KEYWRODS. Layered compound, Vanadium oxide, Hydrothermal Synthesis.

1. Introduction

The supported and unsupported vanadium oxide are used as catalysts in many catalytic reactions [1-2], especially in selective catalytic reduction of NO_x[3]. Kittaka *et al* [4] reported the Synthesis and Characterization of V_2O_5. $1.6H_2O$ layered sphere. The V_2O_5 1.6_2O was prepared by thermal decomposition of NH_4VO_3 at high temperature in the presence of water vapour. The V_2O_5 xerogels Structural studies was done by Abello *et al* [5]. They found that the interlayer distance Shrinks from 11.44A° to 8.75A° for the tranformation from V_2O_5. $1.6H_2O$ to V_2O_5 $0.5H_2O$. We have ended up in a layered structure compound aiming a different structural compound.

2. Materials and Methods

2.1 Materials

For the Synthesis of aluminium-vanadium oxide layered compound analar grade Aluminium foil (SD's, India), Benzytrimethyl ammonium chloride (Fluka) and ammoniummetavanadate (Merck) were used as starting materials.

2.2 Synthesis

The aluminium foil was dissolved in a minimum amount of con.HCl and precipitated as gel using ammonia under hot condition. The gel was washed several times with distilled water and centrifuged. To the purified gel NH_4VO_3 with hot water was added under vigorous stirring condition. Then water sloution of Benzyltrimethyl ammonium chloride (BTMA) was added. The gel was well homogenised then loaded in to 316 type Stainless steel autoclave lined with teflan sealing. The final composition of the gel was 3.5 BTMA $1.0Al_2O_3$ $1.0VO_5$ 300 H_2O.
The autoclave was heated in our air oven at 448K for 40 hrs. The dark green products obtained were washed and dried at 383K for three hrs.

C.A.C. Sequeira and M.J. Hudson (eds.),Multifunctional Mesoporous Inorganic Solids, 107–110.

2.3 Characterization

The X-Ray powder pattern of the products were obtained by Philips CW1050 using CuK$_\alpha$ waves. The samples were scaned in 2Θ angle of 5° to 90° range with scan speed of 2°/min. Thermogravimetric analysis (TGA) of dried sample was analysed by Metler TA 3000 series TGA up to 1173K with temperature rate of 20 K/min. under dynamic air condition.

IR measurements were perfomed with Hitachi 270-0525 instrument. The dried sample were pelletised with KBr and the spectra were recorded.

The dried samples were dissolved in aqua regia and diluted to ppm range. The vanadium and aluminium contents were analysed by ICP minitorch - 3410 instrument.

3. Results and discussion

The XRD pattern (Table 1) obtained did not match with any of the JCDPS file. The pattern was indexed using TREOR program [6]. The program successfuly indexed all the peaks present in the pattern except one with figure of merit 33[7]. Which confirmed that the compound is new one crystallized in tetragonal system with a = 6.08A° and b = 3.899A°. The systamatic presence of (00l) peaks confirmed it is a layered compound with interlayer distance of 8.824A° [8].

TABLE 1. Indexed XRD powder pattern of aluminium - vanadium oxide compound

'd' obsd.	I/Io	hkl
8.8238	100	001
2.9651	34	003
2.7256	30	210
2.6681	23	103
2.2349	23	-
1.9255	25	310
1.3222	7	413

The TGA analysis reports are summarised in Table 2. There are three stages of transformation taking place. The first significant weight loss below 373 K suggests that the compound is having porous structure, the absorbed molecules freely desorbs on heating. The second and third weight loss can be assigned to water and BTMA loss respectively.

From the chemcial analysis and TGA analysis it is found that the compound is having the basic composition.

$$(Al_2O_3)_{0.375}\ (V_2O_5)_{1.0}\ (BTMA)_{0.19}\ (H_2O)_{0.43}$$

In IR the intense band at 996 cm^{-1} and band at 352 cm^{-1} are assigned to V=0 stretching and deformation respectively [9]. The bands at 795 cm^{-1} and 294 cm^{-1} may be assigned to V-O-V stretching and deforming respectively.

TABLE 2. TGA step analysis of layered material

Initial temp. K	Final temp. K	Peak temp. K	% change in mass
306	380	321	-5.3862
380	459	440	-2.7728
459	883	494	-13.673

4. Conclusion

From literature and from our study it is found that the hydrothermal synthesis is one of the route to synthesis layered compound. But this route is not exploited fully for the preparation of layered material. It is expected that the synthesised material will serve and may replace the conventional V_2O_5 and supported catalysts due to its layered and microporous nature as in the case of molecularsieves [10]. Considerable weight loss at 321K and slight variation in structural parameter from hydrated V_2O_5 speculates that the aluminium may be pillared the vanadium oxide layered structure, however further structural study is needed to confirm things.

Acknowledgement. One of the author (N.U) thanks UGC for fellowship.

References

1. Clarebout, G. Ruwet, M. Matralis, H. and Gorage, P. (1991) 'Selective catalytic reduction of nitric oxide with ammonia on vanadium oxide catalysts supported on titania-aluminia', Applied Catalysis 76, L9-L14.

2. Vassileva, M. Andreev, A. Dancheva, S. and Kostev, N. (1989) 'Complete Catalytic oxidation of benzene over supported vanadium oxides modified by Palladium', Applied catalysis 49, 125-141.

3. Handy, B.E. Maciejewski, M. and Baiker, A. (1992) 'Vanadia, Vanadia - Titania and Vanadia - Titania - Silica: Structural genesis and catalytic behaviour in the reduction of nitric oxide with ammonia', Journal of catalysis 134, 75-86.

4. Kittaka, S. Sasaki,S. Ogawa,N. and Uchida, N. (1988) 'Hydrous phase in the crystalline vanadium oxide spheres', Journal of Solid State Chemistry 76, 40-51.

5. Abello,L. Husson,E. Repelin,Y. and Lueareau (1985)'Structural Study of gels of V_2O_5 : Vibrationational Spectra of Xerogels', Journal of Solid State Chemistry 56, 379-389.

6. Werner, P.E. Erickson.L and Westdanl. M (1985) 'TREOR, a semi exhaustive trial and error powder indexing program for all Symmetries', Journal of Applied Crystallography 18, 367-370.

7. De Wolf, P.M. (1968) 'A simplified criterior for the reliability of a powder Indexing', Journal of Applied Crystallography, 1, 108-113.

8. Clearfied, A. (1991) 'Personal communication' dated 29th October.

9. Legrouri, A, Baird. T and Fryer, J R (1988) 'Electron optical studies of the thermal decomposition of ammonium vanadata', Reactivity of solids 5, 53-68.

10. Thomas, J M (1991) 'Curiosity, chance, paradase and perspective in the Chemistry of materials', Journal of Chemical Soceity. Dalton transaction, 555-562.

SHS-THE NEW METHOD OF DIFFERENT INORGANIC MATERIALS PRODUCTION

K. L. PADUKOV
Center of SHS, 164,
Leninsky prospect 4,
Moscow, 117936
Russia

ABSTRACT. The article has its aim to introduce the readers with some characteristic features which make Self-Propagating High-Temperature Synthesis (SHS) very attractive as a unique process to prepare a large variety of different inorganic materials. In this process the strong exothermic reaction propagates spontaneously through the reactant mixture converting it into product. The main advantages of this process are: rapid product formation, energy efficiency, high-purity products and low processing cost. Six main technological types of SHS which are used to produce both materials (sinters, powders and semi-products) and final articles are under consideration.

1. Introduction

Since ancient times the man has tried to make the nature serve him to rule it or at least to understand it better. Step by step the nature reveals its mysteries. Due to the courage, determinedness and hard labour of scientists and researchers the man has learned to understand better different nature phenomena. Mankind is studying nature and as a result of it makes it serve the man. A lot of mysteries of the structure of matter have already been solved. The mysteries of biological energy are under active studying now. Atomic nucleus and nuclear energy are already not an enigma now for scientists, though it is still necessary to treat them with care.

At the end of 60s, the group of the Soviet scientists headed by Alexander G. Merzhanov could open one more nature mystery. The Soviet researchers could make the internal chemical energy serve to mankind. It became obvious that when you are reasonable in using of this internal chemical energy it will help you in a lot of problems (not only scientific problems). Moreover this kind of energy is

C.A.C. Sequeira and M.J. Hudson (eds.),Multifunctional Mesoporous Inorganic Solids, 111–120.

absolutely free of charge.

In particular, it became possible and very promising to use internal chemical energy for the synthesis of different nonorganic materials.

In 1967 at the Institute of Chemical Physics (Soviet Academy of Science) Merzhanov, Shkiro and Borovinskaya studying the combustion of cylindric compacts of the titanium-boron mixture discovered the phenomenon of layer by layer combustion. Chemical conversion of initial components into the final product took place in quite narrow high-temperature zone, but not in the whole volume of the mixture simultaneously. The narrow high-temperature zone was called the front of the reaction. The front of the reaction moved along the sample with a certain rate. The phenomenon in general was called Self-Propagating High-Temperature Synthesis (SHS) [1-3].

So, SHS makes use of the exothermic nature of reactions forming ceramics and similar materials from their constituent elements. In a broad interpretation SHS is a spontaneous frontal propagation of exothermic chemical reaction which is locally initiated by a thermal pulse and takes place in a powder mixture of substances or in a complex heterogeneous system. The result of this reaction is the formation of condensed (solid or melt) products.

For a long time the active research dealing with SHS was held only in the Soviet Union. But starting from 1981 other countries including USA, Japan, South Korea, China, Poland became interested in the problem of SHS too. USA and Japan are most active in SHS research. A lot of well known scientific centers, such as National Livermore Laboratory, California University (Devis), Technology Institute (Georgia), General Electric, Rice University, Los Alamos National Laboratory, Massachusetts Technology Institute, University of Utah are engaged in studying SHS in USA. Osaka University, State Institute of Applied Investigations, Sumitomo Electric Industry, Niipon Steel Corporation, Toyota Motors and some other centers are very active in studying SHS in Japan.

On September 1991 the First International Symposium on SHS took place in Alma-Ata (Kazakhstan). The Second SHS Symposium is supposed to be held on Hawaii (USA) 1993.

Now the question of creation of European center of SHS in Spain is under consideration.

2. The Theory of SHS

2.1. HOW TO REALIZE THE SHS PROCESS

Now, let's find the solution to two questions. What are we to do to realize SHS? What is the subject of studying when we speak about SHS? The SHS process is so simple that everyone can reproduce it. Take the powder of titanium (particle size about 100μm) and the powder of amorphous boron. Mix them. Then you only have to ignite the mixture (for this purpose you may use even matches). After ignition, when the burning wave starts to spread along the mixture, we will see the fast moving of very bright zone. This zone is the front of the reaction. Before the front we have the initial mixture of Ti and B powders and behind the front there is the final product of the exothermic reaction (titanium boride).

The absence of the flame (there is no gaseous products during the reaction between Ti and B) and the fact that the product of combustion is high-quality refractory mixture are the main characteristic features of this process.

2.2. THERMODYNAMICS OF THE SHS PROCESS

And what are scientists and researchers who deal with SHS interested in? The reply is - in everything. Thermodynamics and kinetics of combustion process; structure of the burning wave; the methods of SHS process control; structure and phase transformations which take place with the substances during SHS; characteristics of obtained products; the searching of the new SHS compositions; application of SHS as the method of new substances, semiproducts and final articles production. All these problems are the subject of studying for scientists whose activities deal with SHS.

Now more than 300 different compositions are synthesized by means of SHS. Scientists try to find the new compositions. A number of compositions synthesized by means of SHS are presented in table 1.

Today, practically any reaction can be realized in SHS regime.

Thermodynamic analyses of SHS processes is very fruitful in search for the optimum conditions for the synthesis of the products of desired chemical and phase composition.

The development of the thermodynamic approach is based on the estimation of both temperature and equilibrium compositions of the combustion products. Thermodynamic analyses is especially important and useful for the multicomponent systems, when the combustion product composition is not apparent because of the competing side-reactions and phase transformations.

When you have either two or three component system it is very easy to evaluate temperature of burning (T_b) which is one of the most important characteristics of SHS.

Calculated (adiabatic) and experimental (real) temperatures of burning for a number of simple SHS systems are listed in table 2.

TABLE 1. Different substances'classes and compositions synthesized by means of SHS

Classes	Compositions
Carbides	TiC_x, NbC, HfC, VC, SiC etc.
Borides	TiB, TiB_2, ZrB_2, HfB_2 etc.
Nitrides	BN, TiN, ZnN, HfN, NbN etc.
Silicides	TiSi, $TiSi_2$, $MoSi_2$, ZrSi etc.
Halcogenides	$TiSe_2$, $NbSe_2$, $TaSe_2$, MoS_2 etc.
Solid solutions	TiC-Cr_2C_3, TiC-WC, TaC-TaN etc.
Intermetallics	NiAl, TiAl, TiNi, CuAl etc.
Oxides	$YAlO_3$, $LiNbO_3$, $Y_3Al_5O_{12}$ etc.
Superconductors	$YBa_2Cu_3O_{7-x}$, $LaBa_2Cu_3O_{7-x}$ etc.
Ceramics	AlN-TiB_2, Si_3N_4-TiN-SiC etc.
Cermets	TiC-Ni-Mo, TiC-TiN-Ni-Al-Co, Cr_3C_2-TiC-Ni etc.

TABLE 2. Comparison of calculated and experimental combustion temperatures in some SHS systems

Reaction	Temperature, K	
	Calculated	Experimental
Ni+Al=NiAl	1910	1910
Co+Al=CoAl	1900	1880
Ti+Si=TiSi	2000	1850
Ti+2Si=$TiSi_2$	1800	1770
Nb+2Si=$NbSi_2$	1900	1880
Mo+2Si=$MoSi_2$	1900	1920
5Ti+3Si=Ti_5Si_3	2500	2350
Nb+C=NbC	2800	2650
2Ta+C=Ta_2C	2600	2550

It is clearly seen that accuracy of the thermodynamic calculations are quite high though sometimes there is a difference between adiabatic and real temperatures. The range of temperatures which we have while working with SHS is about 1500 - 4500 K.

Thermodynamics is the most general approach to the solution of most scientific problems. But sometimes it is not possible to find the right solution to a problem using thermodynamics alone. For example, thermodynamics is not able to describe the process of final product formation (structure, phase composition). Therefore SHS theory in general should combine the consideration of some related sciences. Thus, there appeared the new scientific trend - structural macrokinetics.

2.3. STRUCTURAL MACROKINETICS

Structural macrokinetics studies the evolution of the structure (microstructure, macrostructure, crystal structure) in the course of chemical conversions considering heat- and mass exchange phenomena. While analyzing a number of the works concerning the structural macrokinetics one may conclude that the methods of structural macrokinetics are not only perfect for studying structures transformations of substances during SHS. These methods are also of interest for scientists whose activity has nothing to do with SHS but is connected just with the process of phase and structure transformations.

3. SHS: For and Against

The history of SHS is not a long one. But nevertheless SHS has a lot of supporters. A lot of scientists have realized that SHS, in general, reveals great prospectives for studying both fundamental and applied sciences. But at the same time SHS has many opponents too, especially when we speak about new materials production by means of SHS. Opponents prefer the traditional methods of materials production, because they see some significant drawbacks in different technologies based on SHS.

What drawbacks are they talking about? The main of them are:

- no possibility for a perfect control of the spontaneous propagation of the thermal wave and hence the absence of the final structure control;
- a great correlation between the structure and properties of the final products and the raw materials which are used to produce the final articles;
- the result of two above-mentioned drawbacks is the low reproducibility in the characteristics of obtained samples.

Of course, the above-mentioned problems are not the product of the opponents' imagination. These problems do really exist. But the time passes and some things which seemed to be unreal yesterday can be under control and hence find their solution today. A lot of works, which have been carried out by the researchers from the SHS Center staff are aimed to overcome the above-mentioned drawbacks.

The majority of the works in the SHS Center has the applied character. And the main objects of the research are the final product and the influence of different parameters of SHS process on the structure and properties of the final products.

The SHS problem appeared during the study of the combustion process and developed into material science. Today, in spite of some drawbacks, which do really exist different methods of nonorganic materials production based on the SHS process find more and more spheres for application. Sometimes, methods of materials production based on SHS are unique when you want to obtain this or that material. For example, the author et.al. [4] managed to produce by means of SHS diamond-containing materials including hard diamond-containing materials. Due to some unique characteristic features of the SHS process it became possible to produce large-size hard diamond-containing articles (in diameter more than 40 mm). It is practically impossible to produce the same kind of materials by means of any conventional methods.

Practically, any substances may be the products of the SHS reaction see table 1. In order to produce them it is necessary to know how to select the composition of an initial mixture possessing a considerable store of chemical energy and to provide favorable conditions for the process.

The SHS products can be of two types: chemical - metallurgical (chemical products and materials, green products-blanks in the form of sintered materials or ingots, powders) and mechanical engineering (parts and articles of desired shape and size, parts with coatings, welded parts).

4. SHS in Practice

Now, there are six main various technological types of SHS which are used to produce all above-mentioned SHS products. These types are:

- SHS-technology of powders and semiproducts;
- SHS-sintering;
- SHS-densification;
- SHS-metallurgy;
- SHS-welding;
- the technology of gas-transport SHS coatings.

4.1. SHS-TECHNOLOGY OF POWDERS AND SEMIPRODUCTS

SHS-technology of powders and semiproducts is based on the green compact burning in the reactors of up to 30 liters capacity. The SHS products obtained in such reactors are the sinters or ingots of a shapeless form ready for further application. Synthesis of green compact briquettes thermally treated under vacuum yields porous billets of the same form and size. This technological process allows the production of various simple and complex porous articles, including filters, pistons friction pairs, electric contacts, intermetallic catalysts. The initial material for porous articles is chosen in accordance with the further process requirements.

This SHS-technology is most frequently used for powder production.

Such powders as β-Si_3N_4, α-Si_3N_4, TiC, TiB_2, TiN, TiCN, BN, SiC, B_4C, AlN, $SiAlO_xN_y$ etc., are produced at the SHS Center.

4.2. SHS-SINTERING

SHS-sintering of solid products for the manufacture of articles of desired form and size is one of the promising technologies. Conventional methods of sintering ceramic articles (hot pressing, gasostat hot pressing) are characterized by high capital and energy consumption. The production of ceramic materials and articles in the direct synthesis of a metal/nonmetal-gas system in the combustion regime without a preliminary stage of ceramic powder preparation and subsequent densification up to required form and size is of great interest. This a process is realized in special gasostats, which allow the synthesis of final products and simultaneous formation of the articles under active gas pressures up to 8000 atm and in the temperature range of 1000-3000°C attained at the expense of heat evolved in the exothermic reaction.

The main technological parameters of this process are gas agent pressure, briquette thickness and density, powder despersivity, the degree of the green compact dilution by final products. These parameters allow realization of the layer-to-layer regime and the required material porosity, which may vary within 10-50%.

4.3. SHS-DENSIFICATION

SHS-densification applied to the products heated by a combustion wave results in their poreless state.

The method of SHS-densification is of great importance in the creation of new materials, in particular, those of tungsten-free hard alloys of a metastable structure. A number of new compositions of a wide range of physical-mechanical properties have been recently synthesized. SHS alloys possess more perfect grain monocrystallinity and higher purity. These alloys are intensively used as synthetic hard tool materials.

The technology of SHS-densification is used to manufacture hard-alloyed articles, rollers, drawing blocks, pressing equipment, cutting plates, targets etc.

One of the varieties of this technology is SHS-extrusion. Densification is accomplished by pressing of SHS product through a drawing block. It is particularly efficient in the production of long-size articles. Some heat processes have been developed for the optimization of technological regimes of high-quality article production.

SHS-extrusion is used in the production of electrodes for deposition by welding from tungstenless hard alloys, electrodes for arc welding, high-temperature heating elements, drills, microcutters, and stepped rollers.

One of the promising directions in the SHS-densification technology is its combination with various types of rolling. The method of SHS-rolling allows the production of such articles as strips, bars, stepped rotation bodies of ceramic, cermet, and intermetallic materials.

4.4. SHS-METALLURGY

SHS-metallurgy. The areas of SHS application have been extended much by the successful chemico-technological application of combustion processes. various types of combined processes proceeding with the material formation relate to SHS (including SHS with a reduction stage).

The products of SHS with a reduction stage are refractory compounds, ingots or cast coatings.

The technology of SHS with a reduction stage is based on the metal-thermic reaction with rate evolution providing combustion temperatures higher than the melting points of carbides, borides and silicides. The process consists of two stages: melt formation and its treatment by metallurgical procedures. At present there are three technological directions:

- SHS technology of ingots and cast articles;
- centrifugal SHS casting;
- SHS deposition by welding.

4.5. SHS-WELDING

SHS-welding is aimed at the production of strong one-piece connections of refractory parts of both uniform and nonuniform materials in SHS processes. The process is exothermic and its product is a welding material. An SHS welding process is usually carried out in a clearance between the parts welded by synthesizing a high-temperature melt. The heat evolution is enhanced by the electric current passage through the green mixture to initiate a so called "electrothermal" explosion.

The main advantage of the process is the possibility of making a one-piece connection between "materials", which are weakly welded by conventional methods. Among the materials, to which SHS welding is applied, are graphite, titanium, steel, tungsten, molybdenum etc.

4.6. THE TECHNOLOGY OF GAS-TRANSPORT SHS-COATINGS

The technology of gas-transport SHS-coatings is realized by the gas-transport additions to the SHS mixtures with the parts to be coated to initiate the reagent gas-transport transfer to the surface, on which the chemical reaction (similar to that in the mixture) takes place. The SHS product is formed as a coating on the parts plunged into the reaction mixture.

5. Conclusions

Self-Propagating High-Temperature Synthesis being quite a new area of science attracts more and more researchers. The main positive characteristic features of SHS are:

1. Low energy consumption.
2. High productivity.
3. No complicate and expensive equipment.
4. Purity of the final products.
5. The possibility to obtain the final products (articles) in one stage.

These characteristic features make SHS a good competitor to the numerous traditional methods of nonorganic materials production.

Sometimes SHS is the only method which allows to obtain the necessary material or article. And I believe, this method has a great future.

REFERENCES

1. Merzhanov, A. G., Shkiro, V. M. and Borovinskaya, I. P. 'Synthesis of refractory inorganic compounds', Certificate No. 255221, (1967), USSR; (1972), France, Pat. No. 2088668; (1973), USA, Pat. No. 3726643; (1974), U.K., Pat. No. 1321084; (1982), Japan, Pat. No. 1098839.
2. Merzhanov, A. G. (1969) 'The theory of stable homogeneous combustion of condensed substances', Combustion and Flame 13, 143-156.
3. Merzhanov, A. G. (1981) 'SHS-process: combustion theory and practice', Archivum Combustionis 191, No.1/2, 23-48.
4. Padukov, K. L., Levashov, E. A., Kost, A. G. and Borovinskaya, I.P. (1992) 'Self-Propagating High-Temperature Synthesis - the new method of fabricating hard materials. Perspectives of diamond- containing materials synthesis by means of SHS', Industrial Diamond Review, in press.

TREATMENT OF DEHYDRATED Na-Y ZEOLITE WITH THE HETEROPOLYMETALLIC PRODUCTS OF TRANSMETALATION REACTIONS

TAREK M. ABDEL-FATTAH and GEOFFREY DAVIES*
Department of Chemistry and the Barnett Institute,
Northeastern University, Boston, Massachusetts 02115, USA

ABSTRACT. Stoichiometric transmetalation of copper(II) target $(\mu_4\text{-O})N_4Cu_4Cl_6$ (N is a monodentate pyridine) with transmetalators $M(NS)_n$ (NS is a monoanionic S-methyl hydrazinecarbodithioate ligand; n is 2 or 3) containing one of three different metals M, M' and M" in an aprotic solvent at room temperature theoretically gives 35 heteropolymetallic molecules $(\mu_4\text{-O})N_4(Cu,M,M',M")_4Cl_6$ (for example, $(\mu_4\text{-O})N_4CuCoNiZnCl_6$) and co-product $Cu(NS)_2$. These heteropolymetallic molecules easily are thermolyzed to give highly dispersed metals, alloys and mixed metal oxides. Dehydrated Na-Y zeolite reacts with unseparated transmetalation product mixtures at room temperature in methylene chloride to give total mixed metal loadings of 0.4 - 1% by weight. Materials made from molecular families $(\mu_4\text{-O})N_4(Cu,M,M',M")_4Cl_6$ are excellent alternatives to mixed metal catalysts made by conventional methods.

1. Background

Metals and alloys generally are made by high temperature methods. Lengthy processing often is required to obtain single-phase alloys or materials with other special properties. Highly dispersed materials of uniform composition are desirable for catalytic applications. Catalyst lifetime is an important consideration.

Conventional metal catalyst synthesis involves treatment of a support with a metal salt solution followed by drying and reduction. The properties of multimetal catalysts made by treatment of a support with mixtures of different metal salts are difficult to predict and control because each metal has a different affinity for the support and can have a variety of chemical forms in solution, depending on such factors as the solvent, the salt concentration, the anions present, pH and temperature.

Metal oxides also are important catalysts, for example in the synthesis of ethylene oxide. They are made by oxidizing metals with air or oxygen. Bulk, finely divided metal oxide particles are made from metal powders.

1.1 HYPOTHESIS

Putting the metals of the desired multi-metal product into a single, neutral precursor molecule may give more uniform thermolysis products. These precursors might decompose at low temperatures to give highly dispersed materials.

1.2 TRANSMETALATION

Transmetalation is the stoichiometric replacement of the metals in a polymetallic target with other

C.A.C. Sequeira and M.J. Hudson (eds.), Multifunctional Mesoporous Inorganic Solids, 121–126.

metals from reagents called transmetalators[1]. Eq 1 illustrates a series of transmetalation reactions.

$$(\mu_4\text{-O})N_4Cu_4Cl_6 + xM(NS)_2 \rightarrow (\mu_4\text{-O})N_4Cu_{4-x}M_xCl_6 + xCu(NS)_2 \quad (1)$$

Here, N is a monodentate pyridine, M is Co, Ni or Zn, NS is an easily made monoanionic S-methyl hydrazinecarbodithioate Schiff base ligand and x is 1 - 4. Figure 1 shows the target molecule structure. The neutral reactants and products are highly soluble in methylene choride. Reactions 1 are very rapid at room temperature in aprotic solvents. The products are easily separated by gel permeation chromatography, which results in coordination of water by products containing cobalt, nickel and zinc[2,3]. Changing the metal M at each reaction step theoretically gives 35 different heteropolymetallic products with N fixed. An example is $(\mu_4\text{-O})N_4CoNiCuZnCl_6$, which contains four different metals[4].

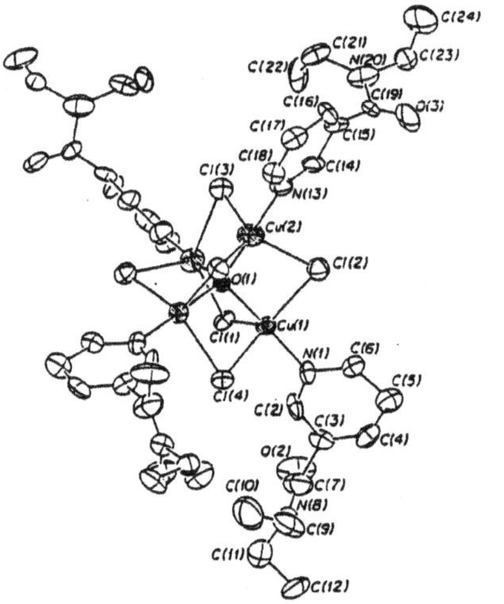

Figure 1. Molecular structure of the target in eq 1[2].

1.3 PREVIOUS WORK

Previous work[5-8] shows that thermolysis of the heteropolymetallic products of eq 1 leads to facile loss of ligands N. Metallic copper and nickel are miscible in all proportions. Bulk precursors from eq 1 that contain copper and nickel are spontaneously converted at 350°C to highly dispersed, single-phase copper-nickel alloys with the same composition as the precursor. For example, solid precursor $(\mu_4\text{-O})N_4Cu_2(NiH_2O)_2Cl_6$ gives the single-phase alloy $Cu_{.50}Ni_{.50}$ on thermolysis in argon. Alloy formation is faster under H_2, and $Cu_{.50}Ni_{.50}$ can be made at temperatures as low as 150°C[6,7].

Hydrogen is needed for conversion of precursors containing cobalt and zinc to the respective metals. Figure 2 shows the product of thermolysis of bulk solid precursor $(\mu_4\text{-O})N_4CoNiCuZnCl_6$ under H_2/Ar. Note the very small spherical uniform particles. Their morphology is due to the involvement of a molten precursor state that does not wet the quartz sample holder, as verified independently. Very highly dispersed products are obtained by spraying a substrate with a solution of the precursor before thermolysis. This gives extremely small supported particles on a substrate, in a polymer matrix or as thin films on platinum[8].

Changing the thermolysis atmosphere to oxygen gives metal oxides at low temperatures. Their catalytic properties may be superior to those of oxides made by conventional methods. Figure 3 shows the beautiful product of thermolysis of $(\mu_4\text{-O})N_4CoNiCuZnCl_6$ under O_2 at 220°C. The individual crystals of Co_3O_4, NiO, CuO and ZnO evidently grew from a melt. The mixed metal oxide products of these experiments are far superior in stoichiometry, homogeneity and morphology to the products

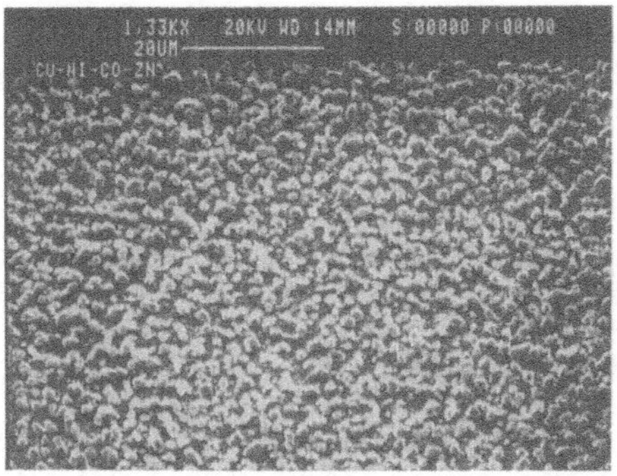

Figure 2. SEM micrograph of (μ_4-O)N_4CoNiCuZnCl_6 reduction product at 350°C under H_2/Ar on quartz.

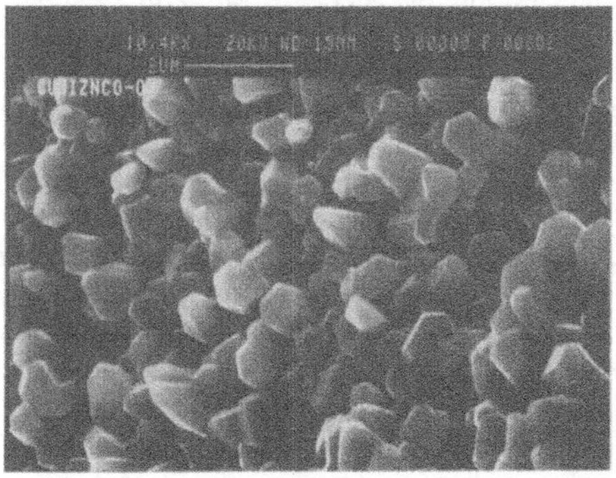

Figure 3. Product of thermolysis of (μ_4-O)N_4CoNiCuZnCl_6 under oxygen on quartz for 20 hr at 220°C[8].

of thermolysis of mechanical mixtures of the carbonates of these same metals, which are incompletely decomposed to the oxides even at temperatures above 300°C[8].

1.4 ZEOLITES

Zeolites have open framework structures. Industrial applications include separation, drying and as catalyst supports. The principal interest in zeolite-supported catalysts has been for shape-selective petrochemical conversions. Zeolites can be made with different sized channels and cavities. Zeolite-Y has 7 Å diameter channels and 13 Å diameter cavities. Zeolites can be loaded with metals by simple ion-exchange or used to encapsulate large metal clusters.

2. Hypothesis

Treatment of a zeolite with discrete molecules containing different metals gives supported products with much more predictable and controllable properties than those made by conventional ion-exchange or mixed salt solution treatment. Heteropolymetallic molecule families come from trans-metalation.

3. Treatment of Dehydrated Na-Y Zeolite with the Heteropolymetallic Products of Trans-metalation Reactions

3.1 EXPERIMENTAL METHODS

Na-Y zeolite (Strem) was completely dehydrated by heating under high vacuum at 350°C for 24 hr, as verified by thermogravimetric measurements. The dehydrated zeolite (1.00 gm) was titrated under dry nitrogen either with 1) an unseparated transmetalation product mixture consisting of a single heteropolymetallic molecule (containing particular fixed combinations of Co, Ni, Cu and Zn) together with co-product $Cu(NS)_2$ from reaction (1); or 2) single heteropolymetallic products that had been separated from transmetalation co-product $Cu(NS)_2$ by gel permeation chromatography[2]. The solvent was anhydrous methylene chloride. In every case, the zeolite sample was allowed to stand in the treatment solution at room temperature for 0.5 hr between each addition of titrant solution. After each waiting period, we separated the zeolite from the treatment solution by centrifugation until the treatment solution from procedure 2) was just visibly colored, indicating a slight excess of heteropolymetallic titrant. With procedure 1) it is necessary to measure the electronic spectrum of the supernatant solution after each waiting period because co-product $Cu(NS)_2$ is not incorporated by the zeolite (see below). After final centrifugal separation, each fully loaded zeolite sample was washed with methylene chloride until the washings were clear. Each fully loaded zeolite sample was dried under high vacuum at room temperature in a rotary evaporator with a liquid nitrogen trap. The dried samples were analyzed by element-specific inductively-coupled plasma spectroscopy and by SEM, XPS and EDAX measurements.

3.2 RESULTS AND DISCUSSION

We found that co-product $Cu(NS)_2$ from reactions (1) in methylene chloride is not encapsulated by dehydrated Na-Y zeolite, as verified by control experiments with only $Cu(NS)_2$ in the treatment solution. The very thermodynamically and kinetically stable[2] co-product $Cu(NS)_2$ is too large to pass through the zeolite channels and is easily washed from the surface. This very important finding means that it is unnecessary to separate the products of reactions (1) before treatment of the zeolite with a heteropolymetallic transmetalation product.

Spectral measurements indicated that the ligands N of precursors $(\mu_4\text{-O})N_4(Co,Ni,Cu,Zn)_4Cl_6$ are left behind in the treatment solution when $(\mu_4\text{-O})N_4(Co,Ni,Cu,Zn)_4Cl_6$ are added to Na-Y zeolite in methylene chloride. This is consistent with the very easy thermal removal of N from $(\mu_4\text{-O})N_4(Co,Ni,Cu,Zn)_4Cl_6$[6] and leads to higher loading levels than otherwise would be the case (the diameter of $(\mu_4\text{-O})N_4Cu_4Cl_6$[2] is close to the channel width). The N-free metal complexes are encapsulated rather than being physically adsorbed on the zeolite surface, as indicated by no effect on the metal loading of washing the treated zeolite samples with methylene chloride.

The loading data (Table 1) for dry, dehydrated Na-Y zeolite indicate maximum total metal capacities from 0.4 to about 1% by weight, with the highest values for precursors containing nickel. The variations with precursor identity suggest that the precursors break down to smaller fragments on encapsulation. Other work suggests the bond strength order $Ni(H_2O)\text{-}Cl$ < Co-Cl < CuCl in $(\mu_4\text{-}$

O)$NCu_{4-x}M_xCl_6$ molecules[9], with the μ_4-oxo group as essentially necessary for integrity of the molecular cores[4]. An extremely important finding is that the molar metal ratios of the loaded zeolite are the same as those of the precursors with which they are treated. Figure 4 shows an SEM micrograph of a fully loaded zeolite sample. Heating such samples in argon, hydrogen or oxygen readily converts them to the metals, alloys and mixed metal oxides, as expected from previous work with bulk heteropolymetallic precursors[5-8].

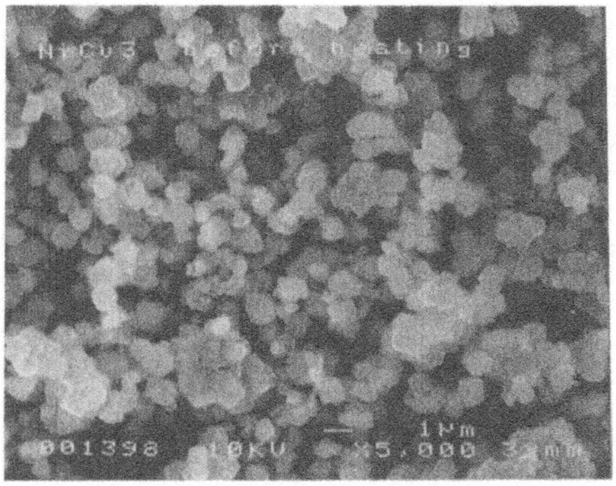

Figure 4. SEM micrograph of a fully loaded Na-Y zeolite sample.

4. Conclusions

Treatment of a zeolite with discrete heteropolymetallic molecules seems to have distinct advantages over conventional approaches to metal loading. We presently are comparing the catalytic efficiency and specificity of our loaded zeolites and their thermal derivatives with those made by conventional methods.

5. Acknowledgements

Our work is financially supported by Grant CHE-8717556 from the National Science Foundation and Grant 324132-AC3-C from the Petroleum Research Fund, administered by the American Chemical Society, which are gratefully acknowledged. This is Contribution Number 521 from the Barnett Institute of Chemical Analysis and Materials Science at Northeastern University.

6. References

1 Davies, G., El-Sayed, M. A. and El-Toukhy, A. (1989) "Transmetalation: A New Route to Heteropolymetallic Molecules and Materials", Comments Inorg. Chem. 8, 203-220; Davies, G., El-Sayed, M. A. and El-Toukhy, A. (1992) "Transmetalation Mechanisms and Applications", Chem. Soc. Rev., in press.

2 El-Toukhy, A., Cai, G. -Z., Davies, G., Gilbert, T. R., Onan, K. D. and Veidis, M. (1984) "Transmetalation Reactions of Tetranuclear Copper Complexes 2. Stoichiometry and

Products of Reaction of $[(DENC)CuCl]_4O_2$, $[(DENC)CuCl]_4(CO_3)_2$, $(DENC)_4Cu_4Cl_6O$ and $[(DENC)CuCl]_4Cl_4$ Complexes (DENC = N,N-diethylnicotinamide) with $Ni(NS)_2$ Complexes (NS is a Hydrazinato-S-methyl Carbodithioate Schiff Base), the Kinetics of Product Isomerization in Aprotic Solvents and Inhibition of Copper-catalyzed Phenolic Oxidative Coupling by Dioxygen through Transmetalation", J. Amer. Chem. Soc., 106, 4596-4605.

3 Al-Shehri, S., Davies, G., El-Sayed, M. A. and El-Toukhy, A. (1990) "Rate Law Variations in the Specific Monotransmetalation of $(\mu_4\text{-O})(N,py)_4Cu_{4-x}M_xX_6$ Complexes with $Zn(NS)_2$ in Nitrobenzene", Inorg. Chem., 29, 1206-1210.

4 Abu-Raqabah, A., Davies, G., El-Sayed, M. A., El-Toukhy, A. and Henary, M. (1989) "Limits of Direct Transmetalation of Polynuclear Copper(II) Complexes with $M(NS)_n$ Reagents", Inorg. Chem., 28, 1156-1166.

5 Marzik, J. V., Carriera, L. G. and Davies, G. (1988) "Copper-Nickel Alloy Formation by Reduction in Hydrogen of a Polymetallic Complex", J. Mater. Sci. Lett., 7, 833-835; Marzik, J. V., Carriera, L. G. and Davies, G. (1990) "Synthesis of Alloys from Heteropolymetallic Precursors", US Patent 4 933 003.

6 Davies, G., Giessen, B. C. and Shao, H. -L. (1990) "Single-Phase $Cu_{.50}Ni_{.50}$ Alloy Preparation by Thermolysis of a Simple Heteropolymetallic Precursor", Mater. Lett., 9, 231-234.

7 Davies, G. and Shao, H. -L. (1991) "Direct Alloy Synthesis from Heteropolymetallic Precursors", US Patent 5 061 313.

8 Shao, H. -L. (1992) "The Synthesis of Metals, Alloys and Mixed Metal Oxides by Thermolysis of the Heteropolymetallic Products of Transmetalation Reactions", Doctoral Thesis, Northeastern University; Davies, G., Giessen, B. C. and Shao, H. -L. (1992), to be published.

9 Al-Shehri, S., Davies, G., El-Sayed, M. A. and El-Toukhy, A. (1990) "Products and Kinetics of the Direct, Specific Transmetalation of $(\mu_4\text{-O})N_4Cu(NiH_2O)_3Cl_6$ (N = N,N-Diethylnicotinamide) by $M(NS)_n$ Reagents (NS = Monoanionic S-methyl Isopropylidenehydrazinecarbodithioate) in Nitrobenzene: A Relationship between Precursor Stabilities and Product Formation Rates", Inorg. Chem., 29, 1198-1205.

TABLE 1. Loading values for dry Na-Y zeolite in dry methylene chloride at 25^oC.

Precursor[a]	%Loading[b]	%Metal Loaded
$(\mu_4\text{-O})N_4Cu_4Cl_6$	2.1	0.46
$(\mu_4\text{-O})N_4Cu_3NiH_2OCl_6$	2.6	0.55
$(\mu_4\text{-O})N_4Cu_2(NiH_2O)_2Cl_6$	4.5	0.93
$(\mu_4\text{-O})N_4Cu(NiH_2O)_3Cl_6$	3.5	0.72
$(\mu_4\text{-O})N_4Cu_3CoH_2OCl_6$	1.9	0.40
$(\mu_4\text{-O})N_4Cu_2Co_2Cl_6$	2.4	0.49
$(\mu_4\text{-O})N_4Co_2(ZnH_2O)_2Cl_6$	2.4	0.50
$(\mu_4\text{-O})N_4Co(NiH_2O)_3Cl_6$	4.9	0.98
$(\mu_4\text{-O})N_4Zn(NiH_2O)_3Cl_6$	5.4	1.1

[a] Ligand N is N,N-diethylnicotinamide; [b] Percent by weight assuming no encapsulation of N from any precursor.

TEXTURAL CHARACTERIZATION OF PILC MONTMORILLONITE PILLARED WITH BINARY OXIDES $LaNiO_x$

S.P.SKARIBAS[1], P.J.POMONIS[1], P.GRANGE[2] and B.DELMON[2]
[1]*University of Ioannina, Department of Chemistry, Ioannina, 45332, Greece.*
[2]*Université Catholique de Louvain,Catalyse et Chimie des Matériaux Divisés, Place Croix du Sud, 2/17, Louvain-la-Neuve, Belgium.*

ABSTRACT. Montmorillonite pillared clay was synthesized by intercalation of binuclear complex of the form $NiLa(fsaen)NO_3$ and subsequent calcination at 500°C. The binary oxides of the form $NiLaO_x$ thus created inside the layers resulted in basal spacings of 14 Å and surface area of 220 m^2/g. Intercalation of more than 2.0 mmoles/g of complex results in a destabilization of the laminar structure and formation of oxidic islands on the external surface of the clay as it can be seen by SEM. The extend of microporosity and the texture of the occuring materials was studied by means of α_s-plots taking as a reference material the pure Na-montmorillonite.

1.Introduction

Recently a whole new class of Angstrong-scale engineered materials has began to be investigated consisted of two dimensional layered aluminosilicates pillared with oxidic pillars of a few Angstroms high.

The introduction of either single or mixed oxidic pillars between the clay layers creates usually fairly stable materials up to 400-500°C. It should be noticed at this point that such materials, may possess, at least in principle, multifold catalytic

C.A.C. Sequeira and M.J. Hudson (eds.),Multifunctional Mesoporous Inorganic Solids, 127–136.

functions and properties toward three different functions : (i) redox catalytic activity due to the oxidic pillars (ii) acidic activity due to the clay and (iii) shape selectivity as a result of the controlled modulation of microporosity and mesoporosity. A cooperation of the above characteristics plays a deterministic role towards production of active and selective catalysts[1]. In this way it is hoped to obtain high performance catalysts for different reactions like cumene synthesis, methanol conversion to hydrocarbons, alkylation and disproportionation of methylbenzene and fine chemical reactions[2-6].

Recently we employed binuclear cationic complexes of the type AB(fsaen)NO_3 (A= transition metal, B= La, H_4fsaen= N,N,'-3-hydroxy-salicylidene-ethylendiamine) for twin purposes. First, we have shown[7,8] that perovskite oxides of the form $LaMO_3$ (M=Ni,Co,Mn) are formed after calcination of the complexes and removal of the organic part of the molecule at relatively low temperatures (500-600^oC). Second it has been also shown that intercalation of MLa(fsaen)$^+$ inside the layers of Na-montmorillonite gives materials with enhanced surface area and d-space[9]. In a recent work[10] the adsorption isotherm of the complex NiLa(fsaen)$^+$ combined with zeta-potential measurements showed that the cationic complex is initially intercalated by an exchange reaction up to the CEC of the clay. Above this loading an intersalation of the complex took place. After calcination and removal of the organic ligand, high surface area microporous/mesoporous materials are formed. XPS studies of the calcined samples showed that by increasing the quantity of LaMOx (M=Ni) there is an increase of the formal oxidation state of M transition metal cation from 2 to 3 which is characteristic of the formation of the perovskitic phase.

One of the major questions addressed in the present work is the textural characterization of the solids that result from intercalation of the binuclear complex and calcination. Thus, the standard isotherm method was used to characterize the porosity and to define the extend of microporosity and/or microporosity of the pillared materials. XRD and SEM microscopy have been also used to investigate the structure of the composite solids formed.

2.Experimental Methods for Textural Characterization

The surface areas and pore volumes of the samples were determined with a volumetric adsorption apparatus (Micromeritics 2000 ASAP Area Analyzer) using nitrogen as adsorbent at liquid nitrogen temperature. Prior to the determination of the adsorption isotherms the samples were degassed at 200^oC until obtaining a static vacuum better than 5.10^{-6} mmHg. The specific surface characterisation of the samples was accomplished applying the BET equation in the linear part ($0.05<p/p_o<0.25$) and assuming a close-packed BET monolayer, with $a_m(N_2)=0.162$ nm^2 at 77°K.

Since no current theory is able to describe mathematically micropore filling[11], the more generalized method of the standard isotherms was used to estimate the extend of microporosity[11-13]. The basic principle of the Standard Isotherm method is to compare the shape of the adsorption isotherm obtained on the test sample with a standard isotherm of a non-porous reference solid of the same type ($\alpha_{(s)}$-plots) or one that is derived using mathematical models (t-plots). Since the validity of the t-plots based on the C value of the BET equation has been criticized, it has been lately[11,14] proposed to use the 'reduced' α_s-plots defined as: $\alpha_s=(n/n_s)_{ref}$ (n_s= amount adsorbed by the reference solid at $p/p_o=s$). According to Sing[15] it is convenient to place $\alpha_s=1$ at $p/p_o=0.4$, since monolayer coverage and micropore filling occur at $p/p_o<0.4$, where capillary condensation takes place at $p/p_o>0.4$. By plotting for a test sample the nitrogen volume, V, adsorbed at a relative pressure, p/p_o, as a function of the α_s value from the reference material at the same p/p_o, valuable informations can be obtained about the porous structure of the adsorbent. Thus as long as the multilayers of adsorbate is formed unhindered, the V-α_s plot is a straight line passing through the origin of axes, the slope of which is a measure of the surface area of the adsorbent according to the equation [16]

$$S_\alpha=2.87.\ dV/\ d\alpha_s$$

At higher relative pressures (higher α_s values) deviations of the straight line might occur. An upward deviation indicates a capillary condensation in the pores of the adsorbent, while a downward deviation is observed when micropores or slit shaped pores are present.In the second case the positive intercept i gives the micropore volume $V_{\mu p}$ after conversion of the gas volume adsorbed in liquid volume of Nitrogen at 77K ($V_{\mu p}=V_{ads(STP)} \cdot 0.001547$).

One problem associated with the application of the $\alpha_{(s)}$-method is the need of a nonporous reference material with the same surface structure.In this work, in order to construct the reference standard isotherm of our series of samples, we made the assumption that the Na-Montmorillonite has no microporosity and was thus taken as a reference material.

A series of film oriented -on glass slides- of the pillared Montmorillonite samples obtained after drying and/or calcination from 125°C to 500°C for 3h under air were studied with a PHILIPS 1130 diffractometer with a Ni filtered CuKα radiation source, from 2θ=2° to 16°

3. Results

As soon as the carboneous material is burned (T>350°C) during calcination, the remaining oxidic pillars create a microporosity by keeping the layers of the clay apart. The nitrogen adsorption isotherms (fig.1) are classified as type IV with H4 hysteresis. They are often associated with narrow slit shaped pores. The parallel shift upwards of these curves until sample NiLa-Mont2.0 also indicates the micropore filling process which takes place at the first steps of adsorption. The high values of the C-BET parameter (table 1) give also an indication of microporosity .

Table 1. BET surface areas and micropore areas from α_s-plots with the corresponding pore volumes for calcined samples at 500°C.

Sample	S_{BET} (m^2/g)	C_{BET}	$S_{\alpha(s)}$ external (m^2/g)	$S_{\alpha(s)}$ μporous (m^2/g)	$V_{\alpha(s)}$ μporous (cm^3/g)	Porous Volume (cm^3/g)
Na-Mont	42	430	42	-	-	0.084
NiLa-Mont-0.2	55	361	23	32	0.025	0.066
NiLa-Mont-0.4	62	478	20	42	0.036	0.072
NiLa-Mont-0.8	139	607	37	102	0.085	0.144
NiLa-Mont-2.0	220	1581	77	143	0.126	0.263
NiLa-Mont-4.0	152	1215	118	34	0.052	0.236
NiLa-Mont-8.0	117	100	117	-	-	0.279

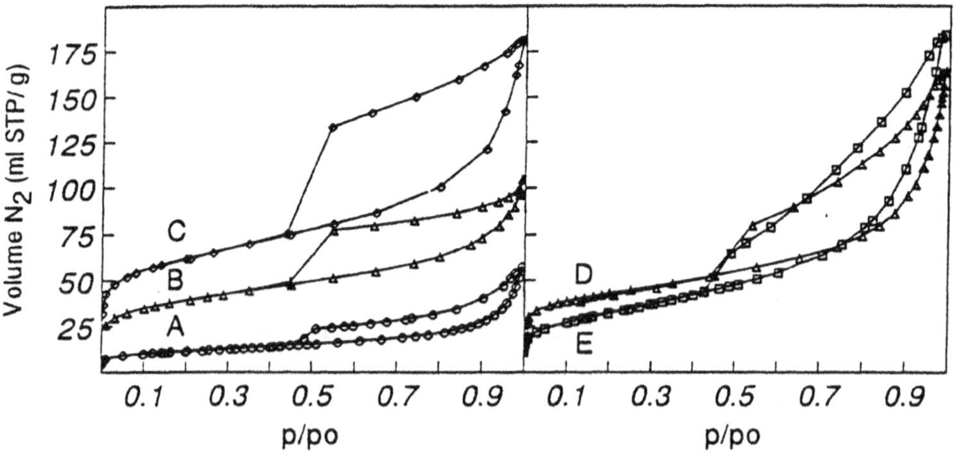

Figure 1. N_2 adsorption-desorption isotherms (77°K) for the samples A)Na-Montmorillonite, B)NiLa-Mont0.8, C)NiLa-Mont2.0, D)NiLa-Mont4.0 and E)NiLa-Mont8.0 after calcination at 500°C.

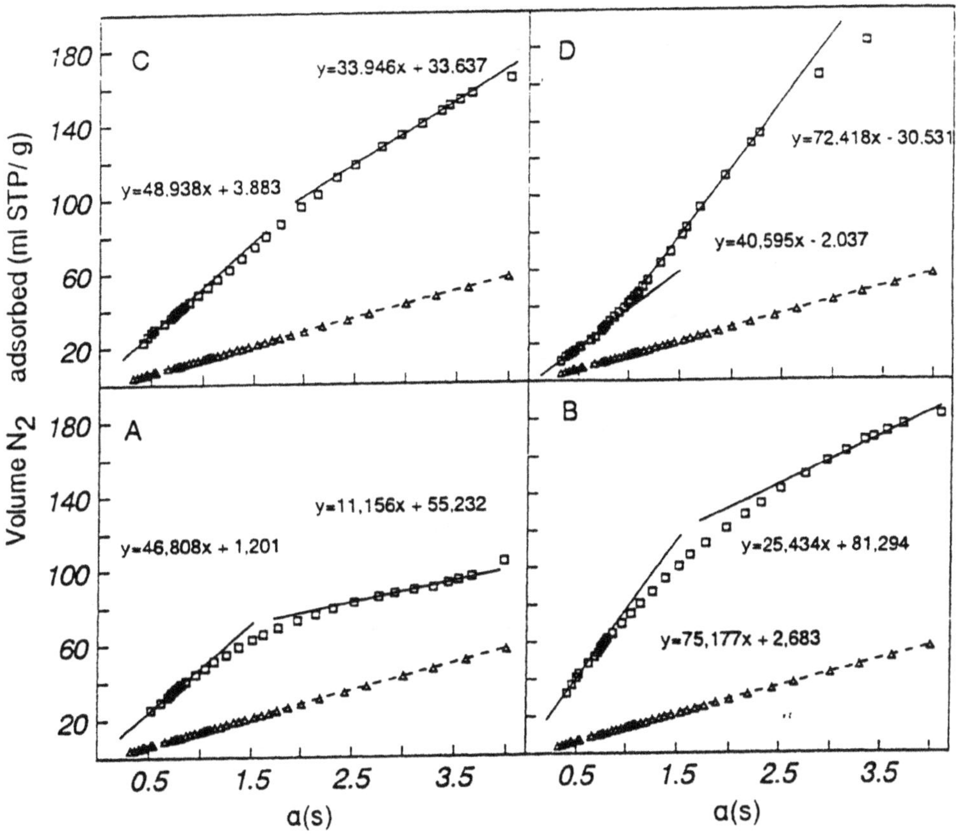

Figure 2. $\alpha_{(s)}$-plots (s=0.4) of the samples A)NiLa-Mont0.8, B)NiLa-Mont2.0, C)NiLa-Mont4.0 and D)NiLa-Mont8.0 after calcination at 500°C.

The α_s plots corresponding to the adsorption isotherms of figure 1 have been calculated as described earlier and drawn in figure 2. The calculated, from these plots, values of the external and micro-porous surface area, as well as the pore volume are included in table 1.

The X-rays patterns (fig.3) for samples until 2.0 mmoles/g loading give only one 001 reflection. This indicates that the calcined samples have a laminar structure[17]. The 001 reflection of the NiLa-Mont4.0 sample was less well defined while the calcined sample NiLa-Mont8.0 did not give any reflection, suggesting that for these two samples a delaminated 'open' structure is created as it has been described by Pinnavaia and Occelli[18,19].

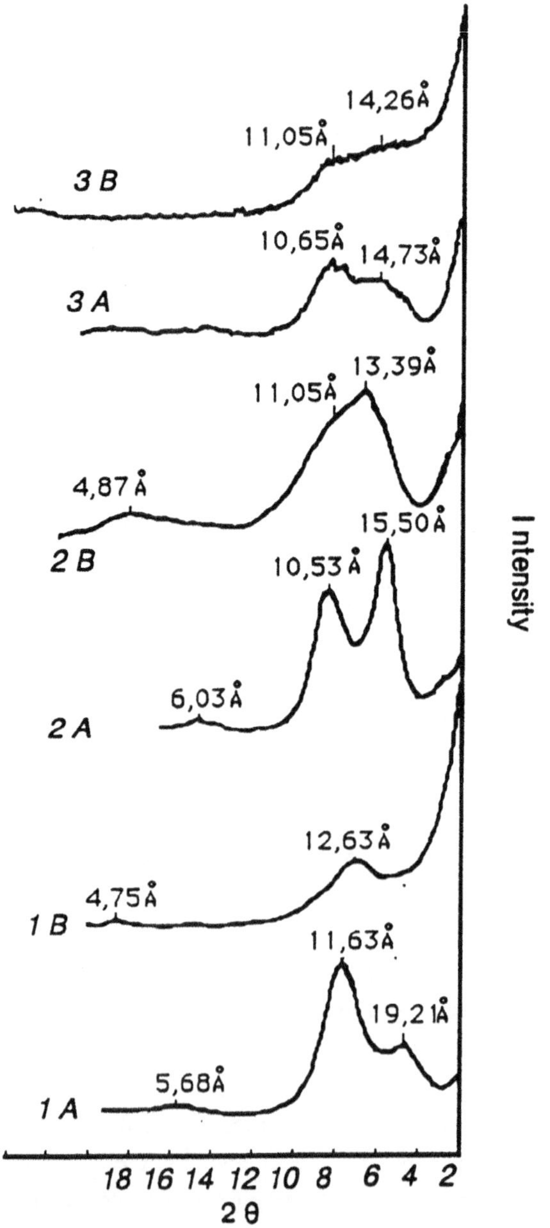

Figure 3. X-rays diffraction patterns of samples 1)NiLa-Mont0.8, 2)NiLa-Mont2.0 and 3)NiLa-Mont4.0 oriented on glass slides, after drying at A)120°C and calcined B) at 500°C.

The above argument is in accordance with the results of the $\alpha_{(s)}$-plot analysis whose shape for the last two samples indicates that less micropore filling is taking place.

The total surface area and the microporosity of the calcined samples (table 1) is increasing monotonically as the ratio Ni+La/clay increases until sample NiLa-Mont2.0 where the maximum surface area (220 m^2/g) and micropore area (143 m^2/g) are attained at 500°C. From the values of normalized surface areas, expressed per mass of pure clay, (table 2) it can be seen that addition of more oxidic pillars does not increase the developed surface area. It seems that the quantity of the precursor complex which is adsorbed after sample NiLa-Mont2.0 does not contribute in the creation of exposed clay layers.

Table 2. Evolution of the reduced BET surface area (surface area / g of pure clay) by increasing the NiLaOx loading.

NiLaOx Loading (mmoles/g)	0.2	0.4	0.8	2.0	4.0	8.0
BET area (m^2/g)	54	66	160	328	301	345

The surface structure of the calcined samples, as the extent of external surface area shows, should be considered as a combination of small particles with laminar structure in random contact. According to Van Olphen[20] the ratio of the crystallographic area (730 m^2/g) to the external area gives the average number of layers per particle. For the samples NiLa-Mont0.8, NiLa-Mont2.0 and NiLa-Mont4.0 this could be converted to 22.0, 10 and 7.3 layers per particle, respectively, on the average.

However it must also be taken in account that a fraction of the formed NiLaOx oxides may be located on the surface of the laminar particles and this fraction is related to the ratio of external to internal area. The above argument is also evident from the fact that, from X-rays diffraction patterns, peaks corresponding to NiO and La_2O_3 were detected for samples NiLa-Mont8.0 and NiLa0Mont4.0 ,less for NiLa-Mont2.0 but not for NiLa-Mont0.8.

The differences found on the textural characteristics of the calcined solids can be interpreted taking account the geometry of the complex precursors as shown shematically in figure 4. Thus, at loading of 0.8 mmoles/g the complex creates a monolayer between the sheets of the clay. In the sample with 2.0 mmoles/g loading the complex can occupy the whole surface of the clay and creates thus after drying and calcination greater distance between the layers increasing the surface which corresponds to micropores.

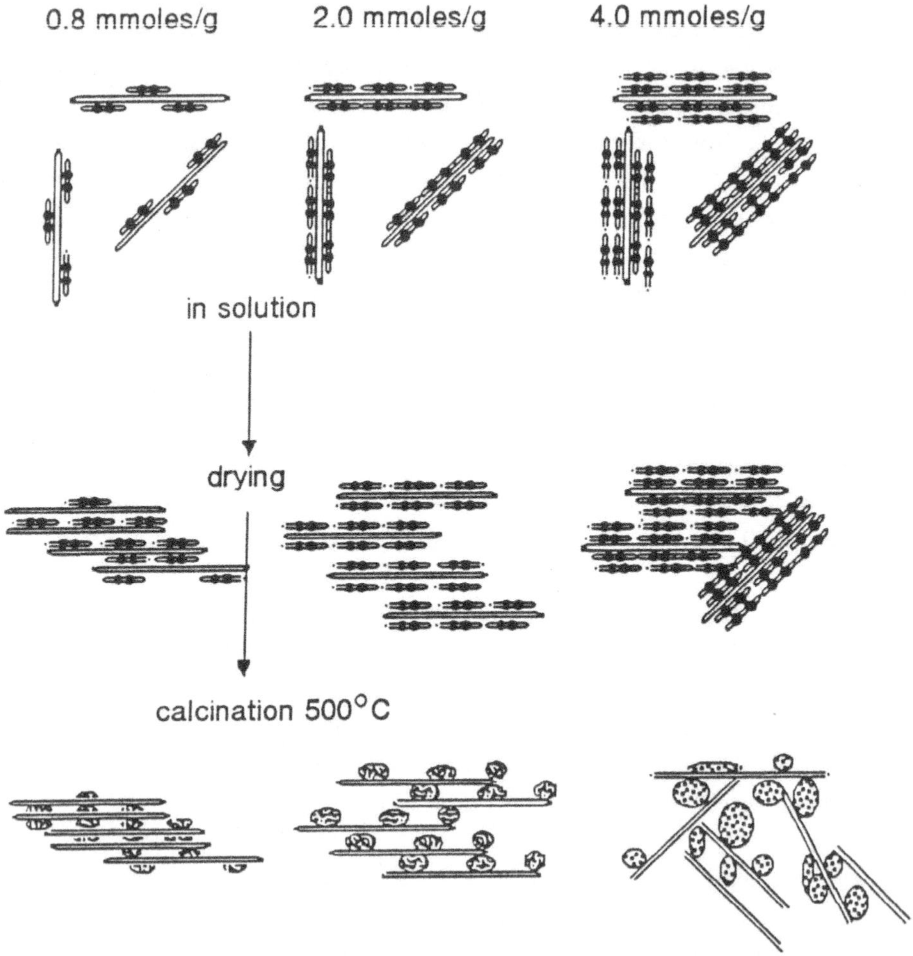

Figure 4. Differentiation of the interlamelar gallery height and texture before and after calcination at 500°C due to the distribution of the number of molecules of $NiLa(fsaen)^{+}$ which form the pillars.

Additional adsorption of complex precursor can not be accumulated inside the layers and is deposited on the outer surface of the clay particles in the form of small islands of bulk oxide unstabilazing at the same time the laminar structure. This distiction can be seen in the SEM photographs shown in figure 5.

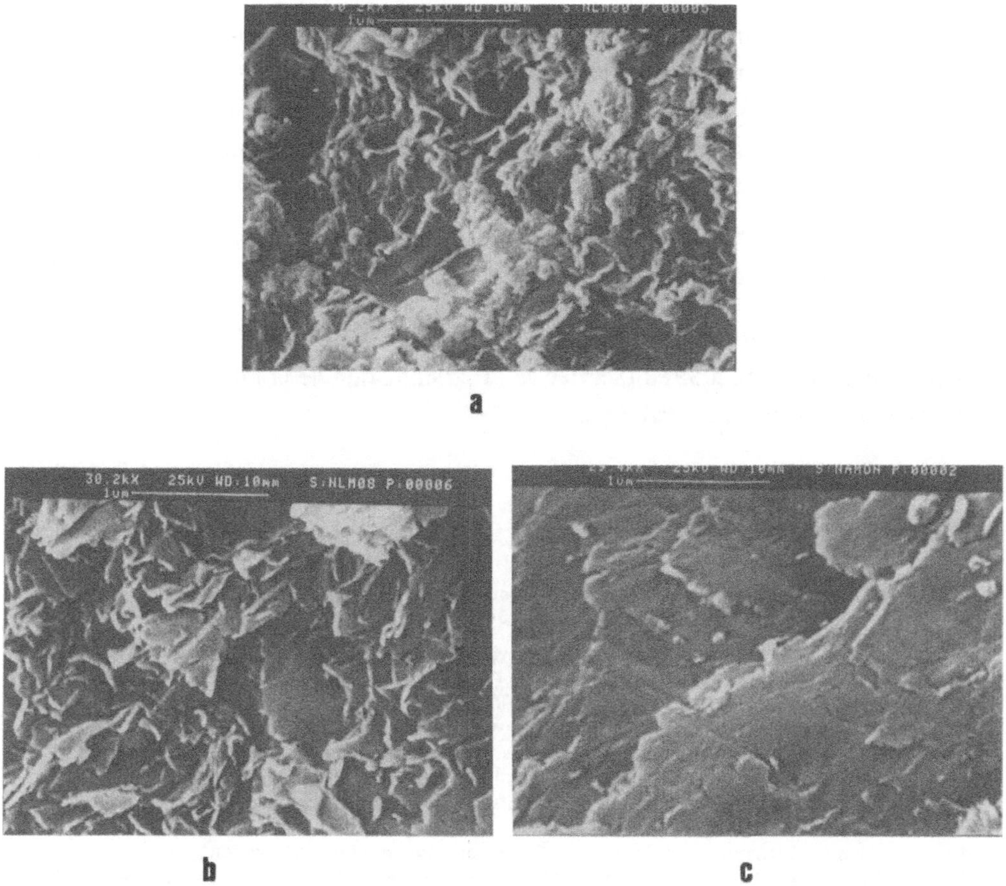

Figure 5. SEM photographs of (a) Na-Montmorillonite, (b) NiLa-Mont0.8 and (c) NiLa-Mont8.0 calcined at 500°C.

As it can be seen in the SEM photographs (magnification ≈ x30000) the non-intercallated Na-Montmorillonite shows a structure solid and without any apparent porosity. The sample NiLa-Mont0.8 shows a quite different structure. In this, groups of clay layers are extensively folded forming thus an apparent macroporosity. The thickness of the bended layers is around 0.03-0.05 μm, or 300-500 A. Considering that the X-rays give a 001 distance of around 15 Å, then each multi-layered should be concidered of around 20-35 grouped aluminosilicate layers. This result is in suprisingly good agreement with the theoretical prediction by Van Olphen[20], cited previously, who suggests, for the corresponding external surface area, 22 layers. Finally the sample NiLa-Mont8.0 shows a two-phase texture : part of the material is consisted of the folded aluminosilicate layers, similar with that of NiLa-Mont0.8, and a second phase of LaNiOx oxide accumulated on the external surface, forming a distict structure.

4.Acknowledgement

We acknowledge the EEC for a Sectoral Grant provided to S.P.Skaribas. Financial support from SPPS (Servies de Programmation de la Politique Scientifique) Belgium is greatfully acknowledged.

5.References

1 B.Delmon and P.Grange, in "New materials in Catalysis", Erdol Erdgas Kohle,1991,9,376.
2 J.J.Fripiat,Clays and Clay Minerals,1986,34,501.
3 P.Laszlo,Science,1987,235,1473.
4 M.L.Occelli, R.A.Innes, F.S.S.Hwu and J.W.Hightower, Applied Catalysis, 1985, 14, 69.
5 H.Mori, T.Kuhara, and M.Misono, Chem.Letters, 1987,2305.
6 D.H.Solomon,Clays and Clay Minerals,1968,16,31.
7 S.P.Skaribas, P.J.Pomonis and A.T.Sdoukos, J.Mater.Chem., 1991, 1, 781.
8 S.P.Skaribas, P.J.Pomonis, T.C.Vaimakis, A.T.Sdoukos, T.Bakas and A.Moukarika, Proceedings of XIV Mendeleev Congress, 11-16 Sept 1989 ,Tashkent,USSR,p554.
9 A.K.Ladavos, P.J.Pomonis and S.P.Skaribas,"Intercalation Compounds"; 6-ISIC,Trans.Tech. Publications, 1992, pp.799
10 S.Skaribas,P.J.Pomonis,P.Grange and B.Delmon, submitted for publication
11 K.S.W.Sing, D.H.Everett, R.A.W.Haul, L.Moscou, R.A.Pierotti and J.Rouquerol,T., Pure and Appl.Chemistry, 1985,57,603.
12 J.Seifert and G.Emig, International Chemical Engineering, 1991, 31, 29.
13 A.Lecloux, Mem.Soc.Royale des Scien.de Liege, 1971,6 Serie, Tome I, Fasc.4, 169.
14 S.J.Gregg, in "Adsorption at the Gas-Solid and Liquid-Solid Interface", eds. J.Rouquerol, K.S.W.Sing, pp153, Elsevier.
15 K.S.W.Sing, in "Surface Area Determination", Eds. D.H.Everett and R.H.Ottewill, p.25.Butterworths,London,1970.
16 A.Lecloux and J.P.Pirard, J.Coll.Interf.Science, 1979,70,265.
17 K.Suzuki,T.Mori,K.Kawase,H.Sakami and S.Iida,Clays and Clay Minerals, 1988,36,147.
18 T.J.Pinnavaia,M.S.Tzou,S.D.Landau and R.H.Raythatha, J.Mol. Catal., 1984,27,195.
19 M.L.Occelli,S.D.Landau and T.J.Pinnavaia, J.Catalysis, 1987,104,331.
20 H.Van Olphen, Proc. Intern. Symposium on Surface Area Determination,1969, Bristol,Butterworths,London,pp255.

STRUCTURE DIRECTION IN ZEOLITE SYNTHESIS

D. E. W. VAUGHAN,
Exxon Research and Engineering Company,
Annandale, New Jersey,
USA, 08801.

ABSTRACT. Developments in instrumental and computational techniques, combined with a major expansion in zeolite synthesis experiments around the world, are promoting a wider understanding of zeolite synthesis mechanisms. This paper explores an approach to structure synthesis by design using several examples of zeolites known to co-crystallize in the same, or similar, synthetic gel compositions, including LTA:SOD:FAU; FAU:BSS; ZSM-5/11; ABC 6-ring series; LTL:OFF:ERI. Extended structures interacting with templates as strong structure directing entities play a major role in specific crystal formation. Particular attention is given to phases which develop at overlapping crystallization field boundaries, where two or more extended building structures may interact to form new zeolites, as in the cases of MAZ: MOR: ECR-1 and MER: RHO: PAU.

1. Introduction

1.1. ZEOLITES

We tend to think of zeolites as being exclusively micro-porous, but over the years there has been much interest in meso-porous, or near meso-porous, derivatives. Initially this took the form of theoretical structure proposals based on extensions of known zeolite structures [1]. Later, despite conflicting theoretical calculations [2], Meier [3] and co-workers [4] proposed that 3-rings were necessary to attain minimum framework densities. Several larger pore zeolites have now been synthesized. VPI-5 [5] is an 18-ring example of the first type (now somewhat complicated by the debate and litigation over its pedigree [6,7,8]), thought to be identical to a theoretical structure expansion of ALPO-5 proposed by Smith [9] (but further complicated, in the absence of single crystal structural data, by the existence of several other similar frameworks [10]). "Cloverite" is a 20-ring example of connected alpha cages [11]. The second type is demonstrated by the recent structure solution for ZSM-18 - the first new structure to be made with 3-ring components and having a 7Å unidimensional 12-ring channel as the main structural feature [12]. Understanding how to control structure expansion of the kind illustrated in Figure 1 is clearly a necessity if large pore zeolites are to be tailored.

C.A.C. Sequeira and M.J. Hudson (eds.), Multifunctional Mesoporous Inorganic Solids, 137–156.

1.2. NON-ZEOLITES

Several other families of materials related to zeolites have stable tetrahedral-octahedral oxide-hydroxide structures and are either microporous or have the potential for mesoporous properties. Particularly important are those with tin [13], titanium [14] and molybdenum [15] in the octahedral sites and Si, Al or P in tetrahedral sites. The recent patent literature claims even larger pore size "crystalline" materials of universal compositions - into the mesopore range - made using surfactant templates such as cetyltrimethyl ammonium [16]. However, rather than being crystalline they are reminiscent of oxide stuffed liquid crystal products [17]. The main feature of all of these materials is that they are prepared in the presence of a large amine or alkylammonium molecule ostensibly templating the structure into a particular structure mode.

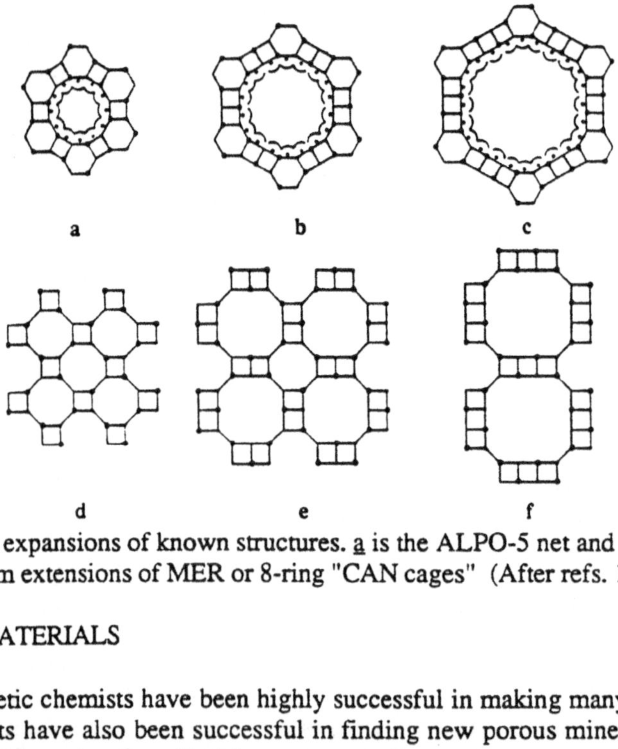

Figure 1: 4-ring expansions of known structures. a is the ALPO-5 net and b the VPI-5 net. d-f obtained from extensions of MER or 8-ring "CAN cages" (After refs. 1 and 9).

1.3. NEW MATERIALS

Although synthetic chemists have been highly successful in making many new materials, the mineralogists have also been successful in finding new porous minerals and solving their structures. Moore has described the structures of a complex mesoporous, zeolite-like, hydroxide minerals - the iron phosphate cacoxenite [18] - and other well known theoretically meso-porous minerals include imogolite [19] and sepiolite [20]. Numerous pillared clays and other layered structures [21], including the layered double hydroxides [22,23,24], have been prepared from layered minerals by exchange with robust inorganic polyions to produce three dimensional structures with pores in the range up to about 25Å. These may be viewed as inorganic extensions of the more numerous but less stable organically intercalated layered minerals[25, 26]. There are many examples of similar tailored structures made by modification of natural or synthetic pre-existing layered materials [27], many of which will be discussed elsewhere in this volume.

1.4. STRUCTURE DIRECTION ISSUES

A major unresolved issue confronting the zeolite synthesis chemist is how to make by design, large pore, three dimensional crystalline target structures, given the present limited state of knowledge of structure building and crystallization mechanisms. This paper focuses on an approach to structure design implied in numerous actual crystallizations; that extended sub-structures exist within a given crystallizing system, and that the ways in which they stack and interconnect may be definable and controllable. Chemists and physicists have yet to define these necessary rules of connectivity and solution chemistry, but improvements in established analytical tools are beginning to provide crucial new information on these processes.The most important are listed below in Table 1.

Table 1: Useful analytical techniques for tracking crystallization mechanisms

Method	Information
29Si-MASNMR	Structure to structure T atom distribution comparisons. Gel/solution speciation.
13C-MASNMR	Template packing, orientation and location.
Laser Raman	Extendedstructure morphology - sheets, rods, etc.
Electron Microscopy	High resolution structure imaging for structural complexity and observation of mixed structures. Gel substructures.
Low angle scattering	Extended structure length scales and morphology.

2. Converting building units into zeolite crystals

2.1. A CONVENTIONAL VIEW

A conventional view of zeolite synthesis mechanisms is based on the observation that most zeolite structures usually comprise only a small number of structural subunits. These were summarized by Meier for a wide range of known zeolites [28] and the term "secondary building unit", or sbu for short, coined to describe them. The recent expansion of known zeolite structures has resulted in the identification of several new sbu [29]. (The primary building unit is the single metal cation coordinated to four oxygen ions - $SiO_4$4-; $AlO_4$5-; $PO_4$3- etc.). These sbu, coordinated to, or clathrating, alkali cations, are thought to be the main solution species which add to the growing zeolite nucleus to form the crystals. The simplified process sequence is shown in Figures 2 - 5, which illustrate the kinds of building units thought to participate in the crystallization process at different stages of growth.

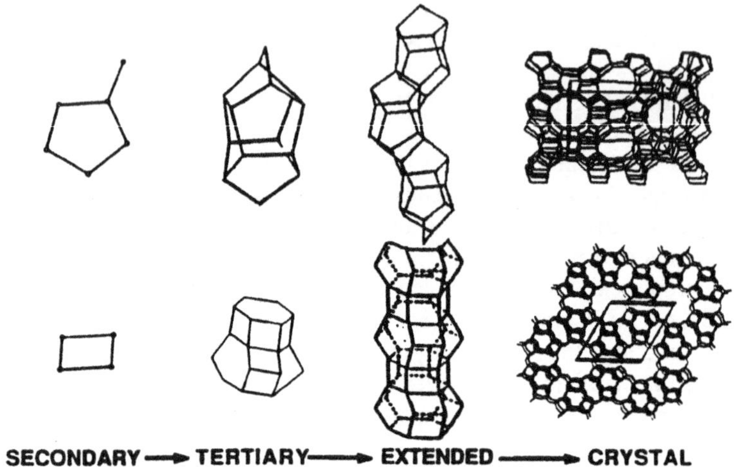

Figure 2: The visualization of a zeolite crystal growth process first polymerizes monomers into a selected group of low molecular weight polymers (sbu) which then clip together to form polyhedral tertiary building units (tbu). A central thesis is that experimental crystallization studies show that the products of structure building require the existence of extended structures. The top scheme is representative of MFI (pentasils) and the bottom structure is that for LTL.

The ability to readily synthesize, in isolation, various silsesquisiloxanes, particularly the double 4-ring $TMA_8Si_8O_{20}$ (TMA= tetramethyl-ammonium), and their Al and Ga substituted forms [30] supports this approach. The developments in solid state NMR in the early 1980's and the attendant interest in Si/ Al/ Ga distributions in several zeolites [31], with a particular focus on sbu components, greatly stimulated this interest. It was further reenforced by the discovery that a zeolite LTA hydrogen bonded molecular structural analogue can be crystallized from the tetrabutyl-ammonium variant ($TBA_8Si_8O_{20}$) [32] shown in Figure 3.

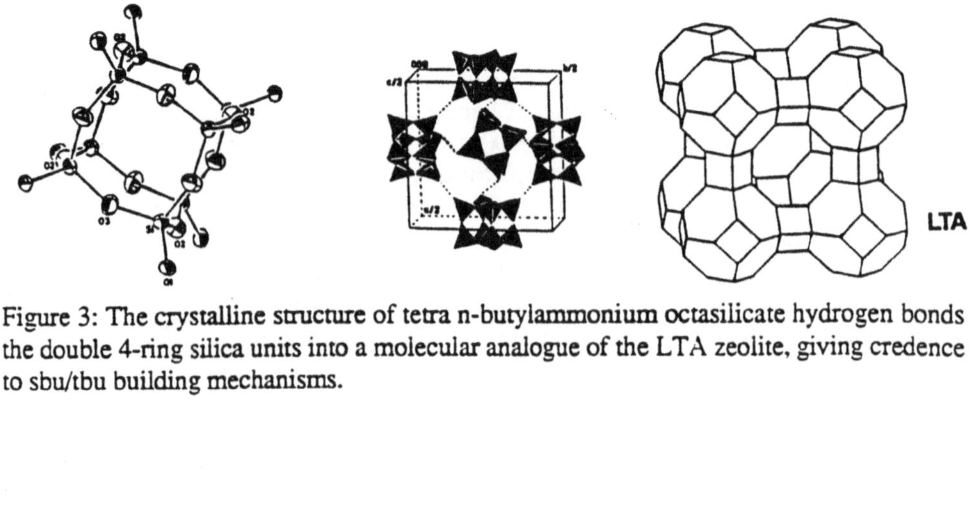

Figure 3: The crystalline structure of tetra n-butylammonium octasilicate hydrogen bonds the double 4-ring silica units into a molecular analogue of the LTA zeolite, giving credence to sbu/tbu building mechanisms.

2.2. CONTEMPORARY RESEARCH

Research interest then turned to characterization of the non-crystalline zeolite gel precursors for clues to the identification of structure building units. Detailed analyses of zeolite synthesis gels using IR [33], NMR [34] and Raman [35] spectroscopies, with increasingly sophisticated instrumentation, indicates that numerous potential sbu are present in the gels. However, as a typical structure only contains one or two sbu, but many more are observed in the precursor gels and solutions, few of the gel species participate in the crystal growth process and most will reconfigure in a dynamic equilibrium process before structural incorporation. Such propositions stimulated our own investigations of structure building mechanisms in Linde A (LTA), faujasite (FAU) and sodalite (SOD) using 29Si- MASNMR [36,37]. These studies also raised the issues of relationships between co-crystallizing zeolites and the tendency for zeolite associations to have common sub-structures, possibly indicating the identities of the real zeolite building blocks. Particularly interesting are the following groups of materials which occur as "boundary phases" or intergrowth structures in relatively simple gel systems;

LTA/ FAU/ SOD
FAU/ BSS/ CsZ-1,3/ ZSM-2,3,20/ ECR-35
LTL/ OFF/ ERI
MAZ/ MOR/ ECR-1/ mazmorites
ZSM-5, 11/ pentasils
"ABC" 6-rings: GME/ CHA; OFF/ ERI/ GME; SOD/ CAN.
PAU/ MER/ RHO

The evaluation of extended larger substructures in crystal building has been examined in detail for silicates by Liebau [38], in which he proposes a classification system based on chain configurations. In a similar approach to zeolites in particular, Barrer [39], at another NATO Workshop in Portugal, considered the many zeolite structures obtained by connecting polyhedra and chains of polyhedra.

3. Zeolite Crystallization Fields

3.1 CRYSTALLIZATION PRODUCTS IN Na/ K - ALUMINOSILICATE ZEOLITES

Consideration of the numerous structures that crystallize in the sodium and potassium aluminosilicate systems at Si/Al ratios in the range of 1 to 5 leads to the conclusion that the most probable primary sub-structure is the Si_{12} unit, shown in Figure 4, configured into several major related structural elements found in the same or similar crystallizations. Evaluation of the bond making and breaking steps needed to reconfigure one unit to another giving a different structure, even 4,6-rings to 5-rings, indicates that few steps are necessary. The implication from many analyses of crystallization sequences is that a particular base cation gel favours a particular multi-component polymer equilibrium, which in turn favours the formation of specific extended structures within the gel selectively depleting the gel in those components, which in turn are replenished by the rapid re-equilibration of low molecular weight species in the gel. That different structures containing the same building blocks co-crystallize gives credence to this thesis, as illustrated in numerous examples described later.

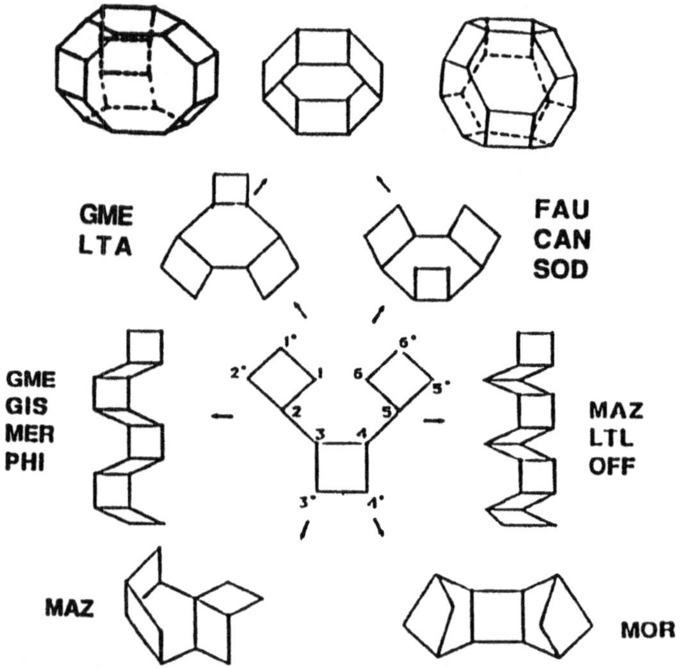

Figure 4: All the many zeolites, listed above, crystallizing in the sodium/ potassium aluminosilicate system at low Si/Al ratios between 80° and 200°C can be built from a re-configured Si_{12} sub-unit as indicated.

Competitive nucleation and polymerization and de-polymerisation processes determine the crystallization sequences within a given composition range. There is considerable experimental evidence that 5-ring containing zeolites (eg. mazzite (MAZ) and mordenite (MOR)) and 4-ring "ladder" structures (eg. phillipsite (PHI), gismondine (GIS) and merlinoite (MER)) have the highest solution stability because they either are the thermodynamically stable species, prevailing at long crystallization times, or that they reorder at a very low rate compared to other structures in reactive mother liquors. Structures containing trapped cations, such as tetramethylammonium (TMA) in LTA - ZK-4, MAZ - ZSM-4 and offretite (OFF), cesium in analcite (ANA, also known as the mineral pollucite) and potassium in Linde L (LTL)), degrade more slowly in solution than structures that do not, implying that enhanced stability is promoted by cations large enough to be clathrated.

The use of large alkyl-ammonium cations as "templates" in the synthesis of large pore, low density, meso-porous framework structures is generally taken as a necessity. Their use in zeolite synthesis is of long standing interest and debate since they were first used to influence zeolite crystallization pathways [40] in late 1950's. Several researchers have reviewed their impact [41, 42] and discussed their roles [43], primarily from the viewpoint of their influence on sub-unit structure and space filling. A common observation in all the zeolite - molecular sieve systems in the general sense, including aluminosilicates, ALPOs, SAPOs, MeAPOs, zeosils etc., is that templates may have more than one function. Many templates can apparently "direct" and be occluded within a given framework; one template may promote the formation of many different frameworks; a template may influence the formation of a framework but not necessarily be occluded within it. The role of the template may therefore be viewed in a variety of ways and in many cases it may seemingly have a multi-functional role in a given synthesis. Apart from detailed studies of tetrapropyl ammonium (TPA) in ZSM-5, the packing and orientation of templates in zeolite cavities and channels has not been extensively studied.

To develop a "structure design" rationale based on several recent synthetic studies on complex synthesis systems seems a possibility. It will attempt to define and differentiate the role of template within a known non-template, "simple" monovalent cation directed synthesis. The repeated indication is that the secondary building units of conventional nomenclature [44] are not the key structures in the formation of extended frameworks, but that the template directed linking of "extended" structures - sheets, columns, large polyhedra - formed from these sbu/tbu is the critical stage in the growth of a specific crystal framework. The recognition in recent zeolite structure evaluations of many defects in numerous crystal structures illustrates the fragility of structural continuity. A phenomenon observed in many other silicate and mineral systems [45,46]. In high purity, near defect free, structures, such as LTA and FAU, the template seems to play a void filling role, in which it forces an increase in the Si/Al framework ratio by reducing the space available for charge balancing cations, but does not change the framework connectivity. In other systems it is common to find multiple structures; some randomly intergrown, others having only occasionally defects present at low percentage levels. In some structures like FAU/BSS, OFF/ERI and other constituents of the ABC 6-ring series [47], almost infinite combinations of two or more end members may randomly intergrow in the same crystal under certain conditions. This reflects the tenuous nature of the crystal growth continuity in the absence of strong structure directing elements within the gel.

3.2. STRUCTURE BUILDING WITH SMALL CLUSTERS

3.2.1. *Co-crystallization of FAU - LTA - SOD*. At Si/Al ratios between 1 and 1.3 in sodium aluminosilicate gels at below 100°C, these three 4-,6-ring structures may be crystallized as either pure phases or as mixtures of two or three components. At long age times hydroxy-sodalite prevails as the thermodynamically stable phase. They all may be viewed as being built from sodalite cages. This commonality of structural unit has stimulated detailed analyses of the structure building mechanism, with proposals of the kind shown in Figure 5. However, detailed 29Si-MASNMR analyses of the components of this crystallization indicates that they do not have the same sbu [36,37], indicating that the zeolite synthesis "mixture" is a complex and dynamic one. Although SOD and FAU probably crystallize from the same sbu, LTA does not, confirmed by the observation that when Ga is partly substituted for Al at Si/(Al+Ga)ratios of unity, co-crystallized SOD and FAU contain Ga but LTA does not [48]. In the pure gallosilicate system LTA is completely suppressed, explaining the ease of making pure gallosilicate FAU(X) at Si/Ga of unity [49] and its near impossibility in the corresponding aluminosilicate gel (although this can be made in the corresponding K/Na gel with poor reproducibility [50]). The present thinking is that FAU and SOD form from the Si_{12} - 6,4-ring unit in the boat configuration and LTA is derived via the same unit in the chair form.

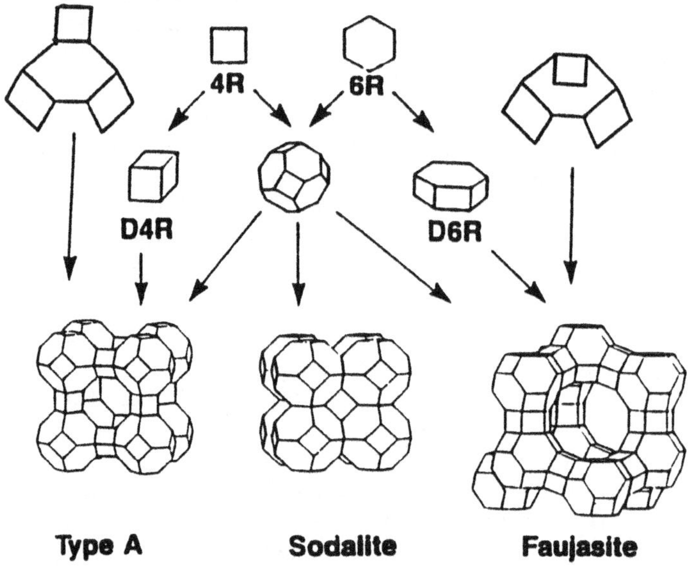

Figure 5: LTA, FAU and SOD all have sodalite cages as constituent units of their structures but are not built from the same building unit.

3.3. STRUCTURES DERIVED FROM EXTENDED SHEETS

3.3.1. *Faujasite, Breck-6 and intergrowths* The FAU/BSS system is one that has become highly complex in the last year in that we can now recognize and synthesize intergrowth materials within a complex matrix of structural and compositional variants [51]. In template free compositions at lower Si/Al ratios (1 to 3) high purity, defect free, faujasite crystallizes (X and Y forms). However, when clathratable cations, such as lithium (with a large hydration shell) and cesium, are included in the gel, planar faults, twins and intergrowth blocks appear at low levels, as seen in CSZ-1 and 3, and ZSM-2 and 3. At higher Si/Al ratios in the presence of templates such as TEA (in ZSM-20 and ECR-35) major intergrowth mixtures occur [52]. The important extended structure in these crystallizations is the sheet of linked sodalite cages shown in Figure 6. By sequential rotation of successive sheets by 60°, three layer orientations are produced A, B and C. An ABC cubic stacking produces a faujasite block of structure. ABAB stacking produces the hexagonal "Breck-6" or BSS structure. Mixtures of these sequences produces a range of structural intergrowth materials, some of which have been synthesized and identified, as indicated on the right of Figure 6. The family of materials can be viewed as a matrix with three end members (AB, ABC or ABABC) [53], as shown in Figure 7, or two end members (AB, ABC) in which ABABC is one of an infinite number of intergrowth possibilities.

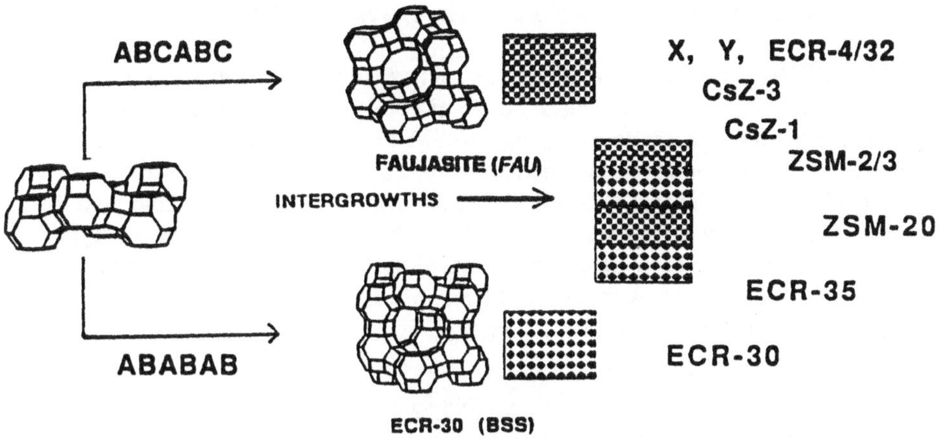

Figure 6: Different stacking of the "faujasite sheet" of double 6-ring (D6R) linked sodalite cages produces two pure phases and an almost infinite number of intergrowth possibilities.

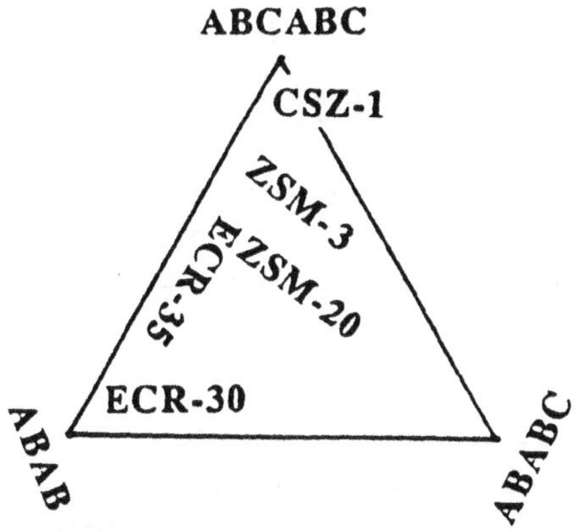

Figure 7: The system of stacked "faujasite" double 6-ring sheets can be viewed as a three phase diagram with pure end members derived from ordered repeat stacking in sequences AB (BSS), ABC (FAU) and ABABC (unknown/ unidentified product).

The system can be further complicated by adding framework compositional variation as the vertical axis - (Si, Ge, P, etc)/(Al, Ga, Fe etc) - to produce a structure/composition triangular prism. Individual structures within a matrix of this type are difficult to recognize and define, but recent developments have greatly simplified this process. Pattern matching with a matrix of theoretical patterns generated with a program such as DIFFAX [54] or by high resolution electron microscopy (HREM) in the structure image mode (preferably in combination) greatly simplifies this process. The approach of proportionally adding and mixing the two end-member X-ray diffraction patterns, though a simple procedure giving functional fits in some cases, is less exact and therefore less satisfactory, particularly as intergrowth block size is always a variable.

3.3.2. *ZSM-5, ZSM-11 and the Pentasils* The structural units of this group are shown in top of Figure 2. Like the FAU/BSS family, these are built from a single sheet (Figure 8), but connected either through a reflection plane in ZSM-11, or an inversion plane in ZSM-5. As successive layers are added to the structure, mistakes may occur to produce occasional intergrowths [55], or if additional strong structure directing variables are operable, induce an ordered mixture of the two stacking modes. These latter have been designated "pentasils" [56]. ZSM-11 is templated with tetrabutyl ammonium (TBA) and ZSM-5 with tetrapropyl ammonium (TPA). Although ZSM-5 is readily made in high purity, ZSM-11 usually contains an appreciable number of defects [57].

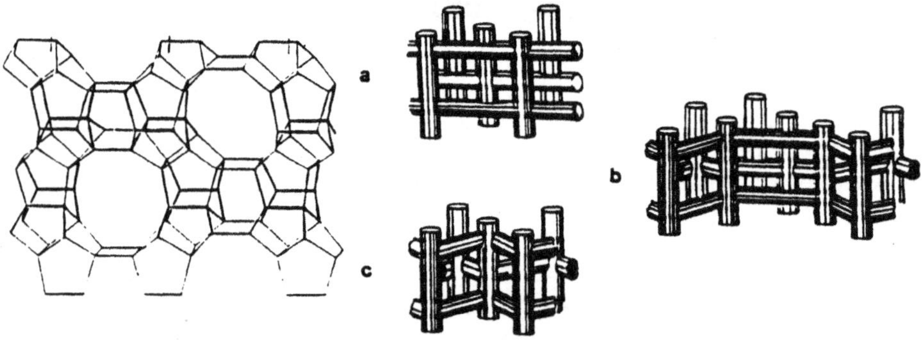

Figure 8: The characteristic sheet (left) is connected to itself either after reflection to form ZSM-11 (a) or after a reflection followed by a 180° rotation (ie. an inversion) to produce ZSM-5 (c). Alternation of these operations with the addition of successive growth layers produce intergrowths of the kind shown as (b).

3.3.3. *ABC 6-ring Sheet Structures* Many zeolites can be derived from the variable stacking of sheets of 6-rings interconnected by 4-rings. 60° rotations of the sheet produce three sheet orientations - A, B and C - which allow sheet to sheet connectivity. Numerous researchers have investigated aspects of this system of materials as they include many natural and synthetic zeolites [,60], as listed in Table 3 after refs. 59 and 60 (there is clearly a much larger number of theoretical stacking sequences). Several complex synthetic intergrowth mixtures contain gmelinite, chabazite and offretite as the dominant components. HREM is the most effective way (possibly the only way) to characterize such complicated inter-relationships [58]. Some sequences have only been observed in the microscope as intergrowth domains or small crystals (indicated by * in Table 3).

Table 3: Members of the ABC 6-ring zeolite family

Structure	Stacking sequence	# layers in repeat	Unit cell parameters a, Å	c,Å
cancrinite	AB	2	12.8	5.1
offretite	AAB	3	13.3	7.6
sodalite	ABC	3	12.6	12.6
gmelinite	AABB	4	13.8	10.0
losod	ABAC	4	12.9	10.9
erionite	AABAAC	6	13.3	15.1
chabazite	AABBCC	6	13.2	15.1
TMA-E(AB)	ABBACC	6	13.3	15.2
liottite	ABABAC	6	12.8	16.1
afghanite	ABABACAC	8	12.8	21.4
levyne	AABCCABBC	9	13.3	23.0
franzinite	ABCABCBACB	10	12.9	26.6
unnamed*	unknown	12	12.9	31.8
unnamed*	unknown	14	12.9	37.3
giuseppettite	ABABABACBABABABC	16	12.9	42.2
unnamed*	unknown	18	12.9	47.9
unnamed*	unknown	24	12.8	63.9
sacrofanite	unknown	28	12.9	74.2

3.3.4. *Other Common Zeolite Sheet Structures* Other structure series based on layers and which involve several structural entities include MOR [59,60] and FER [61,62], both of which involve numerous examples of multiple intergrowths and defect structures. It is usual that even well characterized commercial materials display the variable properties so typical of impure phases. Small and large pore mordenites, showing extremes of molecular sieve behavior (3Å to 7Å) [63], can in some cases be converted one to the other by chemical treatments. The ferrierite family, where the pure type structure may not exist [64] and a range of synthetic commercial products show variable properties [65], does not seem to have similar chemical manipulability.

The model that best fits this mode of structure development is that of sequentially stacking sheets. Particular gel compositions tend to form particular sheet structures or clusters of stacked sheet structures, and specific templates determine the way in which these stack together to form a crystal. The results are families of related structures differentiated by specific layer stacking sequences. Such a model anticipates that in a rapidly crystallizing system stacking errors will occur, producing intergrowths, faults and other defects in the stacking sequence now recognized to occur in numerous zeolite crystals.

3.4 STRUCTURES DERIVED FROM COLUMNAR UNITS

3.4.1. *The LTL/OFF system* In the potassium silicate system at Si/Al ratios between 2 and 10, LTL is the predominant zeolite in the absence of templates, an indication that a columnar structure component may be an important building entity. The building unit sequence for LTL is shown in Figure 2. With the addition of TMA to this system, offretite (OFF) is a cocrystallizing or pure phase. In the mixed phase product LTL may be in the form of separate crystals or as disk like overgrowths parallel to the c-axis of the long OFF crystals. In a similar gel with the addition of sodium but at lower TMA levels, erionite (ERI) intergrows with OFF in the same crystal. The related columnar structures of these three zeolites are shown in Figure 9. In the absence of TMA, ERI increases as an intergrowth component. In other similar syntheses hammer shaped crystals are common wherein the shaft is OFF and the head ERI, with apparent continuity of connectivity between the two. There is not yet evidence for a continuous interface in the former LTL-OFF case. Using models, structural continuity does not seem to be possible in any obvious manner.

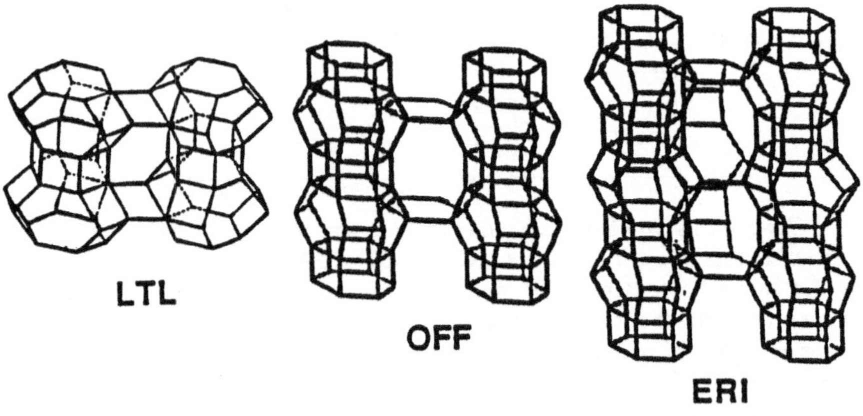

Figure 9: The structures of LTL, OFF and ERI are formed from similar columnar subunits. Specific templates promote the formation of high purity, single phase, products.

In the above described experiments, replacement of TMA with TEA reverts the crystallization to LTL [68]. The evidence in this case supports the presence of an intermediate structure comprising double 6-ring linked columns of "cancrinite" cages. Specific structure direction is determined by preference for the OFF columnar connectivity, which traps TMA in cavities between the columns, when TMA is present. and for the LTL connectivity when potassium is the only dominant trapped species. When the TMA is locally depleted in the gel, the connectivity reverts from OFF to LTL, the latter taking the form of circular disc overgrowths clearly crystallizing later than the OFF (Figure 10). This OFF-LTL arrangement is similar to the well known OFF-ERI co-crystallization, comprising hammer shaped crystals in which the shaft is OFF and the head is ERI.

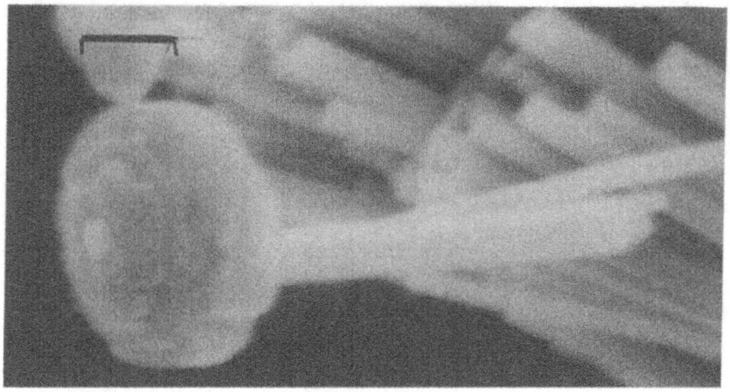

Figure 10: In some co-crystallizations disc like overgrowths of LTL are observed on OFF prismatic crystals. (Bar marker = 1 micron).

3.4.2. *Ladder structures*. LTL, OFF and ERI are usually transient phases in synthetic crystallizations, giving way at long crystallization times to more thermodynamically stable zeolites. These exclude templates and are primarily driven by Na/K cation ratios. Gismondine (GIS) dominates the high Na crystallizations, MER the high K crystallizations and phillipsite (PHI) those in between. As shown in Figure 11, these structures are all built from the same 4-ring ladder structure, derived from the Si_{12} unit (Figure 3), interconnected in different ways. It is closely related to the 4-ring ladder unit which form the columns in LTL and OFF.

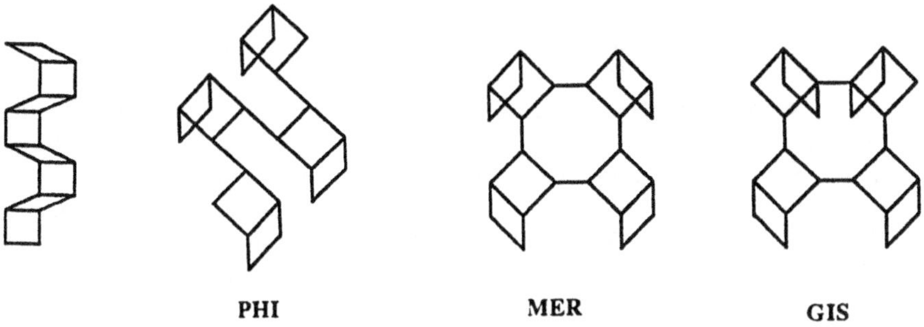

Figure 11: The stable zeolites in this Na/K/template aluminosilicate gel are built from the same 4-ring ladder structure connected in different ways.

4. Crystallization Field Boundary Structures

4.1. OVERVIEW

Although we tend to think of phase or crystallization boundaries as clean demarcations in equilibrium systems, in the kind of fluid, meta-stable, non-equilibrium environments characteristic of zeolite syntheses, they might be expected to be worthy of detailed examination. If zeolites form in the modes outlined in the above sections, mixed building elements will be present at overlapping phase stability fields, with possibilities for incorporating multiple elements into the same crystal. In actuality crystallization boundaries do represent regions of complex multi-phase co-crystallization. Two crystallization boundary regions to be discussed further involve ECR-1 and a synthetic paulingite analogue, ECR-18.

4.2. CASE STUDIES

4.2.1.*Mazzite, mordenite, ECR-1 and the "mazmorites"* One of the most intriguing crystallization regions is the transition from mazzite (MAZ) to mordenite (MOR) with increasing silica levels in the sodium system at temperatures over 100°C. In the presence of even small amounts of TMA, MAZ is the predominant zeolite, MOR being dominant in its absence. However when the template is bisdihydroxyethyl-dimethylammonium (E2M2), a structure comprising an ordered recurrent twin of the MAZ and the MOR sheets crystallizes [67] in a structure designated ECR-1. The crystallization diagram in Figure 12 shows these relationships [68], with the main products being pure MAZ, pure MOR, pure ECR-1, MAZ plus ECR-1 and MOR plus ECR-1, but rarely MAZ plus MOR. The implication of these results is that if the two structural components are present in the gel, ECR-1 crystallizes with E2M2 in the structure, but if this template is depleted before the aluminate component is consumed the excess components form the "end member" phase in addition to ECR-1. (Most zeolites crystallize in the presence of excess silicate, with the crystallization terminating when the aluminate ion is consumed.). Alternatively, the predominance of only one sheet component produces only the pure MAZ (Z) or MOR (M) end member. However, in a reaction system in which two elements M and Z are interconnecting, a pure MZMZMZ ordering will only occur when there is strong structure direction present. This is provided by a preferred template, but it may also be in combinations with other variables (temperature, pressure, reaction mode, Si/Al, other cations, solvent, etc). In the absence of such strong structure direction, alternate ordering sequences may prevail and several have been observed in this system, collectively designated the "mazmorite" family of structures, including MMZMMZ, MZZMZZ and MMZZMMZZ [69]. Special situations exist when a specific template is depleted, in which case an end member may "overgrow" on the earlier crystals, as does MAZ on ECR-1 (MZMZMZMMMM).

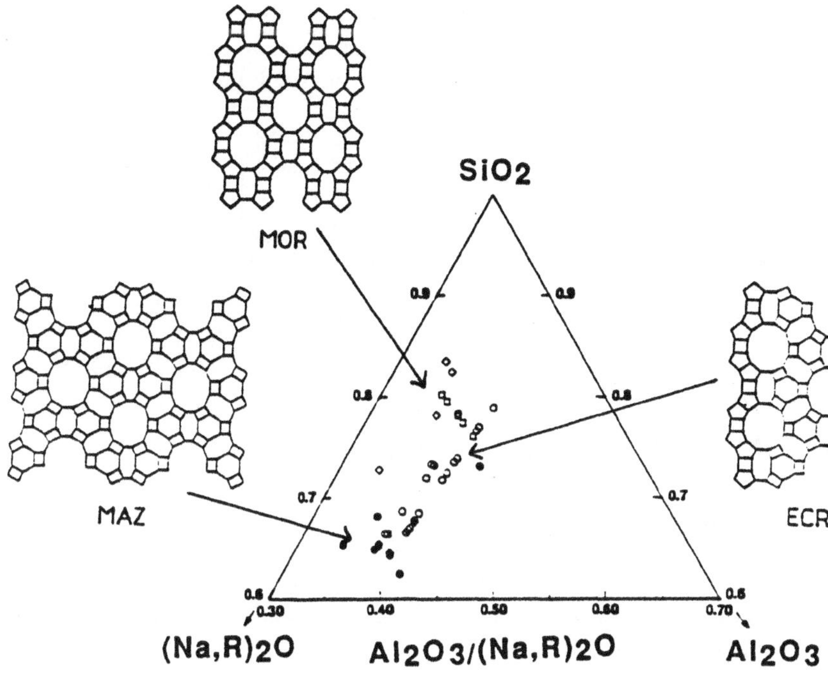

Figure 12: Crystallization composition diagram for ECR-1 showing the structural relationships between the three zeolites and the crystallization boundary between MAZ and MOR. Pure phases are indicated by o (ECR-1), • (MAZ) and ◊ (MOR); mixtures by Δ (ECR-1+MAZ) and □(ECR-1+MOR).

In a similar system MOR has been observed to overgrow on MAZ crystals [71]. The indication in these cases is that both sheet structures are present in the particular Si/Al composition, and the specific template (E2M2 or TMA) determines the way in which the sheets stack together - ECR-1, MAZ or MOR. Variable stacking may also be caused by template impurities, or new templates formed as degradation products of the primary template during synthesis.

4.2.2 *Paulingite, merlinoite and rho.* A similar occurrence of a boundary phase is observed in the crystallization of the templated synthetic analogue of paulingite - ECR-18 [72] - from a multi-cation gel of small (Na), medium (K, Rb) and large cations (TEA). This zeolite crystallizes in a region where MER and RHO may be co-existent phases - the PAU structure being built from the constituent units of both of these zeolites, as shown in Figure 13. In addition other zeolites characteristic of this composition, such as ZSM-20, beta and chabazite, also co-crystallize. With increasing K/Na ratios the dominant product is MER. In the presence of TEA as a template MER first traps TEA then the stable phase switches to PAU. The latter has a narrow crystallization range at the boundary between the several adjoining crystallization fields. TEA-Rho has been made in an adjoining crystallization field. Additional possible structures with this combination of RHO and MER cages include those related to ZK-5 (KFI) and the theoretical structures described by Kokotailo and Maier [72]. One may therefore visualize these zeolites as ones in which large cage units stack together to form specific structures.

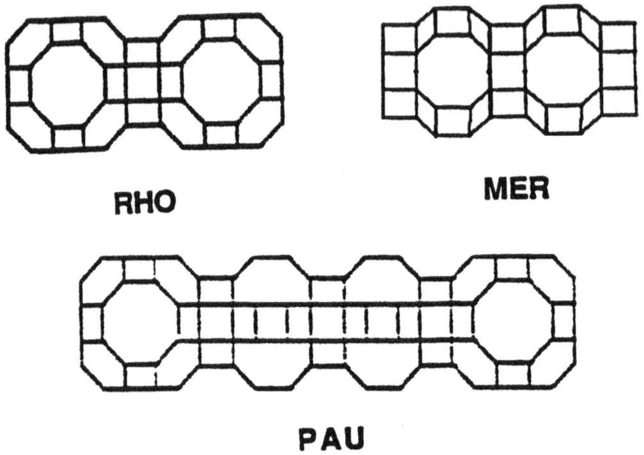

Figure 13: Paulingite contains the structural units of both MER and RHO which have been observed in adjoining systems - all trapping TEA.

5. Anions and Neutral Molecules as Structural Supports

Most discussions of structure direction in zeolite syntheses focus on cations, particularly the alkyl ammonium cations. The thesis developed in this discussion is that the major role of the organic cation is in a secondary structure direction role of directing the way in which the extended structures of sheets, columns or cages, the formation of which is controlled by the smaller alkali cations, are pulled together into specific crystal structures. However, many of the porous phosphates [73] are synthesized in the presence of amines, and clathrasils and zeosils [74] in the presence of relatively non-polar neutral molecules. Although space filling may be a dominant process in such systems, in several cases one structure may be "directed" by many different "templates", indicating either that this is not the sole mechanism of structure direction, or that more flexible structures have wide ranges of template tolerance. Anion "templation" is rarely observed in zeolite crystallizations, although it has been well researched when it does occur, as in the sodalite - cancrinite case [75,76]. The influence of fluoride ions on pegmatites and granites is of long standing interest [77]; similarly with fluorided silicas [78]. Interest in fluoride effects on zeolite synthesis have recently been stimulated with the observation that they facilitate lower pH crystallizations and the structural incorporation of hetero-atoms [79], and that the fluoride ion has a strong preference for locating in double 4-rings, promoting in the $GaPO_4$ system the formation of the LTA framework and a new 3D 20-ring channel structure (designated "cloverite")[11] containing the same F- containing unit. Anions usually have only minor morphological influences, changing crystal size more frequently than crystal shape.

6. Discussion

The formation of a pure zeolite is usually depicted as connecting a particular structural unit together in a repeatable way. Alternate structures obtained by simple symmetry manipulations of the same structural unit commonly co-crystallize or intergrow with the first structure (LTL/OFF; FAU/BSS; OFF/ERI; ZSM-5/11). Miss connections produce defects such as twins of different orientations, or when repeated in an orderly manner, line or block intergrowths. The observed structure relationships in several zeolite crystallizing solutions indicate that two large sub-units may self interconnect or connect with different but similar units in specific ratios and orientations to produce several unique structures. The manner in which the sub-units clip together is largely determined by the templates present in the crystallizing solution.

Although this discussion has focussed principally on the authors own work, numerous other zeolites show analogous behavior. The ZSM-5 and 11 structures are expanded into the "pentasil" family of ordered and disordered structures [56]. ZSM-23 is a recurrently twinned Theta-1 [80]. Beta is an intergrowth of two closely related sheets built from the same building block [81]. Several natural zeolites, including mordenite, ferrierite, gmelinite and chabazite, are also known to have major "structural problems" in mineral and synthetic derivatives, many or which may be related to numerous intergrowth forms not yet fully defined. Many such clusters of structures have been enumerated by Smith and co-workers [82] and Akporiaye [83].

In the last few years the apparent 12-ring pore size limit on zeolite like structures has been broken with the synthesis of 14-, 18- and 20-ring structures, and theoretical studies indicate the possibilities for numerous larger pore stable structures. In instances where structures can be readily expanded by subunit reorientation, it is probable that playing tunes with large template molecules and reaction conditions may induce the formation of new large pore structures. This research clearly indicates that zeolite structures can be extended into the meso-porous pore size range.

References

[1] R.M.Barrer and H. Villiger, Z.Kristallogr., 128, 352-370 (1969).

[2] B.C. Chakoumakos, R.J. Hill and G.V. Gibbs, Amer. Mineral., 66, 1237-1249 (1981).

[3] W.M. Meier, Proc. 7th Intl. Zeolite Conf., Stud.Surf.Sc.Catal. 28, Ed. Y. Murakami, A. Iijima and J.W. Ward, Elsevier/ Kodansha (Tokyo), 13-22 (1986).

[4] G.O. Brunner and W.M. Meier, Nature, 337, 146-7 (1989).

[5] M.E. Davis, C. Saldarriaga, C. Montes, J. Garces and C. Crowder, Zeolites, 8, 362-66 (1988).

[6] D.E.W. Vaughan and R. Sostak, NATO ASI Ser. B 221, Ed. D. Barthomeuf, E. G. Derouane and W. Hölderich, Plenum Press, London, 381-5 (1990).

[7] F. d'Yvoire, Bull.Soc.Chim. France, 1762 (1961).

[8] E. G. Derouane and R. von Ballmoos, Eur.Pat.Appl. 0146389 (1984).

[9] J.V. Smith and W.J. Dytrych, Nature, 309, 607-8 (1984)

[10] J.W. Richardson, Jr., J.V. Smith and J.J. Pluth, J.Phys.Chem., 93, 8212-19 (1989).

[11]A. Merrouche, J. Paterin, H. Kessler, M. Soulard, L. Delmotte. J.L. Guth and J.F. Jolly, Zeolites, 12, 226-232 (1992).
[12]S.L. Lawton and W.J. Rohrbaugh, Science, 247, 1319-1322 (1990).
[13]E.W. Corcoran and D.E.W. Vaughan, Solid State Ionics, 32, 423-429 (1989).
[14]S.M. Kuzniki, K.A. Thrush, F.M. Allen,S.M. Levine, M.M. Hamil, D.T. Heyhurst and M. Mansour, "Molecular Sieves" v.1, Ed. M.L. Occelli and H.E. Robson, Van Nostrand Reinhold (New York), 427-453 (1992).
[15]R.C. Haushalter and L.A. Mundi, Chem.Materials, 4, 31-48,(1992).
[16]C.T.Kresge, M.E.Leonowicz, W.J.Roth, J.C.Vartuli and others U.S. Patents 5,102,643; 5,110,572; 5,112,589 (1992); also PCT WO 91/11390.
[17]J.F. Goodman and J.S. Clunie, in "Liquid Crystals and Plastic Crystals", v. 2, Ed. G.W. Grey and P.A. Winsor, J.Wiley and Sons (London), 1-23 (1974).
[18]P.B. Moore and J. Shen, Nature, 306, 356 (1983).
[19]V.C. Farmer, M.J. Adams, A.R. Fraser and F. Palmieri, Clay Minerals, 18, 1-14 (1983).
[20]A. Preisinger, Proc. 6th Natl. Conf. Clays & Clay Minerals, 61-7 (1959).
[21]D.E.W. Vaughan, Amer.Chem.Soc. Symp.Ser. 368, Ed. W.H.Flank and T.E. Whyte,Jr., 308-323 (1988).
[22]F.Cavani, F.Trifiro and A.Vaccari, Catalysis Today, 11, 173-295 (1991).
[23]G.M.Wolltermann, US Patent 4,454,244 (1984).
[24]T.Kwon, T.J.Tsigdinos and T.J.Pinnavaia, J.Amer.Chem.Soc., 110, 3653-4, (1988).
[25]R.M. Barrer, Clays and Clay Minerals, 37, 385-395 (1989).
[26]G. Legaly, Naturwissenschaften, 68, 82-88 (1981).
[27]M.S. Whittingham and A.J. Jacobson, Eds., "Intercalation Chemistry", Academic Press (New-York), (1982).
[28]W.M. Meier, "Molecular Sieves", Ed. R.M. Barrer, Soc.Chem.Ind.(London), 41-48 (1968).
[29]W.M. Meier and D.H. Olson, "Atlas of Zeolite Structure Types", 3rd Edn., Butterworths/ Intl. Zeolite Assoc., 6 (1992).
[30]G. Engelhard, D. Hoebbel, M. Tarmak, A. Samoson and E. Lipma, Z.Anorg.Allg.Chem., 484, 22-32 (1982).
[31]J.M. Thomas and J. Klinowski, Adv Catal., 33, 200-374 (1985).
[32]G. Bissert and F. Liebau, Zeit.Kristallogr., 179, 357-371 (1987).
[33]G. Harvey and L.S. Dent-Glasser, Amer.Chem.Soc. Symp.Ser. 398, Ed. M.L. Occelli and H.E. Robson, 49-65 (1989).
[34]A.V. McCormick and A. T. Bell, Catal.Rev. Sci.Engrg., 31, 97-127 (1989).
[35]P. Dutta, M. Puri and D.C. Shieh, Mat.Res.Soc. Symp. Proc. 111, Ed. M.M.J. Treacy, J.M. Thomas and J.M. White, 101-106, (1988).
[36]D.E.W. Vaughan, M.T. Melchior and A.J. Jacobson, Amer.Chem.Soc. Symp.Ser. 218, Ed. G. Stuckey and F.G. Dwyer, 233-242 (1983).
[37]M.T. Melchior, ibid, 243-265.
[38]F. Liebau, "Structural chemistry of Silicates", Ch. 7 and 8, Springer-Verlag (Berlin), 90-160 (1985).
[39]R.M. Barrer, NATO ASI Ser. E80, "Zeolites: Science and Technology", Ed. F.R. Ribeiro et al, 35-82 (1984).

[40]R.M. Barrer and P. J. Denny, J.Chem.Soc., 971-982 (1961).
[41]R.M. Barrer, "Hydrothermal Chemistry of Zeolites", Academic Press (London), 157-175 (1982).
[42]B. M. Lok, T.R. Cannan and C,A, Messina, Zeolites, 3, 282-291 (1983)
[43]D.L. Rollmann, NATO ASI Ser. E80, "Zeolites: Science and Technology", Ed. F.R. Ribeiro et al,109-127 (1984).
[44]R. Gramlich-Meier and W.M. Meier, J. Solid State Chem., 44, 41-9 (1982).
[45]P.R. Busek and S. Iijima, Amer.Mineral., 59,1-21 (1974).
[46]C.N.R. Rao and J.M. Thomas, Accts.Chem.Res., 18, 113-119 (1985).
[47]G.R. Millward, S. Ramdas and J.M. Thomas, Proc.Roy.Soc., A399, 57-71 (1985).
[48]D.E.W. Vaughan and M.T. Melchior, (in preparation).
[49]J. Selbin and R.B. Mason, J.Inorg.Nucl.Chem., 20, 222-228 (1961).
[50]H.S Sherry and G. Kuhl, UK Patent 1,580,928 (1982).
[51]D.E.W. Vaughan, K.G. Strohmaier, M.M.J. Treacy and J.M. Newsam, US Patent 5.116,590 (1992).
[52]D.E.W. Vaughan, M.M.J. Treacy, J.M. Newsam, K.G. Strohmaier and W.J. Mortier, Amer.Chem.Soc. Symp.Ser. 398, Eds. M.L.Occelli and H.E.Robson, 544-559, (1989).
[53]J.M. Newsam, M.M.J. Treacy, D.E.W. Vaughan, K.G. Strohmaier and M.T. Melchior, "Molecular Sieves" v.1, Ed. M.L. Occelli and H.R. Robson, Van Nostrand Reinhold (New York), 454-472 (1992).
[54]M.M.J. Treacy, J.M. Newsam and M. Deem, Proc.Roy.Soc, A433, 399-520 (1991).
[55]J.M. Thomas and G.R. Millward, J.Chem.Soc. Chem.Commun., 1380-1383 (1982).
[56]G.T. Kokotailo and W.M. Meier, in "Properties and Applications of Zeolites", Ed. R.P. Townsend, Chem. Soc. Special.Publ. 33, 133-139 (1980)
[57]O. Terasaki, J.M. Thomas, G .R. Millward and D. Watanabe, Chem.Materials, 1, 158-162 (1989).
[58]G.R. Millward, S. Ramdas and J.M. Thomas, Proc.Roy.Soc., A399, 57-71 (1985).
[59]J.D. Sherman and M.J. Bennett, Amer.Chem.Soc. Adv.Chem.Ser. 121, 52-65 (1973).
[60]W.M. Meier, "Natural Zeolites", Eds. L.B. Sand and F.H. Mumpton, Pergamon Press (Oxford), 99-103 (1978).
[61]J.V. Sanders, Zeolites, 5, 81-90 (1985).
[62]R. Gramlich-Meier, W.M. Meier and B.K. Smith, Z.Kristallogr., 169, 201-210 (1984).
[63]L.B. Sand, "Molecular Sieves", Ed. R.M. Barrer, Soc.Chem.Ind.(London), 33-35 (1968).
[64]S.B. Rice and D.E.W. Vaughan, unpublished results.
[65]D.E.W. Vaughan, in "Fluid Catalytic Cracking: Science and Technology.", Ed. J.S. Magee and M.M. Mitchell, Stud.Surf.Sci.Catal., Elsevier (Amsterdam), Ch. 4, in press.
[66]D.E.W. Vaughan, US Patent 4,554,145 (1985).
[67]M.E. Leonowicz and D.E.W. Vaughan, Nature, 329, 819-821 (1987).
[68]D.E.W. Vaughan and K.G. Strohmaier, Amer.Chem.Soc. Symp.Ser. 398, Ed. M.L. Occelli and H.E. Robson, 507-517 (1989).
[69]M.E. Leonowicz, D.E.W. Vaughan and K.G. Strohmaier, U.S. Patent 4,892,721 (1989).
[70]F. Fajula, F. Figueras, C. Gueguen and R. Dutartre, US Patent 4,946,580 (1990).
[71]D.E.W. Vaughan and K.G. Strohmaier, US Patent 5,013,536 (1991).

[72]W.M. Maier and G. Kokotailo, Z.Kristallogr, 121, 211-218 (1965).
[73]E.M. Flanigen, B.M. Lok, R.L. Patton and S.T. Wilson, Proc. 7th. Intl. Zeolite Conf., Stud.Surf.Sc.Catal. 28, Ed. Y. Murakami, I. Iijima and J.W. Ward, Kodansha/Elsevier (Tokyo), 103-112 (1986).
[74]F. Liebau, H. Gies, R.P. Gunarwardane and B. Marler, Zeolites, 6, 272-7 (1986).
[75]R.M. Barrer and J.F. Cole, J. Chem.Soc. A, 1516-23 (1970).
[76]D.E.W. Vaughan, K.G. Strohmaier, I.G. Pickering, G.N. George, Solid State Ionics, 1282-1291 (1992).
[77]P.J. Wyllie and O.F. Tuttle, Amer.J.Sci., 259, 128-143 (1961).
[78]E.M. Rabinovich and D.L. Wood, Mater.Res. Soc. Symp.Proc. 73, "Better Ceramics Through Chemistry", Ed.C.J. Brinker, D.E. Clarke and D.R. Ulrich, 251 (1986).
[79]J.L. Guth, H. Kessler, J.M. Higel, J.M. Lamblin, J. Petarin, A. Seive, J.M. Chezeau and R. Wey, Amer.Chem.Soc. Symp.Ser. 398, Ed. M.L. Occelli and H.E. Robson, 176-195 (1989).
[80]P.A. Wright, J.M. Thomas, G.R. Millward, S. Ramdas and S.A. Barri, J.Chem.Soc. Chem.Commun., 1117 (1985).
[81]J.M. Newsam, M.M.J. Treacy, W.T. Koetsier and C.B. de Gruyter, Proc.Roy.Soc., A420, 375-405, (1988).
[82]J.V. Smith, Proc.8th Intl. Zeolite Conf., Ed. P.A. Jacobs and R.A. van Santen, Stud.Surf. Sci.Catal. 49A, 29-47 (1989).
[83]D.E. Akporiaye, Zeit.Kristallogr., 188, 103-120 (1989).

Note: The structure codes used throughout the text are from Ref. 29 and Zeolites, 12, (June, 1992)(Special Issue).

SECTION 3

PILLARED LAYERED SOLIDS

MESOPOROUS PILLARED LAYERED MATERIALS

Abraham Clearfield
Department of Chemistry
Texas A&M University
College Station, Texas 77843

ABSTRACT. There is a need for catalysts with designed or regular pore systems in the 10-20 Å range for cracking and oxidation of large molecules. The search for such materials has centered on the synthesis of large pore molecular sieves and pillared clays. This work will be briefly reviewed, as well as the authors results on pillaring of layered compounds. A true mesoporous product has been prepared in the synthesis of $Zr(O_3PC_6H_4C_6H_4PO_3)$. B.E.T. measurements and other data indicate an average pore radius of 24 Å.

1. Introduction

Zeolites are microporous naturally occurring or synthetic crystalline aluminosilicates with three dimensional frameworks. The frameworks enclose void space as cavities or channels which contain cations necessary to balance the negative framework charge. The combination of porosity and ion exchange properties have imparted a complexity and versatility to zeolite behavior which has been exploited to satisfy a wealth of commercial applications [1,2]. They are unsurpassed in their ability to crack petroleum and to otherwise catalyze a large number of important reactions [2-4]. For these reasons intense activity, directed towards the synthesis of new zeolites, has been pursued in a number of laboratories. These efforts have been successful in producing many new types of porous or zeolite type materials. Among these we may cite the aluminum phosphates [5], zinc and beryllophosphates [6], gallosilicates [7] and phosphates [8] and pillared clays [9,10].

Until recently the largest cavity in a zeolite are those present in ZeX and ZeY which have cavity diameters of about 13 Å. However, the entrances leading into the cavities are only about 7.4 Å wide. In the aluminum phosphates the largest pores are present in VPI-5 and H1 [11,12]. These compounds have straight channels of about 12 Å diameter running parallel to the **c**-axis. The recently discovered gallophosphate cloverite has an even larger pore than VPI-5 [10]. The latter compound has an 18-

C.A.C. Sequeira and M.J. Hudson (eds.), Multifunctional Mesoporous Inorganic Solids, 159–178.

membered ring whereas the framework of cloverite, shown in Figure 1, contains 20-membered rings opening into a cavity of ~30 Å. Thus, cloverite may be considered as a mesoporous molecular sieve. The formula for cloverite is given as $Ga_{96}\,P_{96}\,O_{372}\,(OH)_{24}\,(RF)_{24}\,(H_2O)_n$ where R is the organic template (quinuclidine) and F is fluoride ion. At its widest extent the entrance into the cavity is 13.2 Å, but the opening is somewhat restricted by hydroxyl groups which point towards the center of the channel (Figure 1) [13]. The role that this restriction will play in catalytic processes is as yet unknown. Cloverite is thermally stable to 700 °C. Its unusual shape of the openings into the large pore restricts the size and shape of molecules that can both be sorbed or desorbed. The molecules must be flat to pass thru the openings, but once inside the catalytic process bulky molecules could be created [13]. However, they would need to be cracked to smaller units in order to desorb. One suggested use of cloverite [13] is the in situ preparation of metal phthalocyanines which, being confined to the cavity, can then be used to mimic enzyme activity in a catalytic behavior [14].

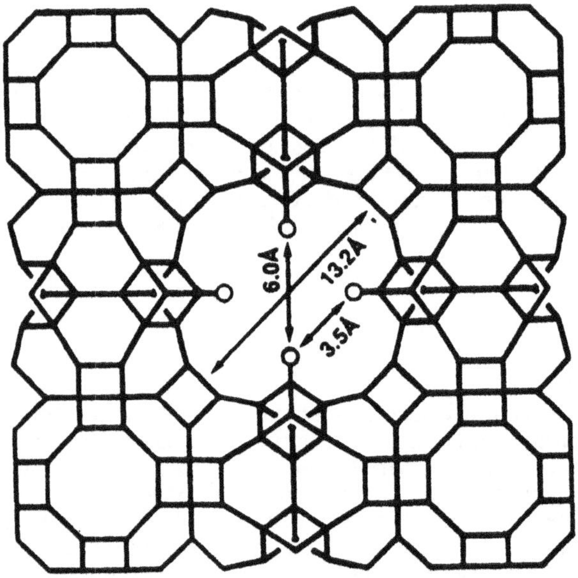

Figure 1. A projection of the pseudo cell of cloverite along the [100] direction showing the cloverleaf-shaped 20-tetrahedral-atom pore opening. The nodes represent tetrahedral atoms (either Ga or P), the circles, terminal hydroxyls. The free dimensions of the pore opening, calculated using an oxygen radius of 1.35 Å, are indicated (from Ref. 13, with permission).

Although VPI-5 has a 12 Å channel with no limiting narrow passageways, as do zeolites X and Y, dynamic argon adsorption measurements suggest that part of the channels are narrowed from 12 Å to about 8.3 Å during dehydration. It is suggested that this narrowing results from a deformation of some 18-membered rings in the anhydrous VPI-5 [15] or to stacking disorders [16]. A recent X-ray diffraction study [17] of the anhydrous VPI-5 and also H1 has shown only slight narrowing due to the change in unit cell dimensions on removal of water. However, the H1 phase exhibited disorder in the position of the phosphate groups in the 6-membered rings which probably accounts for its thermal instability to form $AlPO_4$-8.

Another interesting large pore compound is the naturally occurring mineral cacoxenite, $AlFe_{24}O_6(OH)_{12}(PO_4)_{17}(H_2O)_{24}\cdot 51H_2O$ [18]. A unit of the structure showing the pore system is illustrated schematically in Figure 2. In the long direction the distance is 14.2 Å and in the narrower direction about 12.2 Å. Cacoxenite is not very stable thermally and therefore may have no use as a porous catalyst. The existence of large pore molecular sieves had been predicted earlier by Smith and Dytrych [19]. The largest pore compound, shown in Figure 3, is a member of the series derived from net type 81 and has a predicted pore size of 18 Å. As yet it has not been synthesized in either the aluminum phosphate or silicate systems. The VPI-5 structure is related to the 81b net and efforts are continuing to synthesize additional members of this series. The recent synthesis of cloverite reinforces the notion that zeolites with mesopore cavity sizes may soon become a reality.

Figure 2. A unit of the mineral cacoxenite showing the 15 Å pore system (from Ref. 18, with permission).

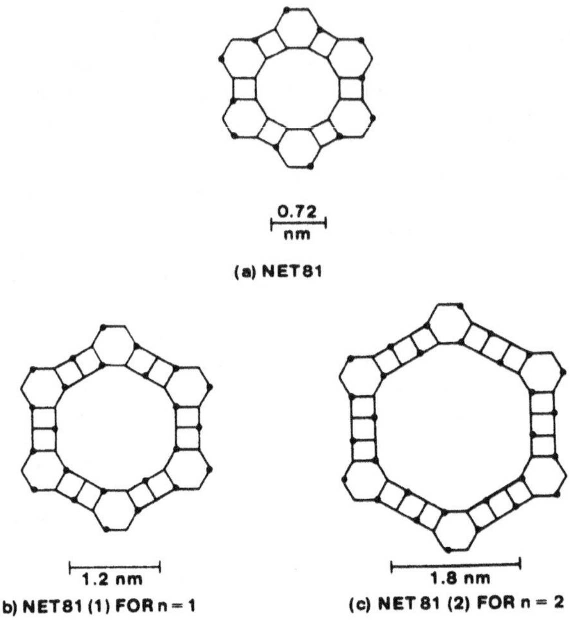

Figure 3. Two-dimensional projections of three members of an infinite series of three-dimensional nets derived from net number 81. The approximate free diameter of each pore is shown below each net (from Ref. 19, with permission).

Mesopores have been shown to exist in dealuminated zeolites [20-22]. When aluminum is removed from the framework by steaming at elevated temperatures, defect sites or holes are created and portions of the framework collapse. A secondary, mesopore channel system is thereby created [22], as shown in Figure 4, having pore diameters in the 3-5 nm range. In the preparation of ultra-stable zeolite Y it was found that those samples which have low amounts of non-framework aluminum also have the maximum free mesopore volume. Such samples were not active for heavy oil cracking and produce high levels of coke. High boiling hydrocarbons are thought to plug the mesopores.

Although great progress has been made in producing large pore molecular sieves, the compounds available are not satisfactory for many of the current needs of industry for catalysts with large pores. They either are too unstable or lack the desired acidity or the requisite combination of pore structure and acidity required for high level performance. This is particularly true in the field of petroleum cracking. The gas-oil, and asphalt fractions of petroleum generally resist normal cracking procedures

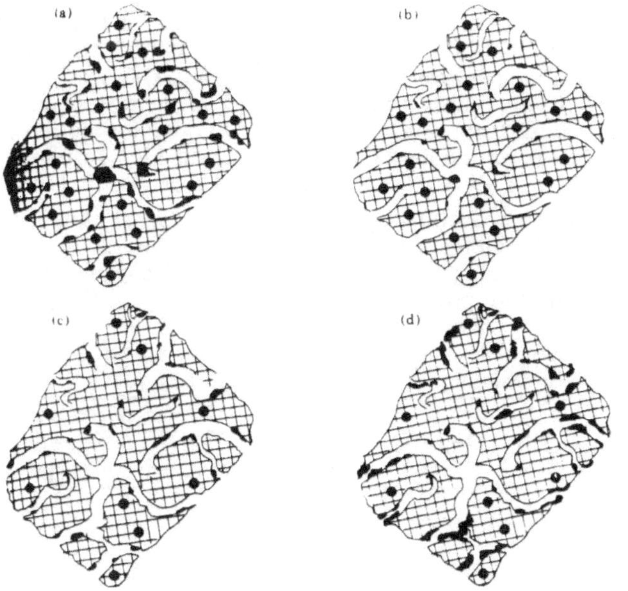

Figure 4. Pictorial representation of mesopores in zeolite-Y and the effect of acid extraction and subsequent steaming on the zeolite non-framework aluminum content (a) ultrastable zeolite-Y (USY), (b) acid extracted USY, (c) USY acid extracted in steam deactivated cracking catalysts, (d) USY in steam deactivated cracking catalysts (from Ref. 20, with permission).

because the molecules of these fractions are too large to enter the zeolite catalysts cavities. Thus, they are first subjected to non-selective precracking to reduce the molecular size of the crude. To avoid the precracking step, it would be advantageous to have porous materials with zeolite-like frameworks in which the pores are large enough to accept the resid molecules and crack them into the gasoline fraction. Thus, the idea was developed to use acidic clay minerals and to prop them open by insertion of large inorganic polymers [23].

Ideally, materials containing pores at the interface of the micro- and mesopore range, 10-20 Å in diameter, are required. In the early studies [23-25] the polymers used as pillaring agents were, the aluminum Keggin ions, $[Al_{13}O_4(OH)_{24}(H_2O)_{12}]^{7+}$, the zirconium tetramer, $[Zr(OH)_2{\cdot}4H_2O]_4^{8+}$ and chromium oligomers. Smectite clays were chosen for pillaring because they swell spontaneously in water and can even be made to exfoliate. These clay minerals are built up of alumina or magnesia octahedral layers sandwiched between silicate tetrahedra. The layers bear a negative charge resulting from substitution of lower valent ions, Mg^{2+} for Al^{3+} and Li^+ for Mg^{2+} in the octahedral layer or, Al^{3+} for Si^+ in the tetrahedra. This charge deficiency is compensated by cations, most commonly Na^+ and Ca^{2+} between layers. A typical smectite clay mineral has layer charges varying from 0.37 to 1.1 electrons per formula weight and a layer thickness of ~9.4 Å. The structures of both the aluminum Keggin ion [26] and the zirconium tetramer [27] are known. Thus, it is possible to calculate the expected distance between pillars from a knowledge of the clay layer charge and the size and charge of the pillaring cation. This has been done for the Al_{13} polymer [28] and gave values of 29 Å for the lowest layer charge and 17 Å for the highest charge. The height of the pillar is ~9 Å. These calculations assumed that the charge on the aluminum species remained constant at 7^+. Subtracting a calcined pillar width, estimated to be 7.2 Å, yields free spaces between pillars of 21.8 and 9.8 Å, respectively. Table 1 presents pore size data obtained by Vaughan and Lussier [23] on two pillared clays. In both pillared products the majority of the pores were in the

Table 1. Comparison of pore size and surface area distributions for Al_{13} Keggin ion pillared clays and zeolites.

			% N_2Surface Area	
Pore Size Range, Å	Pillared Clay	Pillared Clay	NaY	Amorphous Catalyst
>100	1.1	0.6	2.4	1.1
100-80	0.5	0.2	0	3.0
80-60	1.0	0.4	0	19.3
60-40	16.5	19.0	0	47.5
40-30	4.3	7.5	0	23.1
30-20	0.7	0.0	0	6.0
20-14	14.1	18.3	0	0
<14	61.9	64.0	97.6	0
Average pore diameter, Å	23.6	21.9	18.2	51.4
Total surface area, m^2/g	477	373	860	575

Reproduced from Ref. 23, with permission.

micropore region (<14 Å). However, 14-18% of the pores were in the 14-20 Å range and an almost equal amount in the 40-60 Å range. Sorption studies (Table 2) showed a sharp cut off in sorption at low pressures for molecules of 8 Å kinetic diameter and larger. Our own B.E.T. data for an aluminum pillared montmorillonite obtained from LaPorte Industries, Ltd.. is shown in Figure 5. This sorption isotherm is characteristic of a layered material with slit-like pores. The measured surface area was 253 m^2/g of which 83% was derived from pores in the micropore region (<14 Å). From the increase in aluminum content over that of the original clay, it is estimated that the average charge on the aluminum polymer was ~3.25^+. This charge reduction is sufficient to account for the smaller observed pore size. Since the original montmorillonite contained 0.485 moles of Mg^{2+} substituting for Al^{3+}, it would be expected that 0.12 moles of $[Al_{13}O_4(OH)_{24}(H_2O)_{12}]^{7+}$ would exchange for the replaceable Na^+. In fact, 0.2 moles were exchanged for only 0.65 moles of sodium. This level of exchange would require that the pillars be an average of 16.1 Å apart and have a free space between pillars of ~9 Å rather than values of 22.2 and 15 Å, respectively if the full 7+ charge were retained by the Al_{13} polymer. The measured pore sizes are in apparent accord with these lower values.

Table 2. Molecular sieve properties of Al_{13} pillared clay.

Probe Molecule	Size	Sorption Pressue	Amount Sorbed
n-Butane	4.6 Å	600 torr	8.0 wt.%
Cyclohexane	6.1	60	8.4
Carbon Tetrachloride	6.9	106	11.5
1,3,5-trimethylbenzene	7.6	9.4	5.3
1,2,3,5-tetramethyl-benzene	8.0	6.9	0
Perfluorotributylamine	10.4	31.0	0

From Ref. 23, with permission.

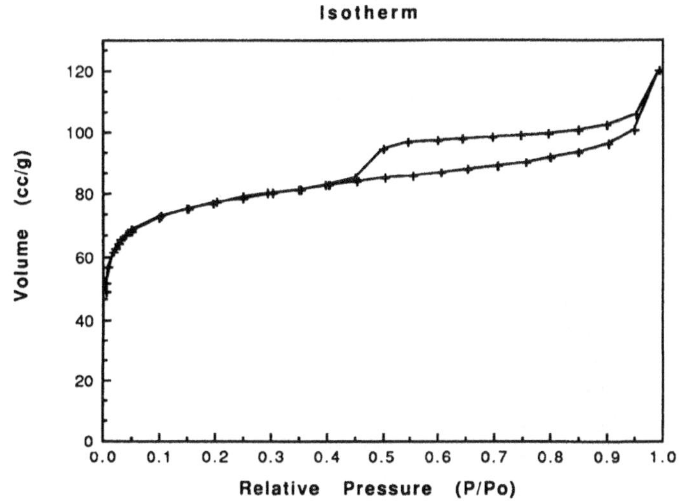

Figure 5. B.E.T. nitrogen sorption-desorption isotherms for an Al_{13} Keggin ion pillared montmorillonite calcined at 550 °C.

Another point to consider is that the charge on the clay layers is not uniform but depends upon the amount of lower valent cation substitution in that layer [29,30]. However, it has been demonstrated that the pillars fill each interlayer region to essentially the same extent, regardless of the layer charge. Thus, the polynuclear hydroxo species apparently adjusts its charge by hydrolysis to yield relatively uniform pores [24,31]. The hydroxy zirconium tetramer is somewhat smaller in length than the Al_{13} species and also is extensively hydrolyzed in solution. Therefore, it would be expected to yield small pores on pillaring. The larger mesopores actually observed in the pillared clays (Table 1) may arise as shown in Figure 6A. Either the layers are distorted (A) or there are gaps left between sheets which do not extend uniformly in length. Mesopores of this type are not built in by design but are the result of the particular way in which the clay layers are brought together by the introduction of the pillaring cation. Another mechanism by which mesopores arise is the formation of a house of cards structure as shown in Figure 6B. This arrangement is a consequence of the strong edge to face attractions of the clay layers. If mesoporous materials are to be prepared in which a uniform large pore is obtained, then new synthetic strategies are required.

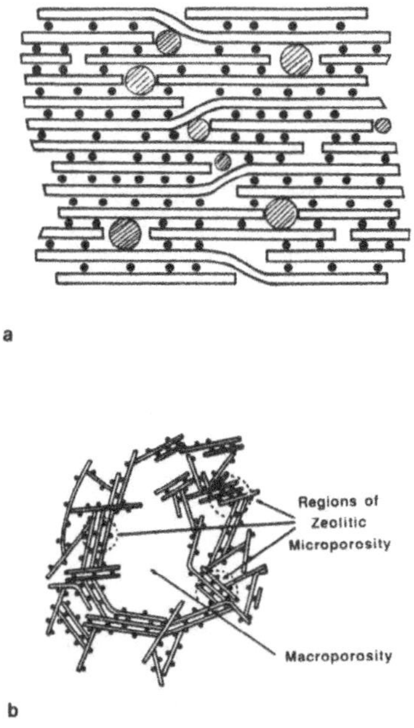

Figure 6. Schematic depiction of proposed origin of mesopores in pillared clays: (a) Faults in pillared clays where filled small circles represent pillars creating microporosity and the shaded circles highlight the regions of mesoporosity, (b) Mesoporosity arising from a "house of cards" structure represented by the medium unlabeled spaces (from Ref. 31, with permission).

A beginning in this direction has been made by introducing larger pillars into the clays. These pillars have been prepared by more extensive hydrolysis of the polyoxycation solutions [32,33] or by use of metal oxide sols [34,35]. in the former technique the polyoxycation is treated hydrothermally and when allowed to react with exfoliated clays achieve interlayer spacings as high as 28 Å. Pores in the 20 Å range were obtained. In the DIMOS methods the exfoliated clays are allowed to react with colloidal silica or titania particles. Basal spacings of 35-40 Å have been observed with accompanying high surface areas. More work aimed towards better characterization of the pores is required, particularly through sorption of large molecules. Sterte [32] also reported hydrothermal treatment of basic aluminum solutions with $LaCl_3$ to obtain large polymer complex cations. These cations produced pillared clays with interlayer spacing of 26 Å and surface areas of 300-500 m^2/g.

1.1 Non-Clay Pillared Layered Materials

There are many classes of layered compounds which potentially could be pillared. These layered compounds have exchangeable cations [36], but generally do not swell in water. Thus, it is necessary to first artificially swell them and this usually can be done by intercalation of alkylamines. Our earlier work had shown [37] that large complex cations could be exchanged for the amine intercalates of zirconium phosphate. The intercalated amines accept protons from the monohydrogen phosphate groups and then exchange as alkyl ammonium ions. The group 4 (14) layered phosphates are excellent candidates for pillaring as they have a clay-like structure. The layers in the α-type structure consist of octahedrally coordinated metal atoms sandwiched between phosphate tetrahedra [38]. The phosphate groups bridge the metal atoms with three of their 4 oxygens with the fourth pointing into the interlayer space as an -OH group. Ion exchange capacities are of the order of 6-7.5 meq./g for these compounds, depending upon the relative mass of the metal. Furthermore, all of this large layer charge is concentrated in $\frac{1}{3}$ as many oxygens as for clays. Straight forward calculations show that the aluminum Keggin ion pillars in these compounds should fill the interlayer space leaving no micropores [28]. This has generally been found to be true for α-zirconium phosphate. However, both tin and titanium did yield higher surface area products in which a proportion of the pores are micropores. This unexpected development was traced to the loss of phosphate from the more hydrolyzable tin and titanium compounds. However, all of the pillared compounds have a mesopore structure associated with them which are spread over a broad range of sizes as shown in Figure 7. Thus, the pores resulting from loss of phosphate essentially produce disordered structures with no regularity. A new approach is needed in order to obtain a more uniform pore structure.

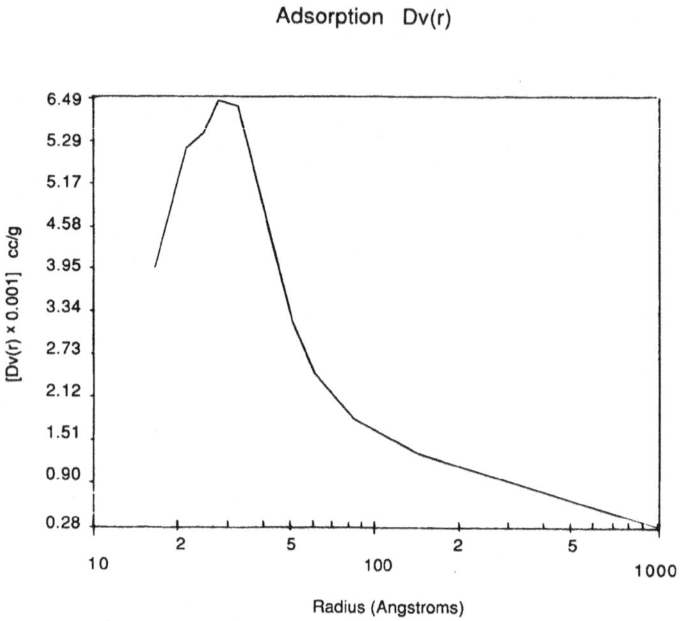

Figure 7. Mesopore size distribution for Al_{13} Keggin ion pillared titanium phosphate from the adsorption branch of the isotherm.

Two procedures which we have examined show excellent promise in this direction. Essentially the process involves the intercalation of propylamine into the four-valent layered metal phosphates to produce thin lamillae or nearly completely exfoliated layers [39,40]. Almost any large cation can then be inserted as a pillaring agent. The other procedure is to allow the polymer to form in situ, whether the layered material is exfoliated or merely propped open with amines. These procedures are illustrated in what follows.

1.2 *Chromium Pillared Phosphates.* We chose to polymerize basic chromium acetate, $Cr_3(OH)_2(CH_3COO)_7$, since the present evidence [41,42] that chromium polymers are one or at most two octahedra wide and therefore that there should be more space between pillars than with the Al_{13} Keggin ion. In our initial experiments we added a hexylamine intercalated zirconium phosphate to a large excess of the basic chromium acetate as a 5.4 M solution and refluxed for 24 h. The air dried product had a d-spacing of 18.7 Å which on calcination reduced to 13.1 Å, an increase of about 7 Å over the α-ZrP layer thickness (6.3 Å). The surface area, obtained from the isotherm shown in Figure 8, was 120 m^2/g of which 61% was attributable to micropores. These pores ranged in size from 8-14 Å in diameter based on an MP-method micropore analysis [43]. The mean radius of the mesopores was 28 Å with a range of 18-50 Å.

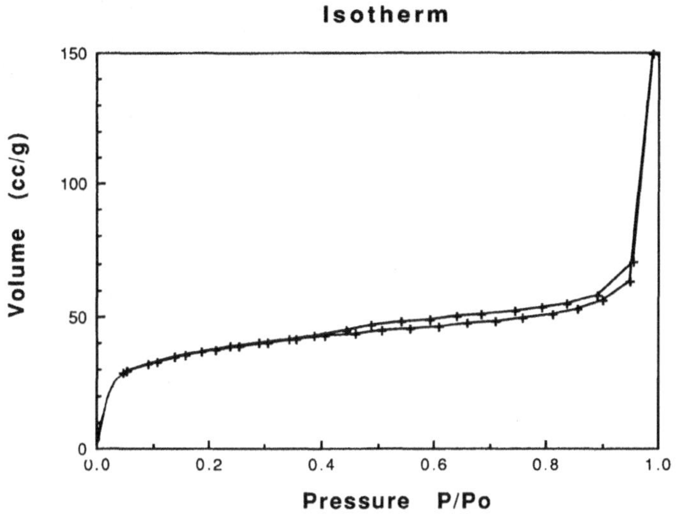

Figure 8. B.E.T. N_2 isotherm of chromium pillared zirconium phosphate (S.A. = $120 m^2/g$).

Antimony phosphate, $HSb(PO_4)_2$, was also successfully pillared by use of basic chromium acetate to obtain porous products. Attempts to prepare Al_{13} pillared materials yielded "stuffed" products where the interlamellar space was filled with the pillars [28]. This result was surprising because antimony phosphate has the same layer structure as zirconium phosphate [44], but with half the exchangeable hydrogen ions. Therefore, one would expect, and calculations show, that porous products would result using the Al_{13} polymer as pillar. However, elemental analysis revealed a high level of aluminum uptake indicative of hydroysis to a lower charged polymer. In the case of antimony phosphate we first prepared a gel by intercalation of propylamine to about half the theoretical maximum uptake. The gel was then added to a dilute basic chromium acetate solution and heated for several days. Two types of products were obtained, those with small surface areas (7-35 m^2/g) and those with larger surface areas (70-220 m^2/g), but all of which contained both meso as well as micropores. It is remarkable how similar the B.E.T. N_2 sorption isotherms are in their shape for the high and low surface area products. The isotherm for the pillared antimony phosphate with S.A. equal to 132 m^2 /g is shown in Figure 9. The higher surface area product (132

m^2/g) had a pore volume of 0.0990 cc/g of which about 48% was attributed to the micropores. The micropores were in the range of 7-12 Å in diameter (MP micropore method) and the average pore diameter for the total sorbed volume was 30 Å. The low surface area pillared phosphoantimonate (29.3 m^2/g) had a total pore volume of 0.0312 cc/g of which 0.0194 cc/g is attributable to the micropores. 90% of the micropores were in the 7-12 Å diameter range while the total average pore diameter was ~42 Å. Thus, the high and low surface area products are not too different in their textural structure. The main difference in the synthetic procedure appears to be the ratio of chromium to layered compound in the initial reactant mix. The higher this ratio in general the lower the S.A. This leads to low chromium uptake and small interlayer spacing.

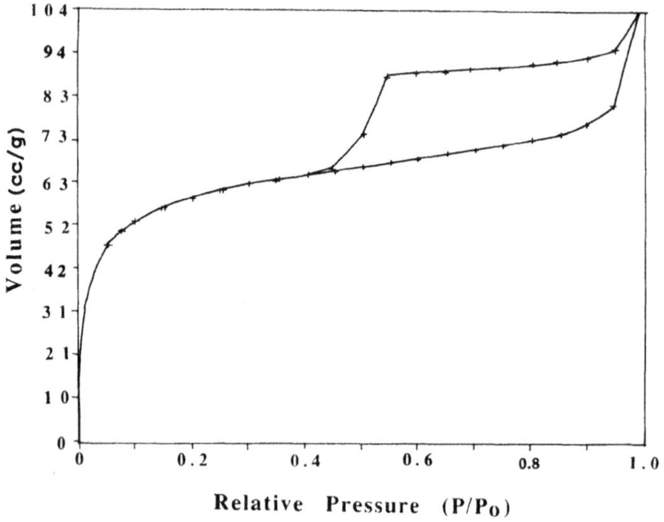

Figure 9. B.E.T. nitrogen sorption isotherm for chromium pillared $HSb(PO_4)_2$ after heating to 400 °C under nitrogen and outgassing at 150 °C.

Tomlinson and coworkers [45,46] have reported on the chromium pillaring of α-zirconium phosphate and α-tin phosphate via the propylamine method. In both cases materials of high surface area (240-386 m^2/g) were obtained. The pore structure was maintained to temperatures in the neighborhood of 500 °C. Their isotherms were similar to ours and are of the Type IV on the BDDT classification. Pore radii were calculated from the adsorption branch of the isotherms using the cylindrical pore model. Pores with radii of less than 15 Å were found to constitute 55-75% of the total. The authors emphasize that the pillared products are predominantly mesoporous with a microporous contribution. Electron micrographs show that the mesopores do not arise from a house of cards structure, but are restricted to lie within the packets of layers.

1.3 *Silica Pillared Zirconium Phosphate.* In a recent study [47] pillaring of α-zirconium phosphate was accomplished by intercalation of 3-[(triethoxy) silyl]-1-propylamine, $NH_2(CH_2)Si(OC_2H_5)_3$, followed by hydrolytic polymerization of the amine. The final product heated to 600 °C exhibited an interplanar spacing of `13.18 Å. Adsorption of n-hexane in a type I isotherm to the extent of 5.6 weight percent was taken as evidence of a porous pillared material.

We carried out a similar procedure with the trimethoxy derivative, but first prepared the zirconium phosphate-propylamine gel. The mixture of gel and silyamine was refluxed for 18 h to yield a product with a 33.2 Å interlayer spacing. On heating to 500 °C this spacing was reduced to 21.5 Å accompanied by removal of the carbon content of the pillars. The isotherm for this heat treated product is shown in Figure 10. It shows a large hysterisis loop with about 76 m^2/g of surface area gain in the range of 8-15 Å in diameter, whereas a well defined mesopore, based on desorption data was observed with a 40 Å diameter. Additional pore size analysis is in progress.

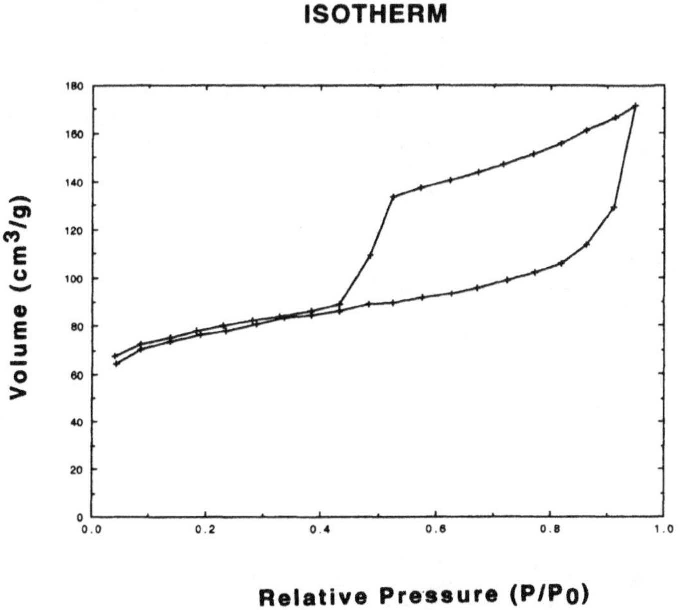

Figure 10. B.E.T. nitrogen isotherm, (both adsorption and desorption branches) for silica pillared zirconium phosphate heated at 550 °C and outgassed at 300 °C.

1.4 *Organically Pillared Zirconium Phosphonates*. We have previously reported the synthesis of compounds of the type $Zr(O_3P\text{-}C_6H_4\text{-}C_6H_4\text{-}PO_3)_x(HPO_4)_{2-2x}$ [48] based on earlier work carried out by M. B. Dines et al. [49]. These compounds, shown schematically in Figure 11 consist of inorganic layers, which are of the type $Zr(O_3P\text{-})_x(HPO_4)_{2-2x}$ and are cross-linked by the biphenyl groups. They have high surface areas in the 200-350 m^2/g range. When these compounds are precipitated from solution slowly, they exhibit a bimodal pore distribution as shown in Figure 12. Since the interlayer distance is 13.6 Å and the distance laterally between phosphorus atoms is 5.3 Å, we might expect oval shaped micropores of ~7 x 10 Å. This is in the range observed since the highest value of $\Delta V/\Delta r$ is at 6 Å. However, the larger pore is unexpected and has a diameter of ~48 Å. In fact Dines et al. [49] has shown that even the fully pillared compound (x=1) had a high surface area. We have found this to be correct although the distance between phenyl ring pillars is too small for even N_2 to penetrate into the interlayer space. The isotherm for the fully pillared compound is shown in Figure 13. The surface area of this preparation is 215 m^2/g, but there are little or no micropores present. This was shown by the almost negligible uptake of tripentylamine (8.1 Å kinetic diameter) at low pressures. Thus, the high surface area must be due to the presence of larger pores since SEM photographs showed the platelets to be in the 0.5-50μ range ruling out particle size as contributing significantly to the overall surface area. Pore size analysis based upon the adsorption brance of the isotherm is shown in Figure 14. It is seen that there exists a range of mesopores centered about an average pore radius of 24 Å. We have initiated a study, via electron microscopy, lattice imaging, and molecule sorption, to elucidate the nature and origin of these mesopores.

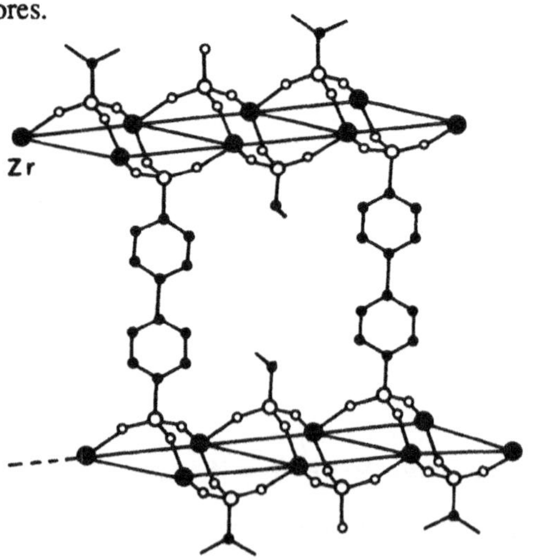

Figure 11. Schematic depiction of mixed derivative zirconium biphenyldiphosphonate phosphate, $Zr(O_3PC_{12}H_8PO_3)_{0.5}(O_3POH)$.

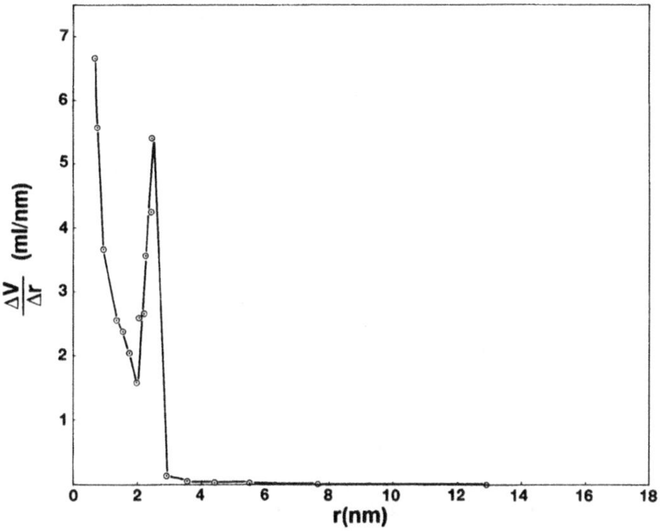

Figure 12. Bimodal pore distribution observed for $Zr(O_3PC_{12}H_8PO_3)(HPO_4)$ as determined from the B.E.T. nitrogen sorption isotherm.

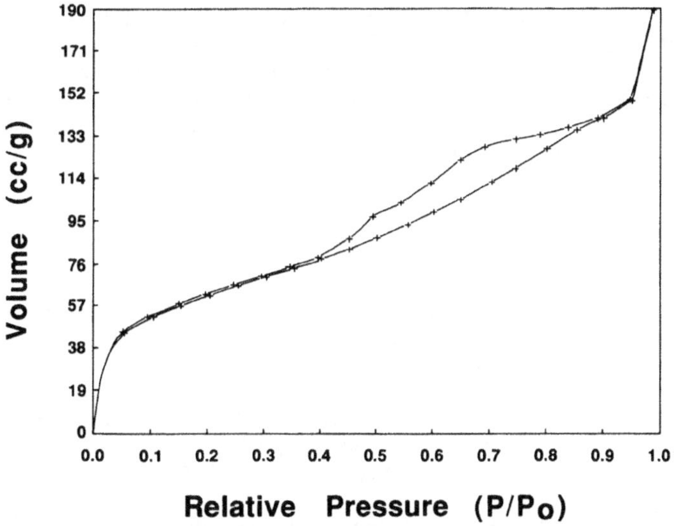

Figure 13. B.E.T. nitrogen sorption isotherm for the fully pillared zirconium biphenyldiphosphonate, $Zr(O_3PC_{12}H_8PO_3)$.

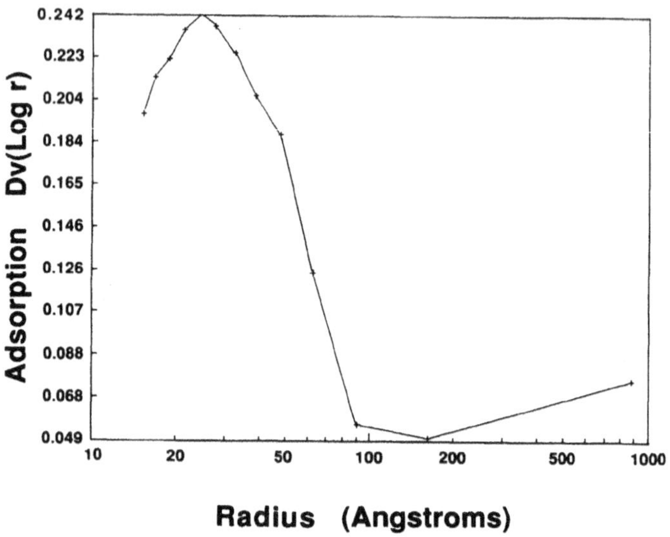

Figure 14. Distribution of mesopores obtained from the isotherm shown in figure 13.

Another group of compounds we have prepared contain pillars of three phenyl rings. The interlayer spacing is 18.2 Å with surface areas of 250-400 m^2/g. Because the triphenyl diphosphonic acid cannot be dissolved, the precipitation of the pillared compounds was rapid and a range of pore sizes from ~10 Å to 50 Å in size was obtained. These organically pillared compounds are interesting from the point of view that better control of the synthetic process may allow us to design materials with specific pore structure. One reaction we have carried out is the sulfonation of the phenyl rings to produce strong acid catalysts. We will report on these compounds in a future publication.

1.5 Conclusions

A variety of clay minerals and layered phosphates have now been pillared. All of these compounds contain mesopores. These large pores may arise in clays by distortion of the layers as shown in Figure 6, but also by a house of cards or edge to face structure. However, this latter mechanism probably is not operative in the phosphate systems. Electron micrographs point towards tightly packed layers with mesopores formed by imperfect joining of packets of layers. Further study of the pore structures of these materials is indicated to establish their type, source and possible control.

Acknowledgment: The author acknowledges with thanks support of this work by the National Science Foundation under grant number DMR-9107715 and the State of Texas Advanced Technology Program.

1.6 References

1. Armor, J. N. (1991) 'New technology commercialized in the USA during the 1980's', Appl. Catal. 78, 141-173.
2. Davis, M. E. (1991) 'Zeolites and Molecular Sieves: Not just ordinary catalysts', Ind. Eng. Chem. Res. 30, 1675-1683.
3. Jacobs, P. A. (1977) Carboniogenic Activity of Zeolites, Elsevier Sci. Publ. Co., New York.
4. Chen, N. Y. (1989) Shape Selective Catalysis in Industrial Applications, M. Dekker, New York.
5. Wilson, S. T., Lok, B. M., Messina, C.-A., Connan, T. R., and Flanigen, E. M. (1983) 'Aluminophosphate molecular sieves: a new class of microporous crystalline inorganic solids', in G. D. Stucky and F. G. Dwyer (eds.), Intrazeolite Chemistry, ACS Symp. Series 218, Wash., D.C., pp. 79-106.
6. Harrison, T. A., Gier, T. E., Moran, K. L., Nicol, J. M., Eckert, H. E., and Stucky, G. D. (1991) 'Structures and properties of new zeolite X-type zincophosphate and beryllophosphate molecular sieves', Chem. Mater. 3, 27.
7. Newsam, J. M., Jacobson, A. J., and Vaughan, D. E. W. (1986) 'Time-of-Flight Powder Neutron Diffraction studies of dehydrated synthetic gallosilicates Zeolite X, XY and Y', J. Phys. Chem. 90, 6858-6864.
8. Kessler, H. (1991) 'Directing effect of fluoride in the synthesis of molecular sieves with new characteristics and of the first twenty-membered ring microporous solid', in R. L. Bedard, T. Bein, M. E. Davis, J. Garces, V. A. Maroni, and G. D. Stucky (eds.) Synthesis/Characterization and Novel Applications of Molecular Sieve Materials, Mat. Res. Soc., Pittsburg, pp. 47-55.
9. Burch, R. (1990) 'Catalysis Today, Special Issue on Pillared Clays', Elsevier Sci. Publ. B.V., Amsterdam.
10. Clearfield, A., Kuchenmeister, M., Wang, J.-D., and Wade, K. (1991) 'New Porous Materials from Layered Compounds', in P. A. Jacobs, N. I. Jaeger, L. Kubelková and B. Wichterlova (eds.), Zeolite Chemistry and Catalysis, Elsevier, Amsterdam, pp. 485-497.
11. Rudolf, P. R. and Crowder, C. E. (1990) 'Structure refinement in the very large pore molecular sieve', Zeolites 10, 163-166.
12. McCusker, L. B., Baerlocher, Ch., John, E., and Bülow, M. (1991) 'The triple helix inside the large-pore aluminophosphate molecular sieve VPI-5', Zeolites 11, 308-313.
13. Estermann, M., McCusker, L. B., Baerlocher, C., Merrouche, A., and Kessler, H. (1991) 'A synthetic gallophosphate molecular sieve with a 2θ-tetrahedral-atom pore opening', Nature 352, 320-323.
14. Parton, R. F., Utterhoeven, L., and Jacobs, P. A. (1991) 'Iron phthalocyanines encaged in zeolite Y and VPI-5 molecular sieves as catalysts for the oxy-functionalization of n-alkanes', Studies Surf. Sci. Cata. 59, 395-403.
15. Martens, J. A., Feijen, E., Lievens, J. L., Grobet, P. J., and Jacobs, P. A. (1991) 'Structure and effective pore size of the dehydrated forms of the molecular sieve VPI-5', J. Phys. Chem. 95, 10025-10031.

16. Sorby, K., Szostak, R., Ulan, J. G., and Gronsky, R. (1990) 'Small pore VPI-5: pore blockage due to thermally induced stacking disorders', Catal. Lett. 6, 209-214.
17. Poojary, D. M., Perez, J. O., and Clearfield, A. (1992) 'The crystal structures of dehydrated VPI-5 and H1 aluminum phosphates from X-ray powder data', J. Phys. Chem., in press.
18. Moore, P. B. and Shen, J. (1983) 'An X-ray structural study of cacoxenite, a mineral phosphate', Nature 306, 356-358.
19. Smith, J. V. and Dytrych, W. J. (1984) 'Nets and channels of unlimited diameter', Nature 309, 607-608.
20. Addison, S. W., Cartlidge, S., Harding, D. A., and McElhineny, G. (1988) 'Role of zeolite non-framework aluminum in catalytic cracking', Appl. Catal. 45, 307-323.
21. Sayari, A., Crusson, E., and Kaliaguine, S. (1991) 'External surface areas of H-ZSM-5 zeolites', Langmuir 7, 314-317.
22. Lohse, V. and Mildebrath, M. (1981) 'Dealuminated Y-type molecular sieves. Porosity of dealuminated molecular sieves', Z. Anorg. Allg. Chem. 476, 126-135.
23. Vaughan, D. E. W. and Lussier, R. J. (1980) 'Preparation of molecular sieves based on pillared interlayered clays (PILC)', in L. V. Rees (ed.), Fifth International Conference on Zeolites, Proceedings, Heyden, London, pp. 94-101.
24. Brindley, G. W. and Semples, R. E. (1977) 'Preparation and properties of some hydroxy-aluminum beidellites', Clay Minerals 12, 229-37.
25. Yamanaka, S. and Brindley, G. W. (1979) 'A study of hydroxy-chromium montmorillonites and the form of the hydroxy chromium polymers, Am. Mineral, 26, 830-835.
26. Johansson, G. (1960) 'On the crystal structures of a basic aluminum sulfate and the corresponding selenate', Acta Chem. Scand. 14, 769-771.
27. Clearfield, A. and Vaughan, P. A. (1956) 'The crystal structure of zirconyl chloride octahydrate and zirconyl bromide octahydrate', Acta Cryst. 9, 555.
28. Clearfield, A., Kuchenmeister, M. E., Wade, K. Cahill, R., and Sylvester, P. (1992) 'Pillaring of layered inorganic compounds: fundamentals', in ACS Symposium Series, Washington, D.C., in press.
29. Stul, M. S. and Mortier, W. J. (1974) 'Heterogeneity in the charge density in montmorillonites', Clays and Clay Minerals 22, 391-396.
30. Peigneur, P., Moes, A., and Cremers, A. (1975) 'Heterogeneity of charge density distribution in montmorillonites as inferred from cobalt adsorption', Clays and Clay Minerals 23, 71-75.
31. Pinnavaia, T. J., (1984) 'Preparation and properties of pillared and delaminated clay catalysts', in Heterogeneous Catalysis, B. L. Shapiro (ed.) Texas A&M Univ. Press, College Station, Texas.
32. Sterte, J. (1988) 'Hydrothermal treatment of hydroxocation precursor solutions', in R. Burch (ed.), Catalysis Today, Elsevier Sci. Publ., Amsterdam, pp. 219-231.

33. Tzou, M. S. and Pinnavaia, T. J. (1988) 'Chromia pillared clays', in R. Burch (ed.), Catalysis Today, Elsevier Sci. Publ., Amsterdam, pp. 243-259.
34. Moini, A. and Pinnavaia, T. J. (1988) 'A new strategy for the synthesis of pillared clays with super galleries: direct intercalation of metal oxide sols (DiMOS)', Solid State Ionics 26, 119-123.
35. Yamanaka, S., Nishihara, T., Hattori, M. and Suzuki, Y. (1987) 'Preparation and properties of titania pillared clay', Mater. Chem. Phys. 17, 87-101.
36. Clearfield, A. (1988) 'The role of ion exchange in solid state chemistry', Chem. Rev. 88, 125-148.
37. Clearfield, A. and Tindwa, R. M. (1979) 'Exchange of large cations and charged complexes with amine intercalates of zirconium phosphate', Inorg. Nucl. Chem. Lett. 15, 251-254.
38. Clearfield, A. and Smith, G. D. (1969) 'The crystallography and structure of α-zirconium bis(monohydrogen orthophosphate) monohydrate', Inorg. Chem. 8, 431-436.
39. Clearfield, A. and Tindwa, R. M. (1979) 'On the mechanism of ion exchange in zirconium phosphates. XXI. Intercalation of amines by α-zirconium phosphate', J. Inorg. Nucl. Chem. 41, 871-878.
40. Alberti, G., Casciola, M., and Costantino, U. (1985) 'Inorganic ion-exchange pellicles obtained by delamination of α-zirconium phosphate crystals', J. Colloid Interf. Sci. 107, 256-263.
41. Stunzi, H. and Marty, W. (1983) 'Early stages of the hydrolysis of chromium (III) in aqueous solution. l. Characterization of a tetrameric species', Inorg. Chem. 22, 2145-2150.
42. Bornholdt, K. Corker, J. M. Evans, J. and Rummey, J. M. (1991) 'EXAFS studies of the formation of chromia pillared clay catalysts', Inorg. Chem. 30, 2-4.
43. Mikhail, R.Sh., Brunauer, S. and Bodor, E. E. (1968) 'Investigations of a complete pore structure analysis 1. Analysis of micropores', J. Colloid Interface Sci. 26, 45-53.
44. Piffard, Y., Verbaere, A., Oyetola, S., Peniard-Courant, S. and Tournoux, M. (1989) 'The layered phosphatoantimonic acid $HSb(PO_4)_2 \cdot XH_2O$: synthesis, structure, thermal behavior and ion exchange properties', Eur.J. Solid State Inorg. Chem. 26, 113-127.
45. Maireles-Torres, P., Olivera-Pastor, P., Rodriguez-Castellon, E., Jimenez-Lopez, A., and Tomlinson, A. A. G. (1991) 'Porous chromia-pillared α-zirconium phosphate materials prepared via colloid methods', J. Mater. Chem. 1, 739-746.
46. Maireles-Torres, P., Olivera-Pastor, P., Rodriguez-Castellon, E., Jimenez-Lopez, A., and Tomlinson, A. A. G. (1991) 'Porous chromia-pillared α-tin phosphate materials', J. Solid State Chem. 94, 368-380
47. Li, L.-S., Liu, X.-S., Ge, Y., Li, L., and Klinowski, J. (1991) 'Intercalation and pillaring of zirconium (monohydrogenphosphate) with $NH_2(CH_2)Si(OC_2H_5)_3$, J. Phys. Chem. 95, 5910-5914.

48. Clearfield, A. (1986) 'New Approaches to the design of materials via preparative inorganic chemistry ', in A. Clearfield and D. A. Cocke (eds.), Design of New Materials, Plenum Press, New York, pp. 121-134.
49. Dines, M. B., Di Giacomo, P. M., Callahan, K. P., Griffith, P. C., Lane, R. H. and Cooksey, R. E. (1983) 'Derivatized Layered M(IV) phosphonate', in J. Miller (ed.), Chemically Modified Surfaces in Catalysis and Electrocatalysis, ACS Symp. Series No. 192, Wash., D.C., 223-253.

PROBLEMS RELATED TO THE PILLARING OF α- AND γ-LAYERED M(IV) PHOSPHATES WITH LARGE CATIONS OR WITH COVALENTLY BONDED DIPHOSPHONATES GROUPS

G. ALBERTI
Chemical Department, University of Perugia, via Elce di Sotto 8,
06100 Perugia, Italy

ABSTRACT. The success obtained in the pillaring of clays has stimulated interest in the preparation of pillared M(IV) phosphates, but controversial results have been obtained up to date in the different laboratories. The main obstacle to a good pillaring of these materials seems to be related to their high density of fixed charges (about 10 times higher than in clays), which makes the pillars too crowded in the interlayer region.

To decrease the density of fixed charges, two different strategies (elimination of some P-OH groups by hydrolytic attack or their partial replacement with neutral P-OR groups) are reported and discussed.

Recent advances in "Molecularly Engineered Layered Structures" (MELS), based on α-Zr phosphite-diphosphonates and γ-Zr phosphate-diphosphonates, are also reported.

1. Introduction

Over the last three decades there has been developed a new class of layered materials having a general formula $M(IV)(RXO_3)_2.nS$, where M(IV) = Zr, Ti, Sn, Pb, Ge, Ce, Th, etc; X = P, As, Sb, ; R = H, OH, organic group; S = molecule intercalated in the interlayer region [1].

The most widely investigated member is zirconium bis(monohydrogen phosphate) monohydrate, $Zr(HPO_4)_2H_2O$; our attention will therefore be essentially focused on this compound and on its organic derivatives. It must however be pointed out that many of the considerations made regarding the properties of zirconium phosphate (ZrP) can easily be adapted to the other members of the class, taking into account only the slightly different distance between the M(IV) atoms in the layer (due to their different atomic radius) and the different tendency to hydrolytic attack (i.e. for M(IV) phosphates (MP) the stability to hydrolysis decreases in the order: ZrP >>TiP > SnP >PbP >> GeP).

The structure of $Zr(HPO_4)_2.H_2O$ was first solved by Clearfield as long as 1964 [2].

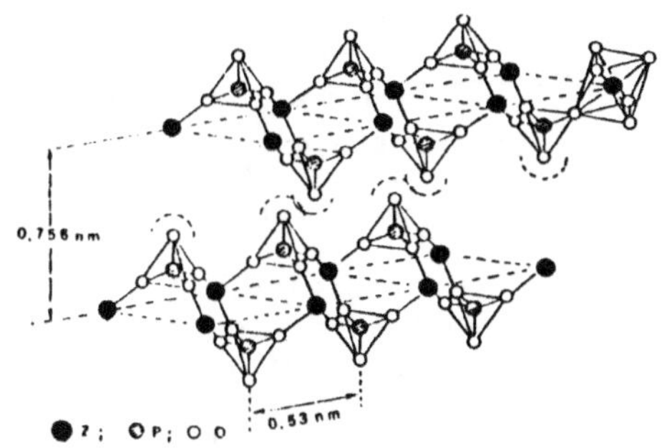

Fig. 1 - A schematic representation of the structure of two layers α-$Zr(HPO_4)_2.H_2O$. The water and hydrogen in the interlayer region are not reported. The dashed lines indicate the Van der Waals size of interlayer P-O$^-$ groups.

C.A.C. Sequeira and M.J. Hudson (eds.),Multifunctional Mesoporous Inorganic Solids, 179–190.

This structure was of fundamental importance for the explanation of many properties of the layered compounds of this class; therefore, although already shown and discussed in many other previous reviews, [3-5] it seemed useful to show here too a schematic picture of this structure (Fig.1).

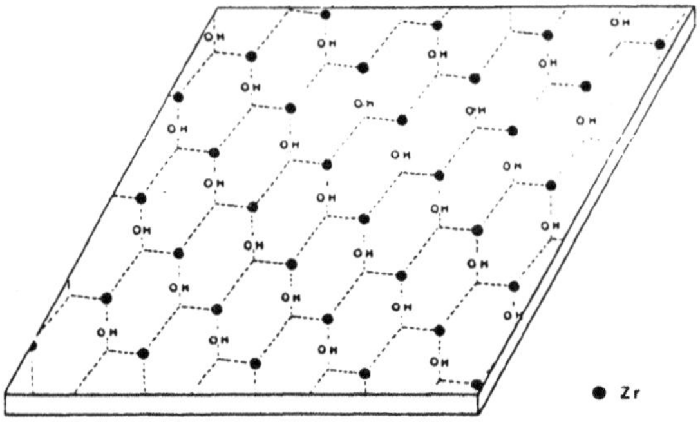

Fig. 2 - Arrangement of P-OH groups on one side of the α -layer.

Each layer can be considered as a planar macromolecule in which acid P-OH groups are present. These groups are arranged in the surface of the layers at alternative vertices of a hexagonal array structure with a distance of 5.3 Å between them (Fig. 2). It can easily be calculated that the number of acid P-OH groups on each side is 4.17×10^{14} per cm^2 and that the corresponding "free area" associated with each group is 24 $Å^2$.

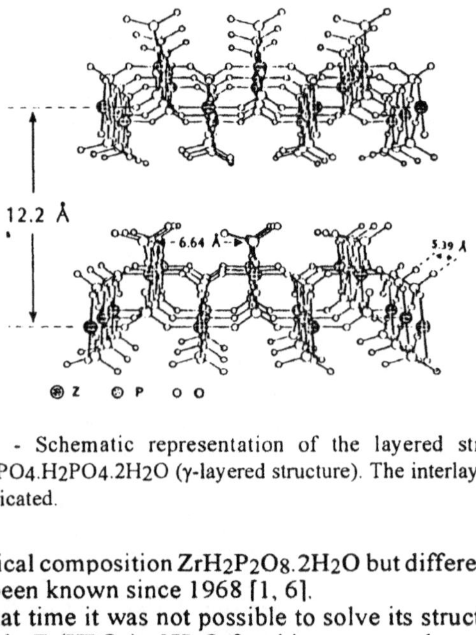

Fig. 3 - Schematic representation of the layered structure of $Zr(IV)PO_4.H_2PO_4.2H_2O$ (γ-layered structure). The interlayer water is not indicated.

The protons of the acid groups (not indicated in the figure for sake of simplicity) can be replaced by other cations and the ion-exchange and intercalation properties of α-layered phosphates of tetravalent metals have also been intensely studied for practical uses [1, 3-5].

It must be pointed out that another zirconium phosphate, having the same chemical composition $ZrH_2P_2O_8.2H_2O$ but different X-ray powder diffraction pattern, has also been known since 1968 [1, 6].

At that time it was not possible to solve its structure; however Clearfield, assuming the formula $Zr(HPO_4)_2.2H_2O$ for this compound, supposed that the difference in the X-ray patterns of the two zirconium phosphates was due only to the different stacking of the layers. Therefore he distinguished the two compounds by the greek letters α and γ (the letter β was assigned to the anhydrous form of the γ-phase). Only recently have we known that these assumptions are not true, the compound being zirconium phosphate dihydrogenphosphate dihydrate, $ZrPO_4H_2PO_4 .2H_2O$ [7, 8]. It is therefore not correct any longer to use the α and γ prefixes to distinguish the two different layered zirconium phosphates. However, to avoid confusion with the previous literature, we still retain these prefixes here to indicate in a concise form the structure of the layers. A schematic picture of two γ-layers is shown in fig.3. Due to their different structure, the possibilities of preparing micro-

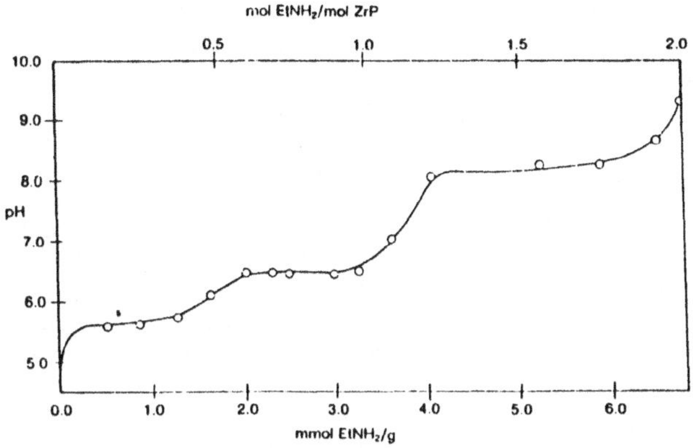

Fig. 4 - Typical monoamine uptake curve of α-$Zr(HPO4)2.H2O$.

porous and/or meso-porous solids from α and γ zirconium phosphate will be examined separately.

2. Pillaring of α-layered M(IV) phosphates

The success obtained in the pillaring of clays [9] has stimulated interest in the preparation of pillared M(IV) phosphates, but controversial results have been obtained in different laboratories [10].

To provide a contribution to these controversies, let us examine the problem of the pillaring of a layered structure from a general point of view.

We start with the "space filling postulate" which, in our particular case , can be adapted as follows:

"In the absence of strong interactions, the most probable host-guest structures will be those in which the most economical use is made of space".

As an example of the application of this postulate to layered α-ZrP, Fig.4 shows the uptake curve of a monoalkylamine as a function of the pH in which three well-defined plateaux can be noted. Since the uptake curve was performed at constant temperature and pressure, it may be expected that along each plateau (where the degree of freedom of the system = 0), there must be the coexistence of two immiscible solid phases [1,3].

In particular, X-ray diffraction analyses of the samples confirmed these expectations. By indicating with the letter a the initial α-ZrP phase, and with b, c, and d the other phases at increasing amine uptake, the three discontinuous phase transitions, corresponding to the three plateaux of Fig. 4, can be simply indicated as: a--->b; b--->c; c--->d.

The arrangement of the alkyl chain in these phases is shown schematically in Fig. 5, where it can be noted that in all the three cases, no free space is present in the interlayer region. This means, for example, that a single phase at 25% amine uptake with free

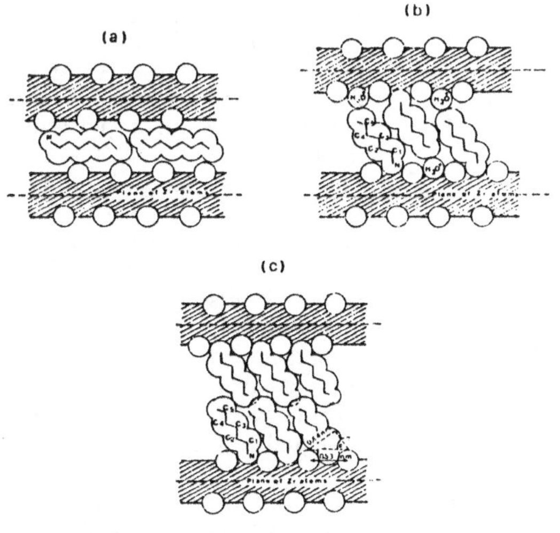

Fig. 5 - The most common orientation of alkylmono-amines in the interlayer region of α-$Zr(HPO4)2$ at various stages of the intercalation: a =10%, b = 50%, c = 100%.

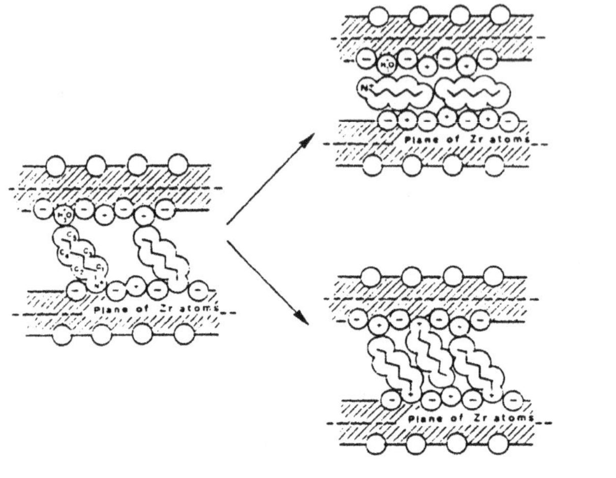

Fig. 6 - A phase with partial conversion in alkylmonoamine and with interlayer porosity tends to segregate in two stable phases without interlayer porosity.

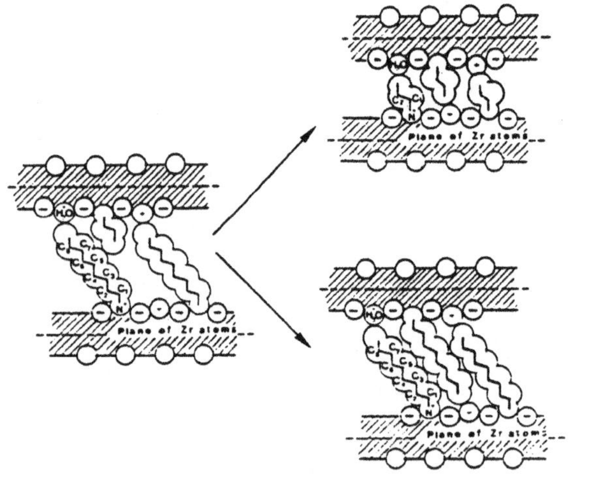

Fig. 7 - A phase containing two alkylamines of different lenght and interlayer porosity tends to segregate two stable phases without interlayer porosity.

space in the interlayer region, as shown in Fig. 6, is not formed. For the "space filling postulate" this phase should be a metastable phase; a more economical use is made of space if this phase is segregated in a mixture of stable phases b and c as shown schematically in Fig. 6. Also porous phases containing two amines of different length are not stable and, if formed, they have tendency to segregate in two stable non porous phases (see Fig. 7).

It must be pointed out however that the "space filling postulate" is not valid in the presence of strong interactions. Thus if the guest species interact strongly with the host (e.g. by covalent bonds or coulombic interactions with fixed charged sites present in covalent structures) the formation of free spaces in the interlayer region becomes possible.In the formation of porous pillared structures the interactions of large polivalent counter-ions with the fixed charges of the layers are very important; in this context the distance between the fixed charges takes on a particular importance.

Let us first examine a simplified model in which: a) the fixed charges are regularly distributed in the surfaces of the layers (the distance between adjacent fixed charges, $d_{f.c.}$, along the axis a and b is the same); b) the free distance between adjacent layers, d_f, is assumed to be equal to the diameter of the counter-ions, that is $d_f = d_i$.

Such situation along one axis is schematically shown in Fig. 8. It is clear from this figure that the lateral distance, d_l, beween adjacent counter-ions will depend on the distance between the fixed charges. A particular simple case is that in which the distance between the fixed charges has a value for which d_f becomes equal to d_l. From Fig. 8 it is easy to see that, for monovalent counter-ions, this occurs when : $d_{f.c.} = 4d_i = 4d_f$. From this simple

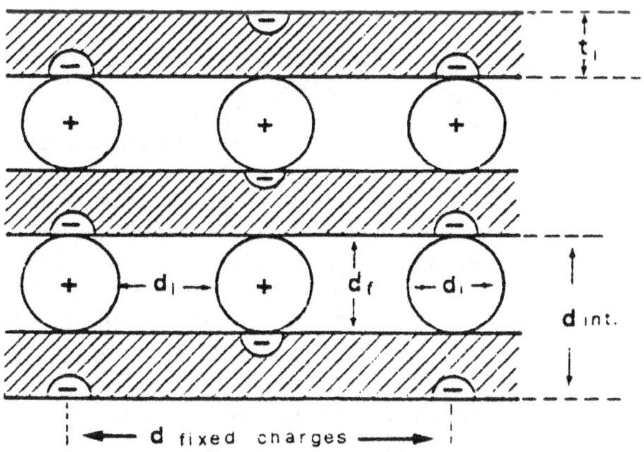

Fig. 8 - A simple model for the pillaring of a layered compound with a cation (see text). It is shown the situation in which $d_i = d_f = d_l$. d_i = ion diameter; d_f = free interlayer distance; d_e = lateral distance; d f.c. = distance between two adjacent fixed charges; d_{int} = interlayer distance; t_l = layer thickness.

equation we can, for example, calculate that if $d_{f.c}$ is 16 Å, the best d_i value for a monovalent pillaring cation is 4 Å.

A maximum free interlayer volume of 4x4x4 Å^3 can be obtained. In fact, if cations with larger diameters are used, higher d_f but lower d_l values are obtained. It easy to see directly in fig.8 that d_l tends to zero as d_i tends to 1/2 $d_{f.c}$.

We can now examine the pillaring of α-ZrP in which $d_{f.c}$ is 5.3 Å and the surface fixed charges are arranged as shown in Fig. 2. In this case we have: d_i = 5.3 /4 = 1.32 Å. It is clear that a microporosity of 1.3 Å^3 is too low for practical applications. The situation is slightly better with polyvalent cations.

The maximum d_l values are: 2.64 Å for divalent cations; 5.3 Å (in one direction) for tetravalent cations. Although the model used is very simple, we believe that no more encouraging data can be obtained with a more refinite model. E.g., taking into account the distribution of the fixed charges in α-ZrP (Fig.2), d_l values greater than 5.3 Å, in two directions, are not expected even with an octovalent pillaring cation.

Keggin Al_{13} cation, which has the shape of a prolate spheroid with a long axis of about 9.5 Å [11], seems to have a right size to act as pillaring agent for montmorillonite ($d_{f.c}$ ~ 15 Å) but is too large for α-M(IV) phosphates.

At least two different strategies can be applied to increase the distance betwen the fixed charges in α-layered phosphates:

1) hydrolytic replacement of some phosphate groups with OH groups;
2) replacement of a large part of phosphate groups with neutral phosphonate groups.

We will examine these two strategies separately.

2.1 REDUCTION OF FIXED CHARGE DENSITY IN α -M(IV) PHOSPHATES BY HYDROLYSIS OF $H_2PO_4^{2-}$ GROUPS.

The monohydrogen phosphate groups present in the surface of α-ZrP can be hydrolysed easily, especially in an alkaline medium and/or at the boiling temperature.

$$Zr(HPO_4)_2.H_2O + nH_2O + 2xOH^- \rightleftharpoons Zr(HPO_4)_{2-x}(OH)_{2x}.(n+1)H_2O + xHPO_4^{2-}$$

The hydrolytic reaction is in practice restricted only to surface groups. The interlayer HPO_4 groups are in fact protected from alkaline attack (it is expected that OH^- cannot enter the interlayer region because of Donnan exclusion of anions in a cation exchanger); furthermore, even if hydrolysed, the phosphate groups could not diffuse out, for steric reasons, from the interlayer region to the external solution.

To obtain a controlled hydrolysis of the interlayer HPO_4 groups, the crystals must first be delaminated so that the hydrolytic attack can occur at the surface of single lamellae,

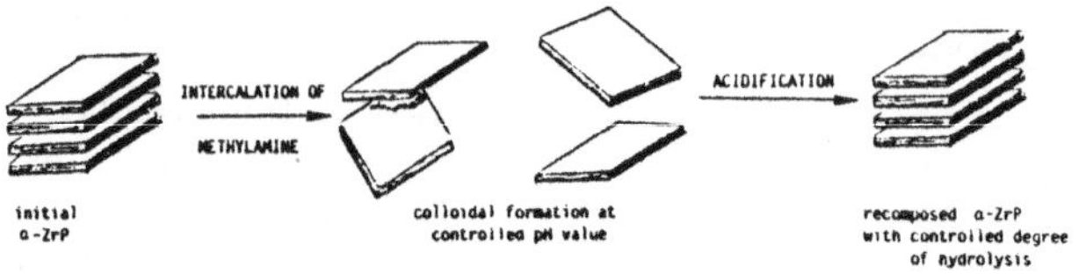

Fig. 9 - A schematic picture of the controlled hydrolysis of - $Zr(HPO_4)_2.H_2O$ by intercalation with methylamine.

and then recomposed again. Since the delamination with the propylamine (50% of conversion) takes place at pH ~5 [12], we found it convenient to carry out the delamination by the intercalation of methylamine (100% of conversion). In this case not only is the intercalation completed in an alkaline medium, but it is also possible to add an excess of amine without provoking the flocculation of the α-ZrP layers [13]. Thus the pH of the solution, and hence the degree of the hydrolysis of α-ZrP, can be fixed at the value desired. The precipitation of the hydrolysed ZrP is then obtained by acidification of the solution. The process is schematically shown in Fig.9. Note that the cycle can be repeated again if very high degrees of hydrolysis are required. In any case, it must be pointed out that for values higher than 30-35% the stability of the α-layer is seriously compromised. Thus, the reduction of the fixed charge density obtainable via the hydrolytic procedure cannot be very high.

The partial success obtained in the pillaring of α-TiP [14] and α-SnP [15] could be due to the fact that these materials are more hydrolysable than α-ZrP and therefore already undergo a considerable hydrolytic attack during the pillaring procedure, especially if this latter is performed after delamination with propylamine.

2.2 REDUCTION OF THE FIXED CHARGE DENSITY IN α -M(IV) PHOSPHATES BY REPLACEMENT OF SOME PHOSPHATE GROUPS WITH NEUTRAL PHOSPHONATE GROUPS.

It was found in our laboratory that M(IV) phosphonates with a layered structure similar to that of Fig. 1 can also be obtained [16].

As an example, Fig. 10 shows the schematic structure of layered zirconium benzene phosphonate. Many other M(IV) phosphonates of general formula M(IV) $(RPO_3)_2.nS$, where R is an organic radical, have already been prepared and investigated [5] (see Fig. 11). The number of the possible compounds was further increased when it was discovered that it is possible to prepare mixed compounds with the general formula $M(IV)(O_3PR)_{2-x}(O_3PR')_x$ nS where R and R' can be an inorganic and/or an organic group. In many of these compounds there was observed the presence of asymmetrical layers with one side richer in O_3P-R groups and the other in O_3P-R' [17].

It is of interest to note that the density of the fixed charges (P-O^- groups) in α-zirconium phosphate is about 10 times higher than that of a clay (e.g. montmorillonite).
By replacing about 90% of O_3P-OH groups with O_3P-R group, the concentration of the fixed charges is reduced to a value similar to that of a clay (Fig. 12).

These compounds can therefore be considered as "clay-like materials"; however, different from clays, their hydrophobicity and the concentration of the fixed charges can easily be modified by changing the nature and the number of the O_3P-R groups. Researches in the preparation and properties of "clay-like" zirconium phosphonate-phosphates, as well as their pillaring with large polyvalent cations, are in progress in our laboratory.

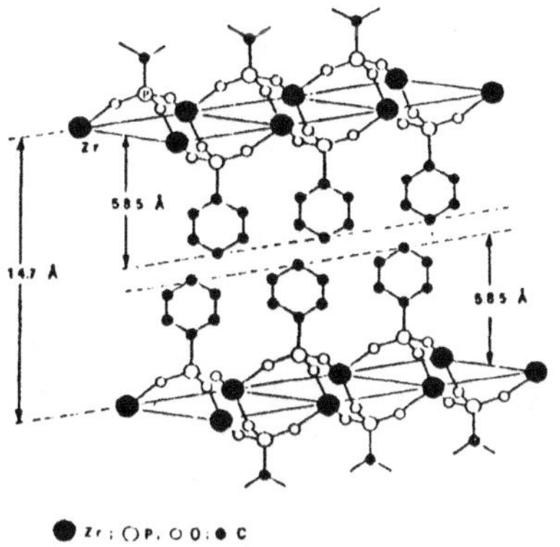

Fig. 10 - Schematic drawing of the layered structure of α-Zr(C6H5PO3)2.

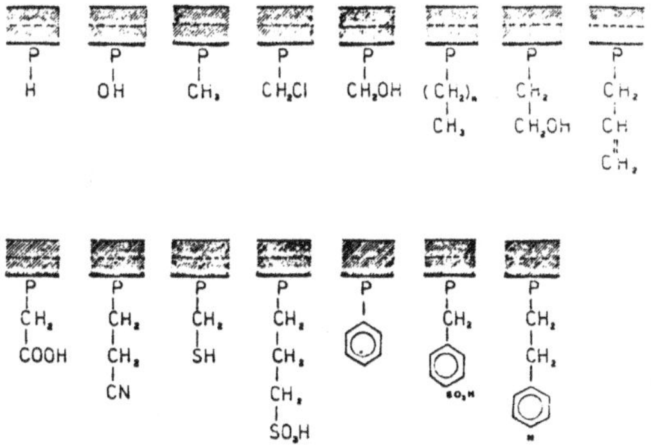

Fig. 11 - Some Zr phosphonates reported in the literature.

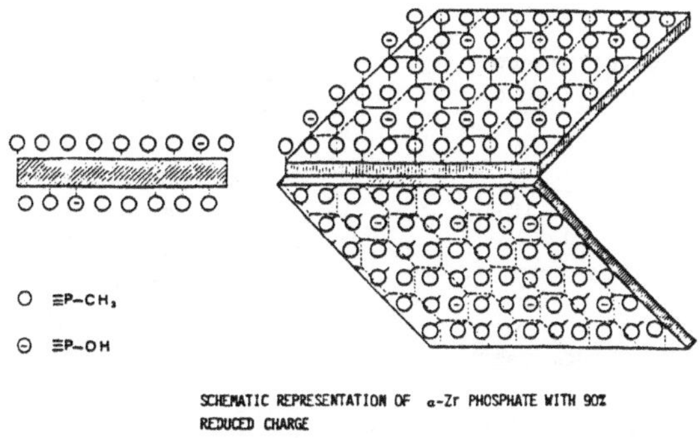

Fig.12 - Schematic representation of α-Zr phosphate-phosphonate with 90% reduced charge respect to the α-Zr(HPO4)2.H2O. This compound can be considered a "clay like" material.

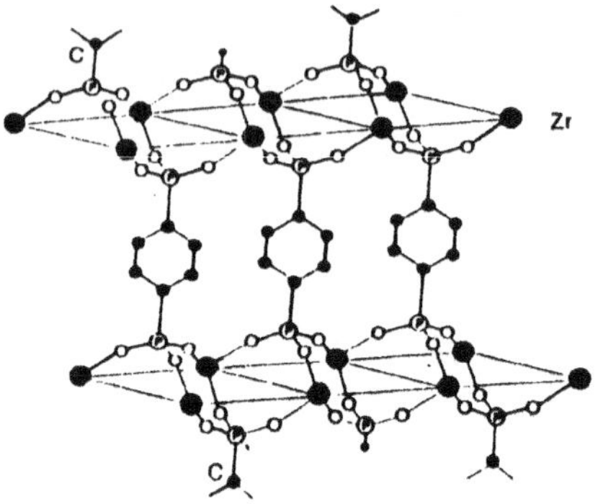

Fig. 13 - Idealized crystal structure of a pillared zirconium diphosphonate, $[Zr(O_3PC_6H_4PO_3)]_n$.

3. Molecularly Engineered Layered Stuctures

Another important advance in the chemistry of layered organo-phosphonates was the preparation of M(IV) diphosphonates of the general formula $[M(IV)R(PO_3)_2]_n$ [18].
As shown schematically in Fig. 13, the layer structure of these compounds is similar to that of the phosphonates already shown in Fig. 3, but with the adjacent layer covalently joined by diphosphonate groups (pillared compounds).

A large number of Zr phosphates and phosphonates have been prepared (see Fig. 14) and many attempts at preparing zeolite-like materials, that is ones containing microcavities in the interlayer region, have been made.

Unfortunately, taking into account that the distance between adjacent pillars, as seen before, is only 5.3 Å, while the average Van der Waals diameter for an aromatic ring is about 3.7 Å, it is possible to deduce that the interlayer region, as shown schematically in Fig. 15, is almost completely filled by the pillars. It has been claimed [18,19] that the formation of a microporosity in the interlayer region was however possible if some O_3P-R-PO_3 diphosphonates were replaced by small O_3P-H groups (see Fig. 16).

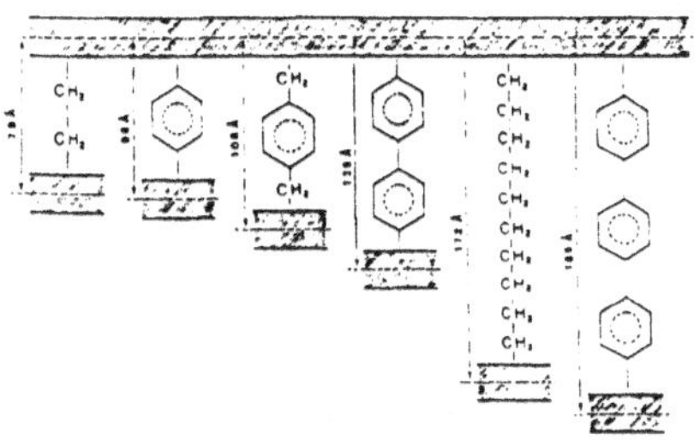

Fig. 14 - Interlayer distance of some zirconium diphosphonates as a function of the pillaring R groups.

Great emphasis was given to these materials [20] that were called with the acronym MELS (Molecularly Engineered Layered Structures).

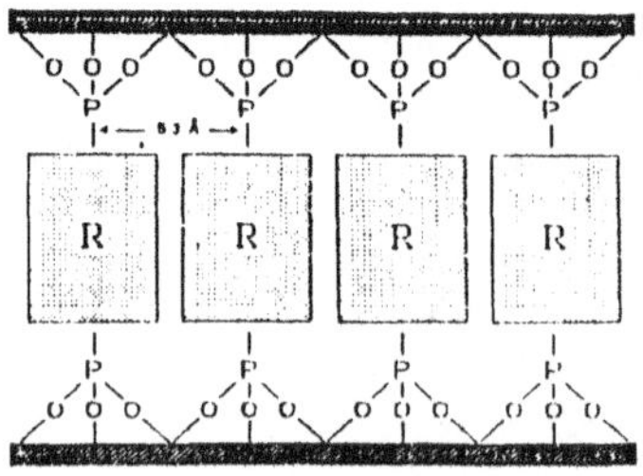

Fig. 15 - A representation of a pillared Zr diphosphate in which the Van der Waals diameter of the aromatic ring is shown. Note that the pillars are crowded so that microporosity in the interlayer region is not present.

The MELS up to now reported however possess a low degree of crystallinity and a low number of microcavities in their interlayer region. Their catalytic properties seem to be due in large part to groups lying in the external surface.

Very recently, it was found in our laboratory that MELS with a high degree of crystallinity and with microcavities of uniform size (and thus useful for shape selective catalysis) can indeed be obtained using particular diphosphonate pillars [21].

It was furthermore found that even Zr phosphite-diphosphonates with a very narrow distribution of intracrystalline

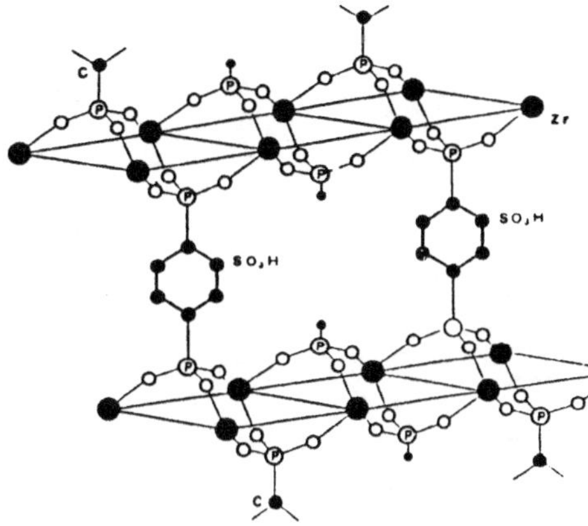

Fig. 16 - Idealized representation of a cavity of a pillared zirconium diphosphonate in which 50% of the diphosphonate groups have been replaced bysmall phosphite groups. Acid functional groups for catalysis applications (e.g. $-SO_3H$) can be inserted in the pillars.

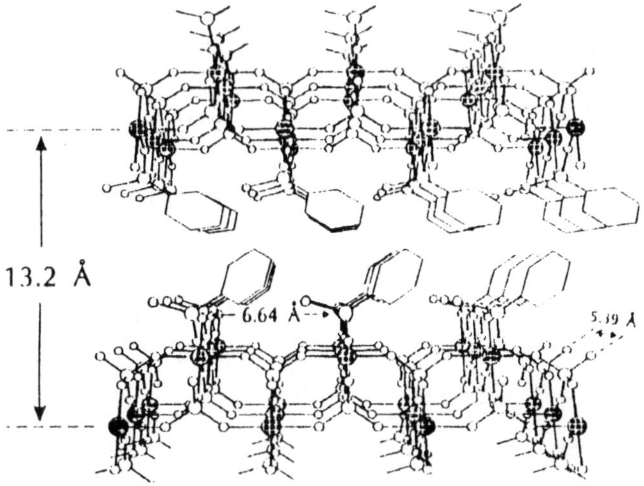

Fig. 17 - Schematic representation of an organic derivative of the γ-phase obtained by topotactic reaction.

mesopores can be prepared [22]. By a suitable choice of the reaction media and of reagents concentrations it seems possible that compounds with the mesoporosity desired may be obtained.

A considerable advance has also recently been obtained in the so-called organic derivatives of γ-M(IV)) phosphates.

The interlayer dihydrogen phosphate groups of γ-zirconium phosphate can be easily exchanged by topotactic reaction with other groups [23-24]. It was therefore possible to prepare many organic derivatives of γ-zirconium phosphate. Fig. 17 shows the schematic

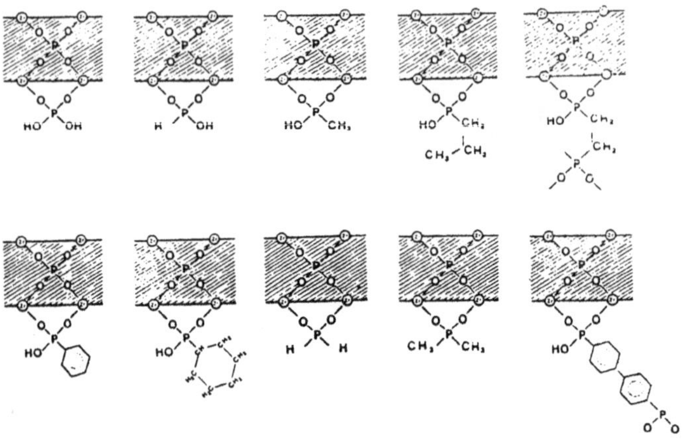

Fig. 18 - Some organic derivatives of the γ-zirconium phosphate obtained by topotactic reactions.

structure of one of these derivatives while Fig. 18 indicates schematically other organic groups that can be attached by topotactic reactions to the γ-inorganic backbone.

Even pillared derivatives (see Fig. 19 as an example) have been obtained recently [25] and this opens another way to the preparation of inorgano-organic lamellar compounds, the properties of which can easily be changed simply by a suitable choice of the organic groups in the interlayer region.

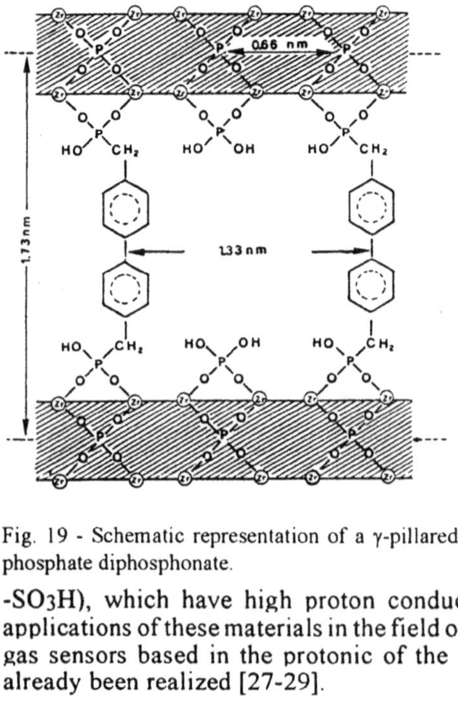

Fig. 19 - Schematic representation of a γ-pillared Zr phosphate diphosphonate.

4. Conclusions

The chemistry of M(IV) phosphates and phosphonates offers the possibility of preparing an enormous variety of phases which range from single planar inorgano-organic macromolecules to crystalline compounds in various ionic forms with a large number of molecules intercalated in their interlayer region. The pillaring of α- and γ- "clay-like" M(IV)phosphates, as well as their pillaring with covalently bonded diphosphonate groups, open the way for the preparation of porous inorganic or inorganic-organic solids with "zeolite-like" structures.

Their applications range from ion-exchange of cations to the intercalation of polar molecules, or from "shape selective catalysis" to ion and molecular sieving.

Finally, the preparation of M(IV) phosphonates containing strong acid functions (i.e. $-SO_3H$), which have high proton conduction (10^{-2} $S^{-1}cm^{-1}$) opens new prospects to applications of these materials in the field of solid state electrochemical devices [26]. Some gas sensors based in the protonic of the zirconium phosphate and its derivatives have already been realized [27-29].

REFERENCES

[1] Clearfield, A. (1982) 'Zirconium phosphates' in A. Clearfield (ed), Inorganic Ion Exchange Materials, CRC Press, Boca Raton (FL) pp. 1-74.
[2] Clearfield, A. and Stynes, J.A. (1964) 'The preparation of crystalline zirconium phosphate and some observation on its ion exchange behaviour' J. Inorg. Nucl. Chem. 26, 117-129.
[3] Alberti, G. (1978) 'Synthesis crystalline structure and ion exchange properties of insoluble salts of tetravalent metals and their salt forms' Acc. Chem. Res. 11, 163-170.
[4] Alberti, G. and Costantino, U. (1982) " Intercalation Chemistry of Acid Salts of Tetravalent Metals with Layered Structure and Related Materials Intercalation Chemistry", in M.S. Whittingham and A.J. Jacobson (Eds.), Intercalation Chemistry, Academic Press, New York, pp. 147-180.
[5] Alberti, G. and Costantino, U. (1991) 'Intercalation of zirconium phosphates and phosphonates', in J.L. Atwood, J.E.D. Davies and D.D. MacNicol (eds.), Inclusion Compounds, Oxford University Press, 5, pp. 136-176.
[6] Clearfield, A., Blessing, R.H. and Stynes, J.A. (1968) "New crystalline phases of zirconium phosphate possessing ion-exhange properties", J. Inorg. Nucl. Chem 30, 2249-2258.
[7] Clayden, N.J., (1987) " Solid State Nuclear Magnetic Resonance Spectroscopic Study of γ-Zirconium Phosphate" J. Chem. Soc. Dalton Trans. 1877-1881.
[8] Christensen, A.N., Andersen, E.K., Andersen, I.G.K., Alberti, G., nielsen, M. and Lehmann, M.S. (1990) "X-ray powder diffraction study of layered compounds. The crystal of structure of γ-$Ti(HPO_4)_2.H_2O$ and a proposed structure for γ-$Ti(H_2PO_4)(PO_4).2H_2O$" Acta Chem. Scand. 44, 865-872.
[9] Pinnavaia, T.J. and Kim, H. (1992) 'Molecular engineering of lamellar solids. Principles derived from the pillaring of smectite clays' in E.G. Deronane, F. Lemos, C. Naccache and F.R. Ribeiro (eds.), Zeolite Microporous Solids: Synthesis, Structure and Reactivity." , Kluwer Academic Publisher, pp. 79-90.
[10] Alberti, G., Costantino, U., Marmottini, F., Vivani R. and Zappelli, P.(1990) " Pillared Compounds of the Zirconium Phosphate, Phosphate-Phosphonate and Diphosphonate type" in I.V. Mitchell (Ed.), Pillared layered Structures, Elsevier Applied Science, pp. 119-126.
[11] Clearfield, A. (1988) "Recent advances in pillared clays and group IV metal phosphates" in J.M. Basset, B.C. Gates, J.P. Candy, A. Choplin, M. Leconte, F. Quignard and C. Santini (Eds.) "Surface Organometallic Chemistry: Molecular Approches to Surface Catalysis", NATO ASI Series vol. 231 Kluwer Academic Publishers, pp. 271-298.
[12] Alberti, G., Casciola, M. and U. Costantino (1985), "Inorganic Ion-Exchange Pellicles Obtained by Delamination of α-Zirconium Phosphate Crystals", J. Colloid. Interf. Sci. 107, 256-263.
[13] Alberti, G. and Marmottini, F., "Preparation of layered α-zirconium phosphate with a controlled degree of hydrolysis via delamination procedure", submitted to J Colloid. Interf. Sci.
[14] Clearfield, A. and Roberts, D.B. (1988) "Pillaring of layered zirconium and titanium phosphates", Inorg. Chem. 27, 3237-3240.
[15] Maireles-Torres, P., Olivera-Pastor, P., Rodriguez-Castellon, E., Jimenez-Lopez, A., Alagna, L. and Tomlinson, A.A.G., (1991) "Porous Cross-linked Materials formed by Oligomeric Aluminium Hydroxides and α-Tin Phosphate" J. Mater. Chem., 1, 319-326.
[16] Alberti, G., Costantino, U, Allulli, S. and N. Tomassini (1978) J. Inorg. Nucl. Chem. 40, 1113-1117.

[17] Alberti, G., Costantino, U., Kornyei, J. and Luciani, M.L. (1985), "Derivatives of α-zirconium phosphate with two different functional groups", React. Polym. 5, 1-10.
[18] Dines, M. and Griffith, P.C. (1983) "Synthesis and characterization of layered tetravalent metal terphenyl mono- and bis-phosphonate" Polyhedron 2, 607-611.
[19] Dines, M., Di Giacomo, P.D., Griffith, P.C., Cooksey, R.E., Callahan, K.P., and Lane, R.H. (1982) "Derivatized Layered M(IV) Phosphonates" in " Chemically modified in Catalysis" ACS Symp. Sez. 192, Washington D.C. pp. 223-240.
[20] Clearfield, A. (1987) 'New approaches to the design of materials via preparative inorganic chemistry' in D.L. Cocke and A. Clearfield (eds.) Design of New Materials, Plenum Press, New York, pp. 121-134.
[21] Alberti G., Costantino U, Vivani R and Zappelli P., "Composizione Solida Cristallina Mesoporosa di Difosfonato-Fosfito di un Metallo Tetravalente con una stretta distribuzione di Mesopori", Italian Patent n 21213 (3/8/1990)
[22] Alberti, G., Costantino, U., Vivani, R. and R.K. Biswas "Preparation and some preliminary investigation of ion exchange and intercalation properties of γ-zirconium phosphate phosphite", React. Polymer, in press.
[23] Alberti, G., Vivani, R., Biswas, R.K. and Murcia-Mascaros, S., "Preparation and some properties of γ-zirconium Phosphate-benzenphosphonate", React. Polymers, in press.
[24] Alberti G., Casciola M. and Biswas R.K., "Preparation and characterization of zirconium phosphate phosphonates, $ZrPO_4(H_2PO_4)_{1-x}(RPO_2)_x.nH_2O$, with γ-layer structure ($R = CH_3, C_3H_7, C_6H_{11}$)" Inorg. Chim. Acta, in press
[25] G. Alberti, R. Vivani and S. Murcia Mascaros, to be published.
[26] Alberti G., Casciola M., Costantino U. and Peraio A., (1992) "Protonic conductivity of layered zirconium phosphonates containing $-SO_3H$ groups. I. Preparation and characterization of a mixed zirconium phosphonate of composition $Zr(O_3PR)_{0.73}(O_3PR')_{1.27}.nH_2O$, with $R = -C_6H_4-SO_3H$ and $R' = -CH_2-OH$", Solid State Ionics 50, 315-322.
[27] Alberti, G. and Palombari, R., (1989), "All Solid State Hydrogen Sensors Based on Pellicular α-Zirconium Phosphate as a protonic Conductor", Solid State Ionics 35, 153-156.
[28] Alberti, G. Palombari, R., De Angelis, L., Maimone, A. and Modica, L., (1990) "A new Hydrogen Sensor with Pellicular Zr Phosphate as Proton Conductor", Sensors and Actuators, B1 121-124.
[29] Alberti, G., Casciola, M., Chieli, S. and Palombari, R.,(1991) "Use of Solid State Protonic Conductors for Oxygen Potentiometric Sensor at Room Temperature", Solid State Ionics 46, 183-186.

THE SYNTHESIS, CHARACTERISTICS AND APPLICATIONS OF LAYERED DOUBLE HYDROXIDES

M Chibwe, J B Valim and W Jones*

Chemistry Department, University of Cambridge, Lensfield Road Cambridge CB2 1EW, UK

ABSTRACT The properties and applications of layered double hydroxides (hydrotalcite-like compounds) are described. Particular areas of application for these materials which are considered include heterogeneous catalysis and photochemistry. The possibility of incorporating large inorganic and organic species to yield pillared solids akin to those studied for clays is outlined. The surface properties - important to catalytic applications - of the parent and calcined materials are described. In particular, results are given with respect to the use of calcined layered double hydroxides in the base-catalysed synthesis of glycol ethers. Their possible use in the controlled solid-state photochemical reaction of cinnamic acid is also discussed.

1. INTRODUCTION

Layered double hydroxides (LDHs - the so-called anionic clays) are a group of materials which are currently of much interest. They have a variety of potential applications as catalysts, pharamaceuticals, ion-exchange materials, sorbents and molecular sieves. A broad range of structural compositions are possible, which give rise to a rich variety of materials with subtle differences in property [1-4]. This review outlines some of the principle features associated with LDHs, illustrating, with suitable examples, the potential uses of these materials.

LDHs consist of positively charged metal hydroxide sheets with intercalated anions and water molecules (Figure 1) [5-7]. In terms of charge they are mirror-images of the much studied family of cationic clay minerals. They may represented by the general formula;

$$M^{2+}{}_{a}M^{3+}{}_{b}(OH)_{2a+2b}(X^{-})_{b}.xH_2O$$

where the identity of M^{2+} involves an extensive range of cations including Mg, Ni, Co, Zn, or Cu and M^{3+} may also be quite varied and includes Al, Cr or Fe. A range of M^{2+}/M^{3+}ratios are possible giving rise to variable anion exchange capacity. The ease of synthesis of a particular LDH is dependent upon the identity of M^{2+} and M^{3+}.

Figure 2a shows the powder X-ray diffraction (XRD) pattern obtained for a synthetic LDH with composition similar to that expected for the mineral hydrotalcite, i.e. $Mg_6Al_2(OH)_{16}CO_3.4H_2O$. The thickness of the interlayer region is approximately 3.0 Å (assuming the thickness of the brucite layer to be approximately ~4.8 Å). This is a value appropriate to a $CO_3{}^{2-}$ ion lying parallel to the Mg-Al sheets, with strong hydrogen bonding.

C.A.C. Sequeira and M.J. Hudson (eds.),Multifunctional Mesoporous Inorganic Solids, 191–206.

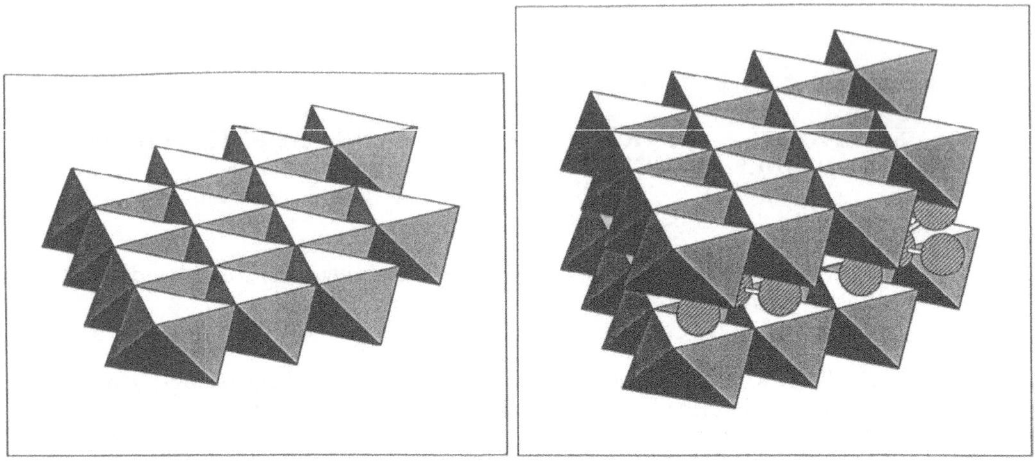

(a) (b)

Figure 1 Schematic illustration of (a) view of brucite [$Mg(OH)_2$] lattice and (b) hydrotalcite $Mg_6Al_2(OH)_{16}CO_3.4H_2O$ - after reference [1].

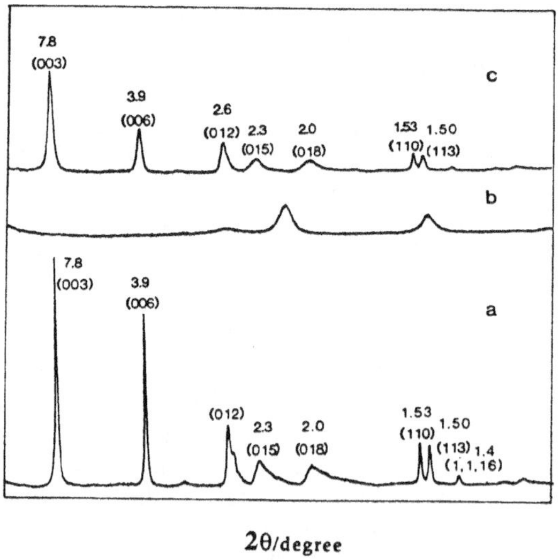

Figure 2 Powder XRD patterns for a Mg-Al-CO_3 LDH (a) as synthesized, (b) after calcination at 450°C for 18 hours and (c) following rehydration in 0.05 M Na_2CO_3 solution.

More generally Cl^-, OH^-, CO_3^{2-} are frequently found as the charge-balancing anions although the intercalation of a great variety of other anions, including organic and inorganic, are possible. Table 1 indicates some of the compositions which are possible as well as the corresponding mineralogical name of the naturally occurring material.

Table 1 Formula for various layered double hydroxides along with mineralogical name for naturally occurring material - after reference 4.

M^{2+}	M^{3+}	Anion	Composition	Name
Mg	Al	CO_3	$Mg_6Al_2(OH)_{16}CO_3.4H_2O$	Hydrotalcite*
Mg	Al	CO_3	$Mg_6Al_2(OH)_{16}CO_3.4H_2O$	Manasseite+
Mg	Fe	CO_3	$Mg_6Fe_2(OH)_{16}CO_3.4.5H_2O$	Pyroaurite*
Mg	Fe	CO_3	$Mg_6Fe_2(OH)_{16}CO_3.4.5H_2O$	Sjögrenite+
Mg	Cr	CO_3	$Mg_6Cr_2(OH)_{16}CO_3.4H_2O$	Stichtite*
Mg	Cr	CO_3	$Mg_6Cr_2(OH)_{16}CO_3.4H_2O$	Barbertonite+
Ni	Al	CO_3,OH	$Ni_6Al_2(OH)_{16}CO_3.4H_2O$	Takovite*
Ni	Fe	CO_3	$Ni_6Fe_2(OH)_{16}CO_3.4H_2O$	Reevesite*
Mg	Mn	CO_3	$Mg_6Mn_2(OH)_{16}CO_3.4H_2O$	Desautelsite*

*Rhombohedral +Hexagonal

2. SYNTHESIS AND STRUCTURAL CHARACTERISTICS OF LDHs

With respect to the laboratory synthesis of LDHs, a variety of routes have been reported. These have recently been considered in a comprehensive review by Cavani et al [4]. Some factors which are important and which are relevant to the potential application of such materials are:

(i) the synthesis conditions (e.g. pH, temperature and atmosphere) required for the generation of phase-pure products
(ii) the thermal stability of the structures obtained and the manner in which this stability is influenced by the nature of the octahedral substituents and the charge-balancing anion and
(iii) the morphology and size of the product crystals - factors which influence the accessible surface area and porosity.

For the co-precipitation method, involving the precipitation of solutions of M^{2+}, M^{3+} with those of the anion salt of interest (whilst controlling the pH) several parameters which may be varied during synthesis have been investigated, including temperature and time of crystallisation as well as applied pressure [1,3,8].

Figure 3 illustrates the PXRD patterns for Ni-Al-CO_3 samples, of approximate composition $Ni_6Al_2(OH)_{16}(CO_3^{2-}).4H_2O$ crystallised at either 65°C or 200°C; the pattern for the 200°C material shows much sharper reflections than the one crystallised at 65°C.

Transmission electron microscopy confirms that crystallisation at 200°C results in well developed hexagonal crystallites whilst the lower temperature route gave somewhat smaller ill-defined crystallites. This variation in crystallite size is in agreement with earlier work on other types of LDHs. There are also significant differences in surface areas for LDHs crystallised at different temperatures. For example, a Mg-Al-CO_3 LDH crystallise at 65°C had a surface area of 45 m^2/g whereas one crystallised at 200°C had a surface area of only 12 m^2/g.

PXRD patterns of Mg^{2+}-Al^{3+} samples with different ratios of Mg^{2+}:Al^{3+}, prepared under similar conditions of temperature and annealing time are shown in Figure 4. At a ratio of 1 and 5, reflections resulting from small amounts of impurities are present. Cavani et al have outlined the compositional ranges possible for an extensive range of M^{2+} and M^{3+} species [4].

Miyata and his colleagues have made extensive studies of the synthesis and properties of LDHs and have reported [10-14] that anion stability in LDHs follows the sequence:

$$OH^- > F^- > Cl^- > Br^- > NO_3^- > I^- \text{ and}$$
$$CO_3^{2-} > SO_4^{2-}.$$

Unlike clays, however, direct ion replacement is not straightforward and a variety of methods have been employed to bring about anion exchange. For example, Bish [15,16] utilised a strategy using the addition of 0.1 N dilute acid HX to release CO_2 and thereby incorporate X^-. An alternative method has been developed by Chibwe and Jones [17,18] based upon the fact that calcination of the LDH at 450°C for 18 hours results in the structure changing to an MgO-like material - see Figure 2b. On exposure of this calcined material to carbonated water the layered structure is regenerated either with CO_3^{2-} restored as the charge-balancing anion or with other anions (inorganic and organic) incorporated (Figure 2c). This approach has been extended by others [19-20]. Calcination of other types of LDHs at 450°C (including Ni-Al, Mg-Fe and Li-Al) leads to products similar to that described for Mg-Al, each consisting of poorly crystalline magnesium-oxide like structures and these may also be reconstructed upon exposure to water. Calcination of a Co-Al LDH at 450°C, for example, results in a product believed to be composed of the spinel type oxides, $CoAl_2O_4$ and Co_3O_4. A detailed summary of the products resulting from calcination has been given by Cavani et al [4].

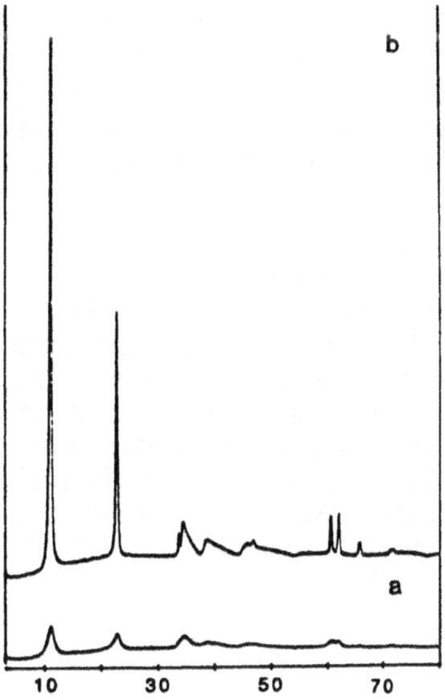

Figure 3 Powder XRD patterns for a Ni-Al-CO_3 prepared by crystallization at (a) 65°C and (b) 200°C.

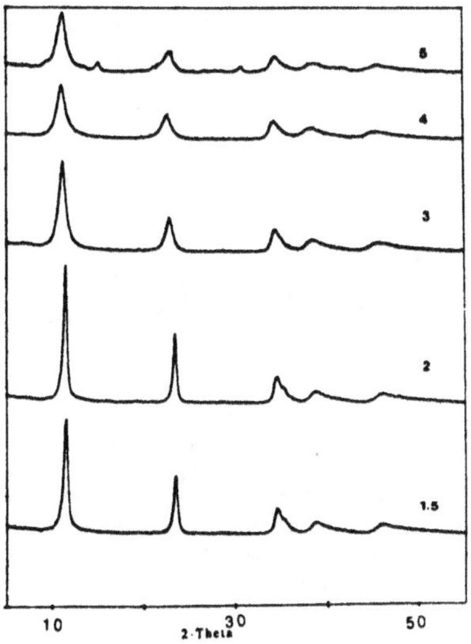

Figure 4 Powder XRD patterns for Mg-Al-CO_3 LDHs with various ratios of Mg/Al - from reference [9].

The calcined precursor method has been used by Chibwe and Jones [17,18] to incorporate a range of inorganic anions, including polyoxovanadate ($V_{10}O_{28}^{6-}$), polyoxomolybdate ($Mo_7O_{24}^{6-}$), ferric cyanide ($Fe(CN)_6^{3-}$) and dichromate ($Cr_2O_7^{2-}$) anions. The spacing for vanadate (d_{003} = 11.8 Å) corresponds to a gallery height of 7.0 Å [21-24]. Various organic anions have also been intercalated using this method including: α-ω dicarboxylates, p-toluene sulphonate, terephthalate, dodecyl sulphate, 2-naphthol-6-sulphonate, phthalate and cinnamate.

Figure 5 shows the PXRD patterns for the regenerated LDHs with dicarboxylate anions as the intercalated species. There is a progressive shift in the observed d_{00l} with increasing chain length. The oxalate dianion (C_2) gives a d_{003} of 9.6 Å and the dodecane dicarboxylate dianion (C_{14}) gives a d_{003} of 24.6 Å. All the products appeared to be monophasic with no interference from carbonate observed using the method. Indeed the data obtained are comparable with those from direct synthesis. In the case of the cinnamate anions PXRD gives for the p-chlorocinnamate derivative a d_{003} of 20.0 Å and for o-chlorocinnamate a d_{003} of 18.6. These values suggest that the cinnamate anions are oriented perpendicular to the sheets - an interpretation supported by studies on pure cinnamate salts [25].

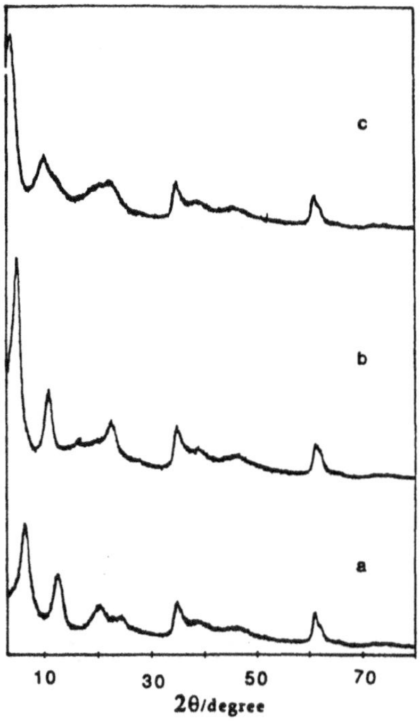

Figure 5 Variation in powder XRD pattern for LDH following intercalation of various dicarboxylate anions. A variation in c-axis repeat following the increasing chain length of the diacid is observed.

Typical surface areas and pore volumes obtained for terapthalate and benzoate-LDH's are given in Table 2 [8,9].

Table 2 BET/N_2 measurements for various hydrotalcite-like materials.

Sample	Surface area (m^2/g)	Pore volume (ml/g)	Average pore diameter (Å)
Calcined hydrotalcite (550°C/18hrs)	262	0.33	50
Terephthalate via calcined precursor	3.9	0.02	190
Terephthalate by direct synthesis	52	0.23	175
Benzoate	125	0.13	42
1-Napthanoic	3.5	0.01	160
9-Anthranoic	3.9	0.08	202

3. APPLICATIONS OF LDHs

A variety of applications for LDHs have been described [1,3,4]. These include: base catalysis, anion exchangers, stabilisers in polymer production, medical and as adsorbents. In terms of catalysis a variety of uses have been outlined [4,26-29]. Some of these use the LDH as the precursor (e.g. calcination followed by H_2 reduction [4]) in others the calcined material serves as the catalyst - such is generally the case for base catalysed reactions [27].

A variety of base catalysed processes have been described in the literature. The vast majority utilise the calcined LDH as the catalyst. Amongst the many reactions studied are:

- polymerization of formaldehyde
- polymerization of β-propiolactane
- polymerization of propylene oxide
- depolymerization of paraldehyde
- methylation of 3,5 xylenol
- aldol condensation reactions

A recent study has been undertaken by Chibwe, Atkins and Jones concerning the use of calcined LDHs in the synthesis of glycol ethers [30-32]. Glycol ethers are important commercial compounds. The current industrial process for manufacturing glycol ethers utilizes a homogenous base catalyst but for a number of reasons heterogeneous catalysts offer added advantages. Acidic clays and pillared clays have been considered [33]. On an acid site 2-ethoxy-1-propanol (i.e. the primary alcohol) is the major product along with a relatively high amount of dioxanes and dioxolanes. On a base site, however, reaction leads to the secondary isomer as the major product with relatively low levels of dioxanes and dioxolanes - Figure 6 illustrates the likely mechanisms. This selectivity of base catalysts towards the secondary isomer as the major product is an important attribute.

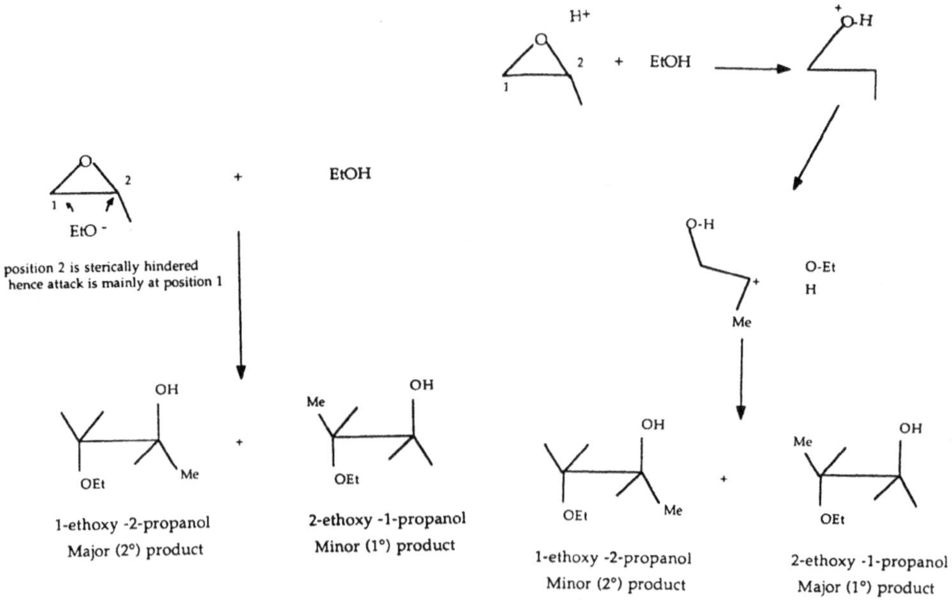

(a) (b)

Figure 6 Schematic illustration of the reaction mechanisms operating in glycol ether synthesis on (a) base and (b) acid sites.

The influence of lattice substitution in calcined LDHs upon conversion (using Mg-Al as the parent material - see Figure 7) has been studied using the following materials [8]:

(i) LDHs in which Mg^{2+} is the M^{2+} ion and in which Al^{3+} is replaced by Fe^{3+}
(ii) LDHs in which Al^{3+} is the M^{3+} ion and in which Mg^{2+} is replaced by Ni^{2+}, Li^{+}, or Co^{2+}.
(iii) LDHs in which both Mg^{2+} and Al^{3+} are replaced by Zn^{2+} and Cr^{3+} respectively.

The activities and selectivites observed as a function of time (Figure 7) for Mg-Al confirm that the material is quite basic with an average selectivity to the 2° product close to 90% - a selectivity which compares well with a homogeneous catalyst under similar conditions (which gave an average selectivity between 92 and 95%). For group (i) the replacement of Al^{3+} by Fe^{3+} results in a significant increase in activity along with a marked decrease in selectivity to the 2° product, see Figure 8. The activity of the Mg-Fe catalyst steadily decreases with time, as a result of coke formation.

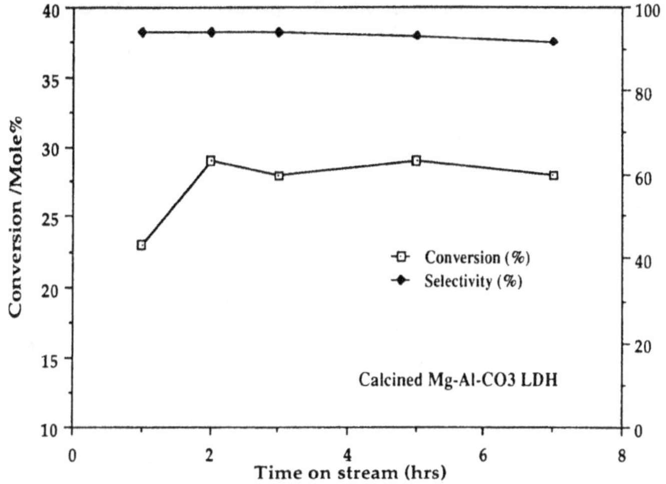

Figure 7 Percentage conversion of propylene oxide and ratio of $2^o/(2^o + 1^o)$ as a function of time on stream for a calcined Mg-Al-CO_3 LDH.

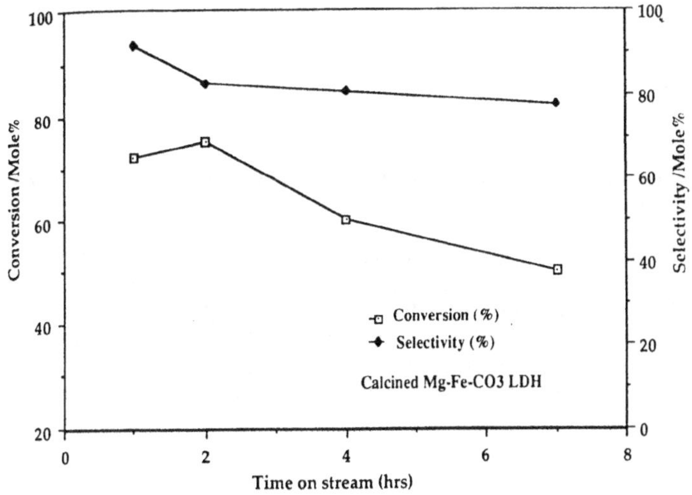

Figure 8 Percentage conversion of propylene oxide and ratio of $2^o/(2^o + 1^o)$ as a function of time on stream for a calcined Mg-Fe-CO_3 LDH.

In group (ii) the relative activities for the catalysts are: Co >>Mg > Li > Ni. Upon calcination at 450°C Mg-Al-CO_3, Ni-Al-CO_3 and Li-Al-CO_3 LDHs behave in similar ways - two general regions of weight loss are observed. The first is attributed to loss of interlayer water and the higher temperature loss is associated with dehydroxylation and loss of CO_2. The Co-Al-CO_3, however, loses interlayer water, structural water and carbonate simultaneously between 100-250°C. Above 200°C the spinel type oxides Co_3O_4 and $CoAl_2O_4$ are formed. The observed activity and the decrease in selectivity (hence the increase in number of acid sites) may be due to the spinel oxides. The calcined Ni-Al LDH is observed to be the least active possibly because of the hydrogenating nature of nickel.

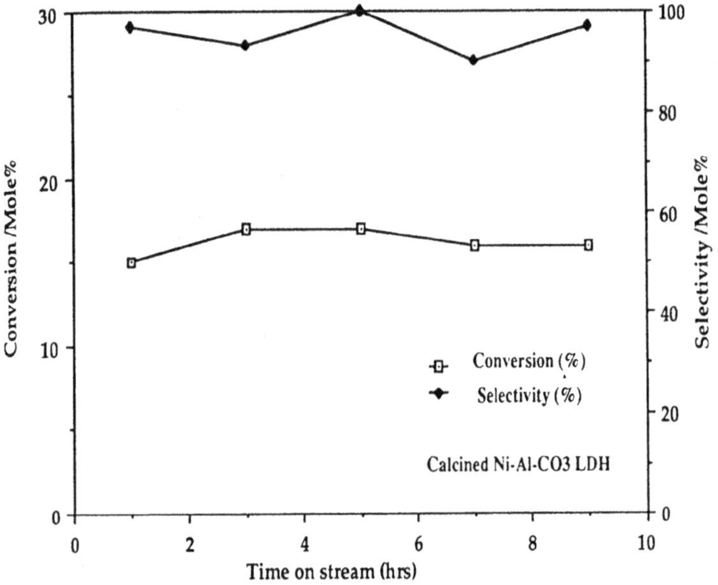

Figure 9 Percentage conversion of propylene oxide and ratio of 2°/(2° + 1°) as a function of time on stream for a calcined Ni-Al-CO_3 LDH.

In group (iii), when compared to Mg-Al, the Zn-Cr catalyst performed better both in terms of activity and selectivity (see Figure 10). The Figure compares the activity of the Zn-Cr with a 330 ppm NaOH sodium hydroxide homogenous catalyst at 120°C. The activity of the Zn-Cr catalyst (34%) and the NaOH (40%) are comparable. The average selectivity was about 97% for Zn-Cr and 94% for the NaOH catalyst. It can be concluded, therefore, that the calcined Zn-Cr LDH is a particularly useful catalyst for this reaction.

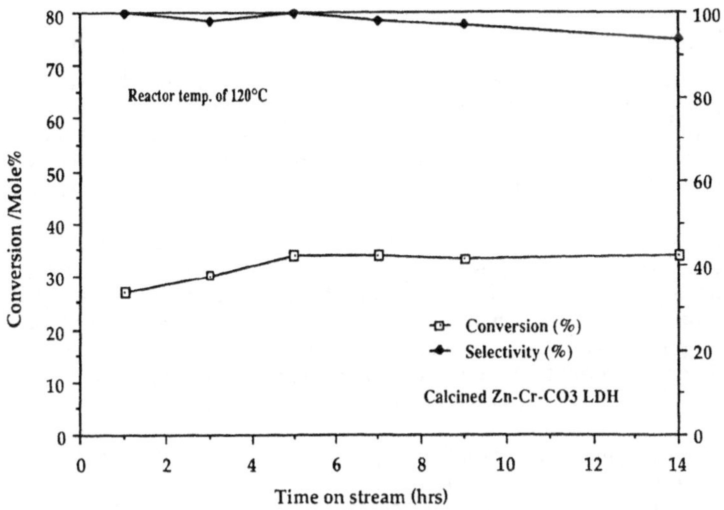

Figure 10 Percentage conversion of propylene oxide and ratio of 2°/(2° + 1°) as a function of time on stream for a calcined Zn-Cr-CO_3 LDH.

Figure 11 shows the variation of activity with calcination temperature for Mg-Al-CO_3. Calcination at 450°C produces a more active material than calcination at 200°C. There is a decrease in catalytic activity for samples calcined at 600°C. There are no significant difference in selectivities, however. Reichle [27] has reported similar observations for other base-catalysed reactions. For example, he found that maximum activity and selectivity for acetone aldol condensation was obtained using a magnesium-aluminium carbonate LDH calcined between 350 and 450°C.

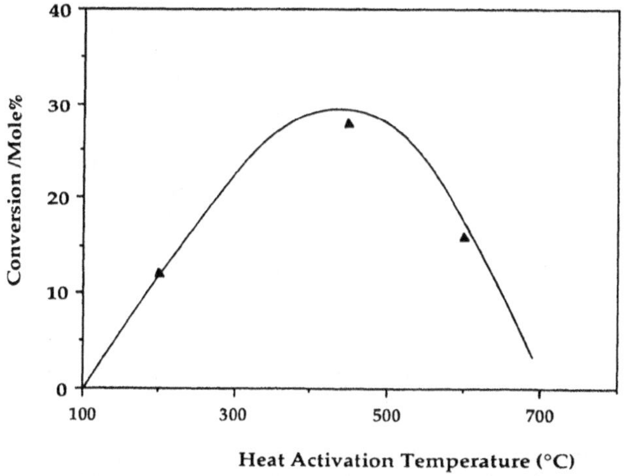

Figure 11 Variation in activity for Mg-Al-CO_3 LDHs calcined to various temperatures, indicating optimum activity for a calcination temperature of 450°C.

Photochemical Applications

Several photochemical processes in LDHs have been described. For example, Giannelis et al [34] have immobilized complex anions of the type $[Ru(BPS)_3]^{4-}$ and have studied the photophysics of such complexes within the LDH. Kwon et al [23] have studied the photo-induced decomposition of isopropanol (to acetone). In a recent paper, Barloy et al [35] have studied the use of LDHs and other layered materials as possible supports for Mn(III)-porphyrins for cycloctene epoxidation and alkane hydroxylation by PhIO.

The study of the photochemistry of cinnamate-LDHs has been described by Valim et al [25] and is based on the well known solid state chemistry of cinnamic acid [36] - see Scheme 1 - and the hope that information concerning the packing and reactivity of the guest molecules within the interlayer region would be obtained. Several sodium and magnesium salts of the cinnamate derivatives were also synthesised in order to help rationalise the formation of the various products. Some interesting comparisons were made between the reactivity of (i) cinnamate LDHs, (ii) the corresponding salts and (iii) the corresponding acid in solution.

Valim et al [25] have described the preparation of the intercalates and the photochemical conditions. They demonstrated that both photodimerization and photoisomerization occur within the gallery region, with the relative amounts of isomerization and dimerization likely to be dependent upon the stoichiometry of the parent layered double hydroxide. By means of a comparison of the powder diffraction data for the intercalates with the full three-dimensional structures of the corresponding salts they were also able to propose an arrangement for the cinnamate anions in the interlayer region.

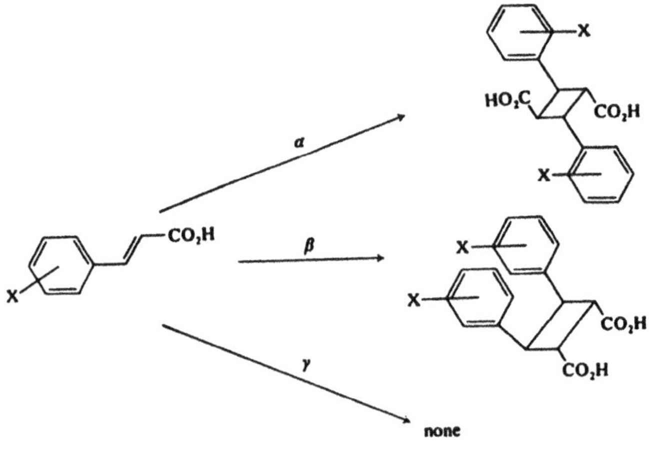

Scheme 1 Schematic illustration of the various reactions possible for cinnamate derivatives - see reference 36.

For example, for the trans-p-chlorocinnamate intercalate a d_{001} of 19.8Å was obtained. In the crystal structure of sodium trans-p-chlorocinnamate monohydrate the molecules are packed as bilayers with a repeat distance of 21.4Å - see Figure 12. This would suggest a comparable bilayer arrangement within the p-chlorocinnamate-LDH.

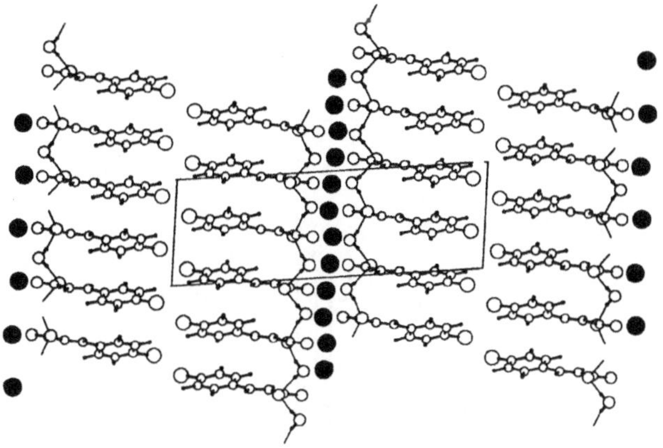

Figure 12 Structure of sodium p-chlorocinnamate hydrate viewed down the c-axis. Photodimeric partners are stacked consecutively along the y axis.

Of particular interest is a comparison of the reactivity of the intercalate compared with the reactivity of the sodium salt and a solution of the acid. Figure 13 shows the ^{13}C NMR spectrum for the UV irradiated trans-p-chlorocinnamate-LDH, after extracting the organic material by acid treatment. One can see additional peaks other than those assigned to the reactant. From these one at 173.9 ppm can be assigned to a carboxyl group and two at 43.9 and 42.5 ppm can be assigned to aliphatic carbons. The fact that only two aliphatic carbon peaks and only one carboxyl peak are present suggests that only one dimer product has been produced and this has a mirror plane or centre of symmetry. Peaks of weak intensities may be considered as indicative of some *cis-trans* isomerisation having also occurred.

Comparison with the UV irradiation of the corresponding sodium salts in the solid state confirms the nature of the product resulting from irradiation of the crystals. Figure 14 shows the spectrum obtained along with the peak assignment.

Calculation of closest double bond distances in the *p*-chlorocinnamate suggests that there are two reasonably short separations, each between two non-symmetry related anions stacked consecutively in the **b** direction and each with a separation in the region of 3.7 Å. Such a separation is well within the range expected to permit photodimerisation [36].

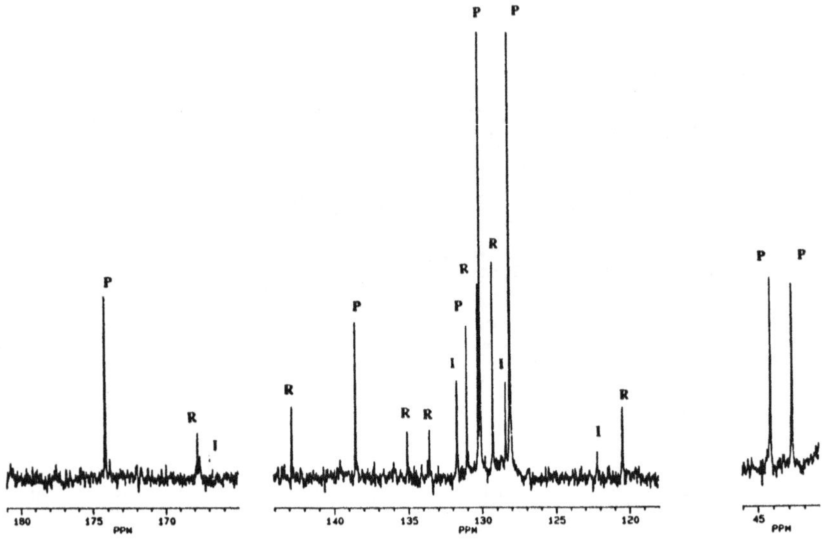

Figure 13 ^{13}C solution NMR spectrum (in CD_3OH) of p-chlorocinnamate after uv irradiation and acid extraction from the LDH. R = reactant, P = product dimer, I = isomer (cis).

CONCLUDING REMARKS

LDHs are proving to be of interest in a variety of applications. This paper has concentrated on only two main aspects i.e. catalysis and photochemistry. It is clear, however, that a thorough understanding of the factors which influence the synthesis of the LDH and the changes accompanying calcination will be important if the full potential of these materials is to be utilised. The extensive review by Cavani et al [4] provides valuable references to many other areas of application.

ACKNOWLEDGEMENTS

We are grateful to the Cambridge Commonwealth Trust, Universidade de Sao Paulo, BP (Sunbury) and the European Research Office (VAL-SLCVA, DAJA45-90-C-0013) for support. Discussions with Drs M. Atkins and C. McAteer are appreciated.

REFERENCES

1. W.T. Reichle (1986) Anionic Clay Minerals, ChemTech, 58-63.
2. K.A. Carrado, A. Kostapapas and S.L. Surb (1988) Layered double hydroxides (LDHs). Solid State Ionics, 26, 77-86.

3. M. Chibwe and W. Jones (1990) The synthesis and catalytic applications of layered double hydroxides in Pillared Layered Solids, Ed. I.V. Mitchell, Elsevier Applied Sciences, 67-78.
4. F. Cavani, F. Trifiro and A. Vaccari (1991) Hydrotalcite-type anionic clays: preparation, properties and applications. Catalysis Today, 11, 173-301.
5. R. Allmann (1968) The crystal structure of pyroaurite. Acta Cryst., 24, 972-977.
6. H.F.W. Taylor (1969) Segregation and cation-ordering in sjogrenite and pyroaurite. Miner. Mag., 37, 338-342.
7. H.F.W. Taylor (1973) Crystal structures of some double hydroxide minerals. Miner. Mag., 39, 377-389.
8. M. Chibwe (1990) Synthesis, characterization and catalytic applications of mixed metal oxides, Ph.D. Thesis, University of Cambridge.
9. W.W. Kagunya (1990) Synthesis and characterization of layered double hydroxides, CPGS Thesis, University of Cambridge.
10. S. Miyata (1975) The synthesis of hydrotalcite-like compounds and their structures and physico-chemical properties - I: the systems Mg^{2+}-Al^{3+}-NO_3^-, Mg^{2+}-Al^{3+}-Ce^-, Mg^{2+}-Al^{3+}-ClO_4^-, Ni^{2+}-Al^{3+}-Cl^-, and Zn^{2+}-Al^{3+}-Ce^-. Clays and Clay Minerals, 23, 369-375.
11. S. Miyata (1980) Physico-chemical properties of synthetic hydrotalcites in relation to composition. Clays and Clay Minerals, 28, 50-56.
12. S. Miyata (1983) Anion-exchange properties of hydrotalcite-like compounds. Clays and Clay Minerals, 31, 305-311.
13. S. Miyata and A. Okada (1977) Synthesis of hydrotalcite-like compounds and their physico-chemical properties - the systems Mg^{2+}-Al^{3+}-SO_4^{2-} and Mg^{2+}-Al^{3+}-CrO_4^{2-}. Clays and Clay Minerals, 25, 14-18.
14. S. Miyata and T. Kumura (1975) Synthesis of new hydrotalcite-like compounds and their physico-chemical properties. Chem. Letts., 843-848.
15. D.L. Bish (1978) Anion exchange in takovite. Bull. Bur. Rech. Geol. Minieres Sect., 3, 293-301.
16. D.L. Bish (1980) Anion exchange in takovite: applications to other hydroxide minerals. Bull. Mineral., 103, 170-175.
17. K. Chibwe and W. Jones (1989) Intercalation of Organic and Inorganic Anions into Layered double hydroxides. J. Chem. Soc., Chem. Commun., 926-927.
18. K. Chibwe and W. Jones (1989) Synthesis of polyoxometallate-pillared layered double hydroxides via calcined precursors. Chem. of Materials, 1, 489-490.
19. E. Narita, P. Kaviratna and T.J. Pinnavaia (1991) Synthesis of heteropolyoxometallate pillared layered double hydroxides via calcined zinc-aluminium precursors. Chemistry Letters, 805-808.
20. E.D. Dimotakis and T.J. Pinnavaia (1990) New route to layered double hydroxides intercalated by organic anions: precursors to polyoxometalate-pillared derivatives, Inorganic Chem., 2393-2394.
21. M.A. Drezdzon (1988) Synthesis of isopolymetalate-pillared hydrotalcite *via* organic-anion-pillared precursors. Inorg. Chem., 27, 4628-4632.
22. T. Kwon and T.J. Pinnavaia (1989) Pillaring of a layered double hydroxide by polyoxometalates with Keggin-ion structures. Chem. of Materials, 1, 381-383.

23. T. Kwon, G.A. Tsigdinos and T.J. Pinnavaia (1988) Pillaring of layered double hydroxides by polyoxometallate anions. J. Am. Chem. Soc., 110, 3653-3654.
24. J. Twu and P.K. Dutta (1989) Structure and reactivity of oxovanadate anions in layered lithium aluminate materials. J. Phys. Chem., 93, 7863-7868.
25. J. Valim, B.M. Kariuki, J. King and W. Jones (1992) Photoactivity of cinnamate-intercalates of layered double hydroxides, Mol. Cryst. Liq. Cryst., 211, 271-281.
26. S. Kohjiya, T. Sato, T. Nakayama and S. Yamashita (1981) Polymerization of propylene oxide by calcined synthetic hydrotalcite. Makromol. Chem. Rapid Commun., 2, 231-233.
27. W.T. Reichle (1985) Catalytic reactions by thermally activated, synthetic, anion clay minerals. J. Catalysis, 94, 547-557.
28. C. Busetto, G. Del Piero, G. Manara, F. Trifero and A. Vaccari (1984) Catalysts for low temperature methanol synthesis. Preparation of Cu-Zn-Al mixed oxides *via* hydrotalcite-like precursors. J. Catalysis, 85, 260-266.
29. W.T. Reichle, S.Y. Kang and D.S. Everhardt (1986) The nature of the thermal decomposition of a catalytically active anionic clay mineral. J. Catalysis, 101, 352-359.
30. M. Chibwe, M.P. Atkins and W. Jones (1991) Process for the preparation of glycol ethers, E.P.A. 0421 679 A1.
31. M. Chibwe, M.P. Atkins and W. Jones (1991) Process for the preparation of glycol ethers, E.P.A. 0421 677 A1.
32. M. Chibwe, M.P. Atkins and W. Jones (1991) Process for the preparation of catalysts.
33. M.P. Atkins (1990) Modified alumina pillared clays as catalysts for glycol ether production, in Pillared Layered Solids, Ed. I.V. Mitchell, Elsevier Applied Sciences.
34. E.P. Giannelis, D.G. Nocera and T.J. Pinnavaia (1987) Anionic photocatalysts supported in layered double hydroxides: intercalation and photophysical properties of a ruthenium complex anion in synthetic hydrotalcite, Inorg. Chem., 26, 203-205.
35. L. Barloy, J.P. Lallier, P. Battioni, D. Mansuy, Y. Piffard, M. Tournoux, J.B. Valim and W. Jones (1992) Manganese porphyrins adsorbed or intercalated in different numeral matrices: preparation and compared properties as catalysts for alkene and alkane oxidation, New J. Chem., 16, 71-80.
36. G.M.J. Schmidt (1971) Photodimerization in the Solid State. Pure Appl. Chem., 27, 647-678.

STRUCTURE AND TEXTURE PROPERTIES OF CALCINED LAYERED Mg,Al DOUBLE HYDROXIDES

F. M. LABAJOS[1], V. RIVES[1*] and M. A. ULIBARRI[2]
[1]Dpto. de Química Inorgánica, Universidad de Salamanca, Facultad de Farmacia, Salamanca, Spain, and
[2]Dpto. de Química Inorgánica e Ingeniería Química, Universidad de Córdoba, Facultad de Ciencias, Córdoba, Spain

ABSTRACT. Mg,-Al double layered hydroxides (Mg,Al-LDH) have been synthetized and characterized by X-ray diffraction, Thermal Analysis, FT-IR spectroscopy, Transmission Electron Microscopy and Surface area and Porosity assessment by nitrogen adsorption at 77 K. The solids have been submitted to thermal and hydrothermal treatments. It has been found that despite any hydrothermal treatment given to the fresh samples, surface properties are the same after calcination at temperatures high enough to destroy the layered structure.

1. Introduction

Layered double hydroxides (LDH), with the general formula $[M_{1-x}M'_x(OH)_2]^{x+}X_{x/m}$ $.nH_2O$, where M is a divalent cation and M' a trivalent cation, have a brucite-like structure and are the only known layered materials with positively-charged layers [1]; these materials exhibit swelling properties and the interlayer space (filled with anions and water molecules) can be modified by intercalation of other anions. When M=Mg(II), M'=Al(III) and X=carbonate, the material formed is hydrotalcite (HT). The interest in these compounds arises not only from the academic point of view, but also because of their application (and that of their partially decomposition products) as catalysts and catalyst supports, their anion exchange capability and formation of pillared structures, etc [2]. The properties and applications of hydrotalcite-like materials have been reviewed recently [3].

Crystallinity of the freshly prepared materials is generally improved by submitting the samples to hydrothermal treatment, but this leads to samples with lower specific surface areas [4]. However, as most of LDH are used after being submitted to calcination at moderate or high temperatures, we have here investigated the structural and surface properties of calcined LDHs that had been (or not) previously submitted to such a hydrothermal treatment, in order to analyze if some sort of *memory effect* persists in the high temperature calcined materials.

2. EXPERIMENTAL

2.1. SAMPLES PREPARATION

The samples were prepared following the so-called "constant pH" method already described [5]. Briefly, a solution containing 0.5 mol $MgCl_2 \cdot 6H_2O$ and 0.25 mol $AlCl_3$ $.6H_2O$ dissolved in 250 ml of bidistilled water was dropwise added to a vigorously stirred

C.A.C. Sequeira and M.J. Hudson (eds.),Multifunctional Mesoporous Inorganic Solids, 207–216.

solution (500 ml) containing 1.7 mol NaOH and 0.5 mol Na_2CO_3. The resulting heavy suspension was further stirred at room temperature for 2 h, and then the solvent was very slowly evaporated at 325 K in a water bath. Two samples were therein obtained: sample A was aged at room temperature for two weeks, while sample B was aged hydrothermally for 13 days at 400 K in a stainless steel Phaxe 2001 bomb. The solids were washed until the absence of chloride and sodium ions was confirmed in the washing liquids by conventional analytical methods. Chemicals were from Probus or Panreac, with adequate quality.

2.2. EXPERIMENTAL METHODS

Instrumental facilities were as follows: (i) chemical analysis for Mg and Al was performed by atomic absorption (ELL-240); (ii) X-ray diffraction (XRD) profiles were recorded on a Siemens D-500 instrument, using Ni-filtered CuKα radiation (λ=154.05 pm); (iii) differential thermal analysis (DTA) and thermogravimetric analysis (TG) were performed in DTA-1700 and TGS-2 instruments from Perkin-Elmer, coupled to a Perkin-Elmer 3600 Data Station, using a heating rate of 10 K/min in air; (iv) Fourier-transform infrared (FT-IR) spectra were recorded by the KBr pellet technique in a Perkin-Elmer FT-IR instrument, with a nominal resolution of 2 cm^{-1} and averaging 100 scans; (v) transmission electron micrographs were obtained on a Jeol 200CX microscope; and (vi) nitrogen adsorption measurements at 77 K were carried out in a conventional, Pyrex high-vacuum system equipped with a MKS pressure transducer, grease-free stopcopcks, an APV-DD4 rotatory pump and a silicon oil diffusion pump (nominal residual pressure *ca.* 10^{-4} N/m^2); analysis of data for specific surface area and porosity was carried out with the assistance of a computer programme described elsewhere [6].

3. RESULTS AND DISCUSSION

3.1. CHEMICAL ANALYSIS

Parameter x for samples A and B was 0.35 and 0.28, respectively; these values lie within the limits reported in the literature for pure HT-like materials ($0.2 \le x \le 0.33$); for x values outside this range either the pure hydroxides or other compounds with different structures are obtained [3]. It should be also noted that this value is fairly close to that existing in the initial solution containing both cations (x=0.3).

3.2. X-RAY DIFFRACTION AND TRANSMISION ELECTRON MICROSCOPY

XRD profiles for both samples indicate a good degree of crystallinity, specially for sample B. Indexing of the peaks recorded has been made on the basis of hexagonal axes and upon comparison with the values reported in the literature for hydrotalcite (file 14-191 OF JCPDS) [6], and are summarized in Table 1. A further confirmation of the lamellar structure of these compounds is concluded upon comparison of the profiles recorded for randomly oriented crystals and oriented aggregates, as shown in Fig. 1 for sample B. In this last case, the intensities of peaks due to diffraction by (00*l*) peaks are enhanced if compared to those of other crystal planes.

From the relative intensities of peaks due to diffraction by the same planes for both samples, it can be concluded that crystal size for sample B should be appreciably larger than that for sample A, probably because of the hydrothermal treatment the former sample has been given, and in agreement with previous studies [4,5]. Parameter c can be estimated from the basal spacing corresponding to planes (003), and corresponds to 22.47 Å for sample A and 22.86 Å for sample B, this last value matching very well with that reported

[3] for hydrotalcite (22.81 Å). Nevertheless, this value is also very close to that reported in the literature [3] for hydrotalcite-like materials with hydroxyl groups in the interlayer space (22.65 Å), but results shown below indicate that carbonate ions, and not hydroxyl ones, really exist in these samples. Finally, reflection (110), recorded close to d=1.5 Å, is independent of the kind of layer stacking and can be used to calculate parameter a ($a = 2d_{(110)}$), obtaining a value of 3.08 Å for sample A and 3.06 Å for sample B (3.05 Å for hydrotalcite [3]).

TABLE 1. Spacings (Å) of diffraction peaks for hydrotalcite samples (randomly oriented crystals)

A	B	(hkl)	[14-191]*
7.49	7.62	(003)	7.69
3.80	3.81	(006)	3.88
2.58	2.60	(009,012)	2.58
2.31	2.32	(015)	2.30
1.96	1.97	(018)	1.96
1.54	1.53	(110)	1.53
1.49	1.50	(113)	1.50)

*from ref. [7].

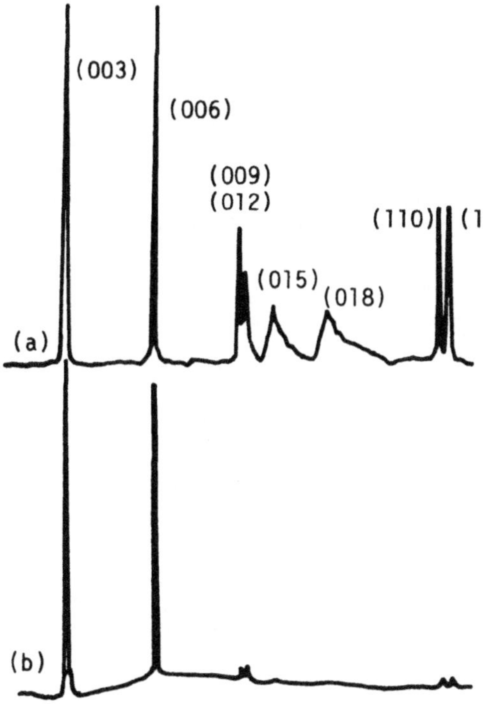

Figure 1. X-ray diffraction profiles of sample B. (a) randomly oriented crystals; (b) oriented aggregate.

Transmission electron micrographs for these two samples indicate rather similarly shaped particles, but with very different size; that for sample B is shown in Fig. 2. The particles appear as stacks of lamellar hexagonal particles, of 100 nm or even larger diameter. As expected, hydrothermal treatment has led to a higher degree of symmetry of sample B (if compared to that for sample A)

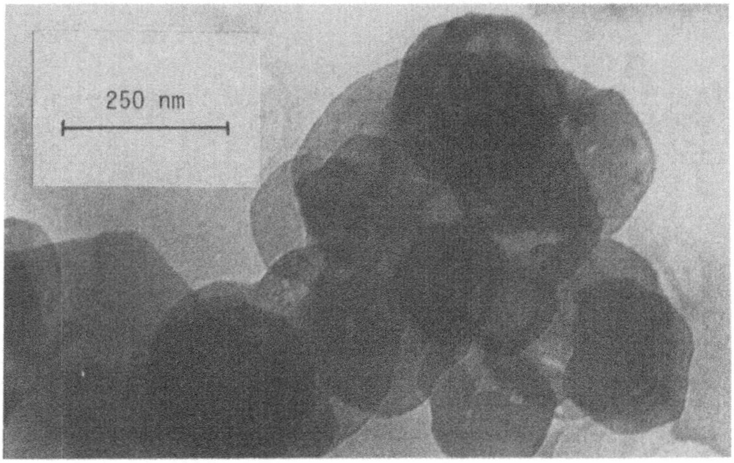

Figure 2. Transmission electron micrograph of sample B.

3.3. THERMAL ANALYSIS

The DTA and TG profiles for both samples were very similar. Three endothermic peaks are recorded in the DTA at 540, 615 and 710 K, these last two more or less overlaped, and correspond to three weight losses (the last two unresolved). Final weight is *ca.* 56-57% of the initial weight. A *plateau* is clearly distinguished between the first and the second weight losses, and so a quantitative analysis can be done; the results are summarized in Table 2.

TABLE 2. Summary of DTA and TG results

DTA		TG[c]	
T/K[a]	ΔT[b]	A	B
≈540	300-550	18.61	13.99
≈615			
	550-1110	24.69	28.49
≈710			

a: position of the minimum in the DTA profile; b: weight loss temperature range; c: percentage of weight loss in the temperature range.

The nature of the gases evolved during thermal treatment at increasing temperatures has been determined [8] by mass spectrometry. It has been found that water is evolved along the first thermal effect, and water and carbon dioxide during the remaining processes. Taking into account the temperatures ranges where these weight losses take place, it can be assumed that the first weight loss corresponds to evolution of molecular water from the interlayer space, while water evolved at higher temperatures comes from condensation of hydroxyl groups from the brucite-type layers. So, from the weight loss coinciding with the first endothermic effect, the number of water molecules existing in the samples could be calculated. The formulae are included in Table 3.

TABLE 3. Composition of the hydrotalcite materials studied

Sample	Formula
A	$[Mg_{0.654}Al_{0.346}(OH)_2](CO_3)_{0.173}\ 0.88H_2O$
B	$[Mg_{0.718}Al_{0.282}(OH)_2](CO_3)_{0.141}\ 0.61H_2O$

3.4. FT-IR SPECTROSCOPY

Positions of the peaks recorded in the FT-IRS spectra of samples A and B are summarized in Table 4. Only minor differences are found between the spectra of both samples, but some other features should be highlighted.

TABLE 4. Main FT-IR peaks for samples of hydrotalcite (cm^{-1})

Mode	Sample A	Sample B	*	*
$\upsilon_{O\text{-}H}$	3441	3470		
$\delta_{H\text{-}O\text{-}H}$	1630	1639		
$\upsilon_3(CO_3^=)$	1364	1368	1360	1415,1365
$\upsilon_1(CO_3^=)$	1060	1060	1050	1055
$\upsilon_2(CO_3^=)$	874	861	865	875,850
$\upsilon_4(CO_3^=)$	671	681	680	
$\upsilon_{Mg\text{-}O}, \upsilon_{Al\text{-}O}, \ldots$	950	934		
	785	776	780	
	556	554	555	
	451	450	450	
	396	396		

*from ref. [9]; ** from ref. [10]

The presence of molecular water can be envisaged from the band at *ca.* 1630 cm^{-1} originated by the bending mode of this molecule. The stretching mode of the hydroxyl groups (both from the brucite-like layers and from the water molecules) give rise to the band centered at *ca.* 3450 cm^{-1}, broadened because of the existence of hydrogen-bonded hydroxyl groups. It can be observed that these two bands shift towards larger wavenumbers from sample A to sample B, probably due to the change in the interlayer space, that would somewhat control the extent of hydrogen bonding and the magnitud of the vibrations.

The band close to 1360 cm^{-1} is due to mode υ_3 of carbonate, the shift form the value for free carbonate (1450 cm^{-1}) being due to restrictions within the interlayer space. The shift towards larger wavenumbers (1364 cm^{-1} for sample A and 1368 cm^{-1} for sample B) is due to larger electrostatic interactions and stronger hydrogen bonding between the hydroxyl groups and the carbonate species [10]. A shoulder is also recorded close to 1400 cm^{-1}, due to splitting of the υ_3 band because of the lower symmetry in the interlayer space, an effect also observed for aragonite [11]. This decrease in symmetry also gives rise to activation of mode υ_1, IR-forbidden for the free carbonate moiety, which band is recorded at 1060 cm^{-1}. This result also indicates that throughtly washing of the sample to eliminate chloride and sodium ions (see above) has not led to substitution of carbonate by hydroxyl in the interlayer space. Other bands due to carbonate ions are recorded at *ca.* 870 cm^{-1} (υ_2, out-of-plane deformation) and 680 cm^{-1} (υ_4, in-plane deformation).

3.5. CALCINED SAMPLES

As mentioned above, the use of LDHs is not restricted to these materials themselves, but also to those obtained after calcination at given temperatures. Such a treatment leads to removal of water moleces, hydroxyl groups and carbonate anions, as confirmed by thermal analysis and mass spectrometry monitoring of the evolved gases. In order to assess the formation of new crystalline phases, as well as their surface properties, samples A and B have been submitted to calcination in air for 2 h at 400, 548, 773, 1023, 1273, and 1373 K, i.e., temperatures below and above the features recorded in the DTA curve. Samples are named A-T or B-T, where T stands for the calcination temperature (in K).

The species identified in the XRD profiles of these samples are indicated in Table 5 for samples obtained from sample A; similar results were obtained for series B-T.

TABLE 5. Crystalline species identified by XRD in the calcined materials

A-400		A-548		A-773		A-1023		A-1273		A-1373	
d/Å	comp.	d/Å	comp.	d/Å	comp.	d/Å	comp.	d/Å	comp.	d/Å	comp.
7.60	H	6.69	H								
								4.66	S	4.64	S
3.79	H	3.37	H								
								2.86	S	2.84	S
2.58	H	2.60	H	2.59	H	2.54	H				
								2.43	S	2.43	S
2.30	H	2.46	H								
				2.09	M	2.09	M	2.10	M	2.10	M
1.97	H										
								1.65	S	1.65	S
1.52	H	1.53	H					1.56	S	1.55	S
1.49	H	1.50	H	1.48	M	1.48	M	1.49	M	1.49	M

H=hydrotalcite (file 14-191); S=spinel $MgAl_2O_4$ (file 10-237); M=MgO (file 4-0829) [7]

While for samples calcined at 400 K lines due to crystalline hydrotalcite are recorded, calcination at 548 K (i.e., after the first DTA endothermic peak due to removal of interlayer water molecules), gives rise to lines shifted towards lower *d* values; ill-defined peaks are also recorded, indicating the existence of more or less amorphous material. Calcination at 773 K (slightly above the main weight loss) gives rise to a highly amorphous material, with only minor diffraction peaks due to poor crystallized MgO, although some residual peaks due to hydrotalcite can be still recorded. Calcination at higher temperatures was performed in order to assess crystallization of the products. So, calcination at 1023 K only gives rise to minor changes in the XRD profiles. On the contrary, upon calcination at 1273 and 1373 K, sharp peaks due to MgO and $MgAl_2O_4$ are recorded.

3.6. SURFACE AREA AND POROSITY

The nitrogen adsorption-desorption curves for all A-T and B-T samples have been recorded and the data have been submitted to analysis [6] to determine the specific surface area of the solids, as well as the pore size distribution and to insight in their surface texture and the changes taking place during calcination and hydrothermal treatments. The values for the specific surface areas., as determined by the BET method, are shown in Table 6.

TABLE 6. Specific surface areas (m^2/g) of the hydrotalcite samples and of the calcined materials.

A-T	S_{BET}	B-T	S_{BET}
A-300	115	B-300	22
A-548	117	B-548	26
A-773	143	B-773	117
A-1023	154	B-1023	156
A-1273	88	B-1273	59
A-1373	25	B-1373	25

All isotherms correspond to type II in the IUPAC's classification, corresponding to unrestricted adsorption on non-porous or macroporous solids. For sample A-300, a type H3 hysteresis loop is found, that corresponds to adsorption of aggregates of platy particles [12]. The presence of hysteresis loops for sample A-1373 can be hardly concluded, due to the much lower adsorption capacitiy of this sample (*ca.* 20% of that for sample A-300). However, isotherms are almost coincident for samples B-300 and B-1373, both correspond to type II and any hysteresis loop is almost undetectable. In all cases, low adsorption capacities lead to low specific surface area values, Table 6. Differences found between samples A-300 and B-300 should be ascribed to the very different particles size that these two samples show, as observed by TEM, due to crystallization and size increase for sample B-300 upon hydrothermal treatment, in agreement with previous results with samples similar to those here being studied [5, 8].

Assessment of changes in the surface texture of both series of samples upon calcination at increasing temperatures has been done following the *f*-plot method due to Gregg [13]. This is obtained by plotting the value of *f*, defined as:

f = [amount of gas adsorbed by sample 1]/[amount of gas adsorbed by sample 2]

vs. the relative pressure; in this way, if both samples are identical, a straight horizontal line with zero intercept equal to one should be obtained; deviations indicate developing or cancellation of pores with a diameter corresponding to the relative pressure where such deviations turn up, as related throughhm the Kelvin equation. The adsorption-desorption isotherm corresponding to sample B-1373 has been taken as a reference (i.e., as "sample 2") to obtain the plots included in Fig. 3. The curves obtained are essentially horizontal, deviating only for P/Po values largere than 0.8, equivalent to diameters with pores larger than 100 Å. It should be noted that, as expected from the isotherms shown in Fig. 3, zero intercept for this plots is close to one for samples B-300 and A-1373, but close to 5 for sample A-300.

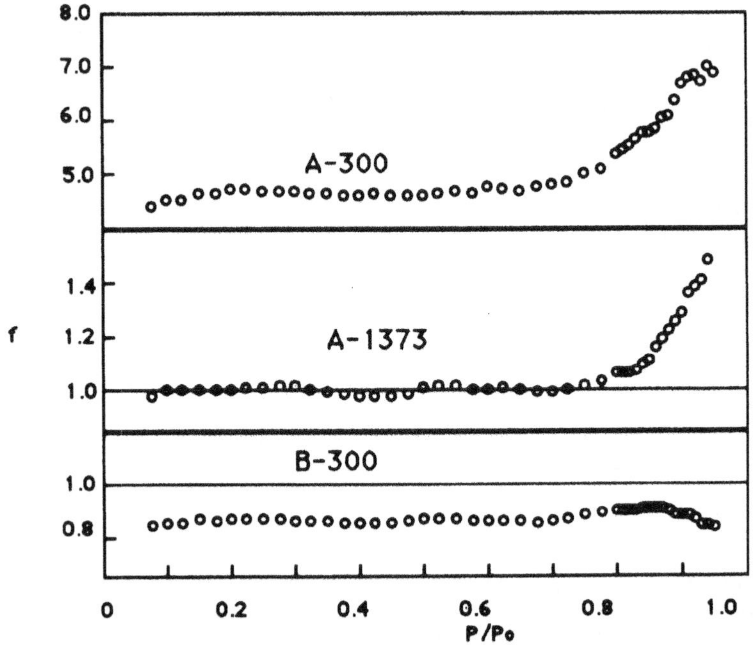

Figure 3. *f*-plots for samples A-300, A-1373 and B-300. Sample B-1373 was taken as reference in all cases.

With regards to changes in SBET values with the calcination temperature, data included in Table 6 are highly informative. So, only minor changes are observed after calcination at 548 K, where the hydrotalcite structure is still maintained (cf. data in Table 5). However, upon calcination at 773 or 1023 K, where the hydrotalcite structure is not longer stable and only MgO and amorphous MgAl2O4 spinel is detected by XRD, a large increase in the specific surface area is nitoceable, leading to essentially *identical* S_{BET} values in both cases, despite sample B had been originally treated hydrothermally. When the calcination temperature is further increased, sintering of the crystals takes place, as concluded from the sharpening of the XRD peaks and the decrease in S_{BET}, finally leading *in both series of samples* to the same surface area development.

4. Conclusions

It has been reported in the literature that the original LDH structure of materials calcined at moderate temperatures (700 K) can be regenerated by hydration in aqueous solutions of the anion to be inserted in the interlayer space [14-16], and this fact has been used in order to intercalate organic or inorganic anions that otherwise would be very unlikely to insert in the interlayer space of these materials [17-19].

From the results reported in this study, it can be concluded that hydrothermal treatment of hydrotalcite materials leads to better crystallized materials, but with a very much lower surface area development (ca. one fifth of the value corresponding to the untreated sample). However, when the samples are calcined at increasing temperatures, once the hydrotalcite structure has been destroyed and formation of MgO and spinel is detected by XRD, the sample looses any *memory effect* and surface area development is not further related to any hydrothermal treatment the original sample had been given, obtaining solids with the same S_{BET} values.

Acknowledgments

Finantial support from CICYT (grant MAT91-0767) is greatly acknowledged.

5. References

1. Reichle, W. T. (1986) 'Synthesis of anionic clay minerals (mixed metal hydroxides, hydrotalcite)' Solid State Ionics 22, 135-141.
2. Jones, W. and Chibwe, M. (1990) 'The synthesis, chemistry and catalytic applications of layered double hydroxides', in I. V. Mitchell (ed.), Pillared Layered Structures: Current Trends and Applications, Elsevier Applied Sciences, London, pp. 67-77.
3. Cavani, F., Trifirò, F. and Vaccari, A. (1991) 'Hydrotalcite-type anionic-clays: Preparation, properties and Applications', Catalysis Today 11, 173-301.
4. Martin Labajos, F. (1990) 'Preparación y caracterización de hidróxidos dobles laminares (Preparation and characterization of layered double hydroxides)', M. Sc. Thesis, University of Salamanca, Spain.
5. F. M. Labajos, V. Rives and M. A. Ulibarri (1992) 'Effect of hydrothermal and thermal treatments on the physicochemical properties of Mg-Al hydrotalcite-like materials', J. Materials Sci. 27, 1546-1552.
6. Rives, V. (1992) 'A computer programme to analyze nitrogen adsorption isotherms on porous solids', Adsorption Sci. Technol., in press.
7. Joint Committee on Powder Diffraction Standards (1977), International Centre for Diffraction Data, Pennsylvania, U.S.A.
8. R. Trujillano Hernández (1992) 'Hidroxicarbonatos de magnesio y aluminio con estructura tipo hidrotalcita (Magnesium and aluminium hydroxicarbonates with hydrotalcite-type structure)', M. Sc. Thesis, University of Salamanca, Spain.
9. Yamaguchi, O., Taguchi, H., Miyata, Y., Yoshinaka, M. and Shimizu, K. (1987) 'Characterization of magnesium-aluminium double hydroxide prepared from alkoxides and its products of thermolysis' Polyhedron 6, 1587-1592.
10. Hernandez-Moreno, M. J., Ulibarri, M. A., Rendon, J. L. and Serna, C. J. (1985) 'IR Characteristics of hydrotalcite-like compounds' Phys. Chem. Minerals 12, 34-38.
11. Rives-Arnau, V., Munuera, G. and Criado, J. M. (1979) 'Raman spectrum of the

split υ_4 mode of $CO_3^=$ ions in aragonite', Spectroscopy Lett. 12, 733-738.
12. Sing, K. S. W. (1989) 'The use of gas adsorption for the characterization of porous solids', Colloids & Surfaces 38, 113-124.
13. Gregg, S. J. (1975) 'A simple method for comparing the shapes of closely related adsorption isotherms', J. Chem. Soc., Chem. Comm. 699-700.
14. Sato, T., Wakabayashi, T. and Shimada, M. (1986) 'Adsorption of various anions by magnesium aluminium oxide ($Mg_{0.7}Al_{0.3}O_{1.15}$)', Ind. Eng. Chem. Prod. Res. Dev. 25, 89-92.
15. Sato, T., Kato, K., Endo, T. and Shimada, M. (1986) 'Preparation and chemical properties of magnesium aluminium oxide solid solutions', React. Solids 2, 253-260.
16. Sato, T., Fujita, H., Endo, T. and Shimada, M. (1988) 'Synthesis of hydrotalcite-like compounds and their physico-chemical properties', React. Solids 5, 219-228.
17. Chibwe, K. and Jones, W. (1989) 'Intercalation of organic and inorganic anions into layered double hydroxides', J. Chem. Soc. Chem. Comm., 926-927.
18. Chibwe, K. and Jones, W. (1989) 'Synthesis of polyoxometalate-pillared layered double hydroxides via calcined precursors', Chem. Mayterials 1, 489-490.
19. Atkins, M. P., Chibwe, M. and Jones, W. (1991) 'Process for the preparation of catalysts' European Patent 0,421,678,A1.

THE MICROSTRUCTURE OF LAYERED DOUBLE HYDROXIDES MODIFIED BY CONTROLLED ANION INTERCALATION

W. Kagunya and W. Jones
Chemistry Department, University of Cambridge, Lensfield Road Cambridge CB2 1EW, England.

ABSTRACT The surface characteristics of some layered double hydroxides before and after thermal treatment are reported. The identity of the anion, and consequently the unit cell parameter *c*, does not appear to significantly affect the observed surface areas. BET analysis reveals that the surface areas observed for the uncalcined samples arise mainly from pores with diameters larger than 20 Å. These pores are generated from inter-plate packing rather than holes or voids within the plates. The opposite is found for those samples calcined at temperatures above 200°C. The results associated with experiments aimed at generating materials with superior porosity by controlling the amount of anions intercalated are also be presented.

1. INTRODUCTION

Different strategies have been reported concerning the tailoring of pore size distributions in pillared clays for applications in the fields of catalysis, selective adsorption and molecular sieving [1]. These strategies include changing the size of the pillar used to separate the clay sheets and controlling of the amounts of intercalate incorporated. In this paper the second of these two strategies was extended to layered double hydroxides (LDHs) - a group of materials known for their high basic properties [2].

LDHs may be represented by the general formula

$$M^{2+}{}_{1-x}M^{3+}{}_{x}(OH)_2.X^{m-}{}_{x/m}.nH_2O$$

As in the case of cationic clays, LDHs have charged sheets, but with the *positive* charge on the sheets being counterbalanced by anions (along with varying amounts of water) within the interlayer region The chemistry of LDHs has been of increasing interest in recent years because of their anion exchange properties, medicinal capabilities and catalytic uses [3-11]. For instance, following thermal activation, LDHs exhibit outstanding basic properties:an aspect which has found use in a number of applications [12].

In this paper we report on the properties of two types of LDH intercalates. In the study it was hoped that by controlling the synthesis of the LDH, and with the use of a combination of BET adsorption and X-ray diffraction analysis to examine the resulting products, a coherent picture of the structure of the two LDH types - namely magnesium aluminium carbonate and magnesium aluminium vanadate, would emerge.

C.A.C. Sequeira and M.J. Hudson (eds.),Multifunctional Mesoporous Inorganic Solids, 217–224.

2. EXPERIMENTAL

The chemicals used were of standard analytical grade, and were supplied by Aldrich. All water used was both distilled and de-ionized.

Synthesis of MgAl- carbonate and MgAl- vanadate intercalates

$MgAlCO_3$ was prepared by dissolving 51.28g of $Mg(NO_3).6H_2O$ and 37.51g $Al(NO_3)_3.9H_2O$ in 140ml water to form an acidic solution. An alkali solution was prepared by dissolving 16g Na_2CO_3 and 24.0g NaOH in 170ml water. The two solutions were then mixed into a container with vigorous stirring so as to maintain the pH close to 10, and the temperature at 25 ± 5°C. The resulting gel was aged for 18hrs at 65 ± 2 °C, followed by cooling, washing and drying at 110°C for 18hrs to yield a product designated as UNC65.

Some of the dry solid was ground into a fine powder, separated into four portions and each portion calcined at either 360, 410, 450 or 550°C for 10 hrs. The products were cooled under vacuum before any further handling. The four calcined samples were exposed to vanadate anions using the following procedure.

5 grammes of the calcined hydrotalcite was added to a three neck round bottom flask containing some thoroughly degassed 0.25M $NaVO_3$ solution (250ml by volume) at 65°C. The mixture was stirred with heavy purging of N_2 for approximately 15 minutes, the actual time being largely determined by the time taken for the pH to stabilise. The pH of the solution was then lowered steadily to 4.5 ± 0.5 using 2M HCl solution. After maintaining this pH for a further 15 minutes, the resulting yellow solid was filtered, washed with hot decarbonated water and allowed to dry at 100°C overnight. Samples were subsequently calcined at 450°C for 18hrs. (Scheme 1 summarises the abbreviations used in this paper).

Scheme 1. Details of samples and treatment.

Acronym	Treatment
UNC65	$MgAl-CO_3$ prepared by crystallisation at 65°C
UNC200	$MgAl-CO_3$ prepared by crystallisation at 200°C
CALC65	UNC65 after calcination at 550°C
CALC200	UNC200 after calcination at 550°C
$UNCV_xO_y$	CALC65 after exposure to vanadate solution
$360V_xO_y$	UNC65 after calcination to 360°C, followed by exposure to vanadate solution, then a 2nd calcination at 450°C
$410V_xO_y$	UNC65 after calcination to 410°C, followed by exposure to vanadate solution, then a 2nd calcination at 450°C
$450V_xO_y$	UNC65 after calcination to 450°C, followed by exposure to vanadate solution, then a 2nd calcination at 450°C
$550V_xO_y$	UNC65 after calcination to 550°C, followed by exposure to vanadate solution, then a 2nd calcination at 450°C

Room temperature X-ray diffractograms were obtained using a Philips diffractometer, with a PW1710 control unit, a PW1050 goniometer, graphite monochromator and Cu K_α radiation (λ =1.5418Å). Continuous scans (from 3 to 80°

(2θ)), were made for samples randomly oriented on glass slides, with a step size of 0.05° and a count time of 1 second.

Surface area measurements were carried out by the nitrogen BET adsorption method on a Micromeritics ASAP 2000 analyser. The samples were cleaned by outgassing any adsorbate until a pressure of the order of 10^{-6} torr was attained (duration of outgassing ranging normally from 2 to 12 hrs), prior to which the space above the sample was measured using helium gas. The temperatures used for outgassing were; 100°C for the uncalcined samples, and the respective temperatures of final calcination for the calcined ones.

3. RESULTS

3.1 Crystallisation temperature

Nitrogen BET surface area data are shown in Table I. The surface area of UNC65 was observed to be approximately four times that of UNC200. Of the total surface areas of these two samples, less than 57% was a contribution from mesopores for UNC200 compared to 81% for UNC65. Though the sorption isotherms for UNC65 and UNC200 were both of type II (BDDT classification) [13], the one associated with UNC65 showed a hysteresis whereas that of UNC200 did not (Figure 1).

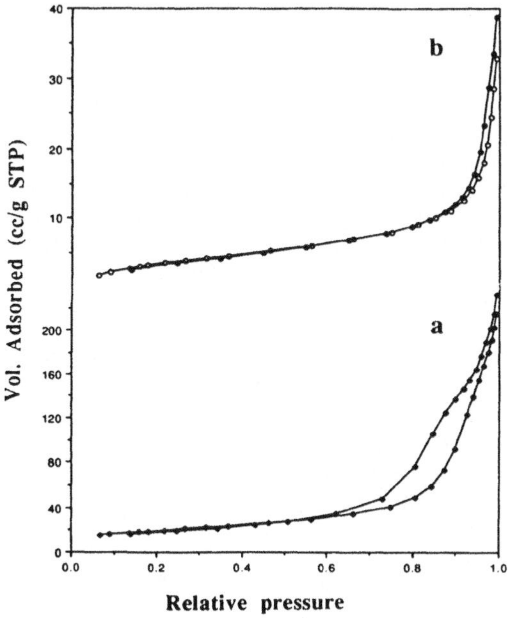

Figure 1. N_2 sorption isotherms for $MgAl\text{-}CO_3$ crystallised at (a) 65 and (b) 200°C.

3.2 Thermal treatment of UNC65

CALC65 had a higher surface area as well as a narrower pore size distribution range compared to CALC200 (Table I). In the preparation of the vanadate samples, a precursor with a high surface surface area as well as a narrow pore size distribution range would present more interesting adsorbtion properties, hence all subsequent investigations focussed on the 65°C crystallised sample. For a range of calcination temperatures up to

550°C the surface areas, as well as the average pore diameters and volumes for UNC65, were found to increase up to 450°C when maximum values were attained (Figure 2).

With the exception of UNC65 and the product obtained from calcining it at 450°C, all other samples exhibited isotherms of type V (A)- de Boer's classification [14]. These isotherms have been associated with mesoporous materials having cylindrical pores open at both sides. The isotherm obtained for the 450°C calcined sample (type IV), is also obtained for reference solids which in addition to being mesoporous, adsorb to form multilayers.

TABLE I. Surface areas and pore size distributions for MgAl-LDH samples.

Sample	Surface area(m^2/g)	Average pore dia. (Å)	% pores micro-	% pores meso-	% pores macro-
UNC65	69	182	0	81	19
UNC200	14	116	6	≈ 57	≈ 37
CALC65	205	142	3	89	8
CALC200	169	40	≈ 33	≈ 56	≈ 11

3.3 Incorporation of vanadate anions

A study on the structure and reactivity of oxovanadate anions has revealed that solutions of pH values between 4 and 6 contain predominantly $V_2O_7^{4-}$ species [15]. It was therefore assumed that since rehydration of the calcined material was performed at pH 4.5, the resulting systems contained mainly this type of anions.

The X-ray powder patterns obtained for UNC65 and CALC65 are shown in Figures 3(a) and (b) respectively. The X-ray powder pattern obtained for the product formed following the exposure of CALC65 to vanadate anions to form a system hereafter referred to as $UNCV_xO_y$, is shown in Figure 3(c). The *c* parameter was observe to increase from 7.6Å (UNC65) to 10.1Å ($UNCV_xO_y$). Assuming a brucite layer thickness of 4.8Å, the observed interlayer space of 5.3Å for the vanadate intercalate, agrees with the height calculated for $V_2O_7^{4-}$ species lying with the O-V-O bond parallel to the brucite-like layer.

No significant differences in surface areas were observed between UNC65 and $UNCV_xO_y$ - the surface area contribution from mesopores and macropores were 81 and 19%, and 90 and 10% respectively (Table II). In addition, the isotherms obtained were both of type V (BBDT), indicating some similarity in surface properties.

TABLE II. Surface areas and pore size distributions for MgAl-LDH samples.

Sample	Surface area(m^2/g)	Average pore dia. (Å)	% pores micro-	% pores meso-	% pores macro-
UNC65	69	182	0	81	19
$UNCV_xO_y$	48	219	0	90	10

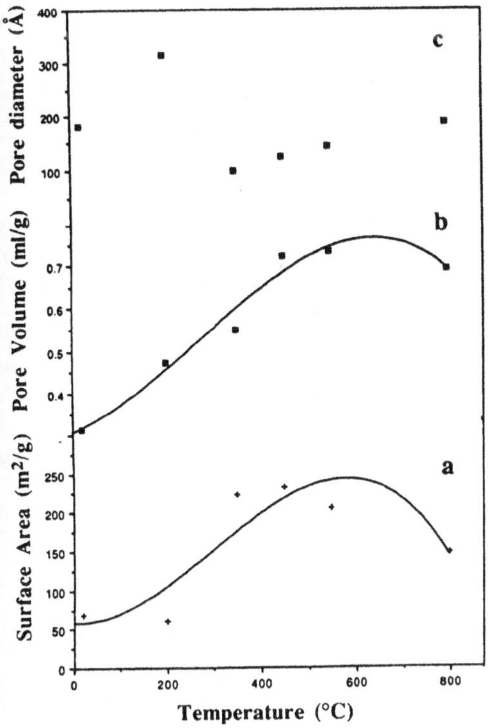

Figure 2. Change in surface properties with calcination temperature for UNC65.

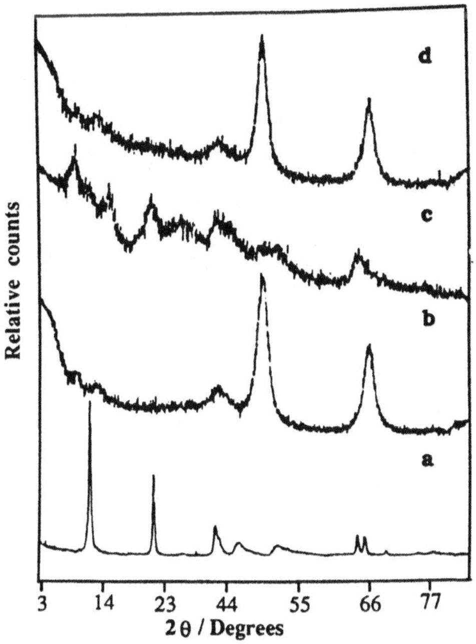

Figure 3. PXRD patterns for (a) UNC65, (b) CALC65, (c) $UNCV_xO_y$, and (d) $UNCV_xO_y$ after calcination at 450°C (18hrs).

3.4 Thermal treatment of $UNCV_xO_y$

Plots showing changes in surface areas, pore volumes and average pore diameters for $UNCV_xO_y$ as a function of calcination temperature, are shown in Figure 4. The surface area increased with temperature, and the pore diameters decreased from an average of 219Å for the uncalcined sample to 86Å for that calcined at 450°C for 18hrs. The temperature at which a rapid decrease in the pore diameter occurred was found to coincide with that at which the surface area increased rapidly. This suggests that the increase in surface area was the result of the formation of small pores due to the loss (in this case) of interlayer water.

For samples heated at temperatures between 250 and 450°C, the isotherms observed were of type V (type B). De Boer has attributed such isotherms to slit shaped pores or spaces between parallel plates. When calcined at 550°C, the resulting sample had an isotherm typical of solids having cylindrical pores open at both ends.

3.5 Controlling the amount of V_xO_y incorporated

The X-ray diffraction patterns of $360V_xO_y$, $410V_xO_y$, $450V_xO_y$ and $550V_xO_y$ were similar, and a representative diffractogram is shown in Figure 3(d). BET results for the various materials are given in Table III. It can be seen that with the exception of $550V_xO_y$, there were no significant differences in the surface areas nor in the average pore diameters. However, the proportion of micropores increased with the calcination temperature used in preparing the precursor.

In the case of $550V_xO_y$, the surface area was significantly less and the average pore diameter double that observed for the other samples. The isotherms for $360V_xO_y$, $410V_xO_y$ and $450V_xO_y$ were of the same type i.e type E, assigned to slit shaped pores or space between parallel plates. That of $550V_xO_y$ had an isotherm of type A, attributed to materials having cylindrical pores open at both ends.

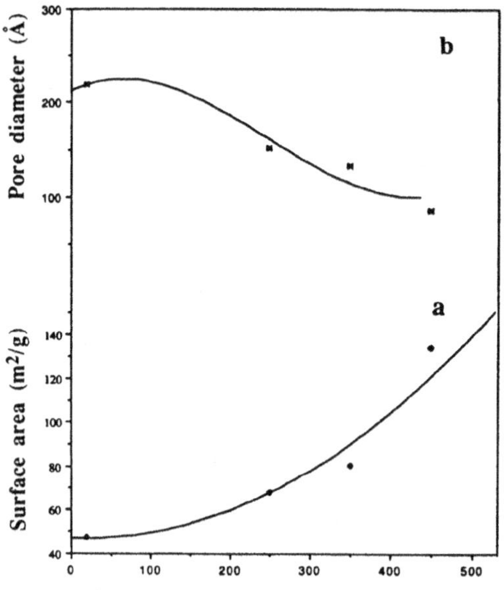

Figure 4. Change in (a) surface area and (b) average pore diameter with calcination temperature for $UNCV_xO_y$.

TABLE III. Surface areas and pore size distributions for MgAl-V_xO_y samples.

Sample	Surface area(m^2/g)	Average pore dia. (Å)	% pores micro-	% pores meso-	% pores macro-
$360V_xO_y$	194	78	2.5	≈ 94.5	≈ 3.0
$410V_xO_y$	193	73	3.8	≈ 95.1	≈ 1.1
$450V_xO_y$	180	75	8.1	≈ 89.4	≈ 2.5
$550V_xO_y$	153	152	1.6	≈ 84.5	≈ 14.0

4. DISCUSSION

The isotherms of both UNC65 and UNC200 were of type II, but they differed in that whereas that of UNC65 had a hysteresis loop, the one of UNC200 did not. Transmission electron micrographs for the two samples indicated that the particles of UNC200 were larger than those of UNC65, and that those of the latter formed agglomerates. The presence of a hysteresis loop in the isotherm for UNC65 could be attributed to the agglomeration of the particles of UNC65 due to their size, an observation which is consistent with microscopy work.

When the interlayer space was expanded from 2.8 to 5.3Å by the replacement of carbonate with vanadate anions, the sorption isotherms were observed to be different. There was no significant change in the pore size distributions, indicating that the variation of the anions had no effect on the surface properties, and more importantly, that the increase in the interlayer space did not result in the accessibility of the interlayer gallery.

The surface areas, average pore volumes and diameters of $360V_xO_y$, $410V_xO_y$, and $450V_xO_y$, differed significantly from those of $550V_xO_y$. These differences could be attributed to the pretreatment temperature used in preparing $550V_xO_y$. X-ray diffraction scans obtained for UNC65 heated to higher temperatures revealed that the LDH framework was destroyed at 550°C. Though the exposure of CALC65 to a solution of vanadate anions resulted in the re-expansion of the lattice, the diffractogram of the resulting product was indicative of a product that was characterised by lattice defects attributable to the temperature used in preparing the precursor.

5. CONCLUSIONS

The synthetic approach described here was aimed at controlling porosity, and possibly the amount of accessible internal surface area. The following conclusions have emerged:

(i) Varying the crystallisation temperature in the preparation of the LDHs with carbonate anions, resulted into a significant difference in the surface areas of the samples, attributable to the small particle size of UNC65. Thermal decomposition of this LDHs, had the effect of increasing mesoporosity for UNC65 and microporosity for UNC200.

(ii) There was no significant difference in the surface characteristics as a result of varying the interlayer anion in the MgAl-host, suggesting that the gallery region was inaccessible even after increasing its height from 2.8 to 5.3Å [16]. However, the pore shapes were observed to be different for the two samples.

(iii) The attempt to influence porosity by using precursors of differing carbonate content in the preparation of the vanadate interlayered LDHs, succeeded in generating materials with increased micro- and mesoporosity.

ACKNOWLEDGEMENTS

We are grateful to the Cambridge Commonwealth Trust for a studentship to WK, and to Dr L. Gladden and Mr P. Grange for the BET measurements. Discussions with Dr M. Atkins and the support of B.P. (Sunbury) are appreciated.

REFERENCES

1. Kikuchi, E. and Matsuda, T. (1988) 'Shape Selective Acid Catalysis by Pillared Clays', Catalysis Today, 2, 297-307.
2. Reichle, W. T. (1986) 'Anionic Clay Minerals', Chem. Tech., 58-63.
3. Bish, D. L. (1980) 'Anion-exchange in Takovite: Applications to other Hydroxide Minerals', Bull. Miner., 103, 170-175.
4. Miyata, S. (1983) 'Anion Exchange Properties Of Hydrotalcite-like Compounds', Clays Clay Min., 31(4), 305-311.
5. Dredzon, M. A. (1988) 'Synthesis of Isopolymetalate- Pillared Hydrotalcite via Organic-Anion-Pillared Precursors', Inorg. Chem., 27, 4628-4632.
6. Sissoko, I. Iyagba, E. T. Sahai, R. and Biloen, P. (1985) 'Anion Intercalation and Exchange in $Al(OH)_3$-Derived Compounds', J. of Solid State Chem., 60, 283-288.
7. Kohjiya, S. Sato, T. Nakayama, T. Yamashita, S. (1981) 'Polymerization of Propylene Oxide by Calcined Synthetic Hydrotalcite', Makromol. Chem., Rapid Commun., 2, 231-233.
8. Cavani, F. Clause, O. Trifiro, F. and Vaccari, A. (1990) 'Anionic Clays with Hydrotalcite- like Structure as Precursors of Hydrogenation Catalysis', Mat. Res. Soc. Extended Abstract, Materials Research Society.
9. Dredzon, M. A. (1989) 'Synthesis And Catalytic Activity Of Pillared Hydrotalcites', Symposium on New Catalytic Materials and Techniques, Petroleum Chemistry, Inc., American Chemical Society.
10. Trifiro, F. and Vaccari, A. (1984) 'Catalysts for Low- Temperature Methanol Synthesis. Preparation of Cu-Zn-Al Mixed Oxides via Hydrotalcite- like Precursors', Journ. of Catal., 85, 260-266.
11. Alzamora, L. E. and Julian, R. H. R. (1981) 'Coprecipitated Nickel- Alumina Catalysts for Methanation at High Temperature', J. Chem. Soc., Faraday Trans. 1., 77, 665-681.
12. Cavani, F. Trifiro, F. and Vaccari, A. (1991) 'Hydrotalcite- type Anionic Clays: Preparation, Properties and Applications', Catalysis Today, 11, 173-301.
13. Gregg, S. J. and Sing, K. S. W. (1982) Adsorption, Surface Area and Porosity, Academic Press Inc., London.
14. Shields, J. E. and Lowell, S. (1984) Powder, Surface area and Porosity, 2nd Edition, Chapman and Hall Ltd, USA.
15. Twu, J. and Dutta, P. K. (1989) 'Structure and Reactivity of Oxovanadate Anions in Layered Lithium Aluminate Materials, J. Phys. Chem., 93, 7863-7868.
16. Kwon, T. and Pinnavaia, T. J. (1989), 'Pillaring of a Layered Double Hydroxide by Polyoxometalates with Keggin-Ion Structures', 381-383.

SHAPE-SELECTIVE INTERCALATION REACTIONS AND CHEMICAL SENSING IN LAYERED METAL PHOSPHATES AND PHOSPHONATES

LOUIS C. BROUSSEAU, KATSUNORI AOKI, MAURIE E. GARCIA, GUANG CAO†, AND THOMAS E. MALLOUK
Department of Chemistry and Biochemistry,
University of Texas at Austin, Austin, TX 78712, and
†Exxon Engineering Co., Annandale, NJ 08801 USA

ABSTRACT. Layered phosphonates containing open metal coordination sites intercalate ligands such as ammonia and amines shape-selectively. As thin films on piezoelectric quartz crystals, these materials form the basis of sensitive and specific gas-phase chemical sensors for these ligands. Selective intercalation of chiral analytes has also been observed in layered metal phosphates that are pillared with chiral organic cations.

1. Introduction

One of the challenges facing both inorganic and analytical chemistry is the preparation of materials that bind small molecules selectively and reversibly. Specific host/guest interactions are used extensively by Mother Nature in processes such as enzymatic catalysis and immune reactions. Molecular recognition has also grown into an important branch of organic chemistry, and several systems of this type now perform unusually complex and often biomimetic functions [1]. In the solid-state, shape-selective reactivity, particularly in microporous solids such as zeolites, forms the basis of a catalytic chemistry of hydrocarbons that is carried out on a huge scale [2]. Nevertheless, it is probably not unfair to say that the state of the art for shape-selective host/guest chemistry in inorganic solids lags far behind that of organic and biochemical systems, and that "designed" solid state materials with true molecular recognition properties are only now beginning to emerge. With this goal in mind, we give examples in this paper that illustrate two strategies for imparting shape selectivity to intercalation processes in layered inorganic materials. One is to prepare the layers in such a way that open ligand coordination sites, which are sterically accessible only to certain molecular shapes, exist within the solid. The second is to incorporate organic pillars, which themselves have functional groups capable of binding specific guest molecules, between layers.

A second important problem to be addressed, particularly if these materials are to be used in analytical applications, is the manner in which they are interfaced to measurement devices. For electronically insulating solids, one of the most readily transduced physical changes is the increase in mass that attends the intercalation process. Transducers that are extremely sensitive to changes in mass, for example surface acoustic wave (SAW) and piezoelectric quartz crystal microbalance (QCM) devices, are commercially available and

C.A.C. Sequeira and M.J. Hudson (eds.),Multifunctional Mesoporous Inorganic Solids, 225–236.

have been widely applied to chemical sensing. Zeolites have been successfully interfaced to these devices by Bein, et al., who have detected at nanogram sensitivity levels molecules that are smaller than the zeolite pores. These sensors are also selective in that only molecules that are reactive at intrazeolitic Brønsted acid or metal coordination sites are detected [3]. In order to bind zeolite particles as thin films to the sensor surface, sol-gel and covalent anchoring techniques have been used. Device response times in these cases are largely determined by the rate of analyte diffusion into the zeolite film.

In principle, lamellar materials have an ideal morphology for application to sensor surfaces as uniform thin films, and therefore a potential (as yet unrealized) for making selective piezoelectric devices with very rapid response times. They offer the additional advantage that their interlamellar spaces may be chemically modified very easily. Layered metal phosphonates and phosphates, for example, may be grown in pellicular form by means of exfoliating reagents such as butylamine [4], and as monolayers or ultrathin multilayer films by sequential adsorption reactions of metal ions and bisphosphonic acids [5]. Two new approaches to growing thin films of these materials on piezoelectric sensor elements are explored in this paper. The first is to prepare colloidal particles by a transmetallation reaction, and to then bind the colloid covalently to the active surface. The second involves chemical generation of metal ions, which comprise the inorganic part of the solid, at the surface. While these methods produce thicker films than the sequential adsorption technique, they are generally applicable to metal salts of simple alkyl- or arylphosphonic acids. The resulting films are used to detect gas phase ammonia and alkylamines shape-selectively.

2. Materials and Methods

2.1. MATERIALS AND SYNTHESIS

Zinc phenylphosphonate monohydrate and zinc methylphosphonate monohydrate were synthesized by literature methods [6,7]. Pure materials were precipitated by combining aqueous solutions of $Zn(NO_3)_2$ and phosphonic acid, and more crystalline products could be obtained by slowly increasing the pH of acidic aqueous solutions, in which the compounds are soluble. Colloidal suspensions of the zinc phosphonates were prepared by adding 0.1 mmol $ZrOCl_2{\cdot}8H_2O$ to a suspension of 0.25 g of the appropriate zinc phosphonate in 100 mL water.

In order to obtain microcrystalline α-zirconium phosphate, a modification of Alberti's procedure [8] was used. The gel prepared by mixing solutions of $ZrOCl_2{\cdot}8H_2O$ and excess phosphoric acid was dissolved in water by dropwise addition of 48% aqueous HF, and the solution was slowly evaporated at 80-100°C by means of an air stream. Water was periodically added to maintain the volume (approximately 50 mL per gram zirconium phosphate), and after one week the suspension was filtered to yield microcrystalline $\alpha\text{-}Zr(HPO_4)_2{\cdot}H_2O$. This solid was suspended in water and titrated to pH 8.5 with tetrabutylammonium hydroxide; typically 0.20-0.25 equivalents were added. The resulting intercalated solid was separated by centrifugation, washed with water and dried, and a layer spacing of 14.6 Å was determined by x-ray diffraction. Preparation of chiral π-acceptor cations and chiral analytes followed procedures described elsewhere [9].

Other compounds were obtained from commercial sources and used without further purification.

2.2. QCM DERIVATIZATION AND MASS MEASUREMENTS

AT-cut, 6 or 9 MHz quartz crystals having Au electrodes on both sides were purchased from International Crystal Manufacturing Co. Two different methods were used to grow thin zinc phosphonate films on the electrode surfaces. In one, the crystal was immersed in an 95% ethanol solution containing 4-mercapto-n-butylphosphonic acid, and then in an aqueous colloidal suspension of the zinc phosphonate salt. In the second method, zinc was electroplated galvanostatically (0.8-1.5 mA/cm^2 x 20 min.) onto one side of the gold electrode from 0.5 M aqueous zinc sulfate. The zinc surface was then immersed for 3-5 minutes in 10 mL of 0.2 M aqueous methylphosphonic acid containing 0.1 g $Zn(NO_3)_2$ and 0.1 g $Zn(O_3PCH_3)\cdot H_2O$. Films prepared by either method were then washed with distilled water to remove soluble ions, and heated to 110°C to obtain the anhydrous form of the zinc phosphonate thin film, $Zn(O_3PR)$, R = methyl or phenyl.

The derivatized QCM's were mounted into a 250 mL three-neck flask, and were connected to a homemade oscillator circuit. Before measuring frequency changes, the flask was flushed with nitrogen, and a measured amount of the appropriate analyte was injected into the flask through a rubber septum. The frequency of the QCM was monitored with a Philips PM6654C frequency counter interfaced to an HP Vectra QS/165 personal computer. At the end of a measurement, the flask was flushed with air or nitrogen, and the crystal heated to 165° to drive off intercalated amines or ammonia. All measurements were conducted at room temperature, 23 ± 1°C.

3. Results and Discussion

3.1. STRUCTURE, TOPOCHEMICAL DEHYDRATION, AND INTERCALATION REACTIONS OF DIVALENT METAL PHOSPHONATE SALTS

Phosphonates of small divalent metal ions (Mg, Mn, Co, Ni, Zn) crystallize from aqueous solutions as layered monohydrate salts, $M(O_3PR)\cdot H_2O$ [10,11]. These compounds are isostructural, and closely related in structure to the inorganic salts $FeNH_4PO_4$ [12]; they are however structurally distinct from compounds of the same formula in which the metal ion is slightly larger (Ca, Cd) [13]. For the small divalent metal ions, the structure allows for alkyl or aryl groups of cross-sectional area 28 $Å^2$ or less (compared to ca. 24 $Å^2$ for the zirconium phosphonates $Zr(O_3PR)_2$), and so simple n-alkyl and phenyl groups are easily accomodated.

One of the most interesting features of these layered compounds is that their water of hydration, which occupies one site in the metal coordination octahedron, can be removed topochemically at relatively low temperatures. While the structure of the dehydrated product is not known precisely, x-ray powder patterns show that the in-plane unit cell dimensions are only slightly altered, and the space group ($Pmn2_1$) is the same after dehydration [6,7]. From symmetry considerations, it can be concluded that the metal must be 5-coordinated in the anhydrous salts, and therefore an open site exists for coordination of an extra ligand. The departing water molecule thus leaves a metal-organic "molecular sieve", in which the chemical nature of the binding pocket is determined by the metal ion; for example, zinc and cobalt salts prefer to bind nitrogen-containing ligands, whereas Mg and Mn have a higher affinity for oxygen ligands. In related work, Johnson and coworkers have shown that similarly prepared microporous vanadyl phosphonates with open coordination sites will bind oxygen-containing ligands, such as alcohols, shape-selectively [14].

The available in-plane area of 28 $Å^2$ in anhydrous $M(O_3PR)$ must accomodate both the R group and the entering ligand, so the coordination site is quite restrictive with respect to ligand size and shape. With R = methyl, primary aliphatic amines readily intercalate the Co and Zn salts, but amines with branching at the α-position are sterically excluded [6]. For R = phenyl, only ammonia is intercalated, and even primary amines are excluded [7].

3.2 DETECTION OF AMMONIA AND AMINES USING DERIVATIZED QCM'S

The shape-selectivity of the intercalation reaction in these solids suggests that they should be usable as active elements in sensors for ammonia and amines. Since the intercalation is irreversible at ambient temperature, these are "integrating" sensors; the anayte can nevertheless be removed thermally, regenerating the active form of the sensor element.

The sensitivity of a QCM device to mass changes is given by the Sauerbrey equation (1) [15], in which Δf is the measured frequency change which accompanies a change in mass

$$\Delta f = -2.3 \times 10^6 \times f^2 \times \Delta m/A \quad (1)$$

Δm. Stoichiometric intercalation of the anhydrous zinc phosphonates produces mass ranges of 5-25%, depending on the molecular weight of the analyte, and the frequency can be measured to an accuracy of a few Hz; therefore, it is sufficient to have as little as a few micrograms per square centimeter, i.e., a film about 100 Å thick, on the surface of the device. In practice, we have used much thicker films, as these are more easily obtained by the methods described below. The frequency change also varies inversely with film area, and we have therefore devised methods for growing films only on the surface of the gold electrodes, which cover about 20% of the entire crystal area.

3.2.1. *QCM Derivatization With Colloidal Particles*. Figure 1 shows schematically the procedure for preparing colloidal particles of zinc phenylphosphonate hydrate and attaching them covalently to the gold surfaces of QCM devices. Initially, the solid is in the form of relatively large crystallites. Since the solubility product of the phenylphosphonate of Zr^{4+} is much smaller than that of Zn^{2+}, the latter is quantitatively displaced from the solid when a substoichiometric amount of $ZrOCl_2 \cdot 8H_2O$ is added to the aqueous suspension. This has the effect of "chopping up" the solid into smaller particles. Also, because the metal ions are in excess, each particle is terminated with a metal ion layer and consequently bears an overall positive charge. A colloidal suspension of small particles is therefore formed.

In order to prevent adsorption of these particles to the exposed quartz surfaces of the QCM, and to promote binding to the gold electrode surfaces, the QCM is first pre-treated

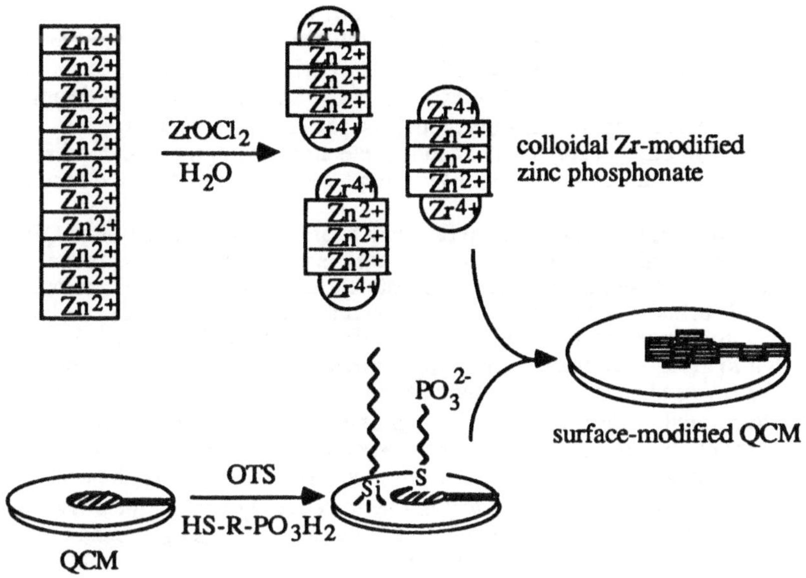

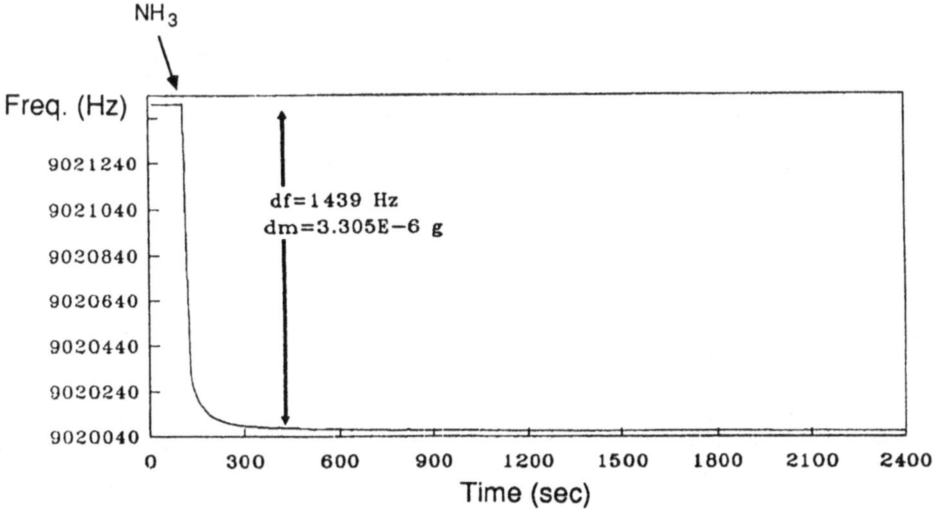

Figure 1. Top: Preparation of a zinc phenylphosphonate colloid, and selective derivatization of the gold electrodes of a QCM crystal. Bottom: Frequency response following injection of ammonia vapor.

with a solution of 4-mercaptobutylphosphonic acid. The thiol has a high affinity for the gold, but not the silica surface. In some experiments the silica surface was also rendered hydrophobic by derivatization with octadecyltrichlorosilane (OTS), in order to prevent binding of particles; this additional step was later found to be unnecessary, as the colloid is not bound at an un-derivatized silica surface. The derivatized gold presents a surface of free phosphonate groups to the suspension, and the colloidal particles bind through strong Zr-phosphonate ionic interactions, much as free Zr^{4+} ions bind to alkylphosphonate multilayer films [5]. Scanning electron micrographs show that a film of randomly oriented micron-size particles is bound, covering only the gold surface.

The response time of QCM's prepared in this way to ammonia vapor is quite rapid. Figure 1 shows a typical plot obtained by exposing the QCM to the vapor above a concentrated aqueous ammonia solution. The intercalation reaction is 80% complete within about 30 seconds, and after 1-2 minutes no more frequency change can be observed. The total mass change, ca. 3 μg, is consistent with stoichiometric intercalation of ammonia into the 2-3 μm thick films. Approximately 20-25% of the observed mass change can be reversed simply by pumping off the gas, and this extra mass may be attributable to reversible adsorption of ammonia at external Zr^{4+} sites. Consistent with external adsorption, we find that larger amines, such as n-butylamine, will cause a reversible response of approximately this magnitude. The remainder of the analyte, which is bound at internal Zn^{2+} sites, can only be removed by heating the device to 165°, whereupon the frequency returns to the baseline, and the crystal is again ready to sense ammonia. The data shown in Figure 1 represent the second run, after thermal regeneration of the $Zn(O_3PC_6H_5)$ film.

3.2.2. *QCM Derivatization From Electroplated Zinc Films.* A second strategy for preparing thin zinc phosphonate films involves creating a locally high concentration of Zn^{2+} and RPO_3^{2-} ions at the surface of the active electrodes. The electrodes are first electroplated with zinc metal, and then exposed to a methylphosphonic acid solution containing zinc nitrate and solid zinc methylphosphonate hydrate. This combination of ingredients ensures that the solution is saturated with zinc methylphosphonate, and is sufficiently acidic (pH ≈ 2) to cause slow corrosion of the metallic zinc film. Locally, a high concentration of Zn^{2+} is created, and at the same time the solution pH increases because protons are consumed in the corrosion reaction. $Zn(O_3PCH_3){\cdot}H_2O$ crystallites nucleate from the locally supersaturated solution, probably at etch pits in the zinc surface, and grow in a "starburst" morphology from these nucleation points. Figure 2 shows an SEM micrograph of these dendridic deposits, in which individual crystallites have grown to an average length of ca. 20 μm.

Films grown in this manner are significantly rougher and thicker, and contain larger individual crystallites, than those prepared from colloidal zinc phenylphosphonate suspensions, and consequently the response to gas phase analytes is slower. Figure 2 shows typical plots obtained upon exposure of the derivatized QCM to 5-10 torr of the four butylamine isomers. *Tert*-butylamine is only adsorbed on the surface of the crystallites, and x-ray diffraction (XRD) patterns show no evidence for bulk intercalation. N-butylamine is intercalated throughout the bulk, and diffraction patterns taken from derivatized zinc surfaces after prolonged exposure show no remnant of the original $Zn(O_3PCH_3)$ phase. *Iso-* and *sec*-butylamines represent intermediate cases. With well-crystallized $Zn(O_3PCH_3)$, we reported previously [7] that iso-butylamine is intercalated, whereas the *sec-* isomer, in which the alkyl chain is branched at the position α to the amino group, is unreactive. In the present case, XRD of similarly prepared zinc

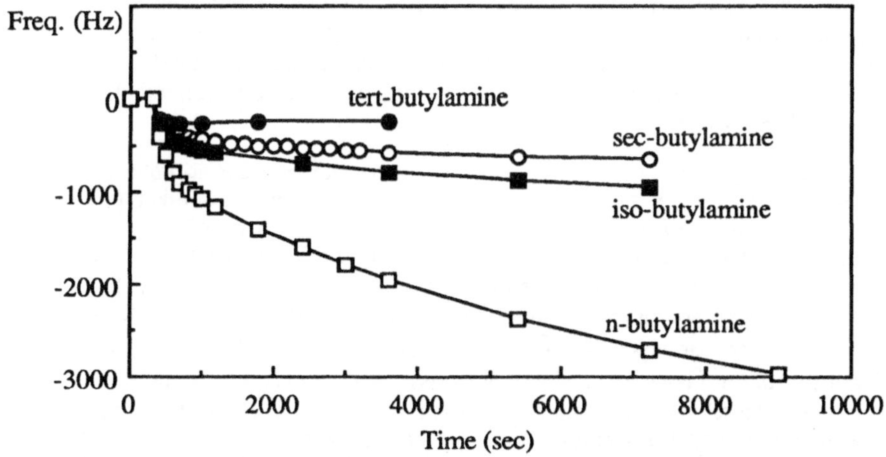

Figure 2. Top: SEM micrograph of $Zn(O_3PCH_3)$ crystallites grown by acid etching of zinc QCM electrodes and thermal dehydration. Bar indicates 10 μm. Bottom: frequency response vs. time following exposure to gas phase butylamines.

surfaces shows evidence for some intercalation by both isomers, the *iso*- isomer being more reactive than the *sec*- isomer. In the case of *sec*-butylamine, the intercalated phase accounts for about 10% of the diffraction pattern after prolonged exposure.

The slowness of these intercalation reactions is consistent with the observed film morphology, and relatively large crystallite size, when zinc methylphosphonate is grown by etching metallic zinc. While physisorption of tert-butylamine is fast, solid-state diffusion of n-butylamine into the intracrystalline volume is slow. As with the zinc phenylphosphonate films prepared from colloidal suspensions, very reproducible frequency/time plots are obtained after repeated thermal activation/intercalation cycles.

4. Chiral Host-Guest Chemistry in α-Zirconium Phosphate

In terms of shape-selectivity, chiral molecular recognition is significantly more demanding than achiral recognition. The process must discriminate two enantiomers that are exactly the same size by virtue of non-covalent interactions in a (necessarily chiral) host binding site. Although chiral microporous solids are known, most notably smectite clays [16] and zeolite beta [17], to our knowledge attempts to prepare pure solid phases of a single enantiomer have met with limited success. An easier approach to this problem is to incorporate chiral organic molecules into microporous hosts, taking advantage of both the considerable body of knowledge which exists on organic chiral molecular recognition and separations [1,18], and the extremely high internal surface area of inorganic intercalation compounds.

4.1 PREPARATION OF CHIRAL ZIRCONIUM PHOSPHATE

α-Zirconium phophate, $Zr(HPO_4)_2{\cdot}H_2O$ is a layered solid which contains acidic, exchangeable protons. Its ion-exchange capacity, 6 meq./gram, represents a density of one proton per 24 $Å^2$ on either side of each layer. This translates to a total internal surface area of approximately 950 m^2/g; put another way, once intercalated, the solid is essentially 100% surface.

We have developed a strategy for pillaring these layers with chiral π-acceptor molecules, which should be easily generalized to other chiral organic receptors of this type. The solid is first titrated with tetra-n-butylammonium hydroxide, TBA^+OH^-, which reacts with some of the acidic protons between layers. In order to achieve charge balance, TBA^+ ions intercalate into the structure to give a gelatinous, colloidal solid. The layers are swollen by water, and once most of the water has been removed by air-drying at room temperature a layer spacing of 14.6 Å (compared to 7.6 Å for α-$Zr(HPO_4)_2{\cdot}H_2O$) is found, consistent with layer expansion to accomodate the bulky TBA^+ cations. Intercalation of TBA^+ is electrostatically unfavorable, since its center of charge is buried within a sheath of nonpolar n-butyl groups; therefore, it acts as a good "leaving group" for subsequent intercalation reactions, being rapidly and quantitatively displaced by either simple alkali, or other quaternary ammonium, cations.

The cationic amino acid derivative **I** is very similar to molecules developed by Pirkle and coworkers for use in chiral chromatography [18], the difference being that **I** contains a

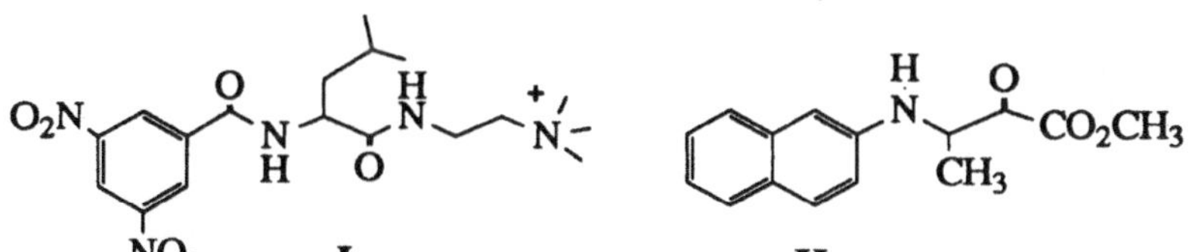

I II

positively charged trimethylammonium group rather than a silanol group. Enantiomerically pure I exchanges readily for TBA^+ in zirconium phosphate to a limiting stoichiometry of ca. 0.18 molecule I per Zr atom, affording a solid with a 19 Å interlayer spacing.

4.2 INTERCALATION OF THE π-DONOR ANALYTE II

The π-acceptor I is known to be a good chiral selector for π-donor analytes such as II. The interaction of these molecules, which occurs most effectively in nonpolar, non-hydrogen bonding solvents, involves three-point non-covalent binding through π-stacking and two hydrogen bonds [19].

Zirconium phosphate intercalated with enantiomerically pure I was reacted with racemic II in acetonitrile, and the resulting supernatant solutions were assayed by chiral HPLC. In these experiments, the mole ratio of II to I was fixed at 4:1, and the concentration of II was varied by addition of more or less solvent. Figure 3 shows the dependence of solution and solid enantiomeric excess (e.e.), which is defined according to equation (2).

$$\text{e.e.} = \left| \frac{[R] - [S]}{[R] + [S]} \right| \times 100\ \% \qquad (2)$$

Because of the mole ratio used, perfect chiral discrimination represents a solution e.e. of 33%, indicated by the dotted line in Figure 3. At high concentration, this limit is nearly achieved, with a ratio of one molecule of nearly enantiomerically pure II intercalating per molecule of the chiral receptor I. The solid acquires the yellow-orange color of the π-donor/acceptor complex, and the e.e. of the solid phase, calculated from the measured concentrations of R- and S-II in solution, approaches unity as the concentration of racemic analyte increases. The intercalation of II is reversible, and the enantiomerically enriched analyte may be recovered by eluting the solid with acetonitrile.

While the intercalation reaction goes nearly to completion at high concentration of II, at low concentration there is essentially no reaction. This behavior may be rationalized in terms of the work needed for layer expansion. When the solid is fully intercalated with II and wet with solvent, a layer spacing of 30 Å is observed by XRD, compared to 19 Å prior to intercalation. The smaller layer spacing is consistent with interdigitation of the π-acceptor cations in $I/Zr(HPO_4)_2$, and formation of a bilayer arrangement upon intercalation by II. This arrangement is shown schematically in Figure 3. The layer expansion requires work, since both the van der Waals attraction of I with its neighbors, and hydrogen bonding interactions between the NO_2 groups of I and the zirconium phosphate protons, need to be overcome. Therefore, a threshold concentration of 70-90

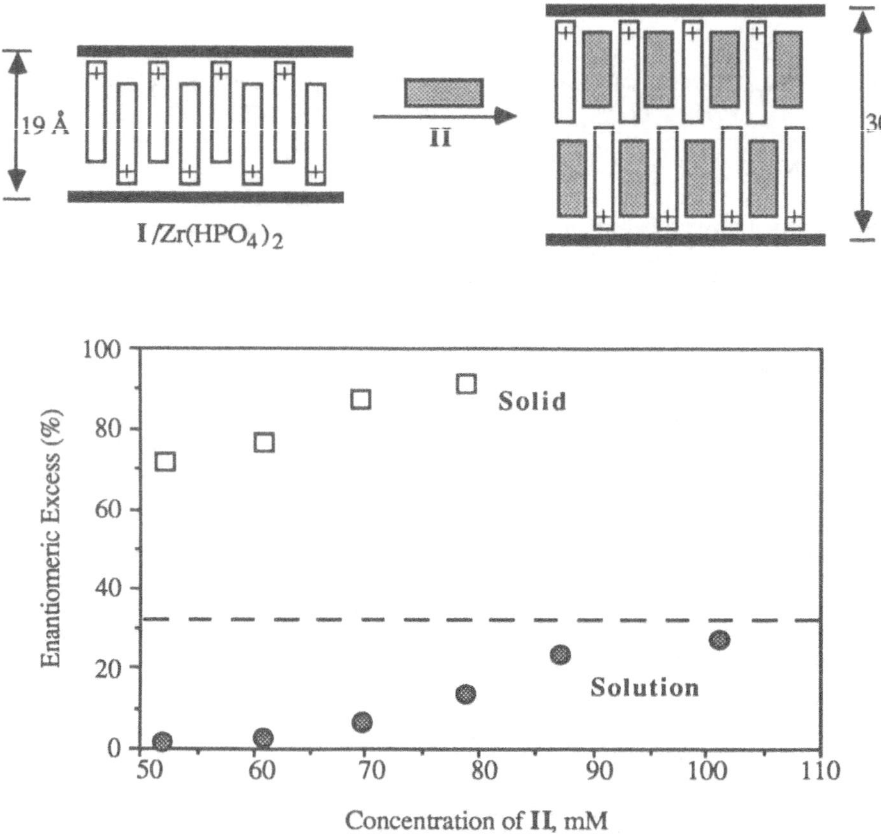

Figure 3. Top: schematic representation of layer expansion in the intercalation of zirconium phosphate/**I** by **II**. Bottom: enantiomeric excess vs. concentration of racemic **II** in acetonitrile. The dotted line represents the maximum theoretical solution e.e. for a 4:1 mole ratio of **II**:**I**.

mM (i.e., 35 - 45 mM of the intercalated enantiomer) is needed for intercalation of the chiral π-donor **II**.

5. Summary and Conclusions

Metal phosphate and phosphonate salts are interesting materials for analytical applications such as chemical sensing and separations. The ease with which the chemical and physical nature of their interlayer volume can be tuned, as well as their very high internal surface area, suggest that they may ultimately be superior to conventional materials in this regard.

We have described two methods for growing zinc phosphonate films on QCM surfaces, and have investigated their potential as sensor elements for ammonia and amines. Because the films deposited by these methods are microscopically rough, significant interferences arise from external surface adsorption. Modification of other film growth techniques [4,5] could in principle yield smoother and thinner films, which should show less interference and faster response times. An interesting new development in candidate materials for sensors is the description of some new copper phosphonates $Cu(O_3PR){\cdot}H_2O$ [20]. These lamellar solids contain 5-coordinate copper(II) ions which are intercalated *reversibly* by amines at ambient temperature; such salts could be active elements in reversible, quantitative QCM-based sensors.

Intercalation of chiral cations into zirconium phosphate results in a material which selectively binds one enantiomer of a complementary racemic analyte. At present, high analyte concentrations are needed. In its present form, this material could be useful in batchwise separation of enantiomers, but not in chromatographic applications where analyte concentrations are low. Pillaring or chemical modification of the interlayer spaces in these solids may in the future help to overcome this limitation.

6. Acknowledgment

We thank Dr. Michael Schmerling, Department of Mechanical Engineering, University of Texas at Austin, for assistance with the SEM experiments. This work was supported by grants from the National Science Foundation (CHE-8657729), the National Institutes of Health (GM43844), and the Texas Higher Education Coordinating Board Advanced Research Program. TEM also thanks the Camille and Henry Dreyfus Foundation for support in the form of a Teacher-Scholar Award.

7. References

[1] Rebek, J., Jr. (1990) 'Molecular Recognition with Model Systems', Angew. Chem. Intl. Ed. Eng., 29, 245-255.

[2] Jacobs, P. A. and van Santen, R. A., (eds.) (1989), *ZEOLITES: Facts, Figures, Future*, Elsevier, New York.

[3] Bein, T., Brown, K., Frye, G. C., and Brinker, C. J. (1989) 'Molecular Sieve Sensors for Selective Detection at the Nanogram Level', J. Am. Chem. Soc., 111, 7640-7641; Yan, Y., and Bein, T., (1991) 'Zeolite Crystal Layers Coupled to Piezoelectric Crystal Sensors: Molecular Recognition Devices', in *Synthesis/Characterization and Novel Applications of Molecular Sieve Materials*, Bedard, R. L., et al., (eds.), Materials Research Society, Pittsburgh, pp. 175-182.

[4] Alberti, G., Casciola, M, and Costantino, U. (1985) 'Inorganic Ion-Exchange Pellicles Obtained by Delamination of α-Zirconium Phosphate', J. Colloid. Interfac. Sci. 107, 256-263.

[5] Lee, H., Kepley, L. J., Hong, H.-G., and Mallouk, T. E. (1988) 'Inorganic Analogs of Langmuir-Blodgett Films: Adsorption of Ordered Zirconium 1,10-Decanebisphosphonate Multilayers on Silicon Surfaces', J. Am. Chem. Soc., 110, 618-620.

[6] Frink, K. J., Wang, R.-C., Colon, J. L., and Clearfield, A. (1991) 'Intercalation of Ammonia into Zinc and Cobalt Phenylphosphonates', Inorg. Chem., 30, 1438-1441.
[7] Cao, G., and Mallouk, T. E. (1991) 'Shape-Selective Intercalation Reactions of Layered Zinc and Cobalt Phosphonates', Inorg. Chem., 30, 1434-1438.
[8] Alberti, G., Costantino, U., and Giulietti, R. (1980) 'Preparation of Large Crystals of α-$Zr(HPO_4)_2.H_2O$' J. Inorg. Nucl. Chem., 42, 1062-3.
[9] Cao, G., Garcia, M. E., and Mallouk, T. E. (1992) 'Chiral Molecular Recognition in Intercalated Zirconium Phosphate', submitted for publication.
[10] Cao, G., Lee, H., Lynch, V. M., and Mallouk, T. E. (1988) 'Synthesis and Structural Characterization of a Homologous Series of Divalent-Metal Phosphonates, $M^{II}(O_3PR){\cdot}H_2O$ and $M^{II}(HO_3PR)_2$', Inorg. Chem. 27, 2781-2785.
[11] Martin, K., Squattrito, P. J., and Clearfield, A. (1989) 'The Crystal and Molecular Structure of Zinc Phenylphosphonate', Inorg. Chim. Acta, 155, 7-9.
[12] Durif, A., and Averbuch-Pouchat, M. T. (1968) 'Metal Ammonium Phosphate Monohydrates III: Crystallographic Data on $M^{II}NH_4PO_4{\cdot}H_2O$ and $M^{II}KPO_4{\cdot}H_2O$', Bull. Soc. Fr. Mineral Crystallogr., 91, 495-6.
[13] Cao, G., Lynch, V. M., Swinnea, J. S., and Mallouk, T. E. (1990) 'Synthesis and Structural Characterization of Layered Calcium and Lanthanum Phosphonate Salts', Inorg. Chem., 29, 3738-3745.
[14] Johnson, J. W., Jacobson, A. J., Butler, W. M., Rosenthal, S. E., Brody, J. F., and Lewandowski, J. T. (1989) 'Molecular Recognition of Alcohols by Layered Compounds with Alternating Organic and Inorganic Layers', J. Am. Chem. Soc., 111, 381-383.
[15] G. Sauerbrey (1959) 'Vermendung von Schwingquarzen zur Wägung dünner Schichten und zur Mikrowägung', Z. Phys., 155, 206.
[16] Yamagishi, A. (1987) 'Optical Resolution and Asymmetric Syntheses by Use of Adsorption on Clay Minerals', J. Coord. Chem., 16, 131-211.
[17] Treacy, M. M. J., and Newsam, J. M. (1988) 'Two New Three-Dimensional Twelve-Ring Zeolite Frameworks of which Zeolite Beta is a Disordered Intergrowth', Nature, 332, 249-251.
[18] Pirkle, W. H., and Pochapsky, T. (1989) 'Considerations of Chiral Molecular Recognition Relevant to Liquid Chromatographic Separation of Enantiomers', Chem. Rev., 89, 347-362.
[19] Pirkle, W. H., and Pochapsky, T. (1987) 'Chiral Molecular Recognition in Small Bimolecular Systems: A Spectroscopic Investigation into the Nature of Diastereomeric Complexes', J. Am. Chem. Soc., 109, 5975-5982.
[20] Zhang, Y., and Clearfield, A. (1992) 'Synthesis, Crystal Structures and Coordination Intercalation of Two Copper Phosphonates', Inorg. Chem., in press.

Micro- and Mesoporous Clays Pillared with SiO_2-TiO_2 Mixed Oxide Sols

SHOJI YAMANAKA
Department of Applied Chemistry
Faculty of Engineering
Hiroshima University
Higashi-Hiroshima 724
Japan

KOICHI TAKAHAMA
Central Research Laboratory
Matsushita Electric Works Ltd.
1048 Kadoma, Osaka 571
Japan

ABSTRACT. The silicate layers of montmorillonite clay were pillared with SiO_2-TiO_2 mixed oxide sols. Though the SiO_2 sols alone are negatively charged, the addition of a small amount of Ti^{4+} converts the sols to the positively charged ones, which can be ion exchanged with the interlayer cations of montmorillonite. The resulting pillared clays have a basal spacing expanded to a value larger than 40 Å. The spacing and the high surface area are maintained on calcination up to 500°C. A structural model for the arrangement of the intercalated sol particles is proposed on the basis of the adsorption studies; the small SiO_2-TiO_2 sol particles are packed between the silicate layers and the micropores with a slit width of 10–12 Å are formed in the interstices between the sol particles and the silicate layers. In an attempt to introduce mesopores between the silicate layers, the arrangements of the sol particles have been modified by the incorporation of organic cations and by using supercritical drying to the sol pillared clays swelled with water.

1. Introduction

Pillared clays can be regarded as layered nanocomposites, in which ceramic oxide particles of nano- and subnanometer scales are incorporated in the interlayer space of two-dimensional silicate layers with a thickness of about 10 Å. The advantage of nanocomposite processing route in developing new functional materials will be found in a variety of combinations of the fundamental oxide particles with the silicate layers, and an easy anticipation of the possible microstructures of the final products [1]. Various kinds of metal oxides such as Al_2O_3 [2], ZrO_2 [3], Cr_2O_3 [4, 5], Fe_2O_3 [6], Bi_2O_3 [7], V_2O_5 [8], ZrO_2-Al_2O_3 [9], Al_2O_3-Ga_2O_3 [10, 11], and Al_2O_3-SiO_2 [12] have been introduced into the interlayer space of clays as pillars to keep the silicate layers apart. Their thermal stabilities, adsorption properties and catalytic activities have been extensively studied. Comprehensive reviews on the recent developments of this field are available [13–18].

In recent years, much attention has been focused onto the new type of pillared clays with large pillar heights, since it will provide larger pores which have not been realized by zeolites

C.A.C. Sequeira and M.J. Hudson (eds.), *Multifunctional Mesoporous Inorganic Solids*, 237–258.

synthesized by a direct crystallization from solutions. Rightor *et al.* [20] prepared iron oxide pillared clays with a basal spacing of 25–27 Å by using hydrolyzed and aged iron solutions. This value is much larger than the spacing for a typical alumina pillared clay of 16–18 Å. Sterte [21] and Yamanaka *et al.* [22] succeeded in introducing TiO_2 sols to make TiO_2 pillared clays with basal spacings of 24–27 Å; the nitrogen adsorption isotherm was found to be Type IV rather than Type I in the classification of BDDT [23], indicating that the pores formed are much larger than those of alumina and zirconia pillared clays. A tubular clay mineral, imogolite with an ideal composition of $(OH)_3Al_2O_3Si(OH)$ and an outside diameter of 23 Å was intercalated into montmorillonite [24]. MaCauley [25] and Sterte [26] used hydrothermally treated pillaring solutions containing aluminum chlorohydrate and cerium or lanthanum salts, and obtained pillared clays with a basal spacing of 25–28 Å. Suzuki *et al.* [27] prepared an alumina pillared clay in the presence of polyvinyl alcohol (PVA). The sample calcined at 400°C showed a sharp pore size distribution centered at 25 Å.

The first and important step in pillaring clay is the ion exchange of the interlayer cations of clays with oligomeric hydroxy-metal cations as pillar precursors, which are converted into the oxide pillars between the silicate layers by calcination. Silica is also an interesting oxide for pillars; its polymeric behavior has been extensively studied, and the size of the sol particle is controllable by the hydrolysis conditions [28]. However, the precursory oligomeric ions available for silica are in the form of silicate anions, which cannot be used for the cation exchange reactions. Avoiding this problem, Endo *et al.* [29] used tris(acetylacetonato) silicon(IV) complex cations, $[Si(acac)_3]^+$ for the precursor cations or neutral silicon compound $SiCl_4$ which was hydrolyzed in the interlayer spaces by *in situ* reactions. The silica pillared clays thus obtained showed relatively low surface areas of 40–190 m^2/g and a maximum basal spacing of 12.6 Å. This spacing implies the existence of the pillars of one oxygen layer thickness. Lewis *et al.* [30] introduced the silica pillar by using polyhedral oligosilsesquioxane compounds, which intercalated into the interlayer spaces by the affinity of the organic groups with clays. Calcination of the products at 170°–700°C yielded silica pillared products with basal spacings of 16.2–19.2 Å and surface area of 140–400 m^2/g. Recently, Moini and Pinnavaia [31] attempted a direct intercalation of a commercial as well as hydrolyzed silica sol into montmorillonite. They found that relatively large surface area solids of 250–460 m^2/g were formed, although those samples were not a regular intercalation phase.

In a series of our studies [32–34], silica sol prepared by the hydrolysis of silicon tetraethoxide has been successfully intercalated into montmorillonite by modifying the sol with the addition of small amount of titania sol. The structures of the interlayer pores have been further modified by the incorporation of organic cations and by using supercritical drying with CO_2.

2. Preparation

2. 1. Sol solutions

A silica sol solution was prepared by mixing silicon tetraethoxide, $Si(OC_2H_5)_4$, 2 N HCl and ethanol in a ratio of 41.6 g/10 ml/12 ml at room temperature. According to Sakka *et al.* [35] this hydrolysis condition gives us three-dimensional or spherical silica sol without having spinnability. Titanium tetraisopropoxide $Ti(OC_3H_7)_4$ was hydrolyzed by adding it to 1 N HCl solution in such a way that the molar ratio of $HCl/Ti(OC_3H_7)_4$ is 4 [22]. The resulting slurry was peptized to a clear sol solution by continuous stirring for 3 hr at room temperature. The silica sol solutions and titania sol solutions were mixed in various ratios.

2. 2. Pillaring with Sol Particles

The clay used in this study was mainly Na–montmorillonite supplied from Kunimine Industrial Co. under the name Kunipia F, the composition of which was $Na_{0.35}Ca_{0.02}K_{0.01}[Si_{3.89}Al_{0.11}]$–$[Al_{1.60}Mg_{0.32}Fe_{0.08}]O_{10}(OH)_2$, and the cation exchange capacity (CEC) was 100 meq/100 g.

The preparation of the sol pillared clays was very simple. The silica and the titania sol solutions thus prepared were mixed in various ratios, and stirred for 30 min at room temperature, which were then added to the clay suspension (1 wt%). The mixed ratios will be hereafter designated like 30/3/1 for example, where the first and the second numbers refer to the molar ratios of silica and titania to the CEC equivalent (the third number) of the clay, respectively. The reaction temperature as well as the mixing ratio were variously changed. After stirring for different reaction times, the products were separated by centrifugation, washed with water several times, and dried in a stream of dry air at room temperature.

2. 3. X–ray diffraction study

When the silica sol solution alone was subjected to reaction with montmorillonite, no swelling of the montmorillonite was observed and the sol solution was clearly separated as a supernatant from

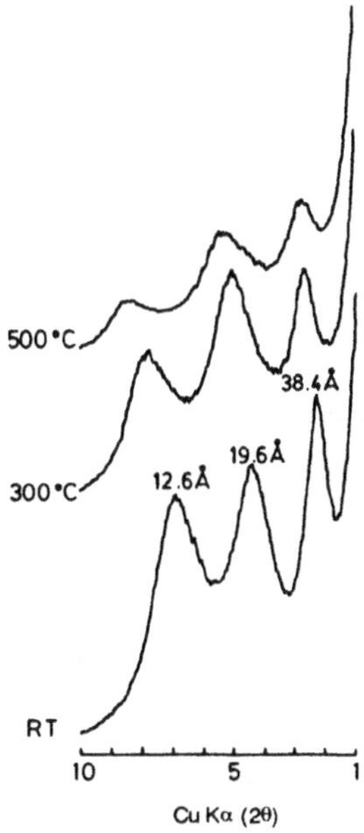

Figure 1. X–Ray powder diffraction patterns of a typical reaction product (30/3/1) air–dried at room temperature and calcined at 300° and 500°C.

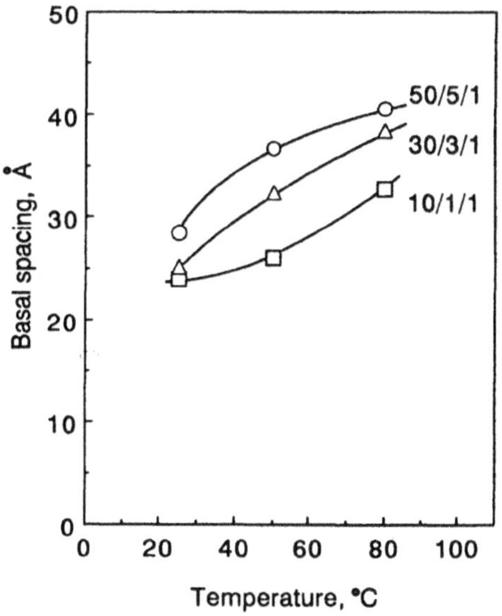

Figure 2. Basal spacings of the products obtained from three different mixtures of SiO_2–TiO_2 sols and montmorillonite as a function of the reaction temperature.

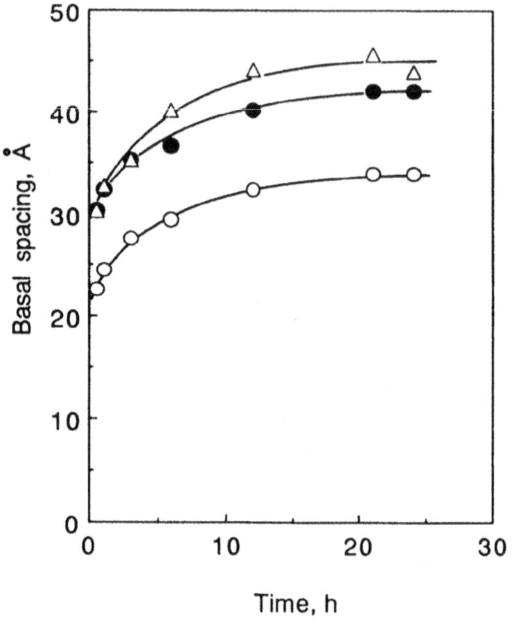

Figure 3. Basal spacing of the products as a function of reaction time at 50°C. Three different mixtures were used: ○, 10/1/1; ●, 30/3/1; △, 100/10/1.

the montmorillonite by centrifugation. However, the addition of even 1 part of the titania sol to 40 parts of the silica sol in mole is effective to increase the basal spacing to about 40 Å. Figure 1 shows a series of typical X-ray powder diffraction (XRD) patterns of the product obtained from a reaction mixture of 30/3/1 at 50°C, and thermally treated at room temperature (air-drying), 300° and 500°C. The basal spacing expanded to a value as large as 38 Å, and the higher order reflections (2nd and 3rd) were observed, suggesting the formation of a regularly intercalated compound. Though the spacing is slightly reduced on heating, the structure appears to be maintained at least up to 500°C. In order to see the effect of the reaction temperature on the basal spacing, three mixtures, 10/1/1, 30/3/1 and 50/5/1, were continuously stirred for 3 hrs at 25°, 50°, and 80°C. Figure 2 shows the basal spacing of the resulting products. Apparently, if larger amounts of sol solution and higher reaction temperatures are used, the products with larger basal spacings are obtained. In order to avoid the precipitation of the silica sol outside the clay at higher temperatures, most of the reactions were performed at 50°C.

Three different mixtures, 100/10/1, 30/3/1, and 10/1/1, were used and the change in the basal spacing was recorded as a function of reaction time, which is shown in Figure 3. The spacing increased rapidly soon after the clay was mixed with the sol solutions and then attained constant values after about 6 hr, though a gradual increase was still observed after that. As seen from the figure, the amount of silica sol 10 times the equivalent of the CEC seems to be not sufficient; the amount 30–50 times the equivalent of the CEC would be required to obtain the fully expanded products at 50°C.

Table 1. Chemical analysis data of the pillared clays prepared from different conditions.

Sol compositions SiO_2/TiO_2/CEC	Reaction conditions	Analytical results, %			Pillar compositions[1)]			
		SiO_2	TiO_2	Al_2O_3	$SiO_2(x)$	$TiO_2(y)$	$x+y$	x/y
24/6/1	50℃, 10h	64.6	15.0	12.7	3.49	1.29	4.78	2.70
15/15/1	50℃, 10h	50.8	31.8	9.43	4.15	3.82	7.97	1.08
6/24/1	50℃, 10h	31.9	51.7	8.29	1.74	6.97	8.71	0.25
10/1/1	50℃, 3h	73.0	5.70	12.4	4.66	0.50	5.16	9.32
30/3/1	50℃, 3h	73.0	8.15	10.9	5.83	0.82	6.65	7.11
50/5/1	50℃, 3h	70.5	10.35	10.2	6.15	1.11	7.26	5.54

1) The compositions (x and y) are given in moles for $O_{10}(OH)_2$ anion basis of the silicate layer.

2. 4. Effect of the sol composition

Sol solutions with a variety of SiO_2/TiO_2 molar ratios were prepared and mixed with the montmorillonite suspension. The pillared products were separated and analyzed on the SiO_2, TiO_2 and Al_2O_3 contents by the atomic absorption method. The reaction conditions and the analytical results are listed in Table 1. The composition of the silicate layers of the clay was assumed to be unchanged. The amount of SiO_2 coming from the silicate layer was estimated from the Al_2O_3 content and subtracted from the total SiO_2 content. The resulting sol formulae, $(SiO_2)_x(TiO_2)_y$, are shown on the $O_{10}(OH)_2$ anion basis of the clay. As seen from the table, more than 6–7 molar equivalent of SiO_2+TiO_2 ($x+y$) are taken up by the clay, and the SiO_2/TiO_2 ratio of the interlayer sol composition (x/y) is always higher than the corresponding ratio of the sol in the solution used. There is a tendency that the interlayer sol composition deviates to a composition richer in TiO_2 than that of the sol solution used. The deviation is larger when a larger amount of the sol to the CEC is used. This can be interpreted in terms of the higher affinity of titania with montmorillonite.

3. Adsorption Properties

3. 1. Nitrogen adsorption

Figure 4 shows a nitrogen adsorption–desorption isotherm of a typical pillared clay (50/5/1). The adsorption isotherm gives a good fit on the Langmuir (Type I) as well as the BET (infinite number of adsorption layers) equation, the regression coefficients being 0.9978 and 0.9989, respectively. A much better fit was obtained on the BET equation for the limited number of nitrogen adsorption layers (Gregg and Sing 1982). The number of the layers was estimated to be in a range of 2–3. Figure 5 shows the basal spacings and the surface areas of the sample (50/5/1) as a function of calcination temperature. The basal spacing and the high specific surface area are maintained at least up to 500°C. For convenience, two kinds of specific surface areas determined by applying the BET (infinite adsorption) and the Langmuir equations, are shown together in the figure. The former surface area is related to the amount of nitrogen required for mono–molecular coverage of

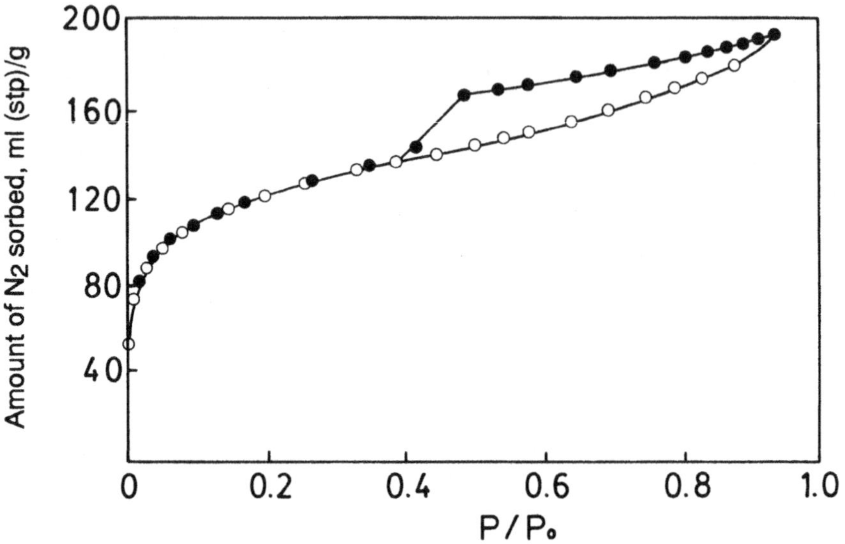

Figure 4. Adsorption–desorption isotherm of nitrogen for a typical reaction product (50/5/1) calcined at 500°C; ○, adsorption and ●, desorption.

the pores, while the latter one is related to the pore filling capacity by liquid nitrogen (1ml of gaseous nitrogen (stp) corresponds to 1.547 x 10^{-3} ml of liquid nitrogen). Figure 6 shows the relationship between the basal spacing and surface areas for the samples shown in Figure 2. The surface areas increases almost linearly with the spacing, suggesting that the porosity of the interlayer spaces are almost constant irrespective of the basal spacing.

The nitrogen adsorption–desorption isotherms and the *t* plots for the samples prepared by using the sol solutions having different SiO_2/TiO_2 ratios are shown in Figure 7. The TiO_2 pillared clay had the adsorption isotherm of Type IV, as reported in a previous paper [22]. With the increase in the SiO_2/TiO_2 ratio of the sol solution used, the isotherm became more Type–I–like, suggesting that the pores formed in the interlayer spaces must be smaller than ~15 Å in diameter. The *t*–plot is given for the adsorption isotherm for each sample in the figure. The pillared clays prepared from the sol solutions with higher SiO_2/TiO_2 ratios showed a downward deviation at high–pressure region. The line extrapolated from the high–pressure branch to the adsorption axis gives a large intercept which is equivalent to the microporous volume [36], though the extrapolations are not very certain for the samples, 0/40/1 and 15/15/1. The nitrogen adsorption isotherm for the pillared clay with a low SiO_2/TiO_2 ratio of 6/24 showed a slight upward deviation. This deviation suggests the existence of a small fraction volume of mesopores. The slit width of the pores can be estimated from the *t* value, where the *t*–plot shows a downward deviation; the value twice this *t* value is assigned to an approximate slit width of the pores. The slit width of the pores thus obtained are shown in Table 2, together with other adsorption data calculated from the isotherms and the *t*–plots shown in Figure 7. The slit width of the pores are in the range of 10–12 Å, and decreases with the increase of the SiO_2/TiO_2 ratio of the sol solutions used.

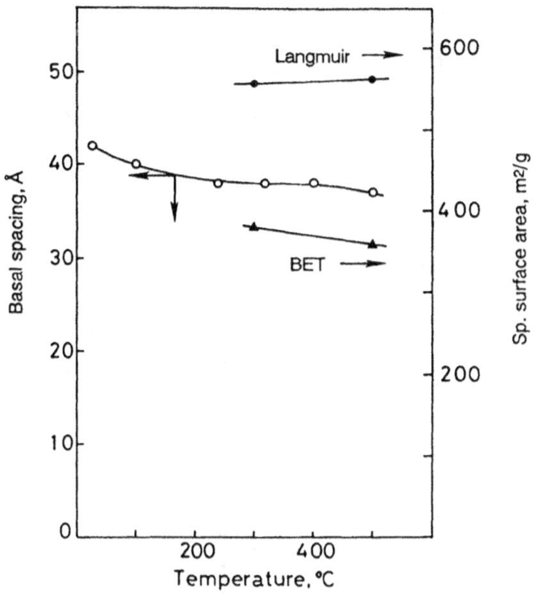

Figure 5. Basal spacing and the specific surface areas of the sample as a function of heat-treatment temperature. Two-kinds of surface areas (BET and Langmuir) were compared.

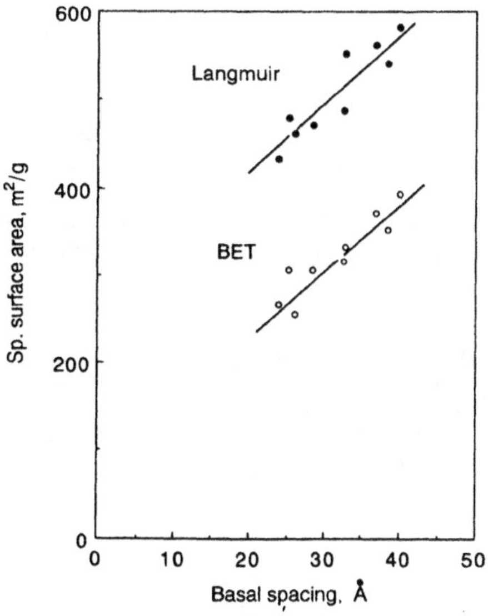

Figure 6. Specific surface aeas as a function of the basal spacing of the samples shown in Figure 2. Two kinds of surface areas (BET and Langmuir) were compared.

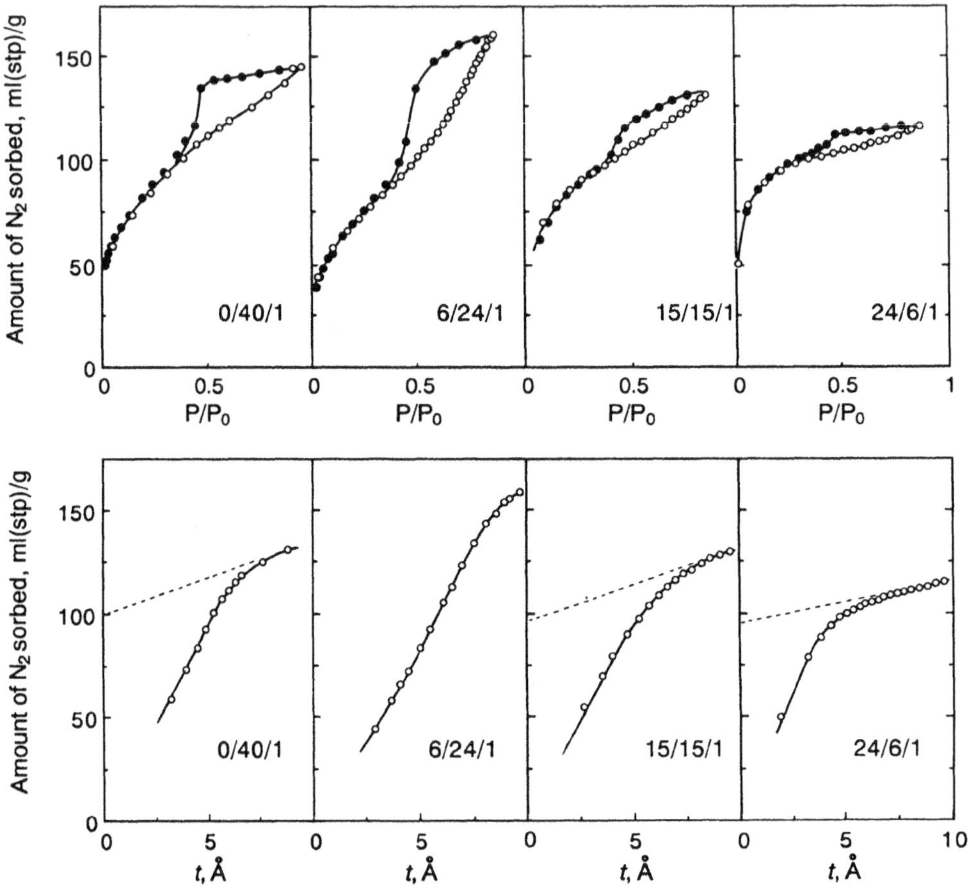

Figure 7. Adsorption–desorption isotherms (upper) and the respective *t*–plots (lower) for the pillared clays prepared by using the sol solutions with different SiO_2/TiO_2 ratios.

Table 2. Adsorption data for the pillared clays prepared from the sol solutions with different compositions.

Sol compositions SiO_2/TiO_2/CEC	Sp. surface area, m^2/g BET	Sp. surface area, m^2/g Langmuir	Micropore volume, ml/g	Slit width 2*t*, Å
0/40/1	284		0.147	14
6/24/1	249			
15/15/1	280	473	0.142	12
24/6/1	286	463	0.142	10

3. 2. Adsorption of solvents

Adsorption–desorption isotherms for various kinds of solvent vapors were measured at 25°C on the sample (30/3/1) and shown in Figure 8. In Figure 9, some of the adsorption data are plotted using Langmuir and BET isotherms. The adsorption isotherms for large molecules fit on the Langmuir linear plot, whereas those for smaller molecules such as water and methanol fit rather on the BET plot. This finding suggests that the pore sizes are in the order of the largest molecular sizes examined (7.6 Å for mesitylene). This is also in good agreement with the finding that the nitrogen adsorption isotherm fit on the BET plot for the limited number of adsorption layers. In the figure, the amount of vapors adsorbed are plotted as the liquid volumes of the solvents. The amount of adsorption near the saturation vapor pressure for each solvent corresponds to the total pore volume to be filled with the liquid. The pore volume (porosity) thus calculated is listed in Table 3. The porosities for the solvents with relatively large sizes are in the range of 0.20 – 0.21 ml/g and those for small sizes are about 0.25 ml/g. The difference in the porosity is attributed to the existence of ultra–micropores which are accessible only by the small molecules such as water and methyl alcohol, but not by the larger molecules. It is interesting to note that the hystereses observed in the desorption of large molecules such as *m*–xylene and mesitylene persist to unusually low P/P_0. Although the reason in not clear, this may not be due to non–equilibrium effects. Further studies would be required for detailed discussions.

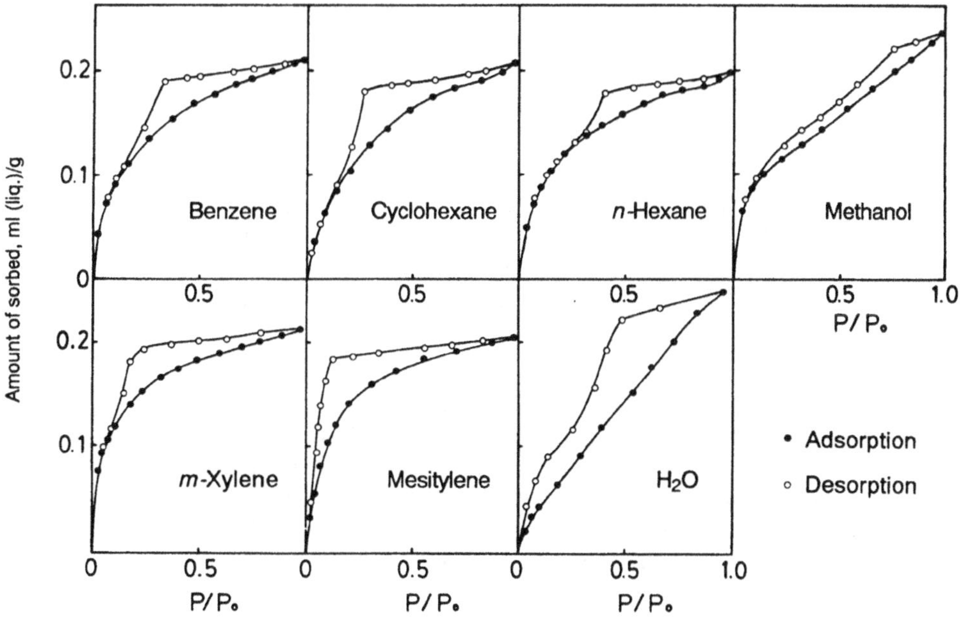

Figure 8. Adsorption–desorption isotherms of the SiO_2–TiO_2 sol pillared clay for different solvents.

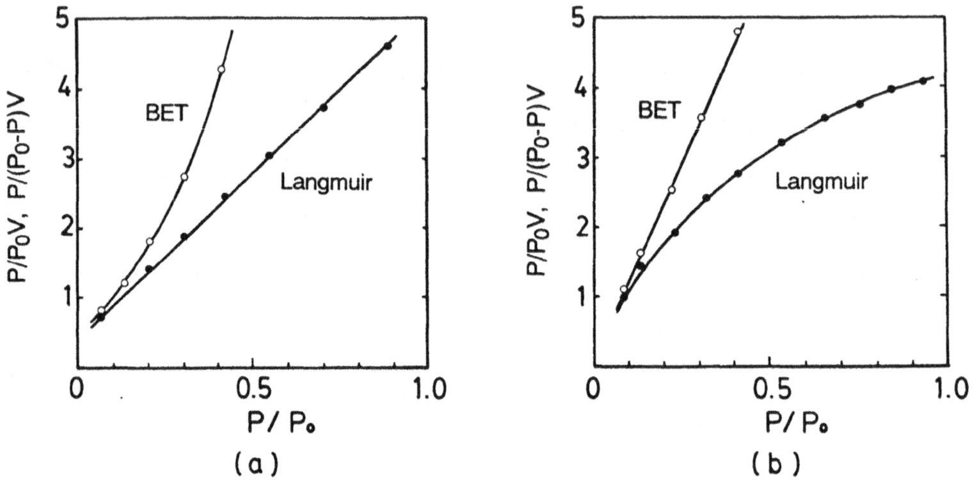

Figure 9. Langmuir (P/P_0V vs. P/P_0) and BET ($P/(P-P_0)V$ vs. P/P_0) plots for the adsorption isothers of mesitylene (a) and methanol (b).

Table 3. Pore volumes of the SiO_2-TiO_2 sol pillared clays for various solvent vapors and the type of isotherm.

Solvents	Pore volumes, ml/g	Isotherm
Benzene	0.21	Langmuir
Cyclohexane	0.20	Langmuir
n-Hexane	0.20	Langmuir
m-Xylene	0.21	Langmuir
Mesitylene	0.20	Langmuir
Methanol	0.24	BET
Water	0.25	BET

The adsorption isotherms of the TiO_2 pillared clay for the same kinds of solvent vapors are shown in Figure 10. Some BET plots for the adsorption isotherms are also shown in Figure 11 [22]. As seen from these figures, all of the adsorption isotherms fitted on the BET isotherm equations, indicating the pore sizes are larger than those of the SiO_2–TiO_2 pillared clays. However, if the pillared clays were prepared using sol solutions with SiO_2/TiO_2 ratios $< \sim 1.0$ they behaved like a TiO_2 pillared clay with large pores.

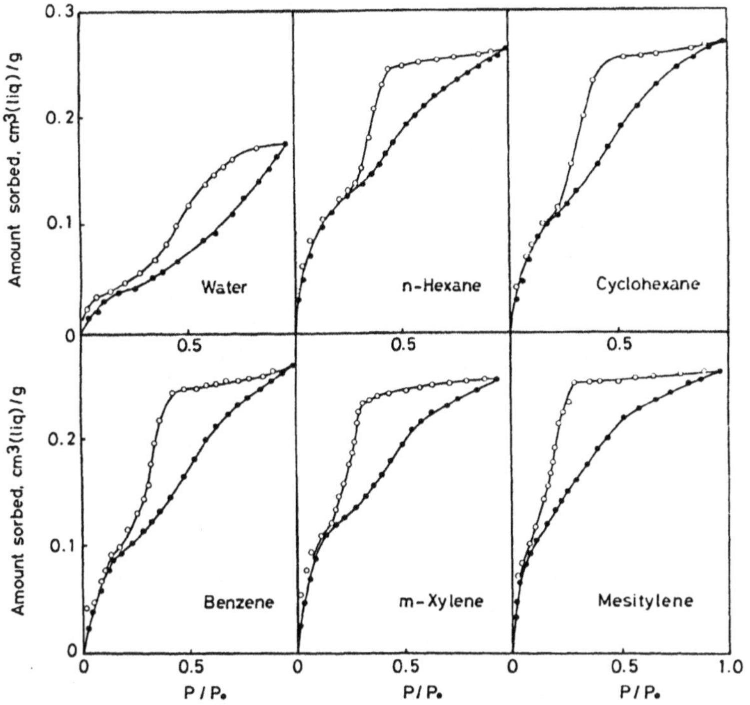

Figure 10. Adsorption–desorption isotherms of various vapors on TiO_2 pillared montmorillonite (calcined at 500°C): • , adsorption; ○ , desorption.

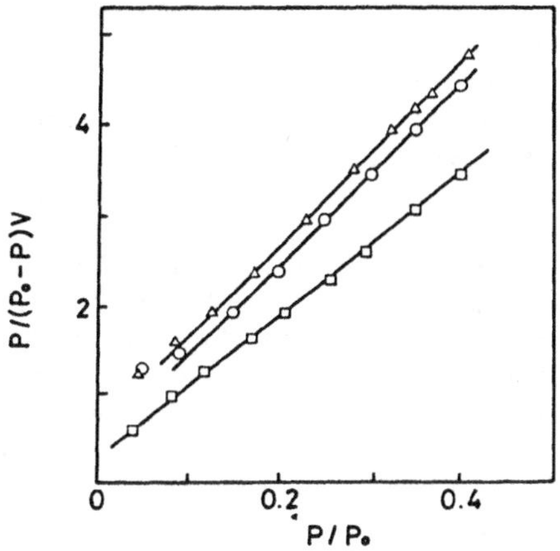

Figure 11. The BET plots of some adsoption data shown in Figure 10: ○ , cyclohexane; △ , benzene; □, mesitylene.

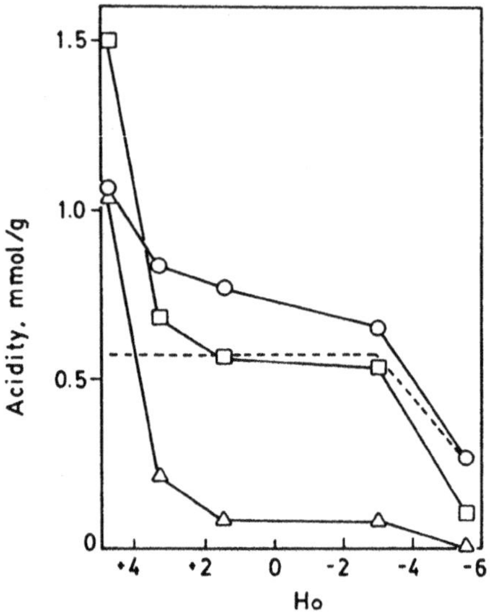

Figure 12. Acid strength distribution for the clays pillared with oxide sols heated at 500°C: ○ , TiO_2; □, SiO_2–TiO_2; △, SiO_2–Fe_2O_3 pillared clays. A dashed line shows a distribution for a binary oxide TiO_2–SiO_2 (Itoh et al. [39]).

3. 3. Acidic properties

The acidity distributions of the pillared clays were determined by the amine titration method [37]. The pillared clays were heated at 500°C and dispersed in *iso*–octane. The dispersion was titrated with *n*–butylamine using various indicators with different pKa. The acid strength is expressed by the Hammett acidity function Ho corresponding to pKa of the indicator. The following indicators were used: benzalacetophenon (pKa = –5.69), dicinnamalacetone (pKa = –3.0), 4–benzenazodiphenylamine (pKa =+1.5), *p*–dimethylaminoazo–benzene (pKa = +3.3), and methyl red (pKa = +4.8). The results are shown, together with those of the clays pillared with TiO_2 and SiO_2–Fe_2O_3 sols in Figure 12 [38]. Although all of the pillared clays showed large acid amounts, the distributions were very different. The TiO_2 pillared clay was characterized by a very strong acidic solid and furthermore, a large portion of this acidity resulted from strong acidic sites ($Ho \leq -3.0$); the portion of the weak acidic sites was very small. The SiO_2–TiO_2 pillared clays was found to be a strong acidic solid and had a similar acidity distribution to that of TiO_2 pillared clay, but it contained a similar amount of weak acidic sites ($Ho \geq 4.8$) as well.

The SiO_2–Fe_2O_3 sol pillared clay could be similarly prepared by depositing a small amount of iron hydroxide onto the SiO_2 hydrolyzed sols, which had a basal spacing as large as 42 Å, and a large BET surface area of 425 m^2/g [38]. The SiO_2–Fe_2O_3 pillared clay also had large acidic amount, but most of the acid sites were very weak ($Ho \geq 4.8$). The acid strength of the pillared clays decreased in the following order, TiO_2 > SiO_2–TiO_2 » SiO_2–Fe_2O_3 pillared clays.

Itoh *et al.* [39] studied the acidic properties of TiO_2–SiO_2 binary oxide, and showed that single component oxides such as SiO_2 and TiO_2 had little acidity, but mixing the two oxides resulted in a remarkable increase in acidity. The acidity for TiO_2–SiO_2 (1:1) at Ho≤1.5 is even larger than those of SiO_2–Al_2O_3. The acidity distribution of TiO_2–SiO_2 (1:1) reported by Itoh *et al.* is compared with those of the sol oxide pillared clays in Figure 12 by a dashed line. It should be noted that the clay pillared with single component TiO_2 sol also had very strong and large acidity. The strong acidity is probably due to the strong interaction between the TiO_2 pillars and the interlayer surfaces of silicate layers. The acidic properties of the SiO_2–TiO_2 pillared clays with high SiO_2/TiO_2 ratio can be also attributed to the strong acidity of the sol surface. Yamaguchi *et al.* [40] have reported that Fe_2O_3 supported on SiO_2 can be promoted to a super strong acid by treatment with SO_3 and can be used as powerful catalyst for isomerization of cyclopropane to *iso*-butane. Since SiO_2–Fe_2O_3 pillared clay has large acidity, it can be converted into an interesting catalyst if it can be improved similarly.

4. Structure of sol pillared clays

The silica sol particles obtained by the hydrolysis of silicon tetraethoxide are negatively charged, which can be converted into positively charged sols by modification with small amounts of titanium ions. The modified sol particles can be ion exchanged with the interlayer cations of montmorillonite. The resulting clays have basal spacing values remarkably larger than those ever reported for the clays pillared with ceramic oxides. As shown in a previous paper [22], the clay pillared with TiO_2 sol particles had a basal spacing ranging from 24–27 Å, and the nitrogen adsorption isotherm was of Type IV. The TiO_2 pillar height was estimated to be 15–18 Å. In the SiO_2–TiO_2 pillared clays, although the basal spacing is much larger than those of the TiO_2 pillared clays, the nitrogen adsorption isotherm is of Type I-like, suggesting that the pore sizes are much smaller than the pillar heights. In addition, as shown in Figure 6, the packing density of the interlayer sol particles seems to be uniform irrespective of the interlayer distance, and pore sizes are about 10~12 Å from the adsorption studies (Figure 7).

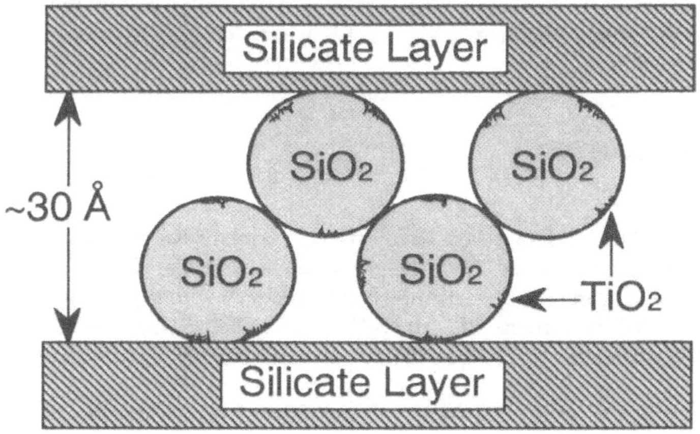

Figure 13. Schematic structural model of SiO_2–TiO_2 pillared clays.

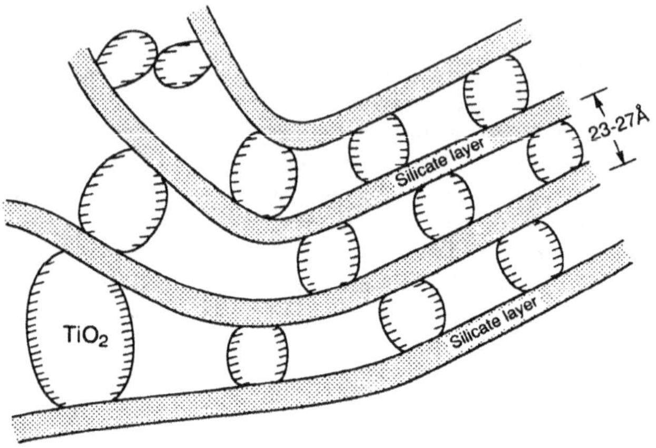

Figure 14. Schematic structural model of TiO_2 pillared clay.

Taking these findings together, we propose a structural model for the SiO_2–TiO_2 sol pillared clays in Fig. 13: the average size of the SiO_2–TiO_2 sol particles must be much smaller than the interlayer spaces and such small sol particles are packed uniformly so as to form small pores in the interstices of the sols and silicate layers. The sols are modified by the TiO_2 on the surface and positively charged. If the pillaring sol solutions with SiO_2/TiO_2 ratios $< \sim 1$ were used, the pores formed in the interlayer spaces were larger. It is likely that the structure is very similar to that of the TiO_2 alone pillared clay as previously proposed as a model shown in Figure 14, where the pore sizes are comparable with the pillar heights.

In the present study, we have used only one hydrolyzed condition in the preparation of the SiO_2 sol solution from silicon tetraethoxide. Different hydrolysis conditions will provide the SiO_2 sols with different sizes and distributions, which will in turn produce the sol pillared clays with different pore structures.

5. Use of Organic Templates

Although the SiO_2–TiO_2 pillared clays have extremely large basal spacings, the interlayers are occupied by the silica sol particles. Only the interstices between the pillars and silicate layers are used for pores. The sol particles were attempted to be replaced with large organic template cations which can be removed by burning off. The SiO_2–TiO_2 pillared clay (4/0.4/1) was treated with octadecyltrimethlammonium (OTMA) chloride as such organic agent in the following two methods:

Process I: OTMA was added to a 0.8% montmorillonite suspension in various ratios and stirred; SiO_2–TiO_2 sols were then added.

Process II: SiO_2–TiO_2 sols were added to a 0.8wt% montmorillonite suspension and stirred; OTMA was then added.

Each mixture was stirred for 1.5 h at room temperature, centrifuged and washed several times.

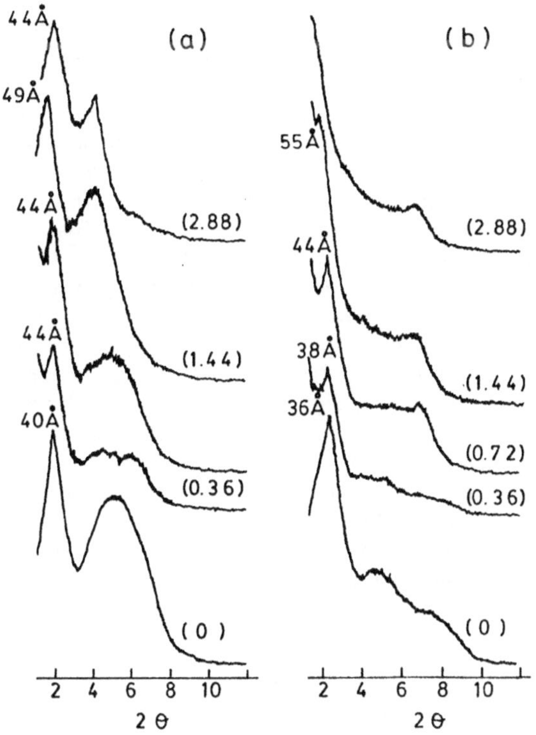

Figure 15. X-ray powder diffraction profiles of the products prepared by the Process II: (a) after air-dried at 60°C and (b) after calcined at 500°C. The OTMA/clay mixed ratios (mol/CEC) are designated in parentheses.

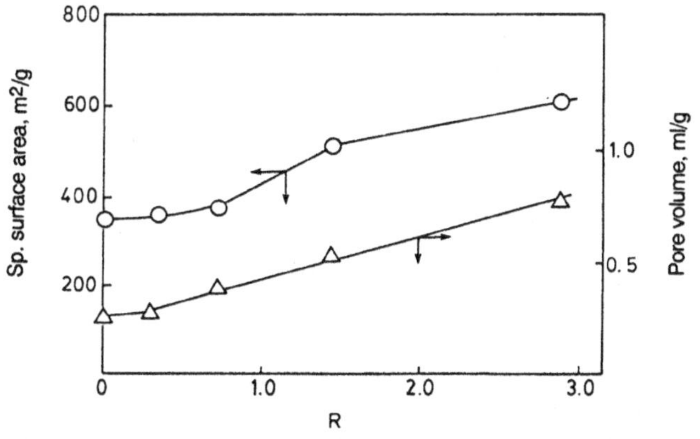

Figure 16. Surface areas and pore volumes of the calcined products prepared by Process II as a function of the OTMA/clay mixed ratio (mol/CEC), R.

The order of the additions of the organic cation and the sol is important. The addition of OTMA first before the ion exchange with sol solutions (Process I) did not have a large effect on the pore structures of the resulting pillared clays. This is due to the large affinity of the organic cations with silicate layers; the OTMA ion exchanged clay held the basal spacing at about 22 Å, and did not open the interlayers for the further ion exchange with sol particles. On the other hand, if the sol pillared clays were first prepared and then treated with OTMA, the sols were partly replaced with OTMA, and rearrangement of the sol particles packed in the interlayer space occurred. Figure 15 shows the change in the basal spacing of SiO_2–TiO_2 pillared clays by the treatment with different amounts of OTMA solutions. The spacing increased from 40 to 44–49 Å by the incorporation of OTMA ions, and the enlarged spacing was maintained by calcination at 500°C. The surface area and the pore volume increased with the increase of the mixed ratio of OTMA/CEC as shown in Figure 16. The most porous sample obtained had a pore volume as large as 0.8 ml/g and a BET surface area larger than 600 m^2/g. Figure 17 shows the pore size distributions of the samples shown in Figure 16, which were calculated on the basis of the adsorption branch of the nitrogen isotherm using the BJH method [41]. As seen from the distributions, the pore volume increased mainly in the part with a mesopore ranging from 20 to 30 Å in diameter. Figure 18 shows a schematic structural model for the formation of mesopores in the interlayer spaces of montmorillonite by Process II.

It will be an interesting subject to control pore sizes of not only mesopores but also micropores and their distribution by changing an arrangement and combination of the kinds of pillaring oxides and organic cations.

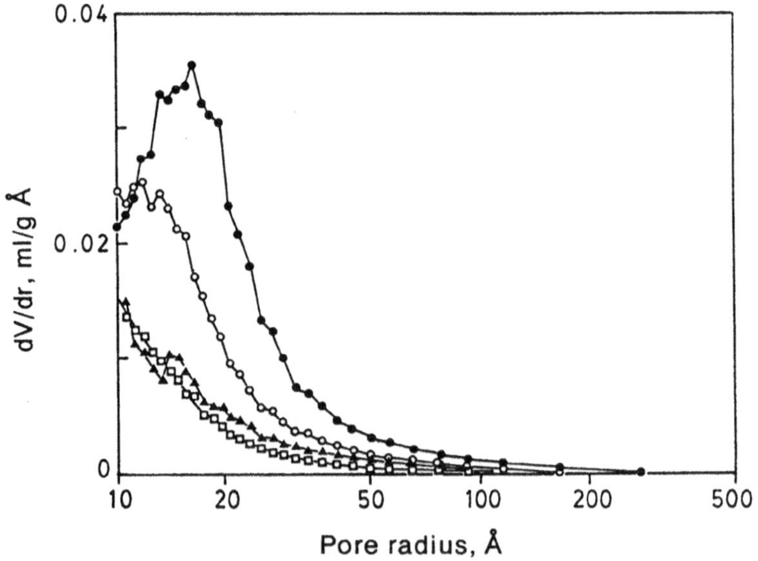

Figure 17. Pore size disribution of the products shown in Figure 16: the OTMA/clay mixed ratios (mol/CEC) were (•), 2.88, (○),1.44, (▲), 0.72 and (□) 0.

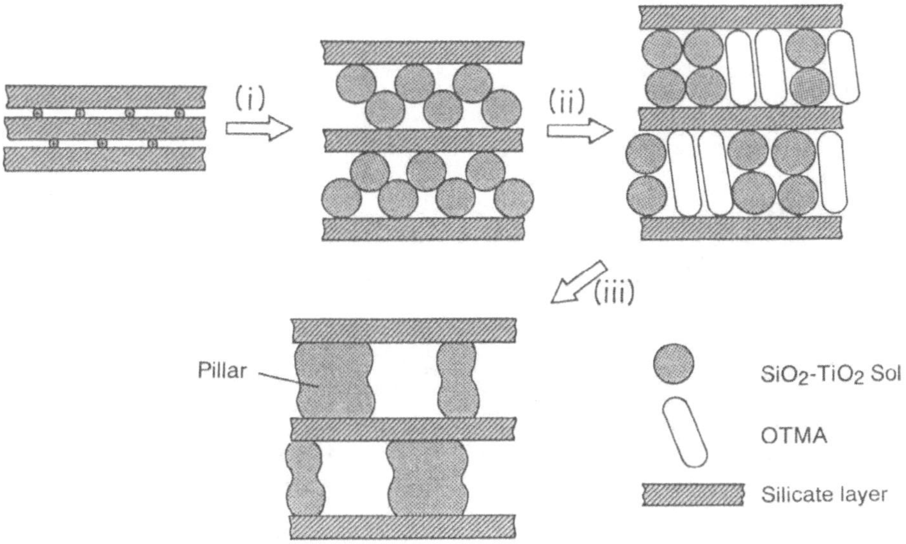

Figure 18. Schematic structural model for the formation of mesopores in the interlayer spaces of montmorillonite by the Process II: (i) ion exchange with sol particles, (ii) uptake of OTMA along with the rearrangement of the interlayer sol particles, (iii) removal of OTMA by calcination and formation of mesopores.

6. Supercritical Drying

The ion exchange of the interlayer cations of montmorillonite with precursory cations for pillars is usually carried out in aqueous media, and the samples separated after the ion exchange are extremely swelled with water. The swelled sample experiences remarkable shrinkage during drying, as in the case of the drying of aqueous gels formed by solution–sol–gel route [42]. The shrinkage of the gel is caused by the large liquid–vapor interfacial (capillary) forces, which act to disrupt the delicate microporous gel structure during an evaporative procedure. The porous structure of the swelled pillared clays must be very different before and after the shrinkage on drying. Recently a new drying technique, supercritical drying, has been developed in which the solvent in the gel pores is substituted by a fluid with a relatively low critical temperature, and the fluid is extracted supercritically avoiding the formation of liquid–vapor interfaces [43]. This technique has been successfully applied to the preparation of highly porous and translucent silica aerogels from alcogel by using a supercritical CO_2 fluid [44]. Supercritical drying technique is applied to the drying of SiO_2–TiO_2 sol pillared clays, and the resulting pore structure was investigated in comparison with that of the sample prepared by conventional air–drying procedures.

The SiO_2–TiO_2 pillared clay (4/0.4/1) was rinsed with ethanol three times by centrifugation in order to replace the water in the product with the ethanol, which was then extracted with supercritical fluid of CO_2 under a pressure of 120 atm at 40°C. The CO_2 was vented for 5 hr.

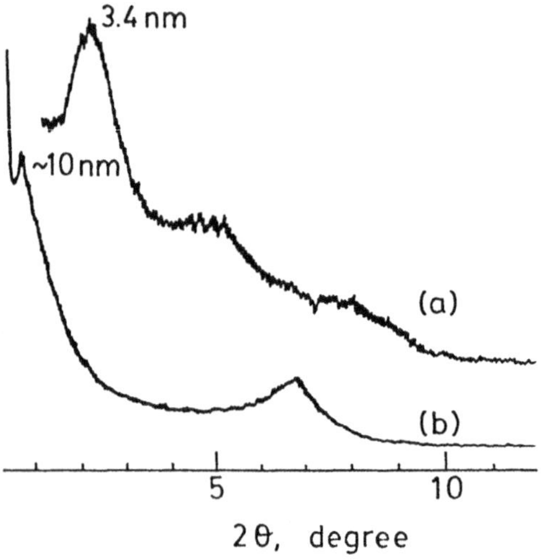

Figure 19. X–ray diffraction patterns of the two pillared clays calcined at 500°C: (a) air–dried, (b) supercritically dried.

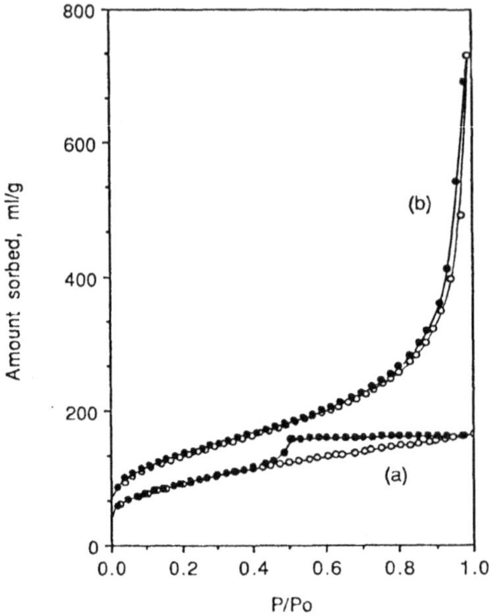

Figure 20. Nitrogen adsorption–desorption isotherms for the two pillared clays calcined at 500°C. (a) air–dried, (b) supercritically dried; ○ , adsorption, ● , desorption.

For comparison, a separate sample of SiO_2–TiO_2 pillared clay was similarly prepared and air–dried at 60°C. The samples thus prepared by the two different drying procedures were calcined at 500°C for 3 hr in air, and supplied for the characterization.

The XRD patterns of the samples prepared by the two different drying procedures are compared in Figure 19. In the preparation of the air–dried sample for XRD measurement, the wet pillared sample was spread on a glass slide before drying so as to attain a preferred orientation, and then calcined at 500°C on the slide. The XRD pattern clearly showed three integral reflection peaks for a basal spacing of 34 Å. For the XRD measurement of the supercritically dried (SCD) sample, the dried and calcined sample was filled in a cavity of a glass slide. The latter sample was therefore, more randomly oriented, and showed only a weak shoulder reflection, corresponding to a basal spacing of about 100 Å and a broad reflection peak for a smaller basal spacing of 12 Å. Since the thickness of the silicate layer itself is 9.6 Å, the appearance of the small spacing can be explained in terms of the occurrence of a partial collapse in the pillared structure during the drying. The two samples compared in Figure 19 do not have similar orientation and thus cannot be compared in a similar manner. However, it appears that the SCD sample has a much larger basal spacing than the air–dried sample.

The nitrogen adsorption–desorption isotherms for the two samples calcined at 500°C are compared in Figure 20. The adsorption isotherm of the air–dried sample is characterized as Type I or Type II, with a restricted number of nitrogen adsorption layers. This indicates that most of the pores formed in the air–dried sample are in a micro– to small mesopore region. The SCD sample also has a similar adsorption isotherm in the lower pressure region, but the isotherm rises steeply in the higher pressure region due to a capillary condensation. This suggests that the SCD sample contains a wide range of meso– as well as micropores. Another characteristic feature of the SCD sample is to show little hysteresis for desorption. The pores in the SCD sample must be very open and have a small number of ink–bottle type necks. The BET surface areas for nitrogen were determined to be 325 and 459 m^2g^{-1} for air–dried and the SCD samples, respectively.

Pore–size distributions in the region larger than about 100 Å in diameter were analyzed by mercury intrusion measurments. The pore volumes measured for the two samples are shown in Figure 21 as a function of pore diameter. It is clearly shown that the air–dried sample has few macropores, while the SCD sample has an extremely large pore volume of 9.8 mlg^{-1}, the most part of which comes from the pores ranging from 5–0.5 μm in diameter. The total pore volumes measured by the nitrogen adsorption study and the mercury porosimetry are summarized in Table

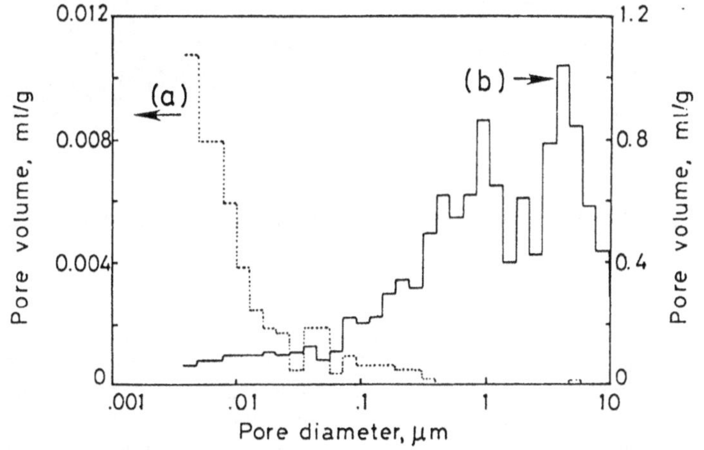

Figure 21. Pore–volume distributions of the two pillared clays measured by a mercury porosimeter. (a) air–dried (dotted line), (b) supercritially dried (solid line).

Table 4. BET surface areas and the pore volumes measured by nitrogen adsorption and mercury porosimetry.

Drying procedure	BET surface area, $m^2 g^{-1}$	Pore volumes	
		N_2 adsorption 1) ml g^{-1}	Hg porosimetry ml g^{-1}
Supercritical drying	459	1.13 (0.09)	9.82
Air drying	325	0.25 (0.19)	0.098

1) Micropore volumes are shown in parentheses.

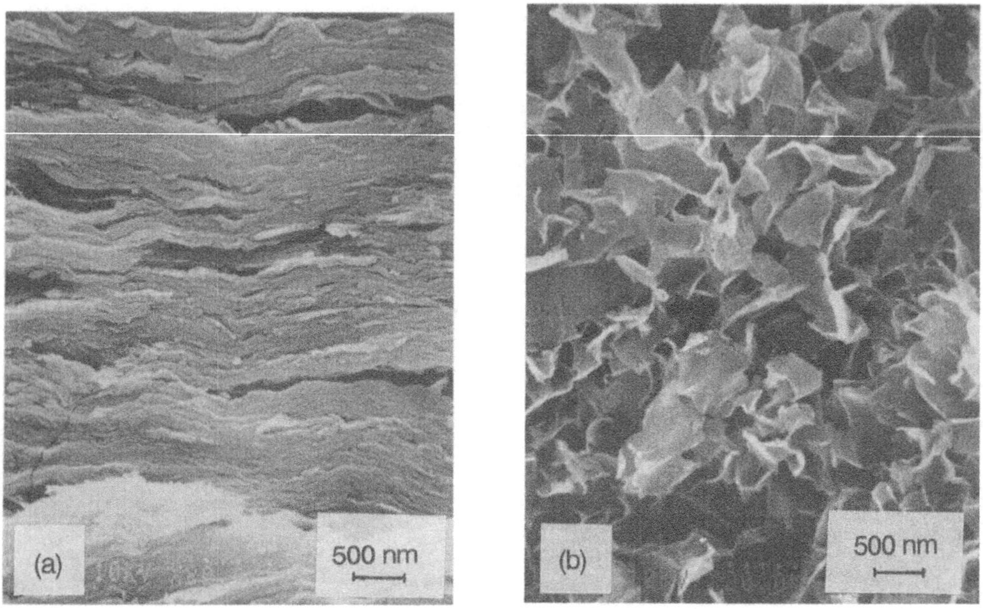

Figure 22. SEM photographs of the two pillared clays. (a) Air–dried, (b) supercritically dried.

4. The micropore volumes were determined by the *t*–plot of the nitrogen adsorption isotherms and are shown in parentheses in the table. Since the micropores are formed in the interstices of the sol particles and the silicate layers, it seems reasonable to consider that the micropore volume will not be influenced by the drying method. The mesopore volume was greatly increased in the SCD sample. It is interesting to note that the total mercury intrusion volume for the air–dried sample is only 1% of that for the SCD sample.

Figure 22 shows SEM photographs of the two dried samples. It is shown that the SCD sample has a well developed card–house structure, and macropores constructed by thin silicate sheet walls. From the photograph, it is also seen that most of the pores are approximately 3–0.3 μm in diameter, in accordance with the data given by the above mercury porosimetry. In contrast to the SCD sample, in the air–dried sample the silicate layers are stacked on each other in a more regular manner, and only a small number of large pores are observed.

As discussed about the structure of TiO_2–SiO_2 pillared clays, the interlayer spaces are packed with small sol particles. The formation of micropores would not be much influenced by the drying method. The macropores which were measured by mercury intrusion had been filled with water before drying. The macropores would disappear in a conventional evaporative drying procedure, because the strong interfacial forces between the swelled silicate layers lead to collapse of the pores on evaporation of the water. In the supercritical drying procedure, however, the water occupying the macropores is replaced with alcohol and then with a supercritical CO_2 fluid. In this case, the formation of liquid–vapor interfaces is avoidable and the delicate card–house structure constructed by the pillared silicate layer can be preserved during drying. The pore–volume data given in Table 4 show that the macropores shrink even to 1% of the original volume on air drying. From a different point of view, it is interesting to note that the sol–pillared clay is in an extraordinarily swollen state before drying.

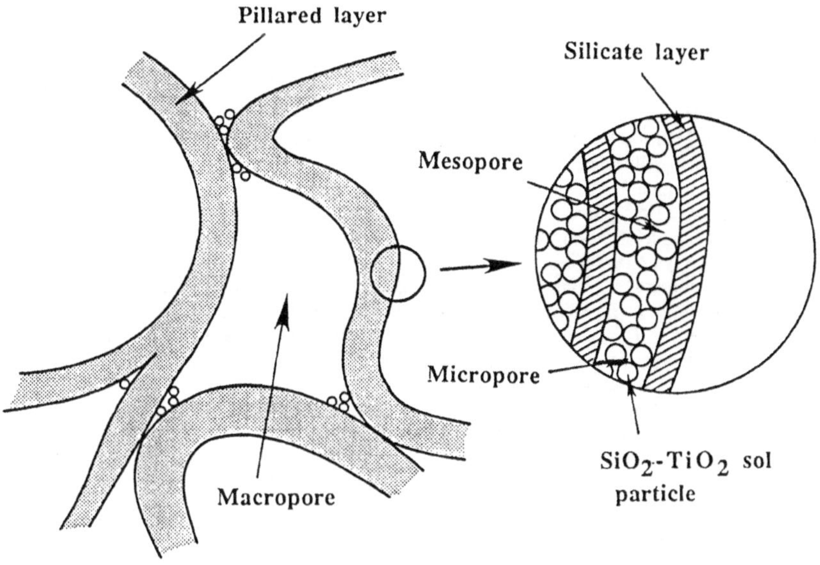

Figure 23. Schematic structural model of the pillared clay supercritically dried.

A schematic structural model showing the dual–pore structure of the SCD sample is shown in Figure 23. The micro– and mesopores are located in between the loosely packed sol particles in the interlayer space of the pillared layers. The macropores are included in the highly porous card–house structure developed by the connection of the pillared layers.

Acknowledgment

This study was partly defrayed by a Grant–in–Aid for Science Research on Priority Area of the Ministry of Education, Science and Culture.

References

1. Yamanaka, S. (1991) Ceramic Bull., **70**, 1056.
2. Brindley, G. W. and Sempels, R. E. (1977) Clays Clay Miner., **12**, 229.
3. Yamanaka, S. and Brindley, G. W. (1979) Clays Clay Miner., **27**, 119.
4. Brindley, G. W. and Yamanaka, S. (1979) Amer. Mineral., **64**, 830.
5. Pinnavaia, T. J., Tzou, M–S., and Landau, S. D. (1985) J. Am. Chem. Soc., **107**, 4783.
6. Yamanaka, S. and Hattori, M. (1988) Catal. Today, **2**, 261.
7. Yamanaka, S., Yamashita, G., and Hattori, M. (1980) Clays Clay Miner., **28**, 281.
8. Choudary, B. M. and Valli, V. L. K. (1990) J. Chem. Soc. Chem. Commun., 1115.
9. Occelli, M. and Finseth, D. H. (1986) J. Catal., **99**, 316.
10. Bradley, S. M. and Kydd, R. A. (1991) Catal. Lett., **8**, 185.
11. Coelho, A. V. and Poncelet, G. (1991) Appl. Catal., **77**, 303.

12. Sterte, J. and Shabtai, J. (1987) Clays Clay Miner., **35**, 429.
13. Pinnavaia, T. J. (1983) Science, **220**, 365.
14. Burch, R., ed. (1988) Pillared Clays, Catalysis Today, **2**, 1–185.
15. Figueras, F. (1988) Catal. Rev. Sci. Eng., **30**, 457.
16. Farfan–Torres, EM and Grange, P. (1990) J. Chim. Phys., **87**, 1547.
17. Yamanaka, S. and Hattori, M. (1991) in Chemistry of Microporous Crystals, eds. T. Inui, S. Namba and T. Tatsumi, Kodansha/Elsevier, Tokyo, p.89.
18. Delmon, B. and Grange, P. (1991) Erdol Erdgas Kohle, **107**, 376.
19. Schoonheydt, R. A. (1991) Stud. Surf.Sci. Catal., **58**, 201.
20. Rightor, E. G., Tzou, M–S., and Pinnavaia, T. J. (1991) J. Catal., **130**, 29.
21. Sterte, J. (1986) Clays Clay Miner., **34**, 658.
22. Yamanaka, S., Nishihara, T., Hattori, M., and Suzuki, Y. (1987) Mat. Chem. Phys., **17**, 87.
23. Brunauer, S., Deming, L. S., Deming, W., and Teller, E. (1940) J. Am. Chem. Soc., **62**, 1723.
24. Johnson, I. D., Werpy, T. A., and Pinnavaia, T. J. (1988) J. Am. Chem. Soc., **110**, 8545.
25. MaCauley, J. R. (1988) Int. Pat. Appl. PCT/US88/00567.
26. Sterte, J. (1991) Clays Clay Miner., **39**, 167.
27. Suzuki, K., Mori, T., Kawase, K., Sakami, H., and Iida, S. (1988) Clays Clay Miner., **36**, 147.
28. Iler, R. K. (1979) The Chemistry of Silica, Wiley, New York.
29. Endo, T., Mortland, M. M., and Pinnavaia, T. J. (1980) Clay Miner., **28**, 105.
30. Lewis, R. M., Ott, K. C., and Van Stanten, R. A. (1985) U. S. Patent 4,510,257.
31. Moini, A. and T. J. Pinnavaia, (1988) Solid State Ionics, **26**, 119.
32. Takahama, K., Yokoyama, M., Hirao, S., Yamanaka, S., and Hattori, M. (1991) J. Ceramic Soc. Jpn. Int. Ed., **99**, 14.
33. Takahama, K., Yokoyama, M., Hirao, S., Yamanaka, S., and Hattori, M. (1992) J. Mat. Sci., **27**, 1297.
34. Yamanaka, S., Inoue, Y., Hattori, M., Okumura, F., and Yoshikawa, M. (1992) Bull. Chem. Soc. Jpn. in press.
35. Sakka, S., Kamiya, K., Makita, K., and Yamamoto, Y. (1984) J. Non–Cryst. Solids, **63**, 223.
36. Gregg, S. J. and Sing, K. S. W. (1982) Adsorption, Surface Area and Porosity: Academic Press, London.
37. Benesi, H. (1957) J. Phys. Chem., **61**, 970.
38. Yamanaka, S., Nishihara, T., and Hattori, M. (1988) Mat. Res. Symp. Proc. **111**, 283.
39. Itoh, M., Hattori, H., and Tanabe, K. (1974) J. Catal., **35**, 225.
40. Yamaguchi, T., Jin, T., Ishida, T., and Tanabe, K. (1987) Mat. Chem. Phys., **17**, 3.
41. Barrett, E. P., Joyner, L. G., and Halenda, P. P. (1951) J. Am. Chem. Soc., **73**, 373–380.
42. Scherer, G. W. (1988) in Ultrastructure Processing of Advanced Ceramics, eds. J. D. Mackenzie and D. R. Ulrich (eds.) Wiley, New York, p.295.
43. Matsonard, D. W. and Smith, R. D. (1989) J. Amer. Ceramic Soc., **72**, 871.
44. Tewari, P. H., Hunt, A. J., and Lofftus, K. D. (1985) Mater. Lett., **3**, 363.

MESOPOROUS CLAYS AND PILLARED CLAYS AS LIQUID PHASE ALKYLATION CATALYSTS

Jean-Rémi Butruille and Thomas J. Pinnavaia
Department of Chemistry and
Center for Fundamental Materials Research
Michigan State University
East Lansing Michigan 48824
USA

Abstract. The propene alkylation of liquid phase biphenyl has been studied at 250°C and 140 psi using two mesoporous clays, namely, acid-treated K10 montmorillonite and alumina-delaminated-Laponite. The mesoporous clays exhibit reactivities and selectivities toward monoalkylated products that are substantially different from those exhibited by microporous alumina pillared clays. The differences in reactivity and selectivity are explained in part by the effects of pore structure on the diffusion controlled reaction.

1. Introduction

Solid acids are promising catalysts for use in chemical industry [1]. Zeolites and chemically modified clays, particularly acid treated and pillared clays, can exhibit acidities close to those of conventional liquid phase catalysts without the drawbacks of corrosivity and effluent pollution. Zeolites have attracted substantial attention as acid catalysts because of their outstanding shape selective properties owing to controlled micropores sizes [2]. However, it is also of great interest to control the mesopores of acid catalysts. Many catalyzed reactions are carried out in the liquid phase, where diffusion control is significant. In this work, we have compared two mesoporous catalysts, namely a commercially available acid-washed montmorillonite (K10) and an alumina-delaminated-Laponite, with two microporous alumina pillared clays. The test reaction chosen was the propene alkylation of liquid phase biphenyl for which the zeolite mordenite is known to be very shape selective [3,4]. The mesopore structure of the clays seems to affect the catalytic activity, as well as the selectivity towards different monoalkylated products.

C.A.C. Sequeira and M.J. Hudson (eds.), Multifunctional Mesoporous Inorganic Solids, 259–272.

2. Experimental

2.1 PREPARATION OF THE CATALYSTS

Wyoming montmorillonite was obtained from the Source Clay Minerals Repository at the University of Missouri, Columbia, Missouri. The sodium-exchanged form was purified by sedimentation. The synthetic clays Laponite (Laporte Industries Ltd.) and fluorohectorite (Corning) were used as obtained from the manufacturer. The pillaring agent was the Keggin-like ion $[Al_{13}O_4(OH)_{24}(H_2O)_{12}]^{7+}$ generated by base hydrolysis of an aluminum chloride solution (OH/Al molar ratio of 2.4). The pillaring procedure described by Pinnavaia et al. [5] was used.

K10 montmorillonite is an acid leached montmorillonite supplied by Aldrich [6]. It was used without further modification.

2.2 CHARACTERIZATION OF THE CATALYSTS

Elemental analyses were performed as described previously [5]. X-ray diffraction patterns (Cu-Kα) of oriented films were obtained with a Rigaku X-ray diffractometer equipped with a rotating anode. Continuous flow N_2 adsorption-desorption isotherms at 77K were obtained on a Coulter Omnisorb 360 CX sorptometer. The samples were outgassed at 150°C for 12 hours under 10^{-4} torr.

The FTIR spectrum for chemisorbed pyridine was obtained on an IBM IR44 spectrometer. Self-supported pellets of the catalyst were outgassed one hour at 300°C prior to pyridine adsorption at 150°C. A minimum of 16 hours of degassing at 150°C was necessary to remove most of the physically adsorbed pyridine. Chemisorbed pyridine was further evacuated at 250° and 350°C for 45 minutes. All the spectra were obtained at 150°C.

2.3 CATALYSIS

The alkylation of biphenyl by propene was carried out in a 300-mL batch reactor (n°4561 Parr Instruments). Propene was CP grade (Matheson Company), and biphenyl was practical grade (Baker company). The catalysts were used without pre-activation.

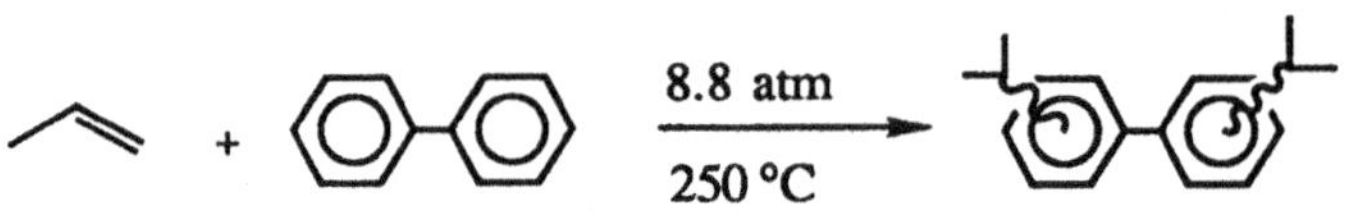

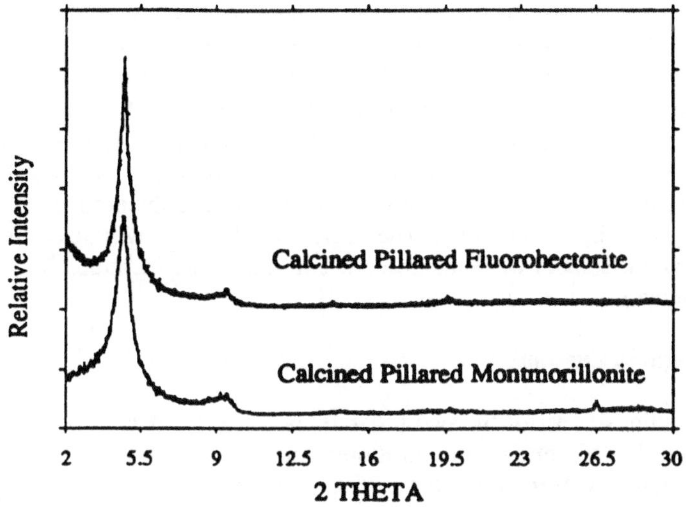

Figure 1: X-Ray diffraction patterns of pillared clays calcined at 350 °C.

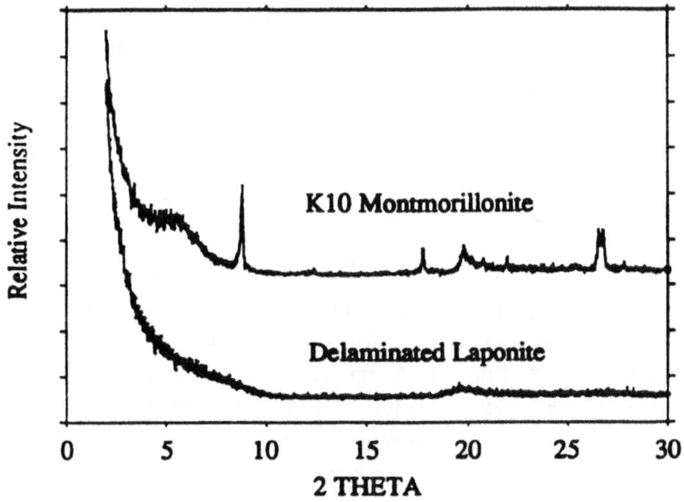

Figure 2: X-Ray diffraction patterns of delaminated clays

Biphenyl (30g, 0.194 mole) and 0.06g to 0.2g of catalyst were heated to 90°C. The system was then purged with propene under stirring until the temperature reached 150°C. The temperature and propene pressure were then carefully increased until they reach the final conditions of 250°C and 140 psi (9.6 x 10^5 Pa). A temperature ramp of 3°/min was used. The reaction was carried out for 20 hours. The reactor was then allowed to cool and the pressure was slowly released. Eventually, the reaction mixture was dissolved in 300 mL toluene, the catalyst was filtered, and the products in the filtrate were analyzed by gas chromatography. The degree of alkylation was determined by GC-MS. The monoalkyl isomers were identified according to the literature [7] on the basis of their retention times.

3. Results and discussion

The calculated compositions for the two pillared clays and for delaminated Laponite are presented in Table I. Wyoming montmorillonite incorporated 1.25 aluminum atom per unit cell and fluorohectorite incorporated 2.45 aluminum atoms per unit cell. Thus, the amount of incorporated aluminum is related to the CEC of the clay. This is not observed with delaminated Laponite.

TABLE I. Alumina-clay catalysts for propene alkylation of liquid biphenyl

Sample	Clay	CEC (meq/100g)	Aluminum atoms incorporated per $O_{20}(OH)_4$ unit
APFH	Fluorohectorite	140	2.45
APWM	Wyoming montmorillonite	75	1.25
ADL	Laponite	30	1.3

The X-ray diffraction patterns of the pillared clays are presented in Figure 1. They show a similar basal spacing (18Å) for both alumina pillared montmorillonite (APM) and alumina pillared fluorohectorite (APF) This demonstrates that the intercalation of the Al_{13} polymer occurred. The X-ray diffraction pattern of the mesoporous clays are presented in Figure 2. With Laponite, an amorphous X-ray diffraction pattern was observed. Laponite platelets are very small (<0.1 μm), and the aggregation mechanism is mainly edge-face rather than face-face [8]. The catalyst obtained was thus called alumina-delaminated-Laponite (ADL). For K10 montmorillonite, the XRD pattern is similarly diffuse with the addition of some sharp peaks owing to impurities (probably muscovite and quartz). Both mesoporous catalysts exhibit a rather amorphous XRD pattern.

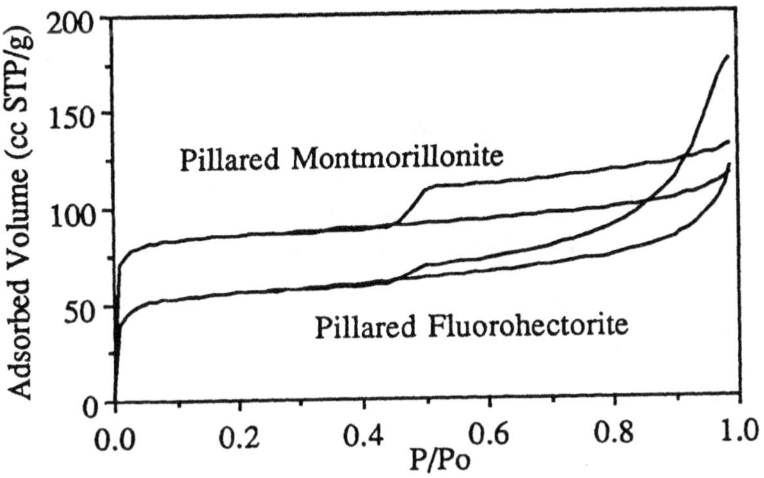

Figure 3: N_2 adsorption-desorption isotherms of pillared clays calcined at 350 °C obtained at 77K.

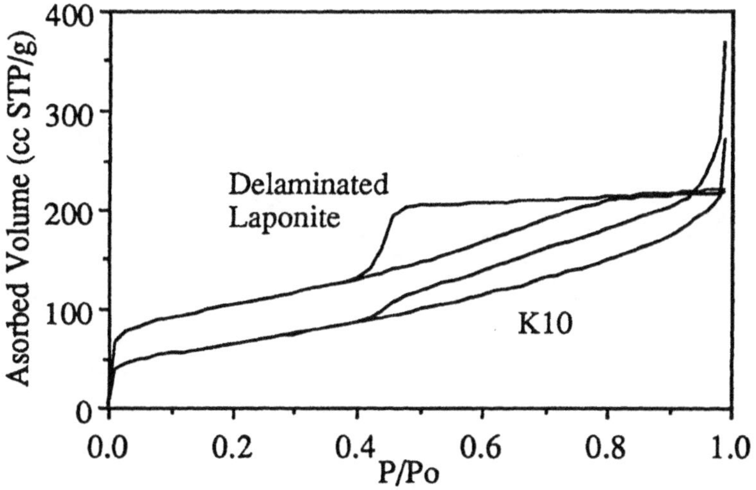

Figure 4: N_2 adsorption-desorption isotherms of mesoporous clays obtained at 77K.

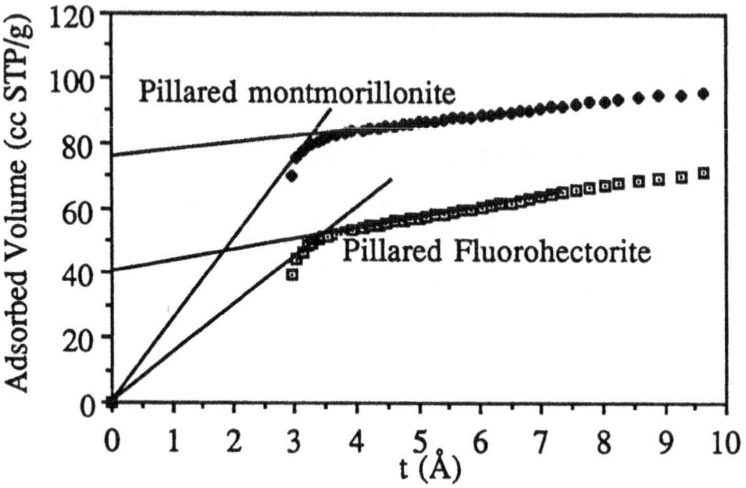

Figure 5: t-plots of pillared clays calcined at 350 °C.

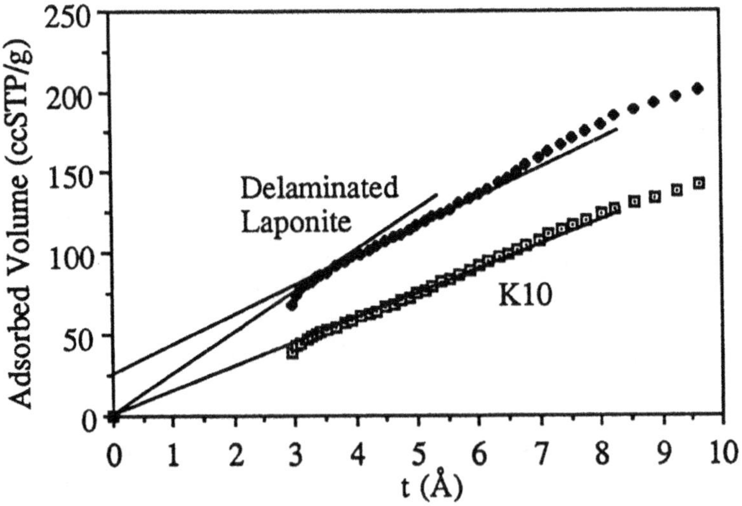

Figure 6: t-plots of mesoporous clays.

TABLE II. Pore analysis summary

	S_{BET} (m^2/g)	S_{total} (m^2/g)	$S_{meso+macro}$ (m^2/g)	S_{micro} (m^2/g)
K10	233	230	230	0
Delaminated Laponite	360	387	294	92
Pillared Montmorillonite	334	377	34	327
Pillared Fluorohectorite	214	234	56	175

The pore structures of the mesoporous clays and pillared clays were studied by N_2 adsorption-desorption measurements. The adsorption isotherms are presented in Figure 3 and 4 for pillared clays and mesoporous clays, respectively. The adsorption data were treated by the BET equation in the range of linearity and by the t-plot method [9]. The t-plots are presented in Figure 5 and 6 for pillared clays and mesoporous clays, respectively. The results are summarized in Table II. As expected, mesoporous clays and pillared clays exhibit very different adsorption properties. ADL and K10 exhibit few or no micropores. This is a consequence of their delaminated structure. Pillared clays (APM, APF) are microporous materials, but the fraction of the total surface area due to the micropores varies somewhat with the clay layer (75% for APF and 87% for APM). The fraction of total surface area due to micropores is probably a consequence of the starting clay morphology. The clay particle size determines the aggregation mechanism and thus the micropore to mesopore ratio.

Mesopore distributions were determined by treating the desorption data using either a slit shape [10] or cylindrical [11] pore model. The data reported in Figures 7 and 8 are those which best fit the t-plot results. For mesoporous ADL and K10 the cylindrical pore model was the more appropriate model, whereas for pillared clays the slit shape model gave better agreement with t-plots. Among the non-microporous clays, K10 exhibited a broad distribution of mesopores, whereas the mesopores of ADL were more or less uniformly centered around 35Å, as shown in Figure 7. Differences in mesopore distribution were also noted for the pillared clays as shown in Figure 8.

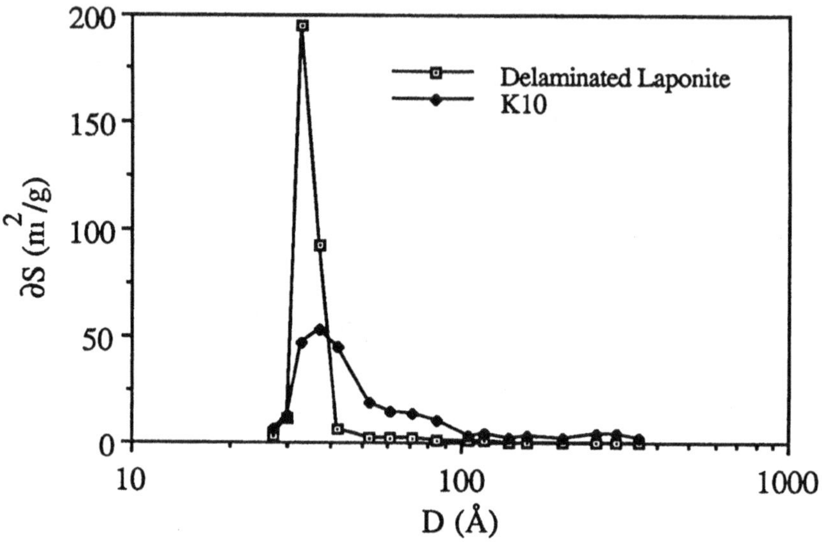

Figure 7: Mesopore size distribution of mesoporous clays obtained from a cylindrical plates pore model. (D = pore diameter)

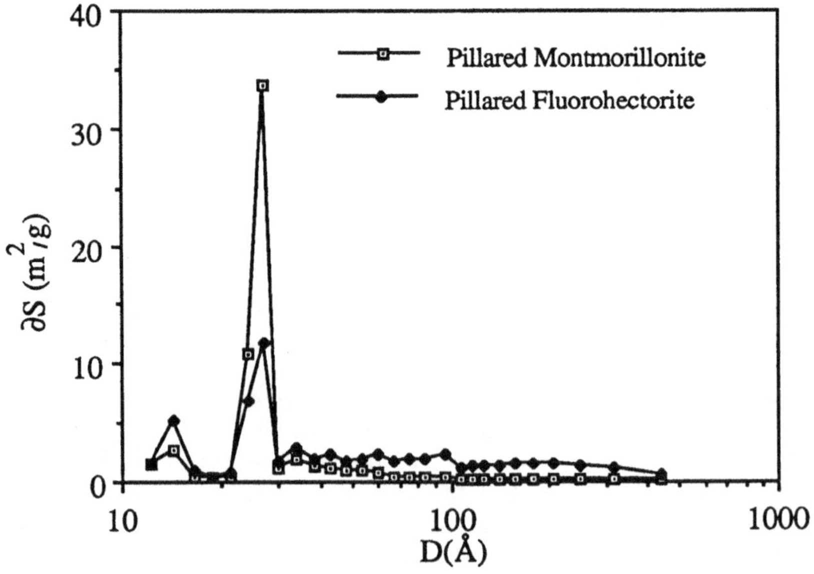

Figure 8: Mesopore size distribution of pillared clays obtained from a parallel plates pore model.

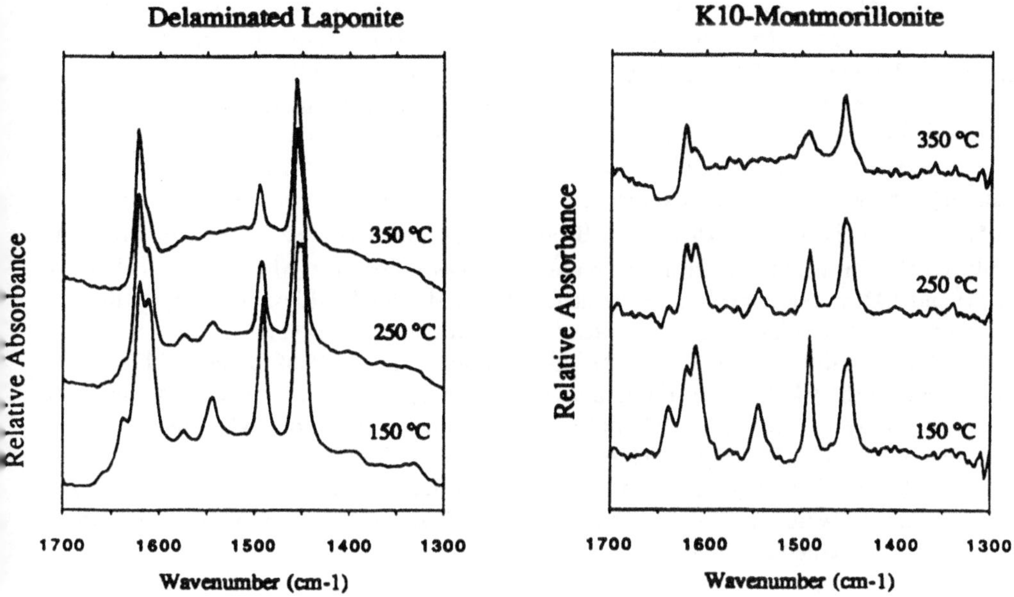

Figure 9: Infrared spectra of pyridine adsorbed on mesoporous clays.

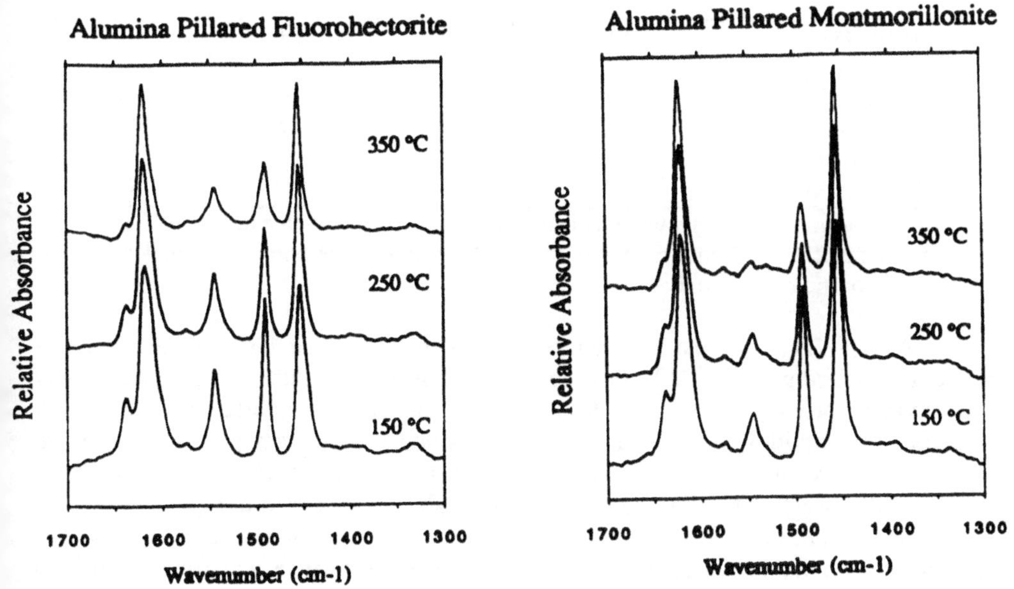

Figure 10: Infrared spectra of pyridine adsorbed on pillared clays.

Alumina pillared montmorillonites exhibited almost no mesopores of diameter larger than 30Å. In contrast, for pillared fluorohectorite, the mesopores of diameter larger then 30Å contributed surface areas of 35 m^2/g.

In order to differentiate the type of acidity for each catalyst, FTIR studies of chemisorbed pyridine were carried out. The results are presented in Figure 9 and 10. Qualitatively, it is easy to distinguish between Lewis and Brönsted acidity for the catalysts studied. Pyridine adsorbed on Lewis acid sites exhibits absorption bands at 1450 cm^{-1} and 1490 cm^{-1} whereas on Brönsted sites absorption occurs at 1490 cm^{-1} and 1540 cm^{-1}. Brönsted sites to Lewis sites ratio was higher for APF and K10 than for APM and ADL. The intensities are related to the acid sites densities. This shows that the number of acid sites increases in the order APM>APF and ADL> K10.

The catalytic activities of the mesoporous catalysts and pillared clays were compared for the alkylation of liquid phase biphenyl by propene at 250°C using the same mass of catalyst (0.2g, 100-200 mesh, 75-150 μm). The biphenyl conversion was reproducible to +/- 5%. Figure 11 represents the evolution of product distribution with time for the reaction catalyzed by APM. These results show that the mono-, di-, tri- and tetraalkylations were successive. The results of the activity tests in order of decreasing activity are presented in Table III.

Table III. Biphenyl conversion and product isomer distribution obtained for different solid acid catalysts [a]

Catalyst	APF	K 10	ADL	APM
conversion (%)	98	96	89	64
% monoalkyl	11	16	29	63
% dialkyl	33	36	39	30
% trialkykyl	38	37	25	7
% tetraalkyl	17	11	7	0

a: The reaction conditions were as follows: 30g biphenyl, 0.2g catalyst, 140 psi propene, 250°C, 20/hours reaction time. All results were obtained with 100-200 mesh particle size catalysts.

APF and K10 were the most active clays, with most of the biphenyl converted to di- and trialkylated products. ADL was less active but still it yielded substantial amounts of dialkylated products. In order to understand better the importance of diffusion limitations on activities, a series of reactions in which the catalysts particle sizes were varied was carried out. Two clay catalysts with very different pore structures, namely, APM and ADL were chosen for this study. The results are presented in Figure 12.

Particle size had a dramatic influence on the catalyst activities. For microporous APM, the conversion increases from 26% to 99% when the particle size decreases from a 75-150 μm fraction to a fraction of particles with a diameter smaller than 45 μm. For mesoporous ADL, the influence of particle size is less dramatic, but it cannot be

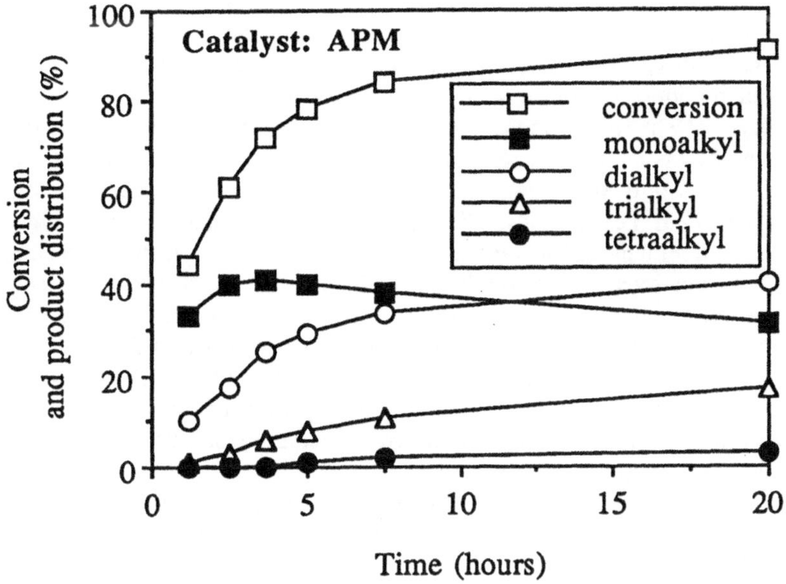

Figure 11: Evolution of product distribution with time for the propene alkylation of biphenyl. The conditions were as follows: 0.6g APM, 30g biphenyl, 140 psi propene, 250 °C.

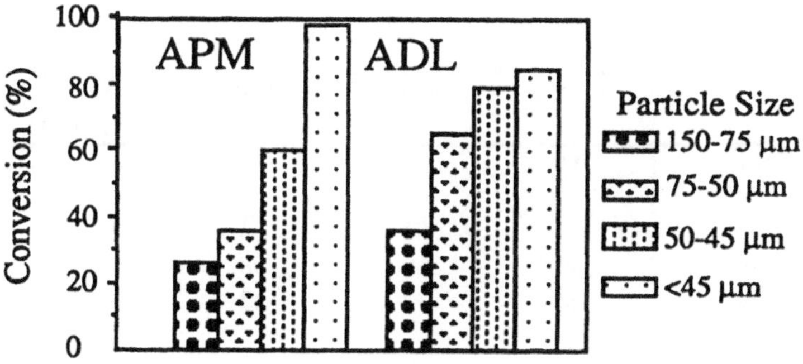

Figure 12: Evolution of the catalytic activity with particle size. The conditions were as follows: 0.1g (APM) or 0.09g (ADL) of catalyst, 30g biphenyl, 250 °C, under 140 psi propene for 20 hours.

disregarded. The reaction was thus probably under diffusion control for all of the catalysts. This may explain why the less microporous catalysts like ADL and K10 are among the most active. For these materials diffusion limitations are low. On the other hand, pillared montmorillonite is very microporous, which could explain its relatively lower activity. The non-microporous catalyst K10, which is a Brönsted acid, is more active than the Lewis acid ADL. Similarly, APF is a Brönsted acid, and is much more active than APM. Solid Brönsted acids might thus catalyze the alkylation reaction more efficiently than solid Lewis acids. The preliminary conclusion is that high catalyst activities are a consequence of low diffusional limitations and high Brönsted acidity.

The product selectivity towards the three possible mono-alkylated isomers was studied in the range 50-65% of conversion. The desired conversions were obtained by using different amounts of a 100-200 mesh (150-75 μm) fraction of catalyst. At this particle size, the reaction was probably under diffusion control. The three monoalkyl isomers have different molecular sizes and probably different diffusivities. The twist angle between the two benzene rings is known to increase from 20° in biphenyl to 60.7° in *ortho*-isopropylbiphenyl in the solid state [11].

The product distribution obtained with an homogeneous catalyst like sulfuric acid in milder conditions (130°C, atmospheric pressure) was, for 50% conversion of biphenyl to monoalkylated products, 44%, 24% and 32% of *ortho*, *meta* and *para* isopropyl biphenyl, respectively [6]. The results presented in Table IV show that for the heterogeneous catalysts studied, the yields of *meta* and *para* isomers are higher than those obtained with sulfuric acid as a catalyst. Although the reaction conditions for the homogeneous and heterogeneous catalysts were not exactly the same, it is clear that the heterogeneous catalysts disfavor the *ortho* isomer. Moreover, catalysts with few micropores such as ADL and K10 afforded a lower yield of *meta* and *para* isomers than the microporous catalysts. There are two main possible explanations for this behavior.

TABLE IV. Distribution of monoalkylated products.

Catalyst:	H_2SO_4[a]	K10	Delaminated Laponite	Pillared Fluorohect.	Pillared Montmor.
weight (g):	0.9	0.09	0.1	0.06	0.2
conversion: (%)	100	61	63	50	64
monoalkyl: (%)	50	60	58	73	63
%ortho:	**44**	**32**	**32**	**23**	**15**
%meta:	24	27	30	34	42
%para:	32	41	37	43	43

a: from D. B. Priddy (1969) t=120-130°C atmospheric pressure

The *ortho* isomer is largest of the three isomers and its formation could be restricted in the micropores. However, we have observed no difference in selectivity by varying the interlayer spacing of the pillared clay from 6 to 15Å [12]. A more favored explanation is that the *ortho* isomer diffusion, being slower than that of the *meta* and *para* isomers, isomerizes or undergoes a second alkylation more readily than the other isomers when the reaction is under strong diffusion control. With very microporous catalysts, diffusion control becomes more important. Thus we expect the yield of *ortho* isomer to depend on the fraction of total surface area due to micropores. This is supported by the fact that the yield of *ortho* isomer decreases as the particle size increases. For example, with APM the *ortho* isomer yield decreases from 21 to 9% of the monoalkylated products when the particle size increases from a fraction less than 40 µm to a 300-150 µm fraction at the same biphenyl conversion of 44%.

4. Conclusion

Mesoporous clays and pillared clays are efficient catalysts for propene alkylation of liquid biphenyl. The results suggest a relationship between the pore structure, the type of acidity and the catalytic activity. In the solid phase, diffusion limitations are severe and the activities and selectivities of the catalysts are very dependent on the particle size used. Shape selective effects due to the micropore structure, (ie, cage effects) are thus difficult to observe under these conditions. The catalytic results are much more dependent on the ratio of microporous surface area to mesoporous surface area. If the mesoporous surface area is relatively low, the molecules cannot readily access the micropores and the activity is lower.

In order to develop new catalysts for liquid phase alkylation reactions, it will be very important to control the pore structure of the solids. Pillared clays are very promising for the alkylation of liquid phase substrates because the diversity of clay properties and wide range of pillars should enable one to design catalysts with the desired pore and acid strength properties.

5. Acknowledgements

The National Science Foundation (DMR-8903579) and Rhône-Poulenc are gratefully acknowledged for the financial support of this research.

References

1. Thomas, J.M. (1992) Scientific American 266, 112.

2. Hölderich, W.,Hesse, M. and Näumann, F. (1988) Angew. Chem Int. Ed. Engl. 27, 226.

3. Lee, G.S., Maj, J.J., Rocke, S.C., and Garcés, J.M. (1989) Catal. Lett. 2, 197.

4. Sugi, Y., Matsuzaki, T., Hanaoka, T., Takeuchi, K., Tokoro, T., and Takeuchi, G. (1991) Stud. Surf. Sci Catal. 60, 303.

5. Pinnavaia, T.J., Tzou, M.-S., Landau, S.D., and Raythatha, R.H. (1984) J. Mol. Catal. 27, 195.

6. Süd-Chemie AG, *K-Catalysts:* Süd-Chemie AG, Munich, Federal Republic of Germany, p. 2.

7. Priddy, D.B., (1969) Ind. Eng. Prod. Rev. Dev. 8, 239.

8. Occelli, M.L., Landau, S.D., and Pinnavaia, T.J., (1987) J. Catal. 104, 331.

9. de Boer, J.H., Lippens, B.C., Linsen, B.G., Broekhoff. J.C.P., Van den Heuvel, A., and Osinga, Th.J. (1966) J. Coll. Inter. Sci. 21, 405.

10. Delon, J.F., and Dellyes, R. (1967) C. R. Acad. Sci. Paris Série D 265, 1161.

11. Barrett, E.P., Joyner, L.G., and Halenda, P.H. (1951) J. Am. Chem. Soc. 73, 373.

12. Suzuki, H., (1960) Bull. Chem. Soc. Jap. 32, 1340, 1350, 1357, 33, 104.

13. Butruille, J.-R., and Pinnavaia, T.J., unpublished results.

OXIDE-PILLARED LAYERED α-METAL(IV) HYDROGEN PHOSPHATES

A. JIMENEZ-LOPEZ* P. MAIRELES-TORRES, P. OLIVERA-PASTOR, and E. RODRIGUEZ-CASTELLON
Departamento de Química Inorgánica
Facultad de Ciencias
Universidad de Málaga
29071 Málaga, Spain.

and

A.A.G. TOMLINSON*
ITSE-CNR
Area della Ricerca di Roma
CP 10 Monterotondo Staz.
00016 Rome, Italy.

ABSTRACT. Oligomeric cations of Al^{3+} or Cr^{3+} are intercalated into α-tin hydrogen phosphate (α-SnP) and α-zirconium hydrogen phosphate (α-ZrP) when colloidal suspensions of these hydrogen phosphates are contacted with partially hydrolysed Al^{3+} solutions or $Cr(OAc)_3$ solutions. Aluminium oligomers are intercalated by ion exchange with tetramethylammonium- α-SnP or *n*-propylammonium-α–ZrP, whereas Cr^{3+}-polyhydroxoacetate cations are intercalated by *in situ* polymerisation. In both cases, highly expanded materials are obtained having interlayer distances between 18 and 33 Å. Calcination of these precursors in air (Al^{3+}) or N_2 (Cr^{3+}) at 400°C leads to porous metal oxide-pillared materials. Although, the more expanded materials tend to amorphise upon further calcination, most materials maintain d_{002} basal spacings in the range 12.3 - 27 Å. The specific surface areas (BET, N_2, 77K) lie in the range 190-386 m^2g^{-1}, which compares well with analogues in clay chemistry (150-400 m^2g^{-1}). Pore size calculations show them to be mainly mesoporous but with micropore contributions. High cation exchange capacities (0.7 - 2 $meq{\cdot}g^{-1}$) confirm that the calcined materials are porous and have more accessible interlayer sites than do the intercalated precursor materials. The pores formed are carbocations generating (like zeolites), permit polymerisation of organic amines (e.g. pyrrole) and can be used to support quantum-well particles.

1. Introduction

The preparation of pillared materials from smectite clays has attracted considerable attention in the past 15 years. The interest was originally centred on their molecular sieving and catalytic properties [1],where, in some cases, pillared clays have been shown to be commercially competitive with zeolite catalysts [2]. The field has now expanded to include other layered materials, such as tetravalent metal salts [3] formulated as $M(IV)(HPO_4)_2{\cdot}H_2O$.

C.A.C. Sequeira and M.J. Hudson (eds.),Multifunctional Mesoporous Inorganic Solids, 273–287.

Pillared layered clays (PILCs) have been known since the pioneering work of Vaughan and collaborators as far back as 1973 [4]. However, doubts still remain as to how interlayer groups are chemically bonded to the layers, and their lateral ordering [5]. It has been demonstrated [6] by ^{27}Al MAS-nmr that, in beidellite at least, a chemical bond exists between AlO_4 groups of the tetrahedral smectite sheet and interlayer alumina tetrahedra and that this cross-linking leads to a stable pillared structure. Other problems derive from sheet instability, in which the sheet degrades owing to protons released upon calcination, which migrate to the octahedral vacancies hence provoking pillar segregation [7] Manipulating the cavity-acidity characteristics in layered phosphates provides materials analogous to smectite clays, with different materials characteristics : higher thermal stability, greater resistance to acids, etc.

Layered hydrogen phosphates may be pillared by several means using classical solution chemistry; the three most important ones reported to date are: i) a polycation is inserted and the material is calcined, the cation being converted to an oxide which holds apart the layers [8]; ii) a ligand, L, is inserted, after which a metal ion is exchanged into the material, which subsequently coordinates to L [9]; iii) the individual basal layer groups are derivatised [10], usually with aryl groups large enough to hold the layers apart (no derivatisation with inorganic groups has been reported).

2. Materials and Methods

2.1 BASE MATERIALS

The initial materials were α-tin hydrogen phosphate monohydrate and α-zirconium hydrogen phosphate monohydrate prepared by standard methods [11,12], α-SnP is virtually isostructural with α-ZrP [13] but its crystallinity is extremely sensitive to details of preparative solutions. In both hydrogen phosphates there is the possibility that manipulations may enhance hydrolysis of the internal surface (α-Snp > α-ZrP).

2.2. DEVELOPMENT OF COLLOID METHODS

The intercalation of robust species usually requires prior delamination of the layered material. This is not necessary in the Na^+ forms of smectites because they lead to stable colloidal suspensions. Alberti *et al* [14] established a method for preparing stable colloidal phosphate suspensions, consisting of intercalation of *n*-propylamine (nPA) at 50% of the saturation capacity. Although suitable for intercalating oligomeric cations in α-ZrP, when α-SnP is used amorphous intercalates are usually obtained, which we attribute to detachment of nPA from the phosphate surface during the exchange reaction instead of $nPAH^+$ ions, owing to differing acidities of the adjacent P-OH groups responsible for exchange and intercalation properties:

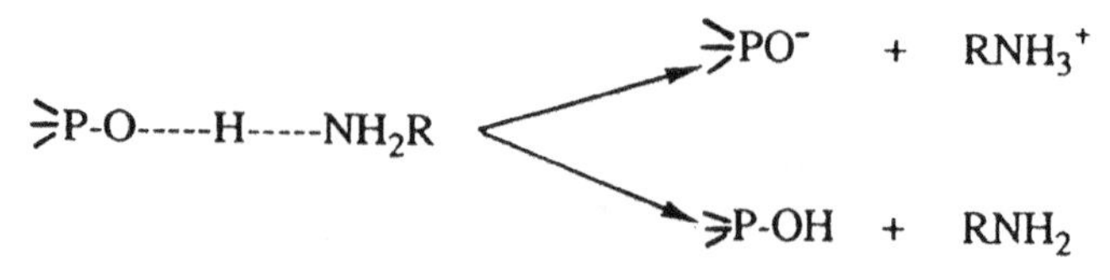

These differences are ultimately due to P-OH bond length differences (currently being clarified *via* Rietveld and neutron diffraction studies). We have developed a new method for preparing colloidal layered phosphates [15], in which the number of interlayer protons is reduced, thus avoiding interferences in the exchange process. The method simply involves exchange of interlayer n-propylammonium by tetraalkyl ammonium ions. α-$Sn[NMe_4]_{0.9-1.0}H_{1.1-1.0}(PO_4)_2 \cdot 4H_2O$, "$NMe_4$-SnP" in particular, has proved to be an excellent exchanger with metal oligomers [16,17]. NMe_4-SnP colloidal suspensions are formed according to the following scheme:

$$\underset{\text{(aqueous suspension)}}{\alpha\text{-SnP}} \xrightarrow[\text{2 hours, 25°C}]{\text{n-propylamine}} \alpha\text{-Sn.}\overline{\text{HnPrA}}\text{.HP}$$

$$\downarrow \text{[NMe}_4\text{]Cl, 1 day, 25°C}$$

$$\alpha\text{-Sn[NMe}_4]_{0.9-1.0}H_{1.1-1.0}(PO_4)_2 \cdot 4H_2O$$

Scheme 1. Preparation of the [NMe_4^+]-SnP colloidal intermediate

3. Results and Discussion

3.1. INTERCALATION OF METAL OLIGOMERS INTO LAYERED PHOSPHATES

We have explored two methods for inserting oligomeric species in the interlayer region of layered phosphates:

3.1.1. Intercalation of polyhydroxyoxo-aluminium cations

Many aluminium species may exist in aqueous solution depending on the conditions of solution preparation [18], although controversy still exists, recent nmr studies have significantly clarified the situation [19]. The species usually used to obtain Al_2O_3-pillared materials is the Keggin-like tridecameric cation $[AlO_4Al_{12}(OH)_{24}(OH_2)_{12}]^{7+}$. However, in solution this is believed to give rise to further species, such as $[Al(H_2O)_6]^{3+}$, $[Al_2(OH)_2(H_2O)_8]^{4+}$, $[Al_6(OH)_{12}(H_2O)_{12}]^{4+}$ and $[Al_{24}O_8(OH)_{46}\text{-}(H_2O)_{24}]^{10+}$, and so on. Fortunately, the nature of the anion also plays an important role in Al^{3+} hydrolysis [20], so that certain ligands (such as acetate) can compete with OH^- groups, faciliting formation of larger species and avoiding precipitation of $Al(OH)_3$. Clearly, the precursor species formed by such cations are expected to lead to several interlayer intercalated species with formation of a variety of alumina species.

3.1.2. Intercalation of Al oligomers from a partially neutralized solution(Al_{13}) and chlorhydrol (Chlor).

Contacting either aged Al_{13}-oligomer solution or chlorhydrol solution with colloidal nPrA-SnP gives only disordered intercalation compounds with a small amount of the polyhydroxyoxo-cation. Conversely, subsequent exchange of nPrA-SnP with NMe_4^+ [16] leads to an ordered material with a large degree of intercalation. The uptake

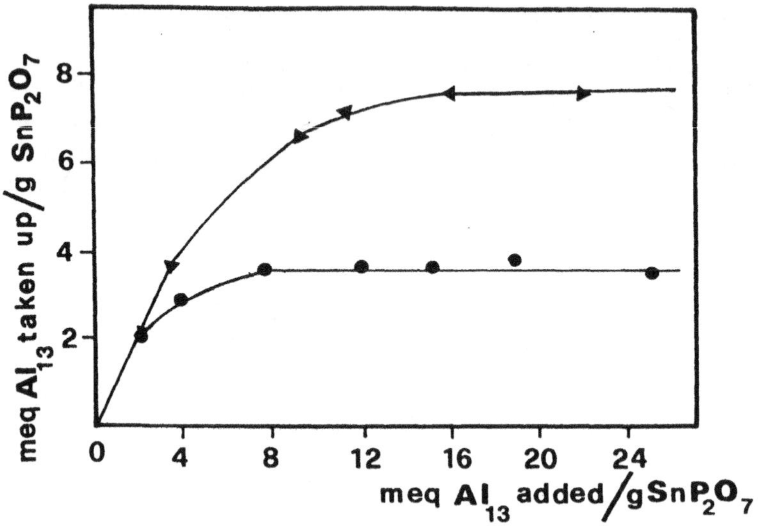

Figure 1. Uptake of "$[AlO_4Al_{12}(OH)_{24}(OH_2)_{12}]^{7+}$" by $[NMe_4^+]$-α-SnP from two different solutions: (●) Al^{3+}/NaOH; (▲) chlorhydrol

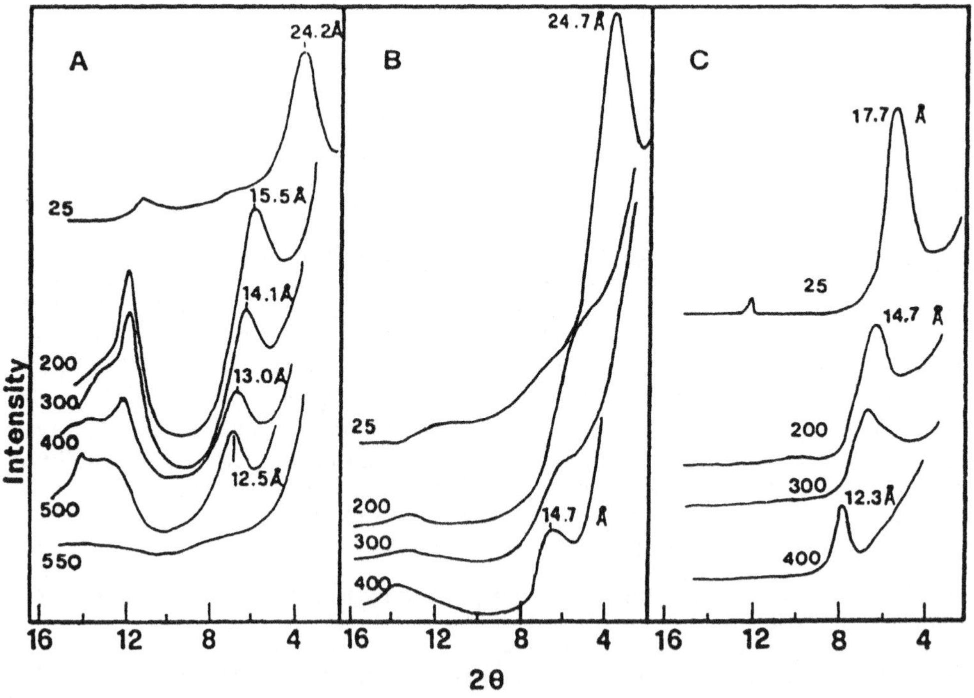

Figure 2. Variation of XRD patterns of chlor-SnP (A, centrifuged; B, dialysed) and Al_{13}-SnP (C, centrifuged) with temperature.

of the two polyhydroxyoxo-cations (Figure 1) into α-SnP is similar at low polyhydroxyoxo-cation: α-SnP ratios, but then diverges at higher ones. Samples prepared under the conditions of the plateaux of the uptake curves show interlayer distances of 17.7 Å for Al_{13}-SnP precursor and 24.2 Å for chlor-SnP precursor, suggesting the presence of different species in each precursor. This was expected since studies of the respective solutions by nmr have revealed the presence of significant differences in structure [18]. Al_{13}-SnP more closely fits with the presence of $[AlO_4(OH)_{24}(OH_2)_{12}]^{7+}$ in the interlayers (this cation has a long axis of ca. 9.5 Å). The chlor-SnP precursor contains more aluminium, indicating that oligomeric species with lower charge per Al ratio are present in this precursor, and that formation of bilayers is possible (a consequence of the high charge density of the phosphate). However, calcination of these two precursors at 400°C leads to materials with similar interlayer distances (about 12 Å) (Table 1), which suggests that similar pillaring species are formed.

N_2 adsorption-desorption isotherms of the pillared materials show that they are mesoporous with a BET surface area of 228 m^2g^{-1} for Al_{13}-SnP(400°C) and 190 m^2g^{-1} for chlor-SnP(400°C). For both calcined materials, most pores (>50% of BET surface area) have radii in the range 15-40 Å (Figure 3). Mesopores are believed to be induced by platelet stacking or end-end interactions, as shown in scheme 2.

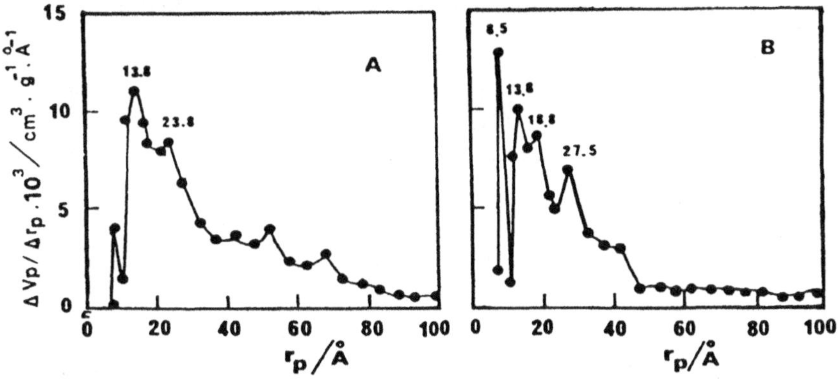

Figure 3. Pore size distributions for Al_{13}-SnP (A) and chlor-SnP (B). Cranston's Method

The interaction of interlayer alumina pillars with the phosphate layers is so strong that no segregation of Al_2O_3 is observed beyond 400°C, SnO_2 and aluminium pyrophosphate being obtained at 1000°C.

3.1.3. Intercalation of aluminium oligomers from buffered solutions.

Throughout the studies to date on PILCs little attention was paid to pH control and the precursors were prepared using ageing techniques. A simple pH control would improve the properties of pillared materials.

A new and faster method [21], which improves the pore regularity consists simply of preparing the cationic aluminium solution oligomer by ading a buffer (acetic acid) to an Al^{3+} solution partially neutralized with NaOH. This solution can then be added immediately (without ageing) to the phosphate colloid. Performing this intercalation reaction at 25°C and under reflux produces two different series of porous materials. Data for two materials separated at the plateaux of the respective uptake curves, are listed in Table 1 in comparison with Al_{13}-SnP and chlor-SnP materials.

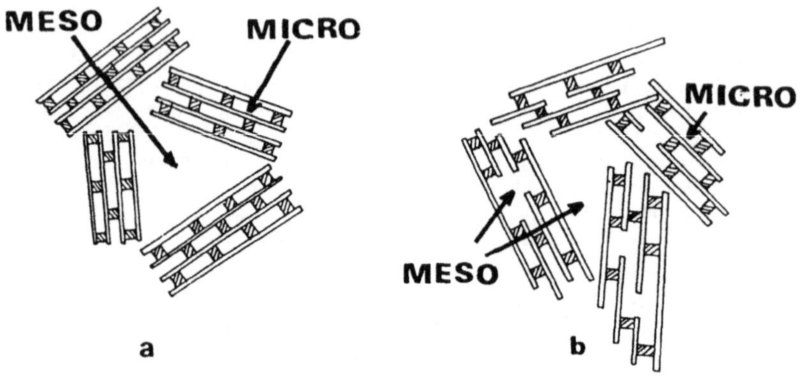

Scheme 2. End-side particle interactions giving rise to mesopores, (a) ideal regular pillaring; (b) with 'cut' layers owing to hydrolysis (or other short-range disorder).

Table 1. Comparison between polyhydroxyoxoaluminium precursors and alumina-pillared final materials prepared from unbuffered (un) and buffered (buff) starting solutions.

Solution	Al_2O_3 (%)	H_2O (%)	d_{002}/r.t. (Å)	d_{002}/400°C (Å)	Δ*(Å)
$OH/Al^{3+} = 2$ (un)	17.1	25.1	17.7	11.7	5.1
Chlor (un)	26.7	30.7	24.6	13.8	7.2
r.t. (buff)	25.2	24.1	16.8	11.9	5.3
Reflux (buff)	49.5	27.6	30.3	22.2	15.6

*Free height, assuming layer thickness = 6.6 Å.

Much higher Al content and much higher interlayer distances are obtained when the reflux method is used (Table 1), and these data well fit the assumption that higher polymers are present in such precursors, presumably *via* coalescence of Al_{13} polymers [22].

The evolution of d_{002} basal spacing with temperature of these four precursors is compared in Figure 4. As shown , the material prepared from refluxed buffered Al solution presents higher basal expansion than all the other phosphate precursors, the free height being higher than those in PILCS.

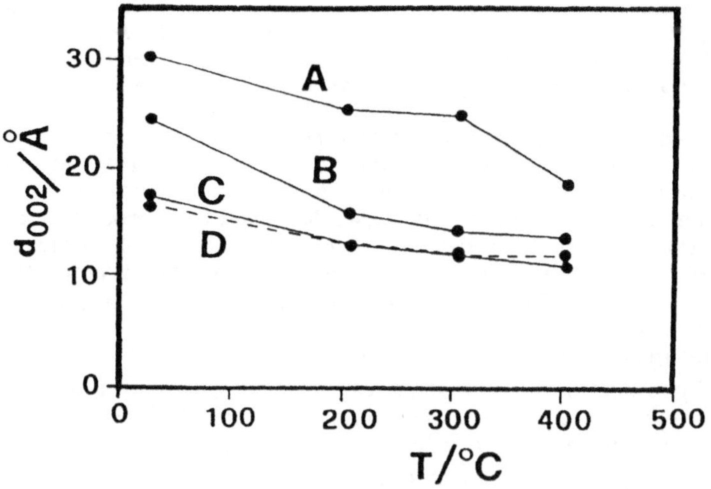

Figure 4. Evolution of d_{002} basal spacings with temperature for four aluminium oligomer exchanged materials prepared from different solutions; A: refluxed buffered, B: chlor., C: r.t. buffered, D: Al^{3+}/NaOH.

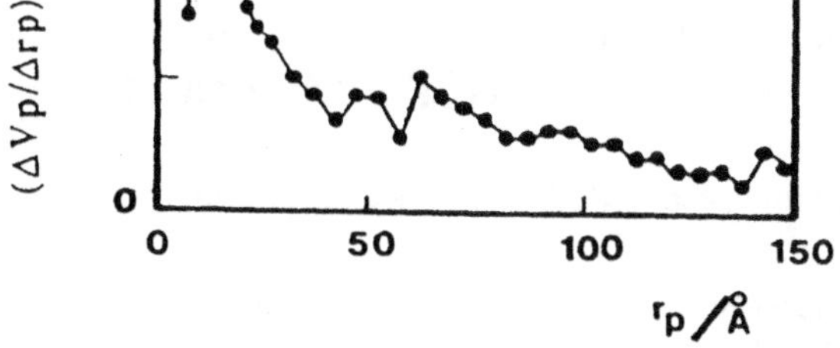

Figure 5. Pore size distribution of a pillared material (400°C) prepared from an Al^{3+} refluxed and buffered solution.

In general, the isotherms correspond to those of mesoporous solids, but there are significant differences between materials prepared from buffered and unbuffered solutions. The former exhibit a higher microporosity (> 20%) and narrower pore size distributions, most of the pores being centred between $r_p = 8.5\text{-}11.3$ Å .

3.1.4. Intercalation of Cr(III) oligomers by "in situ" polymerisation.

The major issue in clay pillaring studies by polyhydroxide precursors is the identification of precursor intercalates. Our experience with polyhydroxyoxoaluminium and polyhydroxychromium(III) formed *ex situ* shows that, apart from problems caused by solution equilibria, such methods can also give rise to large variations in local concentrations of base and side exchange reactions with surface groups (from, for example the Na^+ ions from NaOH used for polymerisation) producing lateral disordering effects during calcination.

Thus, the colloid manipulation methods have been extended to the preparation of chromia-pillared phosphates of α-SnP and α-ZrP by *in situ* polymerisation [17,23]. In both cases, precursors were prepared by direct reaction of aqueous $Cr(OAc)_3$ with NMe_4-SnP or nPA-ZrP suspensions under reflux. Figure 6 compares the uptake curves of hydrolyzed $Cr(OAc)_3$ species by colloidal α-SnP and α-ZrP.

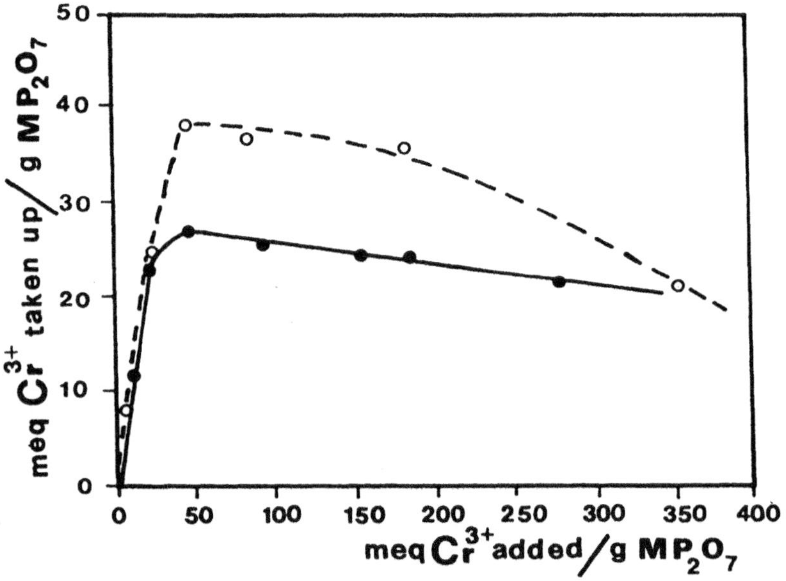

Figure 6. Uptake of hydrolyzed $Cr(OAc)_3$ species by colloidal α-SnP and α-ZrP

In both cases, after initial, rapid, uptake of Cr(III), the amount of Cr(III) exchanged then decreases smoothly. This behaviour is typical of an adsorption by a solid during which a polymeric species forms concurrently in solution [24]. Thus, as the ion exchange reaction proceeds, solution equilibria are set up which lead to formation of large oligomeric Cr(III) species in solution. These are too large to diffuse into the phosphate hosts and leave fewer smaller Cr(III) oligomers available to do so. The intercalated species are linear Cr(III)-polyhydroxyacetate moieties as deduced from chemical analyses and IR-uv-vis spectroscopy. Acetate ligands coordinate to Cr(III) ions in such way as to preclude formation of high connectivity-clusters, which are usually found in solution [23].

Different orderings of the precursors appear to be present in α-SnP and α-ZrP, the former probably containing monolayers and considerable amounts of water on reforming from the colloidal solution and the latter a bilayer structure.

Surface characteristics, such as the charge density, are obviously crucial for *in situ* polymerisation of Cr(III) hydroxyacetate, and it appears that the higher charge density of phosphate surfaces than those of smectite clay permits more orderly growth of polyhydroxyacetate-Cr(III) oligomers. Spectroscopic data and chemical analyses indicate that, in most cases, single linear oligomeric species are intercalated into the phosphate hosts in contrast to montmorillonite, where more than one species are present in the interlayer region [25]. Table 2 compares empirical formulations and d_{002} basal spacings of Cr(III) polyhydroxyacetate phosphates and montmorillonite intercalated obtained in similar conditions.

Table 2. Polyhydroxyacetate phosphates and montmorillonite intercalates

Empirical formulations	d_{002}/r.t. (Å)	d_{002}/400°C (Å)	Δ(Å)*
$[Cr_3(OH)_6(OAc)]_{0.8}(H_2O)_{5.8}SnH_{0.4}(PO_4)_2$	26.5	14.0	7.4
$[Cr_5(OH)_{10}(OAc)_2]_{0.65}(H_2O)_{8.8}Zr(PO_4)_2$	39.3	26.9	20.3
$[Cr_{3.7}(OH)_{9.4}(OAc)](H_2O)_{4.8}\cdot$Mont.	20.4	15.6	6.1

*Free heights, assuming the layer thickness to be 6.6 Å (phosphate) and 9.5 Å (montmorillonite).

Importantly, calcination of the materials under N_2 up to 400°C gave, in most cases, polycrystalline products with a wide range of basal typically expansions, 6 to 21 Å. The absence of a pyrophosphate phase until 1000°C and chemical resistance against dilute acid solutions (0.2 M) is a consequence of the chromium pillars holding the layers apart. However, calcination of these precursors in air gives rise to a triple process of oxidation-segregation-reduction causing the layer structure to collapse. Oxidation of Cr(III) and organic matter occurs between 250-390°C; the CrO_3 formed is subsequently segregated from the interlayer region and reduced at higher temperatures.

Conversely, Cr(III)-polyhydroxyacetate oligomers, intercalated in montmorillonite using the same method, only form interlayer chromia pillars if calcination is carried out in ammonia. In this way, the protons released during calcination are taken up by ammonia and NH_4^+ ions subsequently retained in the interlayer space. This ion is too large to diffuse into the octahedral vacancies in the silicate layer[22].

N_2 adsorption-desorption isotherms of chromia-pillared phosphates are all characteristic of mesoporous materials (with micropore contributions of 0.05-0.12 cm^3g^{-1}) and present high specific surface areas of 250-390 m^2g^{-1} with pore radii ranging from 8.5 to 15 Å and very narrow pore size distributions. Differently from the alumina pillared phosphates, some of the pores in the internal surfaces correspond to mesopores, possibly a consequence of the high free height. Calcination beyond 500°C leads to amorphous chromia-pillared phosphates, with virtually the same pore size and pore size distributions as the crystalline form (Figure 7).

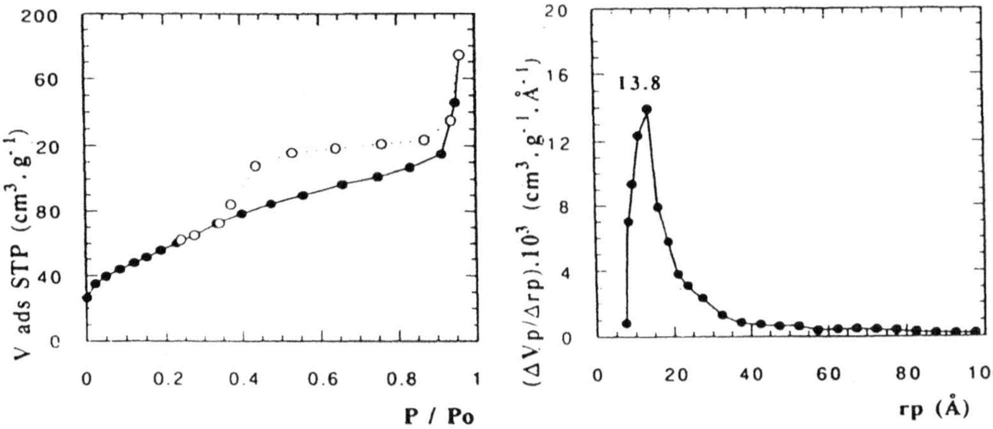

Figure 7. N_2 adsorption-desorption isotherm and pore size distribution of a chromia pillared zirconium phosphate material after calcination at 800°C.

This poses the problem of whether XRD alone is a good measure of porosity induced by pillaring. The amorphous nature of the porous chromia pillared phosphate may be rationalised by assuming that it arises from end-to-end sintered particles which are so small (< 50 Å) as to give only ill-defined ('amorphous') XRD patterns but yet do not preclude long-range order. Beyond 800°C, such materials decompose to pyrophosphate and Cr_2O_3.

4. Pore Characteristics

All the alumina-pillared phosphates retain ion exchange properties (see Figure 8, for ion exchange with Co^{2+}, Ni^{2+} and Cu^{2+}). These change with type of oligomer used and amount of Al^{3+} inserted. As for clays, prior ammoniation enhances exchange capacities, but to a much greater degree: they lie between 0.5 - 1.5 meq g^{-1}.

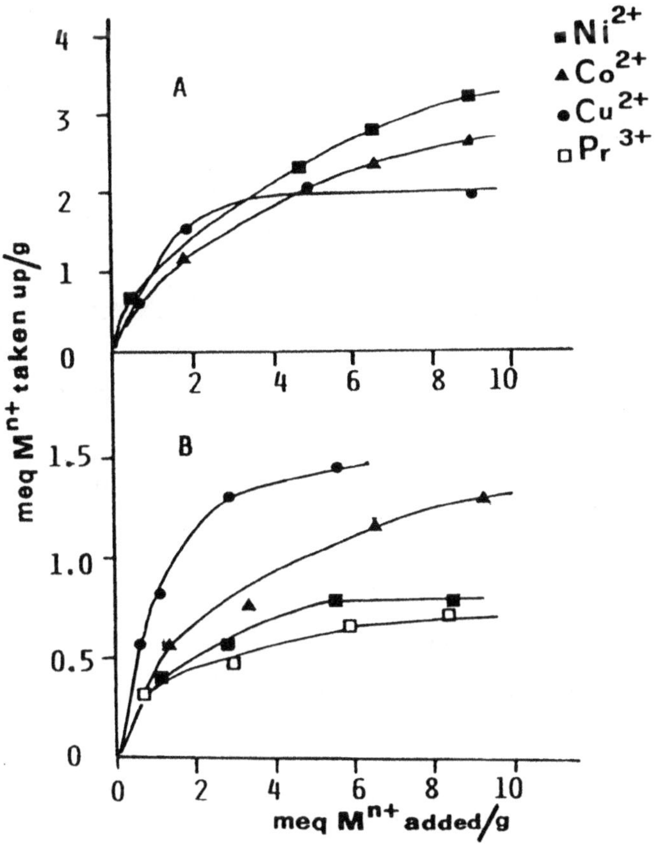

Figure 8. Uptake curves of metal ions by alumina-(Al_{13}) and chromia-pillared tin phosphate materials after prior ammoniation: (A) Al_{13}-SnP; (B) chromia-SnP.

Measurement of the pore characteristics is slow in coming owing to uncertainties as to the type of material obtained after calcination. Figure 9 shows the effect of exchanging Cu^{2+}up to the fully exchanged material on alumina-pillared (unbuffered) and chromia-pillared α-tin phosphates. In both series, the surface areas decrease, as expected, but do not go to zero, and only certain types of pores are filled, suggesting that specific internal exchange sites exist. The chromia-pillared materials are well-behaved, but the alumina-pillared ones are not. Surface areas decrease to less than half their original values, and the pore-size distribution is destroyed. Mesopore distributions are now centred at *ca* 50 Å (Cu^{2+}/Al_{13}-SnP and 25 Å (Cu^{2+}/Chlor-SnP, see Figure 9) indicating not only break-up of card-stacks but also filling of micropores. Since the XRD do not change on Cu^{2+} exchange, this suggests that the "packet lengths" of Scheme 3 are greater than 50 Å.

Presumably, exchange may occur at hydrolysed positions in the layers, i.e. it may occur at Sn-OH groups (hydrolysis has been estimated to be *ca.* 6% for the zirconium analogue). If so, hydrolysis is still not sufficient to lead to total amorphicity. These results suggest that retention of pore uniformity on exchange is a further criterion for "pillaring".

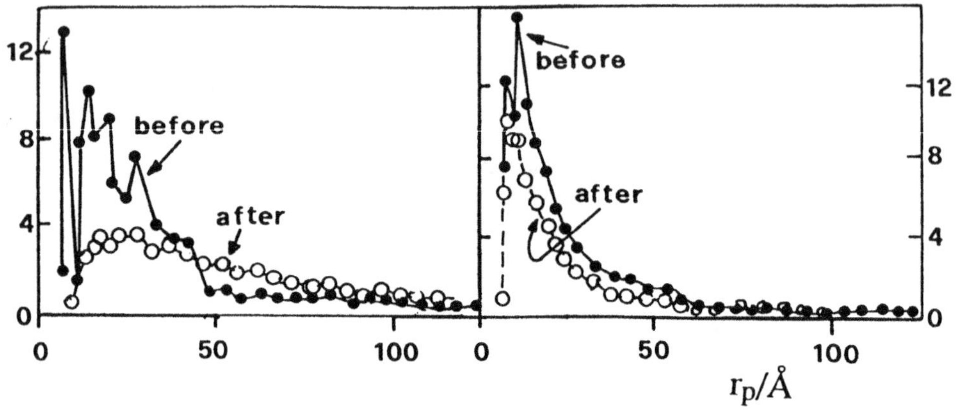

Figure 9. Effect of Cu^{2+} exchange on pore size distribution in chlor-SnP (left) and a typical chromia-SnP material (right).

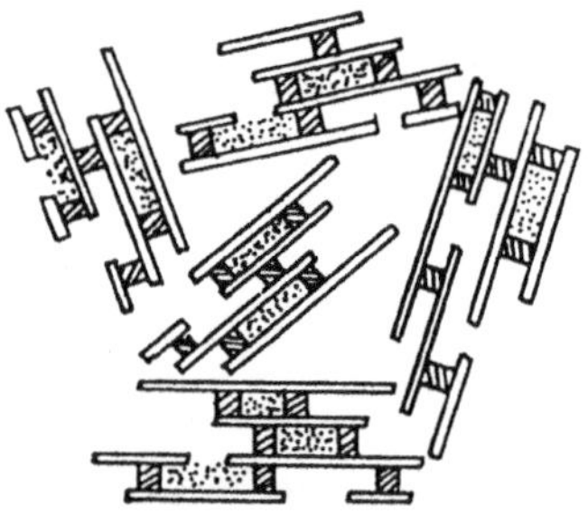

Sheme 3. Suggested model for alumina-pillared tin phosphates after cation exchange; (⁘) exchanged pores.

The porosity remaining after exchange in these materials is sufficient to permit polymerisation of pyrrole [26] (kinetic diameter 5.7 Å) to high levels, such that "molecular wire" nanocomposites are formed, similar to those found in zeolite Y and mordenite [27]. Like the latter, they are non-conducting ($\sigma < 10^{-9}$ Scm^{-1}). In some chromia pillared materials, the polymerisation occurs without prior Cu^{2+} exchange, indicating that carbocation formation ability is present in both cases. The polypyrrole is present as mixed neutral and polaron oligomers, so that assignment of polaron states is difficult.

5. Optical Properties

Another means of assessing pore size is to form quantum wells (QW) inside the pores, e.g. by first exchanging Cd^{2+} and then reacting tha exchanged materials with H_2S (prepared in a "clean" fashion, to avoid sulphur deposition). The shift in band gap threshold (the "Berstein shift") is related to the quantum well size, as shown in the correlation of figure 10 [28].

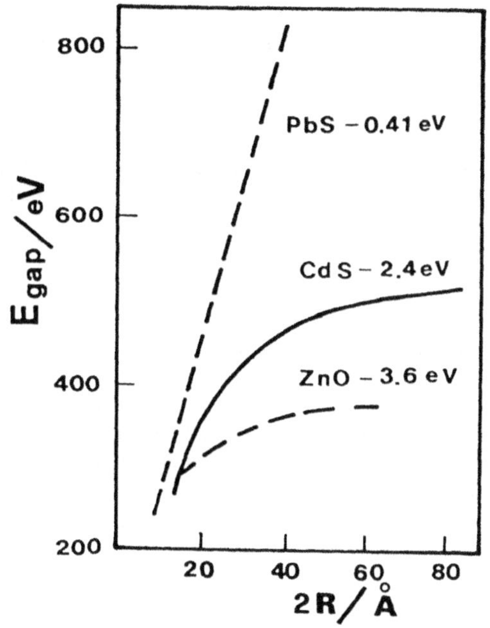

Figure 10. Band gap threshold vs particle size for some common semiconductors (from ref. 28).

For CdS quantisation, particles having mean particle size (2R) > 60 Å (i.e. larger than an exciton in bulk CdS) start to absorb close to 515 nm.(2.4 eV) the band gap of bulk CdS. As 2R decreases, the threshold shifts to lower λ, so that CdS particles with mean particle sizes 2R < 2.2 nm are colourless. Henglein [28] has also separated colloids having structured optical spectra,and suggested that discrete states in clusters of defined geometry.may be present. Figure 11 shows the optical spectra (DRS) of two such materials prepared from differently loaded Al_{13}-SnP.

Unfortunately, there is no clear edge shift, which we ascribe to the presence of CdS clusters sufficiently large as to give bulk CdS characteristics, generated from extended surfaces of groups of "packets" like those in Scheme 1. These give rise to the first derivative peaks at 479 (25% Cd^{2+}) and 484 nm (5% Cd^{2+}), both with threshold shift at 313 nm. Nevertheless, a crude factoring out of this absorption does suggest edge shifts to *ca* 400 nm in the 25% Cd^{2+} exchanged material and to ca 450 nm in the 5% Cd^{2+} exchanged one. Clusters having 2R *ca* 25Å appear to be present in the former and with 2R *ca* 35 Å in the latter. There is also some evidence for structure on the absorption (shown by asterisks in Figure 11; the band at < 300 nm is due to matrix) which differs between the two materials. This suggests that differently sized discrete CdS clusters are

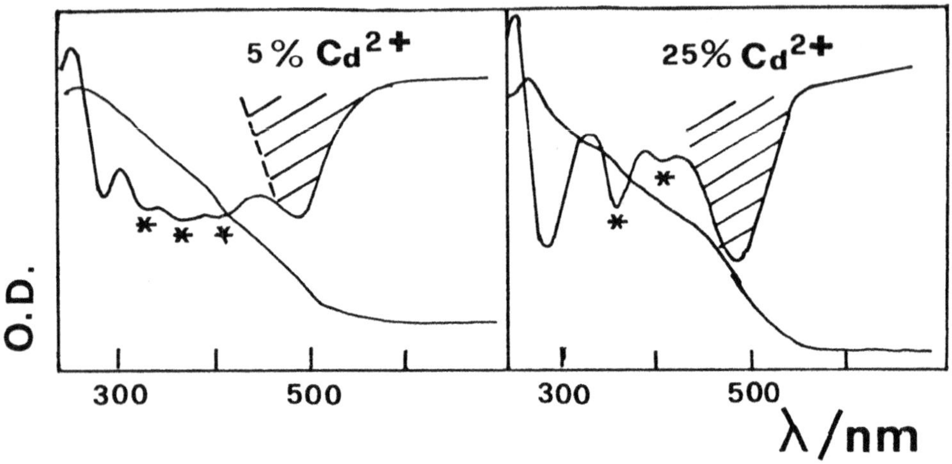

Figure 11. Optical spectra (including first derivative) of *in situ* prepared CdS clusters in Al_{13}-SnP.

present. Confirmation of these preliminary results is under way by X-ray absorption spectroscopy studies and optical measurements on the uniform alumina-pillared materials [29]. It appears that oxide-pillared phosphates may be useful as "bottled quantum well reagents", and with more accessible pores than zeolite-quantum well analogues [30].

Acknowledgements

We thank the E.C. Euram programme (project 0027-I) and Spanish C.I.C.Y.T. (project MAT90-298) for financial support. A.A.G. Tomlinson thanks the Spanish Ministry of Education for a sabbatical period at the University of Málaga.

References

1. Barrer, R.M. (1986), J. Inclusion Phenom., 456; Vaughan, D.E.W. (1988) in R. Burch (ed.), Catalysis Today. Pillared Clays, Elsevier, Amsterdam, p. 187.
2. Pinnavaia, T.J. (1987) in A.P. Legrand and S. Flandrois (eds.) NATO ASI Ser. B, Vol 172, Plenum Press, New /York, p. 271.
3 Tomlinson, A.A.G. (1990) in I.V. Mitchell (ed.), Pillared Layered Structures, Current Trends and Applications, Elsevier Applied Science, London, p. 91.
4. Vaughan, D.E.W., Lussier, R.J. and Magee, I. US P. 4,176,090; 4,248,739.
5. Purnell, J.H. (199) in I.V. Mitchell (ed.) Pillared Layered Structures. Current Trends and Applications, Elsevier Applied Science, London, P. 107.
6. Plee,D., Borg, F., Gatineau, L. and Fripiat,J.J. (1985), J. Am. Chem. Soc., 107, 2362.
7. Poncelet, G. and Schutz, A. (1986) in R. Setton (ed.), Chemical Reaction in Organic and Inorganic Constrained Systems, Riedel Publ. Co., p. 186.
8. Clearfield, A. and Roberts, B.D. (1988), Inorg. Chem. 27, 3237.
9. Ferragina, C., Massucci, M.A., Patrono, P., La Ginestra, A. and Tomlinson, A.A.G. (1986), J. Chem Soc., Dalton Trans., 265.

10. Alberti, G., Costantino, U., Allulli, S. and Tomassini, N. (1978), J. Inorg. Nucl. Chem., 40, 1113.
11. Costantino, U. and Gasperoni A. (1977), J. Chromatog., 51, 289.
12. Alberti, G. and Torraca, T. (1968), J. Inorg. Nucl. Chem., 30, 317.
13. Troup, J.M., and Clearfield, A. (1977), Inorg. Chem., 16, 3311.
14. Alberti, G., Casciola, M. and Costantino, U. (1985), J. Colloid Interface Sci., 107, 256.
15. Olivera-Pastor, P., Jimenez-Lopez, A., Maireles-Torres, P., Rodriguez-Castellon, E., Alagna, L. and Tomlinson, A.A.G. (1989), J. Chem. Soc., Chem. Comm., 751.
16. Maireles-Torres, P., Olivera-Pastor, P., Rodriguez-Castellon, E., Jimenez-Lopez, A. and Tomlinson, A.A.G. (1991), J. Mat. Chem., 1, 319.
17. Maireles-Torres, P., Olivera-Pastor, P., Rodriguez-Castellon, E., Jimenez-Lopez, A. and Tomlinson, A.A.G. (1991), J. Solid State Chem., 94, 368.
18. Jones, S.L. (1988) in Burch, R. (ed.), Catalysis Today, Pillared Clays, Elsevier Amsterdam, p. 209.
19. Fui, G., Nazar, L.F. and Dash A.D.(1991), Chem. of Mater., 3, 602.
20. Akitt, J.W. and Farthing, A. (1978), J. Magn Res., 32, 345.
21. Maireles-Torres, P., Olivera-Pastor, P., Rodriguez-Castellon, E., Jimenez-Lopez, A., Tomlinson, A.A.G., Perez. G. and Keheyan, Y. (1992), Ital. Pat. Applic. No. RM92A000376.
22. Vaughan, D.E.W. (1988), in ref. 18., p. 187.
23. Maireles-Torres, P., Olivera-Pastor, P., Rodriguez-Castellon, E., Jimenez-Lopez, A. and Tomlinson, A.A.G. (1991), J. Mat. Chem., 1, 739.
24. Giles, C.H., McEwan, T.H.,Nakhwa, S.N. and Smith, D. (1960), J. Chem. Soc., 3973.
25. Jimenez-Lopez, A., Maza-Rodriguez, J., Olivera-Pastor, P. and Rodriguez-Castellon, E. (1992), Clays and Clay Miner., submitted.
26. Maireles-Torres, P., Olivera-Pastor, P., Rodriguez-Castellon, E., Jimenez-Lopez, A. and Tomlinson, A.A.G. (1992), J. Incl. Phenom., submitted.
27. Bein, T. and Enzel, P. (1989), Angew. Chem. Int. Ed. Engl., 28, 1692; Bein, T. (1991) in "Electron Transfer in Biology and the Solid State", ACS Series, p. 493.
28. Henglein, A. (1989), Chem. Rev., 89, 1861.
29. Tomlinson, A.A.G., in preparation.
30. Wang, Y. and Herron, N. (1991), J. Phys. Chem., 95, 525.

SYNTHESIS OF POROUS AND CROSSLINKED LAYERED PHOSPHATES BY THE INTERCALATION OF ORGANOMETALLIC PRECURSORS

DEBORAH J. JONES, THIERRY CASSAGNEAU AND JACQUES ROZIERE
Laboratoire des Agrégats Moléculaires et Matériaux Inorganiques,
URA CNRS 79, Université Montpellier 2,
Place Eugène Bataillon,
34095 Montpellier Cédex 5, France

ABSTRACT. The hydrolytic condensation of alkyltrialkoxysilanes of general formula $ZSi(OR)_3$ leads to the formation of oligomeric species, the nuclearity and structure of which depend upon specific conditions of hydrolysis and ageing of the solutions, and which can be identified by ^{29}Si NMR. Synthetic conditions conducive to the formation of the siloxane octamer were adopted, with Z chosen as aminopropyl or 2-aminoethyl-3-aminopropyl. The presence of basic functions facilitates the intercalation process with α-$Zr(HPO_4)_2.H_2O$, $Sn(HPO_4)_2.H_2O$ and γ-$ZrPO_4(H_2PO_4)_2.2H_2O$ by making accessible an acid-base reaction. Each series of compounds demonstrates particular uptake characteristics for aminosiloxane, depending on the nature of the phosphate layers (Sn, Zr; α, γ). The [Si]:[P] ratio at which a single phase is first observed is crucial to the preparation of porous pillared - and not simply crosslinked - materials. Single phase pillar precursor materials have interlayer distances of 17 - 30 Å, representing an increase in free height of between 8 and 21 Å. Calcination at 600 °C removes the organic moieties, liberating the space they occupied laterally and anchoring silica fragements to the phosphate layers. After calcination, silica-pillared phosphates have interlayer distances of 12 - 18 Å, and specific surfaces areas up to 250 m^2/g, with a mean pore diameter of 13 Å. The phase prepared from γ-ZrP/aminopropyltriethoxysilane represents the first inorganic-pillared γ-zirconium phosphate. This is an adsorbent for basic molecules e.g. NH_3, $C_6H_5NH_2$. All the silica pillared phosphates have a high residual ion exchange capacity (Cu(II), Zn(II), Cd(II) etc.) and the pores can be used to size-quantize metal sulphide particles. Preliminary structural characterisation of silica pillared materials has been made using X-ray absorption spectrosocopy, which has allowed identification of the local environment around silicon.

1. Introduction

The insertion of inorganic cationic polyhydroxometallic species such as $[Al_{13}O_4(OH)_{24}(H_2O)_{12}]^{7+}$ and $[Zr_4(OH)_8(H_2O)_{16}Cl_x]^{(8-x)+}$ into a solid in which the layers carry a negative charge has been one of the most usual routes to crosslinking layered compounds [1]. Several examples are given in the chapters of this volume, including the pillaring of smectite clays and metal hydrogen phosphates, as well as silicic acids. Layered titanates and niobates [2] have also been crosslinked by ion-exchange with inorganic oligomeric cations. A similar strategy is used for layered double hydroxides where the inorganic layers carry a positive charge [3]. Use may also be made of the properties of an organic sol-modifier to stabilise colloidal solutions of metal ions, such that insertion into a two-dimensional host of a composite sol

C.A.C. Sequeira and M.J. Hudson (eds.), Multifunctional Mesoporous Inorganic Solids, 289–302.

system can be achieved. The organic template serves to space and uniformly distribute the inorganic cations, and concomitant pore size control is possible through the nature of the organic molecule, subsequently removed by calcination.The insertion of organometallic oligomers into lamellar compounds will also lead to inorganic pillared materials, although the method has been relatively much less used. The synthesis of silica-crosslinked zirconium (both α and γ forms) and tin phosphates by intercalation of octa(amino/diaminoalkyl)siloxanes prepared by the hydrolytic condensation of the corresponding alkyltrialkoxysilanes, will be described in this paper.

Modulation of pore size in directions parallel and perpendicular to the layers is one of the most important synthetic challenges in pillared layered solid chemistry. Optimisation of this process, and the possible control of the distribution of the pillaring species, will lead the state of the art in this area to one of "materials by design".

2. Lamellar host materials: smectite clays and non-clay substrates

Whereas crosslinking layered solids by the insertion of an inorganic or organometallic oligomer followed by thermal degradation to form the (covalent) link between layer and interlayer species requires favourable energetic parameters brought about by the chemical nature (nature of the components, Lewis/Brønsted acidity, steric considerations) of the solid and of the intercalant, their pillaring necessitates, in addition, a spacial distribution of crosslinking ions so as to leave empty certain regions of the interlayer volume. The most important difference in this respect between smectite clays and non-clay substrates concerns the charge density on the layers which is perhaps 8-10 times higher in metal phosphates for example, than in a smectite clay. This low layer charge density in clays has several consequences favourable to the formation of porous solids, and the creation of a similar situation in non-clay materials requires a more elaborate chemical strategy. The spontaneous swelling of smectite clays in water is not generally observed for their non-clay counterparts, for which a more thermodynamically favourable reaction must precede the polyoxo-ion insertion step, such as the use of polar organic molecules to expand the layers.

In addition, high layer charge density tends to lead to close spacing of the inserted species, even when the metal ion is highly charged. If inducing porosity is synonymous with overcoming the high layer charge density on non-clay lamellar matrices, then in addition to the use of multifunctional organic "templates", or of organometallic oligomeric species where the organic moiety is, as above, expendable, modification of the layer structure in such a way that the charge density is reduced can also be considered. Taking the example of metal phosphates, it has been shown that a significantly high percentage of active sites in an α-zirconium phosphate can be deactivated by modification during the synthesis such that the structure contains less highly-charged groups than phosphate [4].

3. Formation of silica-crosslinked and pillared α and γ-structured metal phosphates from organosilicon precursors

The insertion of an octameric siloxane $[ZSiO_{1.5}]_8$, where Z is an alkylamino function, into α-zirconium phosphate and the tin congener was carried out as an initial route to pillaring layered phosphates. Subsequently, attempts were made to increase the internal surface area of the silica-pillared solids formed after calcination, by modifying the nature of the phosphate substrate (limited ion exchange characteristics) and the nature of the organic group Z (more sterically hindering), whilst retaining the same synthetic approach. In each case, uptake of the siloxane by the layered host was measured, and the evolution of the X-ray diffraction pattern was followed. The pH varies throughout the reaction, in particular as a function of the metal phosphate:aminosiloxane ratio, and causes a certain degree of hydrolysis of the phosphate substrate. The extent of this hydrolysis was evaluated by the gravimetric determination of liberated phosphate.

3.1 STRUCTURE OF $Zr(HPO_4)_2.H_2O$ (α-ZrP) AND $ZrPO_4.H_2PO_4.2H_2O$ (γ-ZrP)

The structures of these host materials are described in other chapters of this volume, and only points salient to the intercalation process will be recalled here. The α-form can be considered as a solid monobasic acid with one type of hydrogen phosphate group [5]. The hydrogen atom on HPO_4 projects into the interlayer zone and 100% exchange is commonly observed in the absence of dominating steric effects. The γ-structure is more complex, and it is only recently that structural information has been obtained on this form using Rietveld profile refinement techniques on a neutron powder diffraction spectrum of the anhydrous titanium analogue [6]. Here, two types of phosphate group are associated with each metal atom, one of which carries two acid functions, the other bearing none. This arrangement introduces particular behaviour into the intercalation process since generally only one proton is transferred although, more rarely, a maximum of around 1.5 hydrogen atoms per zirconium atom has been observed [7].

3.2 OCTAMERIC ORGANOSILICON SPECIES AS SILICA-PILLAR PRECURSORS

The hydrolysis of alkylalkoxysilanes of general formula $ZSi(OR)_3$ leads to the formation of oligomeric species, the nuclearity and structure of which depend upon the particular conditions of hydrolysis (solvent) and ageing of the solutions. The oligomers are all built from a polyhedral silicon oxygen skeleton bearing organic side arms, Z, attached to the silicon atoms, figure 1, and whereas the even numbered members carry only one organic group per silicon atom (so-called T3 silicon atoms), the odd-numbered members each have one silicon atom bearing two organic groups (T2).

The chemistry of the siloxanes is well-known, and fundamental studies on the polymerisation process and the spectroscopic and structural characterisation have been reported [8]. For the purpose

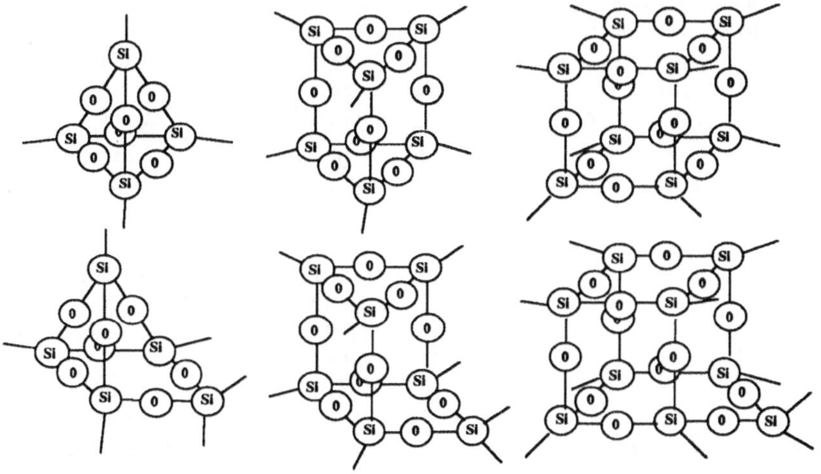

Figure 1. Polyhedral oligosiloxanes formed (above) by the hydrolytic condensation of trifunctional monomers $ZSiY_3$ with, in this work, Z = alkylamino, Y = OCH_3, OC_2H_5 and (below) by cohydrolysis of tri- or di- functional monomers Z_2SiY_2

of intercalation into a two-dimensional host, the siloxane octamer $[ZSiO_{1.5}]_8$ is particularly attractive, since it is has the same lateral dimension is all three directions, and the orientation of the polyhedron is thus unambiguous. Furthermore, its ^{29}Si n.m.r. spectrum is characteristic, enabling its identification with reasonable certainty, in the sol, prior to the intercalation reaction.

The octameric siloxane formed by hydrolysis of aminopropyltriethoxysilane has an approximate lateral dimension, including the organic side arms, of 11.2 Å, fig 2. The organic moieties of this oligomer play two important roles. Firstly, the eight basic end groups avoid the necessity for an additional expansion step, since in a one-stage process they allow expansion of the layers and entry of the organometallic system in an acid-base reaction. Secondly, the organic groups bearing the amino function are chosen to be bulky, to enable pore formation on calcination of the pillar precursor system, the pillar itself being represented by the inorganic Si-O-Si skeleton.

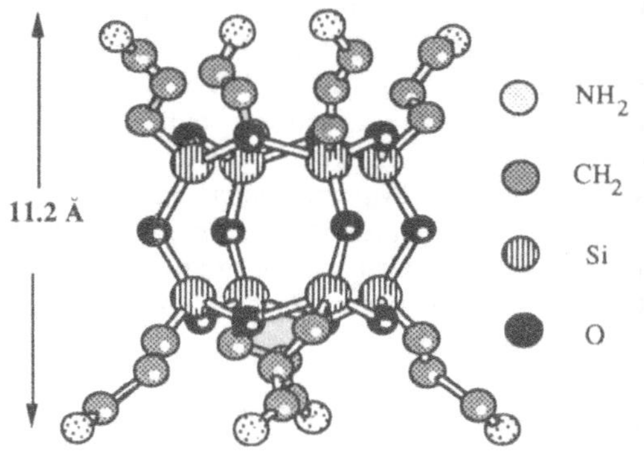

Figure 2. octa(aminopropyl)siloxane

3.3 CROSSLINKING AND PILLARING

3.3.1 *Octa(aminopropyl)siloxane and α-ZrP*. The insertion of siloxanes into zirconium phosphate was first reported by Lewis [9]. Whilst also using conditions for sol preparation conducive to the formation of the $[ZSiO_{1.5}]_8$ octamer, the temperature at which the intercalation compounds were calcined was limited to ~ 400°C. Octa(aminoalkyl)siloxanes are thermally stable species and the temperature reported is too low to remove all organic matter from the composite system. Furthermore, the host matrix was limited to α-ZrP which can, as will be shown, lead only to crosslinked, and not porous, solids. In another study, also using α-ZrP, a sol containing a mixture of dimeric and trimeric siloxane species was used, and although no surface area data were given, the resulting calcined material was reported to adsorb hexane [10].

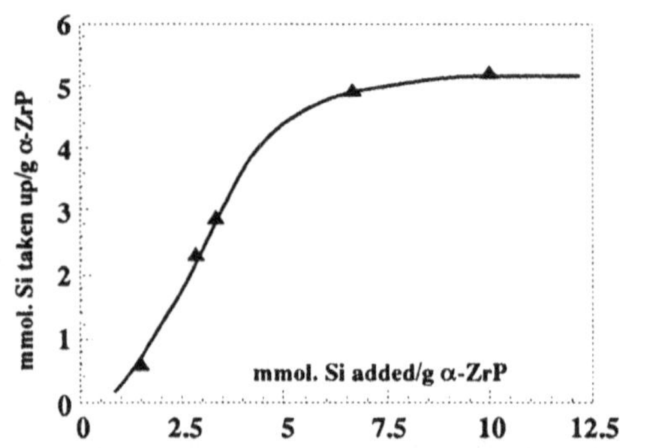

Figure 3. Uptake of $[H_2N(CH_2)_3SiO_{1.5}]_8$ by α-ZrP from aqueous ethanolic soution

In the present work [11], expanded α-ZrP was synthesised by contacting the solid, prepared according to published methods and suspended in deionised water, with aliquots of octa(aminopropylsiloxane) solution. After contact for 15h to 5 days, the materials were recovered by centrifugation, washed and air dried, and the silicon remaining in the supernatant liquid determined as quinoline silicomolybdate. The uptake isotherm is given in fig. 3, where it may be seen that a plateau region is reached when 5.5 mmol Si/g α-ZrP are intercalated, corresponding to a [Si]:[P] mole ratio of 0.8. X-ray diffraction (XRD) patterns (Cu-Kα radiation) indicate that a biphasic system subsists almost up to the plateau region, when a single phase, layered material of interlayer spacing d_{002}=17.7 Å is observed, figure 4. This corresponds to an increase in interlayer distance of 11.1 Å, or the lateral dimension of the siloxane octamer. Furthermore, the siloxane:α–ZrP ratio determined experimentally can be rationalised in terms of the ratio of the surface area of the active sites -POH and of that of the octameric "cube".

Thus the interaction of each face parallel to the internal surfaces of α-ZrP must involve four acid sites -POH on the phosphate layer above and on that below. The surface area of each active site in α-ZrP is known to be 24 Å², and proton transfer from four acid sites requires the effective use of 96 Å² on the layer above and that below. The siloxane octamer occupies a surface area of 125 Å² [(11.2 Å)²], from which it may be deduced that the "covering effect" prevents stoichiometric intercalation of the organosiloxane, which is limited in fact to 96/125 or 77%, in agreement with the experimental result.

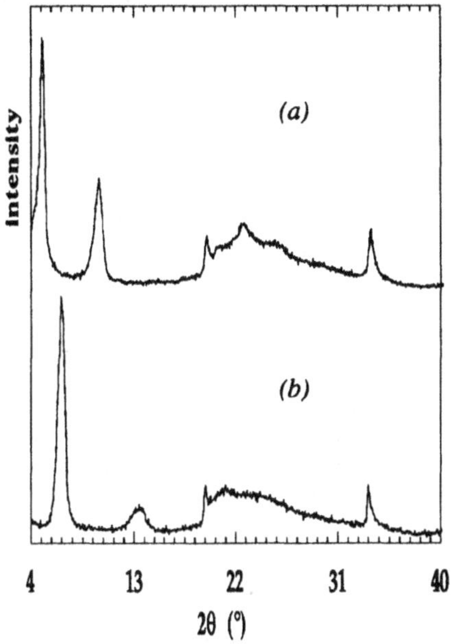

Figure 4. Intercalation of octa(aminopropyl)-siloxane into α-ZrP: evolution of the X-ray diffraction pattern
(a) $\alpha\text{-}Zr(PO_4)_2H_{0.4}\{H_3N(CH_2)_3SiO_{1.5})_8\}_{0.2}$
(b) as (a) after calcination at 500°C

Thermogravimetric analysis (figure 5) of the intercalation compound formed at maximum siloxane uptake indicates a total weight loss of ~ 25% up to 600°C, corresponding to elimination of cointercalated water up to 100°C and loss of organic matter at higher temperature. Concomitantly, the interlayer distance contracts, and a slight loss in crystallinity is observed. At 500°C, the basal spacing is 12.5 Å, and it may be considered that the phosphate layers are crosslinked through silica fragments. Despite occupancy of only 80% of the interlayer volume, the specific surface area is only slightly increased with respect to the starting α-ZrP (α-ZrP ~ $5 m^2g^{-1}$; $\alpha\text{-}Zr(PO_4)_2H_{0.4}\{H_3N(CH_2)_3SiO_{1.5})_8\}_{0.2}$ 30 m^2g^{-1}) Obtaining high surface area, porous solids requires modification either of the layered substrate, or of the pillaring organosiloxane.

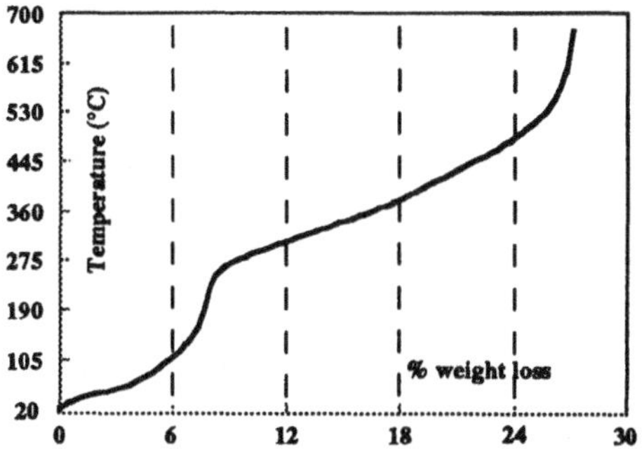

Figure 5. Thermogravimetric analysis in air of $\alpha\text{-}Zr(PO_4)_2H_{0.4}\{H_3N(CH_2)_3SiO_{1.5})_8\}_{0.2}$

3.3.2 *Octa(aminopropyl)siloxane and α-SnP [11].* The structure of α-SnP is entirely analogous to that of α-ZrP, but the compound shows a much higher propensity to hydrolysis, a property that can be exploited since it leads to the replacement of some of the hydrogen phosphate groups by hydroxyl, and a lowering of the overall layer charge density. One of the consequences of this partial hydrolysis is the formation, under the same experimental conditions as described above for α-ZrP, of a siloxane bilayer in the interlayer region, inferred from the uptake curve (identical to that observed for α-ZrP) and the evolution of the X-ray diffraction pattern of the intercalated phases. The phase diagram shown in fig. 6 depicts the approximate percentage of each phase present as a function of the concentration of octa(aminopropyl)siloxane per gram of α-SnP.

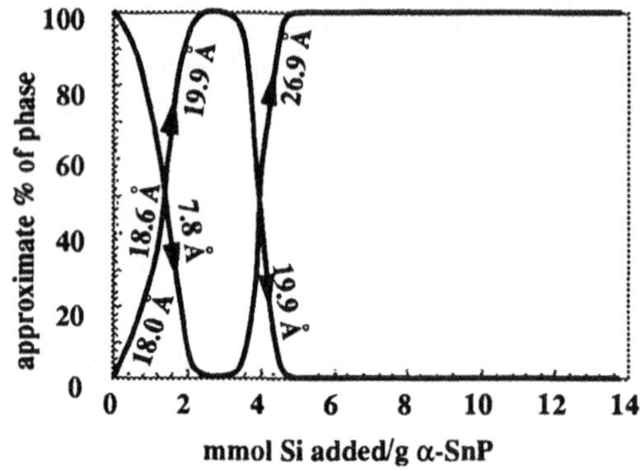

The system first goes through a monolayer arrangement of interlayer distance 19.9 Å, followed by a region where both single and double layers coexist, before a purely bilayer arrangement is observed. During the transition from a single to a bilayer of octa(aminopropyl)siloxane, phosphate ion is released to reach ~ 3 % of the total phosphate present, fig 7. Hydrolysis occurs more rapidly once intercalation is complete {[Si]:[P] ratio 0.8} to reach a maximum of 10% before reprecipitation, probably of amorphous SnP, occurs.

Figure 6. Phase diagram in the octa(aminopropyl)siloxane /α-SnP system : approximate percentage of each phase (from XRD) as a function of the number of mmol silicon present in solution/g α-SnP

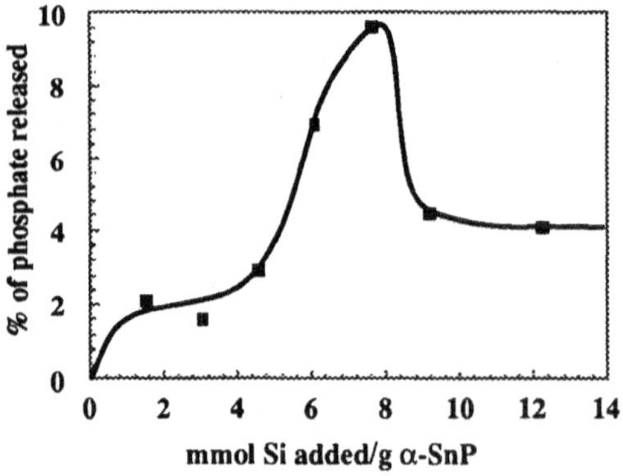

Figure 7. Intercalation of octa(aminopropyl)-siloxane into α-SnP: % PO_4 released as a function of mmol Si added/g α-SnP

Bilayered α-$Sn(PO_4)_2H_{0.4}\{H_3N(CH_2)_3$-$SiO_{1.5})_8\}_{0.2}$ has a basal spacing of 26.9 Å, two harmonics (004, 006) are seen in the X-ray diffraction pattern, fig 8. Calcination at 600°C in air or under vacuum gives the crosslinked material of interlayer distance 18 Å, indicating that the bilayer is conserved after expulsion of the organic groups.

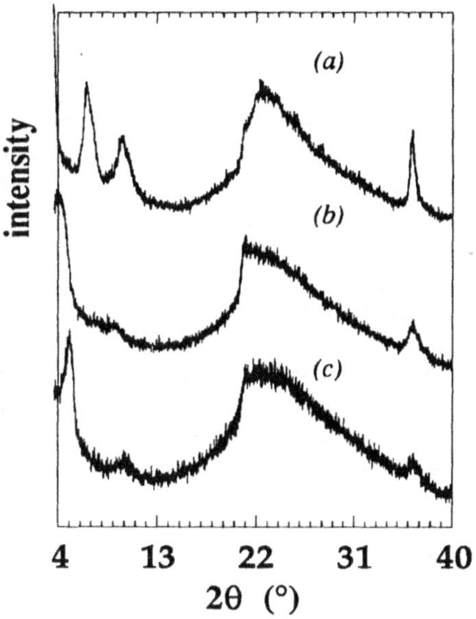

Figure 8. Effect of temperature on the interlayer distance in bilayered α-$Sn(PO_4)_2H_{0.4}\{H_3N(CH_2)_3SiO_{1.5})_8\}_{0.2}$ (a) 25 (b) 380 (c) 460 °C

The BET adsorption isotherm is of type IV, indicating a mesoporous structure with some contribution of micropores, and the pore distribution derived from these data (fig. 9) demonstrates that the majority of pores have a radius in the 13Å range. The specific surface area is 230 m²/g.

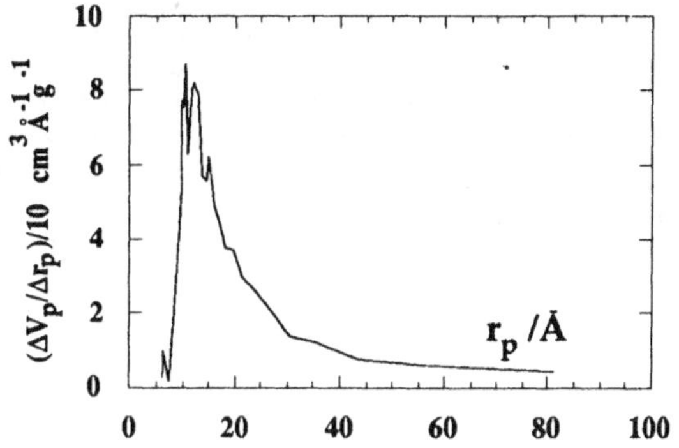

Figure 9. Pore size distribution in bilayered α-$Sn(PO_4)_2H_{0.4}\{H_3N(CH_2)_3SiO_{1.5})_8\}_{0.2}$

3.3.3 *Octa(aminopropyl)siloxane and γ-ZrP*. The intercalation of siloxane can be restricted in a more controlled fashion by the use of a substrate of more limited ion exchange properties. Using the same route to preparation of the siloxane octamer, and a similar room temperature intercalation reaction, the uptake curve shown in fig. 10(a) was obtained, which shows a linear dependence on the concentration of siloxane, and a practical total uptake of the siloxane in solution up to 5.5 mmol Si/g γ-ZrP. However, unlike intercalation into α-ZrP or SnP, a single phase system is observed before the maximal uptake, fig 10(b). At the point where a monophasic system is first observed, the interlayer distance is 17.4 Å, when the uptake of siloxane (3.2 mmole Si/g γ-ZrP) corresponds to exactly half the number of

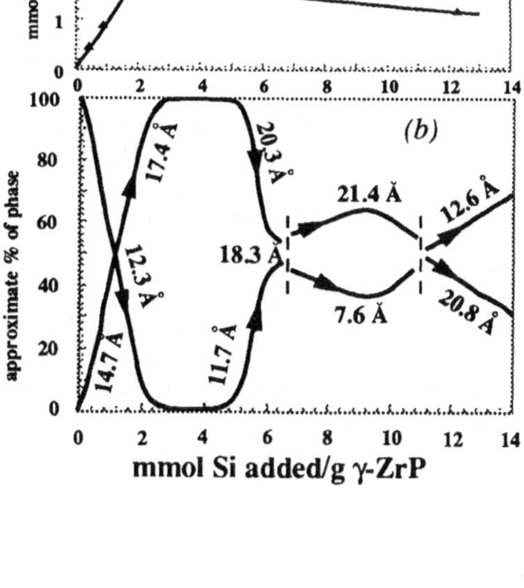

Figure 10. Intercalation of octa(aminopropyl)siloxane into γ-ZrP (a) uptake isotherm (b) phase diagram

protons in this matrix, or exchange of one proton per H_2PO_4 group. However, intercalation does proceed further, and the basal spacing gradually increases to 20.3 Å when the [Si]:[P] ratio is 0.8.

At higher concentrations still, complex behaviour is observed, with dissolution of phosphate, as shown by gravimetric analysis of the supernatant solution, and reprecipitation of zirconium phosphate of both α and γ forms, as shown by X-ray diffraction. γ-ZrP is a metastable phase in the Zr^{4+} - H_3PO_4 - H_2O system, and a transformation of the γ-type zirconium phosphate layers into those of the α-type has been observed in the presence of strong acids, through a process involving complexation of zirconium. A similar mechanism is assumed to operate here, with the formation of complexed Zr^{4+} in the presence of base as the pH rises on continued addition of octa(aminopropyl)siloxane.

The maximal basal spacing observed, 20.3 Å, is consistent with that expected from the sum of the thickness of the zirconium phosphate layers in the γ-form (9 Å) and the lateral dimension of the octamer, 11.2 Å. At lower uptake of siloxane, when each H_2PO_4 is half exchanged, the interlayer distance of 17.4 Å would seem to imply that the flexible organic arms can be deformed. In general, maximum use is made by an intercalated molecule of available interlamellar space; a slight deformation of the organic groups on siloxane octamers thus implies that, at this level of intercalation (half exchange), some additional surface area is available to them. At this point, the internal surface of γ-ZrP can be schematically represented as the chessboard-like arrangement shown in fig 11, where proton transfer from H_2PO_4 to alkylamino groups and the HPO_4---siloxane interactions are represented in terms of hydrogen bonding. The surface area associated with each active site in γ-ZrP can be estimated from crystallographic studies [12] as 6.6 x 5.3 = 36 Å². The use of one proton on four H_2PO_4 sites gives a total available surface area of 144 Å², or an average maximum possible dimension for the face of the octamer of 12 Å, being 0.8 Å longer than the lateral dimension of the siloxane cube, so rationalising the probable deformation of the aminopropyl groups at low uptake.

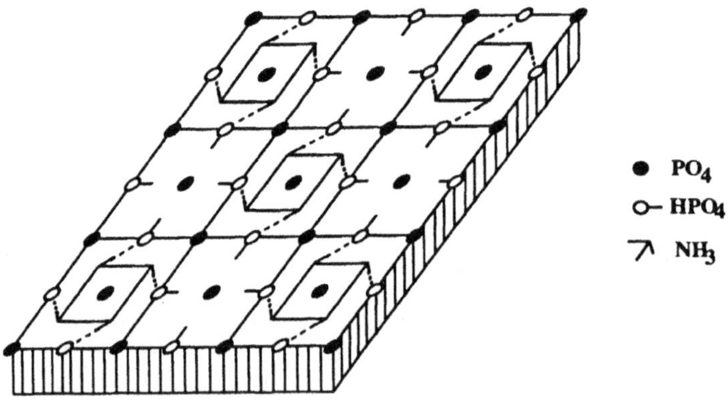

Figure 11. "Chessboard arrangement" of siloxane octamers among the internal surfaces of a γ-ZrP layer

3.3.4 *Octa[N-(2-aminoethyl-3-aminopropyl)]siloxane and α-ZrP*. In a third approach, an aminoalkyl-trimethoxysilane bearing a bulkier organic group was used as pillaring agent in α-ZrP. The diamino functions on octa[N-(2-aminoethyl-3-aminopropyl)]siloxane should promote interaction of each of the two cube faces parallel to the phosphate layers with eight active sites -POH and not four, and the uptake of siloxane should be limited accordingly. Fig. 12(a) expresses the uptake of Si/g α-ZrP and fig. 12(b) the corresponding phase diagram. After stirring at room temperature for 5 days, a monophasic system

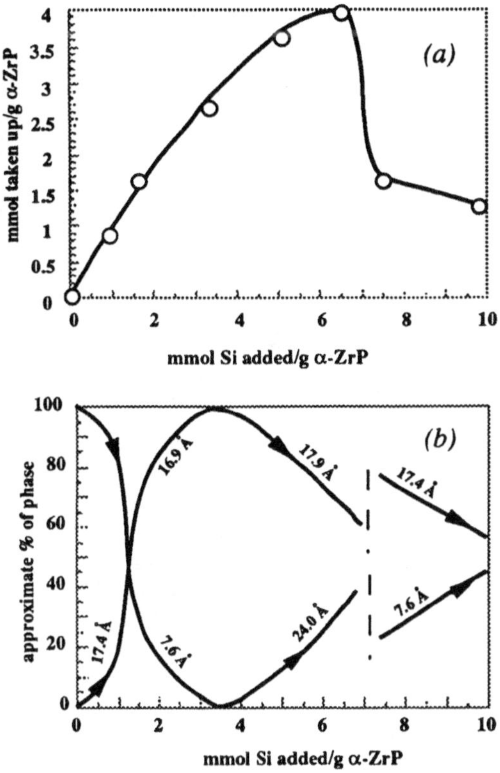

Figure 12. Intercalation of octa[N-(2-amino-ethyl-3-aminopropyl)]siloxane into α–ZrP
(a) uptake isotherm (b) phase diagram

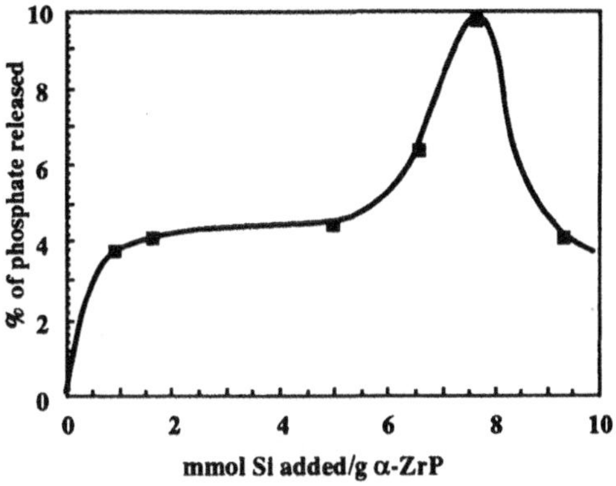

Figure 13. Intercalation of octa[N-(2-amino-ethyl-3-aminopropyl)]siloxane into α–ZrP : % PO_4 released as a function of the concentration of Si/g α-ZrP

of interlayer distance 16.9 Å is first obtained when 3.4 mmol Si/g α-ZrP is present in solution (a biphasic system consisting of the intercalation compound and α-ZrP existing up to then). This gives an uptake of silicon of 2.7 mmol/g α-ZrP, corresponding to a [Si]:[P] ratio of 0.4, exactly half the value observed for intercalation of the octamer derived from aminopropyltriethoxysilane with the same host lattice, but with use of the same number of acid functions, in view of the diamine nature of the organic groups. At higher organosiloxane concentrations, the system is biphasic , In fact intercalation proceeds further to values exceeding the use of the total ion exchange capacity of α-ZrP, with the 16.9 Å phase giving way to one of interlayer spacing 24 Å. This is assumed to adopt a bilayer arrangment of diaminosiloxane units, a situation which permits intercalation without additional proton transfer from the phosphate layers. However, the bilayer phase is never observed alone and, above intercalation of 4 mmol Si/g α-ZrP, delamination occurs, and reprecipitation of non-intercalated α-ZrP is detected by X-ray diffraction. Determination of liberated phosphate confirmed these results, fig. 13.

In the last two systems - octa(aminopropyl)siloxane/γ-ZrP and octa[N-(2-aminoethyl-3-aminopropyl)]siloxane/α-ZrP - thermal degradation at 600°C leads to materials which recognisably retain their starting structures. In the first case (fig. 14), the 020 and 200 reflections characteristic of the γ-type structure can be seen at 3.31 and 2.67°, and in the second (fig. 15), a characteristic line of the α-type structure. The interlayer distances, 13 and 12 Å respectively, indicate that the organic matter has been expelled, and that fragments of silica anchor the phosphate layers together.

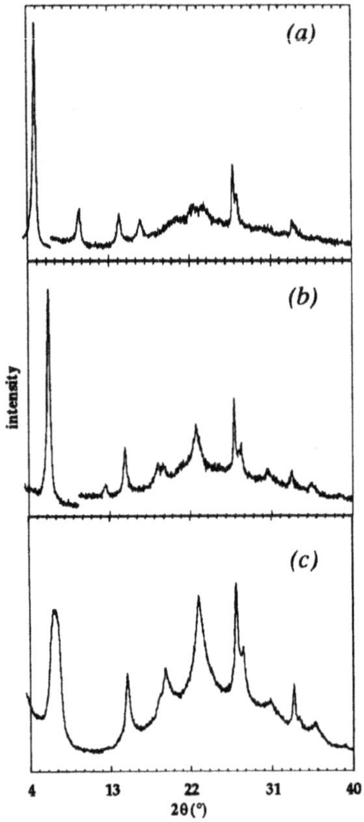

Figure 14
γ-$Zr(PO_4)_2H_{0.8}\{(H_3N(CH_2)_3SiO_{1.5})_8\}_{0.1}$ (a) as prepared, d_{002} = 17.4 Å and after thermal degradation at (b) 400 and (c) 600°C, d_{002} = 13.1 Å

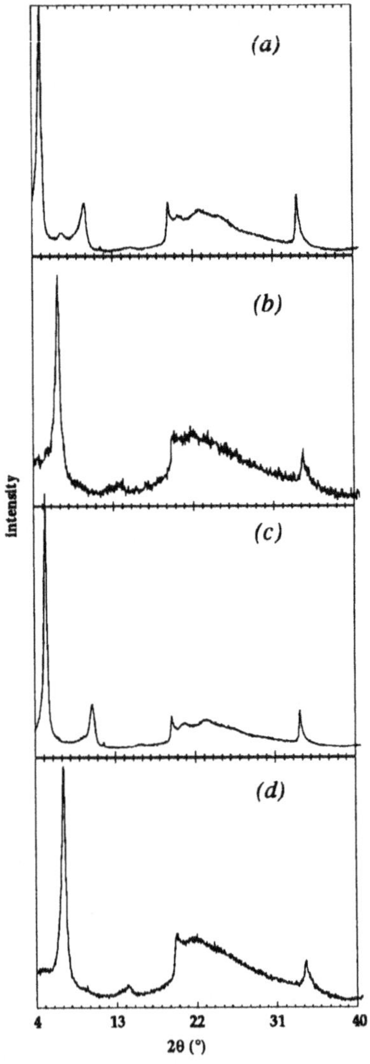

Figure 15 α-$Zr(PO_4)_2H_{0.4}\{(H_3N(CH_2)_2H_2N(CH_2)_3SiO_{1.5})_8\}_{0.1}$
(a) as prepared, d_{002} = 16.9 Å and (b) after calcination at 540°C, d_{002} = 12.4 Å
(c) α-$Zr(PO_4)_2\{(H_3N(CH_2)_2HN(CH_2)_3SiO_{1.5})_8\}_{0.25}$ as prepared, d_{002} = 17.9 Å and 24 Å and (d) after calcination at 600 °C, d_{002} = 13.0 Å

3.3.5 *Summary*. Insertion of alkylaminotrialkoxysilanes is a route to all-inorganic crosslinked phosphates. New silica pillared materials having a range of pore characteristics can be prepared depending on the nature of the phosphate and the identity of the functional groups of the siloxane precursor. In the case of SnP, partial hydrolysis of the layer allows the formation of bilayers and the introduction of porosity not possible in the analogous zirconium phosphate system, where the material is crosslinked and the interlayer region completely filled [11]. In the other two cases, either by using a bi-functional precursor or by exploiting the particular ion-exchanged host substrate, limited insertion of the siloxane is possible, allowing the synthesis of mesoporous phosphate-derived nanocomposite solids. The resulting compounds are depicted schematically in fig. 16.

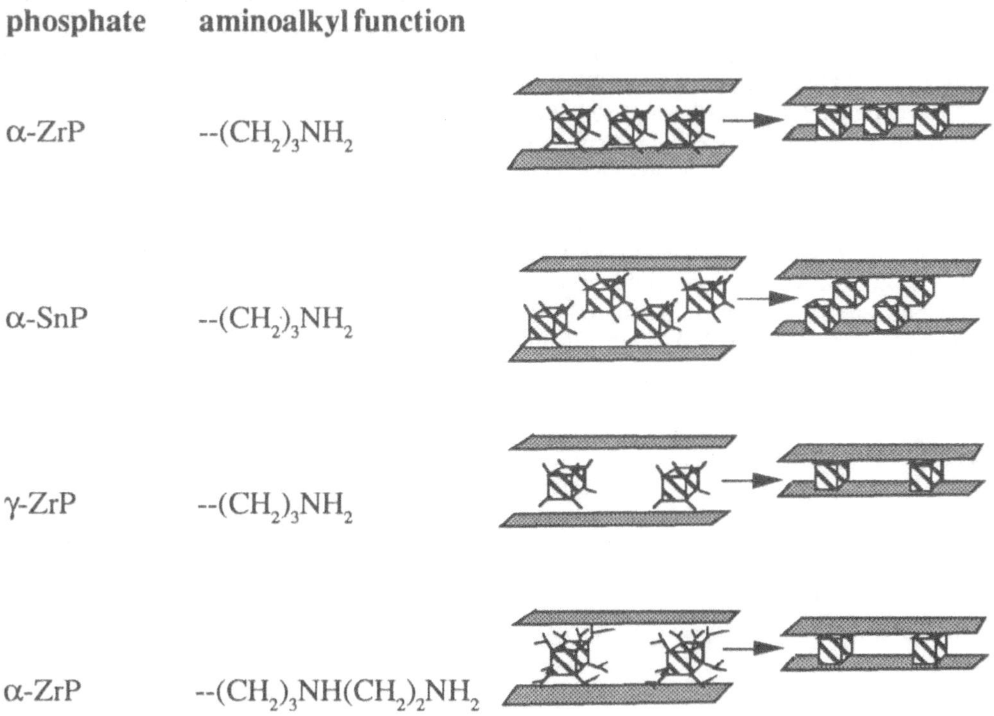

Figure 16. Schematic representation of the arrangement of intercalated octa(aminoalkyl)siloxanes and silica fragments before and after calcination respectively in metal(IV) phosphates

4. STRUCTURAL ANALYSIS

Despite much synthetic work on the insertion of inorganic oligomers into layered substrates (including clays etc.), almost all evidence for assuming that a covalent bond effectively exists between the pillaring species and the consistuent groups (AlO_4, PO_4, etc.) of the layers after calcination is indirect, coming mainly from powder X-ray diffraction and the deduction of interlayer free height and geometrical considerations. MAS NMR has an important role to play here, but only in one case to date - pillaring of beidellite with the aluminium Keggin ion - has the technique been used to demonstrate, by means of ^{27}Al MAS NMR, structural transformation interpreted as the growth of a tridimensional network grafted on to the clay sheets [13].

X-ray absorption spectroscopy is able to provide even more detailed structural information than this, including the identification of nearest neighbours to an absorbing atom, the number of atoms in this coordination sphere, and their distance from the absorbing atom. Furthermore, the development of synchrotron rings capable of furnishing soft X-rays means that determination of the environment of the lighter atoms is also accessible, including those used in pillar chemistry such as aluminium and silicon.

Preliminary studies of the near edge (XANES, x-ray absorption near edge structure) and EXAFS (extended X-ray absorption fine structure) regions of two series of silica pillared compounds, and their precursor intercalation compounds containing octa(aminopropyl)siloxanes, have been carried out at the silicon edge (1839 eV) using the SOXAFS beam line 3.4 at the Daresbury Laboratory, UK. In all cases, the powdered samples were suspended with graphite (approximately 25 % by volume) in acetone, and the suspension pipetted on to stainless steel slides to form a thin, homogeneous layer of dimensions 1 x 1 cm². The X-ray beam was monochromatised using an InSb(III) double crystal and its intensity detected by total electron yield.

Figure 17 shows the XANES region of octa(aminopropyl)siloxane intercalated SnP (siloxane double layer) and zirconium arsenate, $Zr(HAsO_4)_2.H_2O$ (ZrAs). This latter has an α-type structure analogous to that of ZrP, to which it was preferred because of potential interference in the EXAFS of the phosphorus absorption edge at some 300 eV above that of silicon. The XANES region of α-$Zr(AsO_4)_2H_{0.4}\{H_3N(CH_2)_3SiO_{1.5})_8\}_{0.2}$ after thermal treatment at 470 and 540 °C is also shown.

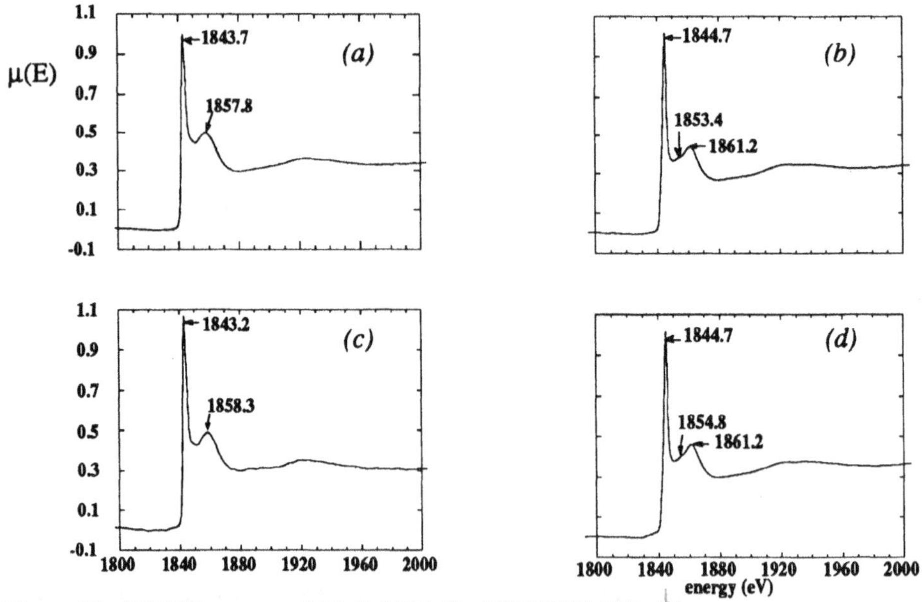

Figure 17. XANES spectra of (a) $Sn(PO_4)_2H_{0.4}\{H_3N(CH_2)_3SiO_{1.5})_8\}_{0.2}$
(b) $Zr(AsO_4)_2H_{0.4}\{H_3N(CH_2)_3SiO_{1.5})_8\}_{0.2}$ as (b) after calcination at 470 °C (d) as (b) after calcination at 540 °C

The positions of the maximum of the absorption edge, the first oscillation and of the resonance which is visible in some spectra are summarised in Table 1.

	edge	resonance	first oscillation (eV)
pillar precursors (intercalation compounds)			
$ZrAs\{[H_3N(CH_2)_3SiO_{1.5}]_8\}_{0.2}$	1843.2		1858.3
$SnP\{[H_3N(CH_2)_3SiO_{1.5}]_8\}_{0.2}$	1843.7		1857.8
after calcination			
ZrAs-"SiO_2"/470 °C	1844.7	1853.4	1861.2
ZrAs-"SiO_2"/540 °C	1844.7	1854.8	1861.2

Table 1. Positions of maxima in the XANES spectra of pillar-precursor $M(XO_4)H_{0.4}[H_3N(CH_2)_3SiO_{1.5})_8]_{0.2}$, M = Sn, Zr; X = P, As and for the corresponding pillared compounds when M = Zr, X =As.

The intercalation compounds derived from both ZrAs and SnP are characterised by an edge position at 1843.5 ± 0.3 eV and a rounded first oscillation of maximum at 1857.9 ± 0.3 eV. On the other hand, the absorption edge of the calcined samples ZrAs-"SiO_2" is at higher energy (1844.7 eV), as is the first oscillation (1861.2 eV) than in the precursors, and an additional feature superimposed on this first oscillation is probably indicative of multiple scattering and shows the presence of longer range ordering (second shell contribution). The edge position provides some useful information on the nature of the atoms surrounding silicon. Bulk silica, for example, lies at 1844.3 eV [14], sufficiently close to the values in the calcined samples examined here to conclude that, in the pillared materials, the silicon environment is probably a fragment of that in bulk silica. Furthermore, it may simply be calculated using the edge positions of bulk silica and bulk silicon carbide (1842.5 eV), that a system containing silicon in a "1 carbon-3 oxygen" environment is expected to have an absorption edge at 1843.8 eV, practically identical to that observed for both the intercalation compounds (derived from SnP and ZrAs), and in agreement with the environment expected in an oligosiloxane.

The difference in energy between the positions of the absorption edge and first oscillation (ΔE) can be used to calculate the second shell distance through the relationship $\Delta E{*}R^2$ = constant = 165.7 eVÅ² [14], providing a silicon to second nearest neighbour distance of 3.17 Å.

The Fourier transformed EXAFS spectra of the ZrAs intercalation compound and corresponding pillared phase after calcination at 540 °C are identical; that of the latter is given in fig. 18. In this radial distribution-like function, the peaks correspond to shells of atoms at a distance R from silicon (the phase correction for the silicon-oxygen pair has been applied). A contribution from only one

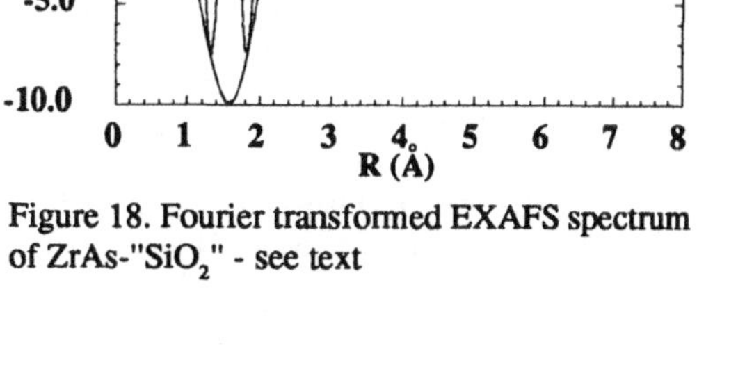

Figure 18. Fourier transformed EXAFS spectrum of ZrAs-"SiO_2" - see text

shell of atoms is observable, and this at the same distance, R(Si-O) = 1.58 Å in the precursor and the silica pillared compounds. Observation of a second shell would have allowed identification of the backscatterer producing the effect in the XANES and also confirmation of the Si-second neighbour distance. Further experiments are programmed. However, if it is assumed that the second neighbour is layer arsenic, and by making use also of the known As-O bond length, 1.76 Å, then the Si-O-As angle can be calculated as 143 °. The complementary nature of the information that can be derived from the XANES and EXAFS spectra is well illustrated in this example.

5. Conclusion

Ion exchange, intercalation or insertion reactions represent an approach to nanoscale chemistry in two-dimensional substrates. In the present case, the intercalation of organometallic precursors in layered substrates has been described as a route to pillared phosphates. The modulation of pore characteristics is potentially accessible by the appropriate choice of the organic fragment. Furthermore, the range of organic moieties can be exploited to enable partial retention of grafted organic matter, the reactivity of which can be used to lead to the preparation of true inorganic-organic composites. The porous nature of the solids is demonstrated, in addition, by their high ion exchange capacity, comparable to that of the metal(IV) phosphates, and their ability to adsorb basic molecules. The use of silica pillared phosphates as template solids for reactions in a constrained environment, including particle size quantization, is one aspect currently being investigated.

Acknowledgement
We acknowledge the use of the synchrotron source at Daresbury Laboratory made available through the EEC Large Facilities Programme.

References

[1] Vaughan, D. E. (1988) A. C. S. Symp. Ser. 368 pp. 308 - 323; I. V. Mitchell (ed.), Pillared Layered Structures, Current Trends and Applications, Elsevier Applied Science, London, 1990
[2] Cheng, S. and Wang, T-C. (1989), Inorg. Chem. 2, 1283; Hardin, S., Hay, D., Millikan, M., Sanders, J. V. and Turney, T. W. (1991) Chem. Mater. 3, 977
[3] Kwon, T. and Pinnavaia, T. J. (1989) Chem. Mater. 1, 381
[4] Alberti, G., Costantino, U., Marmottini, F., Vivani, R and Zappelli, P. (1990) in I. V. Mitchell (ed.), Pillared Layered Structures, Current Trends and Applications, Elsevier Applied Science, London, pp. 119 - 126
[5] Troup, J. M. and Clearfield, A. (1969) Inorg. Chem. 8, 431; Albertsson, J., Oskarsson, A., Tellgren, R and Thomas, J. O. (1977) J. Phys. Chem. 81, 1574
[6] Christiansen, A. N., Krogh Andersen, E., Krogh Andersen, I. G., Alberti, G., Nielsen M. and Lehmann, M. S. (1990) Acta Chem. Scand. 44, 865
[7] Costantino, U. (1981) J. Inorg. Nucl. Chem. 43, 1895
[8] Voronkov, M. G. and Lavrent'yev (1982) Top. Curr. Chem. 102, 199
[9] Lewis, R. M., van Santen, R. A. and Ott, K. C. (1985) Eur. Pat., 0159756 B1
[10] Li, L., Liu, X., Ge, Y, Li, L. and Klinowski, J. (1991) J. Phys. Chem. 95, 5910
[11] Rozière, J., Jones, D. J. and Cassagneau, T. (1991) J. Mater. Chem. 1, 1081
[12] Alberti, G., Bernasconi, M. G. and Casciola, M. (1989) React. Polym. 11, 245
[13] Plee, D., Borg, F., Gatineau, L. and Fripiat, J. J. (1985) J. Am. Chem. Soc. 107, 2362
[14] Laffon, C., Flanck, A. M. and Lagarde, P. (1991) in S. S. Hasnain (ed.), X-ray Absorption Fine Structure, Ellis Horwood, pp. 603 - 605

SECTION 4

SOL-GEL METHODS

SOL-GEL SYNTHESIS OF MULTIFUNCTIONAL MESOPOROUS MATERIALS - I. METAL-ORGANIC PRECURSORS

J. LIVAGE
Chimie de la Matière Condensée
Université Pierre et Marie Curie
4 place Jussieu, 75252 Paris
France

ABSTRACT. Inorganic oxide materials can be synthesized from molecular precursors such as metal alkoxides. Hydrolysis and condensation then lead to the formation of an oxide network. Alkoxides react with a large variety of nucleophilic species allowing a molecular design of ceramic materials at a nanoscale. Moreover metal organic precursors can lead to the formation of hybrid organic-inorganic materials. These new nanocomposites exhibit the properties of both the organic and the inorganic species. They can be used as multifunctional smart materials.

1. Introduction

The synthesis of mesoporous materials requires a precise control of the size and shape of solid particles. This can be conveniently achieved via the sol-gel route. The sol-gel synthesis of metal oxides is based on the hydrolysis and condensation of molecular precursors. This chemical route offers unique advantages for the preparation of monodispersed powders, multicomponent ceramics, coatings, membranes or fibres. The oxide network is progressively built via inorganic polymerization reactions. The design of molecular precursors and the control of processing parameters such as hydrolysis ratio, solvents or catalysts allow the production of tailored microstructures. A gel is obtained at room temperature in which a liquid fills the pores of the oxide network. Drying leads to a porous solid and densification can be performed upon heating at higher temperature [1].

Inorganic gels exhibit high surface area and small pore sizes. These unique properties may be exploited in applications such as filtration, separation, catalysis and chromatography. Many porous materials have been made via the sol-gel route. The so-called aerogels for instance are solid materials with porosities up to 99.8%. Such a high porosity is achieved via supercritical drying of wet gels in an autoclave. They can be used

C.A.C. Sequeira and M.J. Hudson (eds.), Multifunctional Mesoporous Inorganic Solids, 305–319.

as catalytic substrate [2] or thermal insulant [3]. Colloidal particles are needed to generate mesopores less than 100 nm in diameter. The first inorganic ultrafiltration membranes based on Al_2O_3, TiO_2 or ZrO_2 oxides have been synthesized via the sol-gel route [3]. They are produced for the microfiltration (macropores), ultrafiltration (mesopores) and nanofiltration (nanopores) of gases and liquids [4]. Porous coatings are also useful as catalytic surfaces. Titania based sol-gel coatings have been employed as catalysts for the hydrogenation of olefines [5]. Antireflective coatings are deposited from colloidal solutions at the surface of glasses, plastics or silicon. These coatings have interconnected pores whose dimensions are controlled by the composition, size and morphology of colloids in solution prior to deposition. The index of refraction is then varied smoothly from the free surface to the substrate [6]. Sol-gel chemistry is a low temperature process. Organic species can be embedded into porous oxide matrices or even bonded to the oxide backbone leading to the formation of hybrid organic-inorganic materials [7]. New functionalities can then be introduced giving rise to multifunctional materials. These smart materials can be used for sol-gel optics [8], as chemical sensors [9] or even for the encapsulation of enzymes [10] or proteins [11].

Two sol-gel routes are currently used for the production of mesoporous solids. The metal-organic route is based on the hydrolysis-condensation of alkoxide precursors in organic solvents. The inorganic route involves the formation of condensed species from aqueous solutions of metal salts via pH modifications [1]. This paper adresses the basic concepts of sol-gel chemistry, hydrolysis, condensation and complexation of molecular precursors. It shows how the chemical control of these reactions can be used for the synthesis of multifunctional mesoporous materials.

2. Hydrolysis and condensation of metal alkoxides

Sol-gel chemistry is based on the hydrolysis and condensation of metal alkoxides [12]. These reactions can be described as follows :

$$\text{hydrolysis : } -M\text{-}OR + H_2O \longrightarrow -M\text{-}OH + ROH \quad (1)$$

$$\text{condensation : } -M\text{-}OH + RO\text{-}M- \longrightarrow -M\text{-}O\text{-}M- + ROH \quad (2)$$

Condensed species are then progressively formed giving rise to oligomers, oxopolymers, colloids and gels or precipitates. At the end of the process, an hydrous oxide or hydroxide is obtained.

Actually these reactions correspond to the nucleophilic substitution of alkoxy ligands by hydroxylated species XOH as follows :

$$M(OR)_z + x\ XOH \longrightarrow [M(OR)_{z-x}(OX)_x] + x\ ROH \qquad (3)$$

where X stands for H (hydrolysis) or M (condensation)
They can be described by a S_N2 mechanism [12] :

```
H                        Hδ+                         Hδ+
 \                        \                          /
  Oδ- + Mδ+-OR —1—>   O-M-Oδ-R  —2—>  XO-M-O   —3—> XO-M+ ROH
 /                        /                          \
X                        X                            R
```

1) nucleophilic addition of a negatively charged $HO^{\delta-}$ group onto the positively charged metal atom $M^{\delta+}$. This first step leads to an increase of the coordination number of the metal atom in the transition state.
2) transfer, within the transition state, of the positively charged proton toward a negatively charged OR group.
3) departure of the positively charged protonated alkoxide ligand ROH.

As a result of these reactions, OR groups are replaced by OX ligands. The chemical reactivity of metal alkoxides toward hydrolysis and condensation mainly depends on the positive charge of the metal atom δ_M and its ability to increase its coordination number "N". As a general rule the electronegativity of metal atoms decreases, their size increases and the chemical reactivity of the corresponding alkoxides increases when going down the periodic table (Table 1). Silicon alkoxides are not very reactive. Gelation occurs within several days after water has been added. Hydrolysis and condensation rates have to be increased via acid or base catalysis [1]. The hydrolysis rate of $Ti(OEt)_4$ ($k_h = 10^{-3} M^{-1} s^{-1}$) is about five orders of magnitude greater than that of $Si(OEt)_4$ ($k_h = 5.10^{-9} M^{-1} s^{-1}$) [12]. Cerium alkoxides are very sensitive to moisture. They must be handle with care under a dry atmosphere otherwise precipitation occurs as soon as water is present. Alkoxides of highly electronegative elements such as $PO(OEt)_3$ ($\chi(P)$=2.11, $\delta(P)$=+0.13) cannot be hydrolyzed under ambient conditions [13].

TABLE 1 : Electronegativity "χ", partial charge "δ", ionic radius "r" and maximum coordination number "N" of some tetravalent metals (Z=4).

alkoxide	χ	δ	r(Å)	N
$Si(OPr^i)_4$	1.74	+0.32	0.40	4
$Ti(OPr^i)_4$	1.32	+0.60	0.64	6
$Zr(OPr^i)_4$	1.29	+0.64	0.87	7
$Ce(OPr^i)_4$	1.17	+0.75	1.02	8

Acid and base catalysis of silicon alkoxides $Si(OR)_4$ not only increase gelation rates, they also lead to completely different silica materials [14].

. Inorganic acids reversibly protonate negatively charged alkoxide ligands and increase the reaction kinetics by producing better leaving groups. The partial charge of OR groups increases as the electron-providing power of alkoxy, hydroxo and oxo ligands increases. The ease of protonation of OR groups therefore decreases as the connectivity of the adjacent Si atom increases. Acid-catalyzed condensation is directed preferentially toward the ends of oligomeric species, resulting in chain polymers (Fig.1a).

. Base catalysis provide better nucleophilic OH^- groups for hydrolysis while deprotonated silanol groups $Si\text{-}O^-$ enhance condensation rates. The positive partial charge δ_{Si} increases with its connectivity so that nucleophilic addition of $Si\text{-}O^-$ is directed preferentially toward the middles of oligomers leading to more compact, highly branched species. Monodispersed silica spheres can be produced via the hydrolysis of silicon alkoxides in the presence of ammonia (Fig.1b).

. Hydrolysis and condensation rates can be also increased via nucleophilic activation using highly nucleophile species such as fluoride. It has been shown that the addition of NaF to TMOS-sols leads to an increase in average pore size and the pore size distribution becomes narrower. The micropore area decreases from $300 m^2/g$ down to $100 m^2/g$

whereas the mesopore area increases from $120m^2/g$ to $720m^2/g$ when about 10^{-3} NaF is added to the hydrolyzing solution [15].

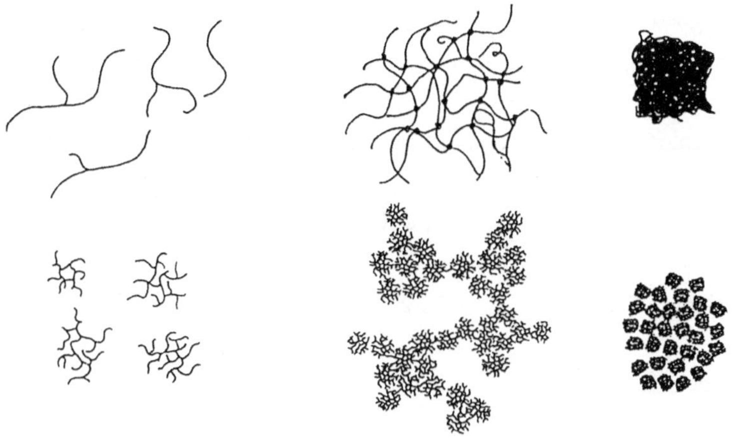

Figure 1. Growth and dessication of silica gels for (a) acid-catalyzed gels, (b) base-catalyzed gels, according to ref.[14].

Drying Control Chemical Additives (DCCA) such as formamide or oxalic acid have been used by L. Hench *et al.* to tailor the pore structure of sol-gel derived silica [16]. Without a DCCA a wide range of pore sizes and diameter of solids network are produced when gelation occurs whereas addition of a DCCA produces a sol-gel network with uniform pores. Transpiration cooled porous silica rocket windows have been prepared in which pores few nanometers in diameter allow the diffusion of He gas at velocities needed for transpiration cooling [17].

3. Molecular structure of metal alkoxides

Table 1 shows that the coordination number N of most metal ions in oxides is usually larger than their oxidation state Z. As a consequence, the chemistry of metal alkoxides is dominated by their tendency to increase their coordination via nucleophilic addition. This currently arises via oligomerization or solvation. Therefore the molecular structure of metal alkoxides is not obvious. Several monomeric or oligomeric species are often

simultaneously in equilibrium. Their molecular complexity depends on physical parameters (concentration, temperature) as well as chemical factors (solvent, oxidation state of the metal atom, steric hindrance of alkoxide groups) [18].
Complexity increases with the size of the metal atom and the "N-Z" value (Table 2). Titanium (r_{cov}=1.32Å, N-Z=2) for instance gives monomeric $Ti(OPr^i)_4$ species while dimeric $[Zr(OPr^i)_4]_2$ oligomers are observed for zirconium (r_{cov}=1.45Å, N-Z=3) [18].

Table 2 : Molecular complexity of tetravalent metal alkoxides $M(OEt)_4$

alkoxide	$Ti(OEt)_4$	$Zr(OEt)_4$	$Th(OEt)_4$
covalent radius (Å)	1.32	1.45	1.55
N-Z	2	3	4
molecular complexity	2.9	3.6	6.0

The steric hindrance of alkoxide ligands appears to be a major parameter. Bulky secondary or tertiary alkoxide groups tend to prevent oligomerization whereas primary alkoxides are usually oligomeric. Therefore most metal methoxides are comparatively non-volatile and insoluble in organic solvents. Oligomeric species $[Ti(OEt)_4]_n$ (n=2,3) have been evidenced for titanium ethoxide (Fig.2b) while titanium iso-propoxide only gives monomeric $Ti(OPr^i)_4$ (Fig.2a). Therefore titanium alkoxides with bulky OR groups, in which Ti has a lower coordination number, are more reactive toward hydrolysis. The reverse was observed with $Si(OR)_4$ which remains fourfold coordinated so that steric hindrance factors prevail. Spherical monodispersed TiO_2 powders are obtained via the controlled hydrolysis of diluted solutions of $Ti(OEt)_4$ in EtOH while $Ti(OPr^i)_4$ leads to polydispersed precipitates [19].

The choice of the solvent is also very important. Coordination expansion via solvate formation occurs when alkoxides are dissolved in their parent alcohol. The stability of such solvates increases with the size and the electropositive character of the metal, *i.e.* when going down the periodic table. At room temperature monomeric $Ti(OPr^i)_4$ is not solvated while solvated dimeric species $[Zr(OPr^i)_4(Pr^iOH)]_2$ have been observed in solution and even isolated as single crystals (Fig.2c).

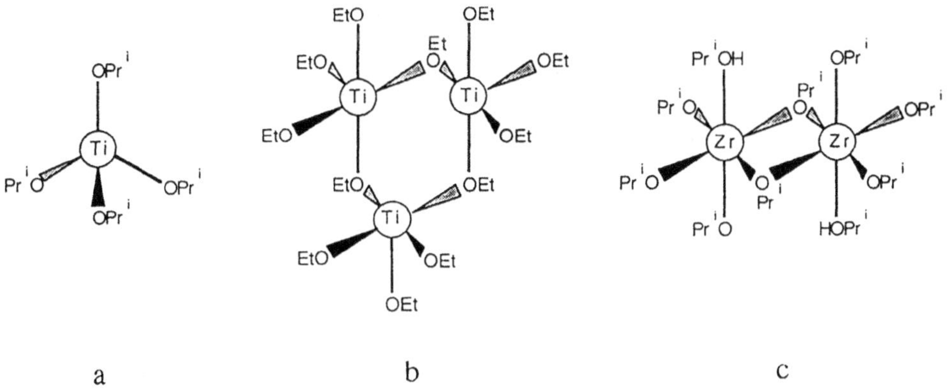

Figure 2. Molecular structure of metal alkoxides. a = $Ti(OPr^i)_4$, b = $[Ti(OEt)_4]_3$, c = $[Zr(OPr^i)_4,Pr^iOH]_2$.

Coordination expansion occurs via alkoxide bridging or solvation. The molecular complexity of metal alkoxides can then be controlled by an appropriate choice of the solvent [20]. $[Zr(OPr^n)_4]_n$ oligomers are formed ($n \leq 4$) in non polar solvents such as cyclohexane allowing slow hydrolysis rates and the formation of clear gels. Less condensed solvates are formed in n-propanol, hydrolysis becomes much faster and leads to precipitation .

Alcohol interchange reactions can also occur when metal alkoxides are dissolved in non parent alcohols [18]. These reactions are rather slow with silicon alkoxides while they can be very fast with transition metal alkoxides. They lead to the formation of a whole range of mixed alkoxides. Five different species for instance are readily formed when $Ti(OPr^i)_4$ is dissolved in Am^tOH :

$$Ti(OPr^i)_4 + x\ Am^tOH \longleftrightarrow Ti(OPr^i)_{4-x}(OAm^t)_x + x\ Pr^iOH$$

With x ranging from 0 to 4. Again it is possible to control hydrolysis and condensation by varying the value of x. Precipitation occurs readily with x=0, stable colloids are obtained for x=4 while polymeric gels are formed around x=2 [21].

The choice of the molecular precursor and the solvent allows to tailor the pore structure of the gel. This was clearly evidenced by J.D. Mackenzie *et al.* for silica gels as shown in Table 3 [22].

Table 3. Physical properties of silica prepared from different precursors and solvents after 600°C, 24 h heat treatment (from ref.22).

precursor	solvent	porosity (%)	surface area (m^2/g)	pore diameter (Å)
$Si(OMe)_4$	EtOH	27	169	36
$Si(OEt)_4$	EtOH	28	180	36
$Si(OBu)_4$	EtOH	35	773	13
$Si(OMe)_4$	MeOH	34	305	29
$Si(OEt)_4$	MeOH	25	160	34
$Si(OBu)_4$	MeOH	38	673	16

4. Molecular engineering of alkoxide precursors

Chemical additives such as carboxilic acids or β-diketones are currently employed to control the sol-gel synthesis of oxide materials. Most metal alkoxides are very prone to nucleophilic reations and react readily with such additives [18]. These reactions are also nucleophilic substitutions and can be described as shown in eq.3 where X stands for a complexing ligand L. A new molecular precursor $[M(OR)_{z-x}L_x]$ is formed which exhibit different molecular structure, chemical reactivity and functionality.

The chemical modification of titanium or zirconium alkoxides by acetic acid AcOH or acetylacetone acacH has been extensively studied.

Carboxylates $R\text{-}COO^-$ are known to behave as bidentate ligands. Acetic acid for instance reacts readily with most metal alkoxides giving oligomeric species in which acetate bridges have been evidenced. A large variety of oxopolymers is formed depending on the amount of acetic acid. Small oligomers $[Ti(OPr^i)_3(OAc)]_n$ (n=2 or 3) are first obtained when acetic acid reacts with titanium iso-propoxide in a one to one ratio (Fig.3a). Acetate ligands are less easily hydrolyzed than alkoxy groups therefore slowing down condensation reactions. The functionality of the new precursor decreases leading to the anisotropic growth of oxo-polymers rather than the precipitation of oxide particles. Esterification reactions between acetic acid and alcohol also occur when more acetic acid is added (x>1). Fibrous oxo-acetate polymers are obtained in the presence of a large excess of acetic acid.

Acetylacetone (acacH = CH_3-CO-CH_2-CO-CH_3) is often empoyed as a stabilizing agent for highly reactive metal alkoxides. Its enolic form contains OH groups which react readily with metal alkoxides. This strongly chelating ligand (Fig.3b) cannot be removed upon hydrolysis (unless pH <2) even in the presence of a large excess of water. Condensation is then prevented and only small oligomers are formed [23].

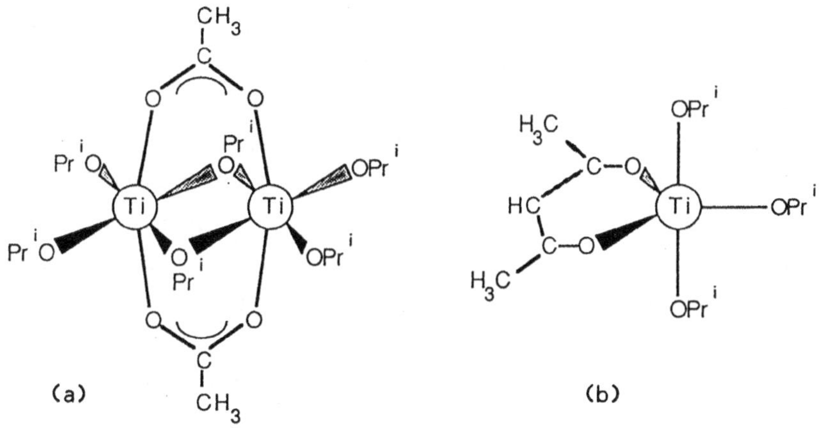

Figure 3. Chemical modification of $Ti(OPr^i)_4$ by (a) acetic acid and (b) acetylacetone.

Metal alkoxides react with water molecules (hydrolysis) and nucleophilic species (complexation). The first reaction is followed by condensation and lead to the formation of larger oligomeric or polymeric species whereas complexing ligands prevent condensation and favor the formation of smaller species. It becomes therefore possible to design sol-gel derived materials by controlling the two following chemical parameters : the nature and amount of complexing additive XOH (x=X/M) and the hydrolysis ratio (h=H_2O/M). A large variety of oligomeric species can then be obtained upon hydrolysis and condensation. As a general rule, more condensed species are obtained, as x decreases and h increases [23].

The hydrolysis and condensation of metal alkoxides $M(OR)_z$ lead to oligomers in which metal atoms tend to acquire their maximum coordination numbers. The formation of these species can be controlled by the hydrolysis ratio $h=H_2O/M$.

In the presence of a large excess of water ($h >> z$) all alkoxide groups are removed and highly cross-linked tridimensional polymers are formed. They lead to hydrous oxides $MO_{z/2},xH_2O$ similar to those synthesized from aqueous solutions. The adsorption and dissociation of water molecules at the oxide-solvent interface leads to the formation of charged particles.

All alkoxide groups are not removed when $h \leq z$ and chain oxo-polymers are formed. They can be advantageously employed for drawing fibers or making coatings. The dielectric constant of the organic solvent is rather low, the surface charge of polymeric particles is very small and van der Waals interactions prevail.

For very low hydrolysis ratio ($h<1$), condensation is mainly governed by the formation of μ-oxo and alkoxide bridges. Solute molecular oxo-alkoxides are then formed and can often be isolated as single crystals from the solution [24]. Their structure appears to be closely related to that of the corresponding polyanions. The first hydrolysis product ($h=0.6$) of titanium ethoxide for instance, $Ti_7O_4(OEt)_{20}$, contains $[Ti_7O_{24}]$ units isostructural with $Mo_7O_{24}^{6-}$ [25].

The chemical reactivity of metal alkoxides toward hydrolysis and condensation can be modified by complexation. Less electronegative alkoxide ligands are hydrolyzed preferentially whereas more strongly bonded complexing groups are more difficult to removed. They prevent condensation and gelation or precipitation rates decrease upon complexation. Complexing ligands behave as termination reagents. They cause the effective functionality toward condensation to be reduced, resulting in less condensed and more anisotropic polymeric species. Precipitation can then be avoided and gelation is promoted.

Esterification occurs when more than one mole of AcOH is added to one mole of titanium alkoxide. Acetic acid in excess reacts with the alcohol molecules released during complexation providing an *in-situ* generation of water giving rise to clear transparent gels. Only solute oligomeric species are obtained in the presence of small amounts of water. Single crystals of hexameric $Ti_6O_4(OPr^i)_{12}(OAc)_4$ have been obtained upon aging an equimolar mixture of AcOH and $Ti(OPr^i)_4$ in a closed vessel (Fig.3). Water, provided via esterification reactions arising from acetic acid in excess ($x=4/6=0.66$), leads to the slow hydrolysis of alkoxide groups which are replaced by oxo bridges [26].

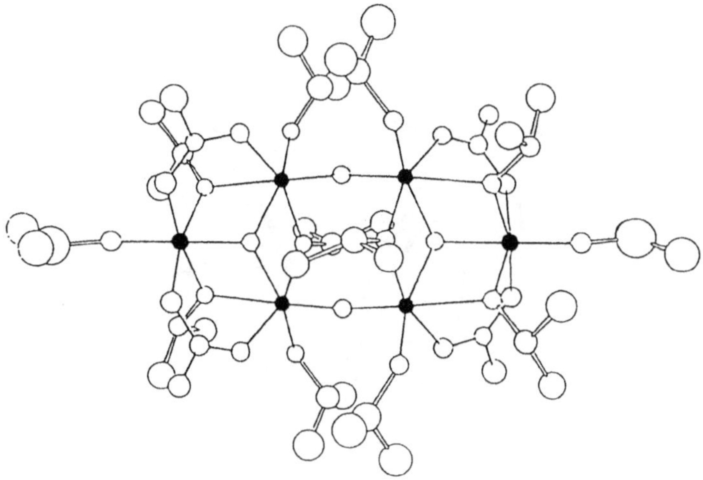

Figure 4. Molecular structure of $Ti_6O_4(OPr^i)_{12}(OAc)_4$, (●=Ti, O=other atoms)

The reaction of titanium alkoxides with acetylacetone leads to several molecu compounds. $TiO(acac)_2$ is formed in the presence of an excess of acetylacetone or up hydrolysis of $Ti(acac)_2(OR)_2$. Single crystals have been isolated and X-ray diffract experiments show dimers with sixfold coordinated Ti atoms linked through oxygen ato Strongly complexing acac ligands cannot be hydrolyzed easily and condensation does not any further. Larger molecular species can be obtained for smaller amount of acetylaceto Single crystals of $Ti_{18}O_{22}(OBu)_{26}(acac)_2$ molecular species have been obtained recently. Th are made of eighteen $[TiO_6]$ octahedra sharing edges or corners and correspond to x=(h=1.2. Complexing acac ligands remain outside of the $Ti_{18}O_{22}$ core of the molecule (Fig. These few examples show that condensation can be taylored via complexation and hydroly Gels are obtained in the presence of an excess of water and for small values of x ($x \leq 0.3$, $h \geq 1$ Clear sols are obtained when $Ti(OBu^n)_4$ is hydrolyzed in the presence of acetylacetone. T mean hydrodynamic diameter of colloidal particles was measured by quasi-elastic li scattering. It appears to increase from 2nm to 40nm as the hydrolysis ratio increases from h to h=4 (for x=0.3). It decreases from 40nm to 3nm when the amount of acetylacetone increa from x=0.3 to x=1 (for h=4) [27].

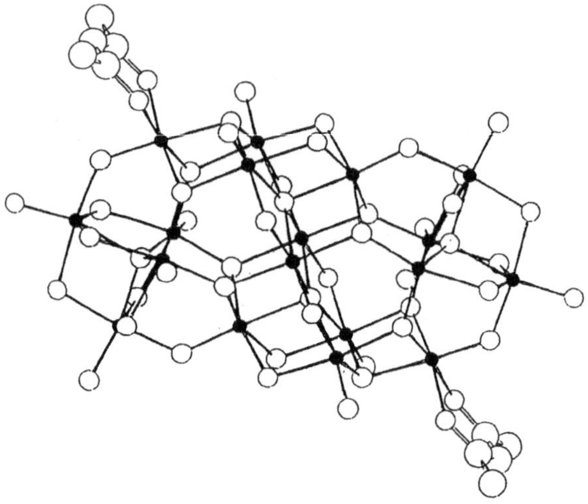

Figure 5. Molecular structure of $Ti_{18}O_{22}(OBu)_{26}(acac)_2$, (●=Ti, O=other atoms)

5. Sol-gel synthesis of hybrid "organic-inorganic" materials

The sol-gel synthesis of metal oxides is based on metal-organic molecular precursors. Hydrolyis and condensation reactions are performed at room temperature in organic solvents such as alcohols. One of the major advantage of the sol-gel process has been the ability to dope oxide glasses with organic molecules. The basic idea is simple, adding a solution of the desired dopant to the alkoxide precursor [28]. The encapsulation of active molecules into porous sol-gel glasses allows the fabrication of chemical sensors. The analyzate diffuses into the pore network and reacts with reagent molecules which, although immolized by trapping, are exposed to the pore volume [9]. Even biological species such as enzymes [29] or proteins [11] can be encapsulated into porous sol-gel glasses.

Sol-gel chemistry makes it possible to run simultaneously inorganic and organic reactions in order to form hybrid "organic-inorganic" materials. This kind of chemistry has already been widely developped in the case of silicon compounds. Many silicon precursors such as $R'_xSi(OR)_{4-x}$ are commercially available. They contain non hydrolyzable Si-C bonds so that the organic moieties are not removed during the hydrolysis-condensation process. These precursors lead to the so-called ORMOCERs

(ORganically MOdified CERamics) [7][30]. Depending on the nature and amount of organic and inorganic components it becomes possible to tailor microstructure and obtain multifunctional materials. Aryl-gels have been developped from aryl-bridged polysilsesquioxanes precursors [31]. Organic spacers (aryl groups) are introduced at regular intervals in the silicate framework leading to the formation of a very porous network that resembles lightweight, brittle glass. They have surface areas as high as $800m^2/g$ and micropores with a mean diameter below 2nm [31]. The introduction of organofunctional substituents like carboxy and amino groups multiplies the possibilities of such materials. Polymerizable organic ligands such as epoxides allow the formation of an organic network in addition to the inorganic one. These hybrid materials, in which organic groups behave as network modifiers or network formers, exhibit unique properties and can be employed as soft abrasives, coatings or optical lenses [7]. It is also possible to trap organic molecules which exhibit specific optical or chemical properties for making optical devices or sensors [32].

Organically modified gels can be used for the production of "smart membranes". Active sites responsible for specific interactions can be randomly distributed in the membrane at a nanoscale. It becomes therefore possible to go beyond the classical sieve concept for which separation is performed on size criteria only. Specific interactions could discriminate chemical species according to their chemical function, electric charge, hydrophobicity or hydrophilicity [33].

Organically modified ceramics are more difficult to synthesize when metals other than silicon are involved. The more polar M-C bond would be broken upon hydrolysis. Strongly complexing organic ligands such as β-diketonates must then be used in order to obtain a chemical bond between the metal atom and the organic species. Zirconium-oxo-PAAEM copolymers have been synthesized by chemical modification of zirconium alkoxide precursors with polymerizable chelating ligands AAEM [34]. Both polymerization reactions have been managed simultaneously leading to hybrid organic-inorganic polymers intimately interpenetrated at a nanoscale. The zirconium oxo core is made with oxo-alkoxo AAEM modified species where zirconium is likely in seven fold coordination as in monoclinic zirconia. The zirconium oxo species are chemically bonded to polymeric methacrylate chains via the β-diketo complexing function. The complexation ratio (AAEM/ Zr) seems to be the key parameter which controls the structure and texture of these hybrid materials. A careful adjustment of this parameter leads to zirconium oxo species with more or less open structures. Organic and inorganic polymerization processes are not independent and such a system exhibits probably the same behavior as the so-called interpenetrating polymer networks.

References

1. Brinker C.J., Scherer G.W. (1989), Sol-Gel Science, Academic Press, New York.
2. Teichner, Rev. Phys. Appl. 24 (1989) C4-1
3. Fricke J., Emmerling A., Structure and Bonding, 77 (1992) 38.
3. Larbot A., Julbe A., Guizard C., Cot L., J. of Membrane Sci., 44 (1989) 289
4. Klein L.C., Giszpenc N., Ceramic Bull. 69 (1990) 1821.
5. Carturan G., Facchin G., Navazio G., Gottardi V., Cocco G., in "Ultrastructure Processing of Ceramics, Glasses and Composites" (1984) Hench L.L. and Ulrich D.R. Ed., Wiley and Sons Pub. pp.197.
6. Uhlmann D.R., Boulton J.M., Teowe G., Weisenbach L.W., Zelinski B.J., in "sol-gel optics", J.D. Mackenzie and D.R. Ulrich Ed. S.P.I.E. Proc. Ser. (1990) 1328.
7. Schmidt H., in "Inorganic and Organometallic Polymers with Special Properties" Kluwer Pub., Netherland (1992) 297.
8. Nogues J.L., Moreshead W.V., J. Non-Cryst. Solids 121 (1990) 136.
9. Zusman R., Rottman C., Ottolenghi M., Avnir D., J. Non-Cryst. Solids, 122 (1990) 107.
10. Braun S., Rappoport S., Zusman R., Avnir D., Ottolenghi M., Materials. Lett. 10 (1990) 1.
11. Ellerby L.M., Nishida C.R., Nishida F., Yamanaka S.A., Dunn B., Valentine J.S., Zink J.I., Science, 255 (1992) 1113.
12. Livage J., Henry M., Sanchez C., Progress in Solid State Chem., 18 (1988) 259.
13. Livage J., Barboux P., Vandenborre M.T., Schmutz C., Taulelle F. J. Non-Cryst. Solids (in press).
14. Brinker C.J., Scherer G.W., J. Non-Cryst. Solids, 70 (1985) 301.
15. Chan J.B., Hua D.W, Winter R., Jonas J., J. Mater. Res. (in press).
16. Hench L.L., West J.K., Chem. Rev. 90 (1990) 33.
17. Fosmoe A.G., Hench L.L., Mat. Res. Soc. Symp. Proc. "Better Ceramics through Chemistry IV", 180 (1990) 843.

18. Bradley D.C., Mehrotra R.C., Gaur D.P. (1978) "Metal Alkoxides", Academic Press, London.
19. Barringer E.A., Bowen H.K., J. Am. Ceram. Soc., 65 (1982) C-199.
20. Sanchez C., Livage J., New J. Chem., 14 (1990) 513.
21. Nabavi M., Doeuff S., Sanchez C., Livage J., J. Non-Cryst. Solids, 121 (1990) 31.
22. Chen K.C., Tsuchiya T., Mackenzie J.D., J. Non-Cryst. Solids, 81 (1986) 227.
23. Livage J., Sanchez C., J. Non-Cryst. Solids (in press).
24. Sanchez C., Toledano P., Ribot F., Mater. Res. Soc. Symp. Proc. "Better Ceramics through Chemistry IV" 180 (1990) 47.
25. Watenpaugh K., Caughlan C.N., Chem. Comm., 2 (1967) 76.
26. Doeuff S., Dromzee Y., Taulelle F., Sanchez C., Inorg. Chem., 28 (1989) 4439.
27. Toledano P., In M., Sanchez C., C.R. Acad. Sci. Fr., 313 (1991) 1247.
28. Avnir D., Levy D., Reisfeld R., J. Phys. Chem. 88 (1984) 5956.
29. Carturan G., Campostrini R., Dire S., Scardi V., de Alteriis E., J. Mol. Cat., 57 (1989) L13.
30. Schmidt H., J. Non-Cryst. Solids, 73 (1985) 681.
31. Shea K.J., Loy D.A., Webster O.W., Chem. Mater. 1 (1989) 572.
32. Sakka S., Yoko T., Structure and Bonding, 77 (1992) 89.
33. Guizard C., Julbe A., Larbot A., Cot L., Key Engineering Materials, 61 (1991) 47.
34. Sanchez C., In M., J. Non-Cryst. Solids (in press).

SOL-GEL SYNTHESIS OF MULTIFUNCTIONAL MESOPOROUS MATERIALS - II. INORGANIC PRECURSORS

J. LIVAGE, M. HENRY and J.P. JOLIVET
Chimie de la Matière Condensée
Université Pierre et Marie Curie
4 place Jussieu, 75252 Paris
France

ABSTRACT. The precipitation of metal cations from aqueous solutions can be preformed via pH modifications. As for metal alkoxides hydrolysis and condensation reactions are involved in the formation of solid materials and the structure or chemical reactivity of inorganic precursors can be tailored via complexation. Complexing anions change the morphology, structure and even chemical composition of condensed phases. These reactions are described in the frame of the so-called Partial Charge Model.

1. Introduction

Inorganic precursors are much cheaper and easier to handle than metal alkoxides. Therefore the industrial production of oxide powders for ceramics and catalysts is mainly based on the precipitation or coprecipitation of inorganic salts from aqueous solutions. Gibbsite $Al(OH)_3$ is precipitated from aluminate solutions. TiO_2 powders are made via the controlled hydrolysis of titanium salts. Stabilized zirconia is coprecipitated from aqueous solutions of $ZrOCl_2$ and $Y(NO_3)_3$.

Most mesoporous inorganic solids described in this NATO ASI are prepared from aqueous solutions. Layered zirconium phosphates α-$Zr(HPO_4)_2.H_2O$ are formed by refluxing phosphoric acid and zirconium oxychloride solutions [1]. Double hydroxides such as $[Zn_2Al(OH)_6]NO_3.2H_2O$ are precipitated from zinc nitrate with freshly prepared aluminium hydroxide at a pH 6.2 [2]. Pillared clays are made via the intercalation of cationic $[Al_{13}O_4(OH)_{24}(OH_2)_{12}]^{7+}$ or anionic $[H_2W_{12}O_{40}]^{6-}$ polyoxometalates made from aqueous solutions [3].

However during the sol-gel synthesis water behaves both as a solvent and a chemical reagent. Hydrolysis and condensation reactions are therefore more difficult to

C.A.C. Sequeira and M.J. Hudson (eds.),Multifunctional Mesoporous Inorganic Solids, 321–338.

control. Several monomeric or oligomeric solute species are formed simultaneously when a metal salt is dissolved in the aqueous solution. These species are in rapid equilibrium and it is not easy to know which one is going to nucleate the solid phase. Literature provides many data describing the hydrolysis of metal cations in dilute solutions but very little is known about their condensation [4]. Moreover anions are known to play an important role even when they do not seem to be chemically involved in the hydrolysis and condensation reactions. As for the chemical modification of metal alkoxides by organic ligands, anions are able to form complexes with cationic precursors allowing a chemical control of the hydrolysis and condensation reactions [5].

This paper presents a theoretical description of inorganic polymerization reactions of metal cations in aqueous solutions. It is based on the so-called "Partial Charge Model" and does not apply only to the sol-gel synthesis of metal oxides, but to any wet chemical method from aqueous solutions such as precipitation, coprecipitation or thermohydrolysis [6].

2. The Partial Charge Model.

The Partial Charge Model (PCM) is based on the electronegativity equalization principle stated by R.T. Sanderson as follows : "when two or more atoms initially different in electronegativity combine, they adjust to the same intermediate electronegativity in the compound" [7]. Parr and coworkers showed that the electronegativity χ is related to the electronic chemical potential μ ($\chi = -\mu$), providing a theoretical basis to Sanderson's assumption [8]. The main consequence is that both the electronegativity χ_X of a given atom X and its partial charge δ_X vary when the atom is chemically combined. These two parameters must be related and a linear relationship is usually assumed as follows :

$$\chi_X = \chi_X^{\circ} + \eta_X \delta_X \quad (1)$$

where η_X is the hardness of atom X as introduced by Pearson. Hardness is related to the softness $\sigma_X = 1/\eta_X$ which provides a measure of the polarisability of the electronic cloud around X. Softness increases with the size of the electronic cloud, i.e. with the radius r of X. Therefore hardness varies as 1/r. According to the Allred-Rochow scale, electronegativity is proportional to Z_{eff}/r^2 [9]. Hardness may then be approximated as :

$$\eta = k\sqrt{\chi^\circ} \qquad (2)$$

where k is a constant that depends on the electronegativity scale. It can be shown that k = 1.36 when Pauling electronegativities are expressed in the frame of Allred-Rochow's model [6].

The total charge "z" of a given chemical species is equal to the sum of the partial charges of all individual atoms $z = \Sigma\delta_i$. This together with equations (1) and (2) leads to the following expressions for :

the mean electronegativity
$$\chi = \frac{\Sigma_i\sqrt{\chi_i^\circ} + 1.36\,z}{\Sigma_i 1/\sqrt{\chi_i^\circ}} \qquad (3)$$

and the partial charge
$$\delta_i = (\chi - \chi_i^\circ)/1.36\sqrt{\chi_i^\circ} \qquad (4)$$

Eq. (4) can also be writen as
$$\delta_i = \sigma_i(\chi - \chi_i^\circ) \qquad (5)$$

where
$$\sigma_i = (1.36\sqrt{\chi_i^\circ})^{-1} \qquad (6)$$

The Partial Charge Model provides an easy way to work out the mean electronegativity of chemical species and the charge distribution on each atom. In the case of the water molecule for instance equations (3) and (4) lead to $\chi(H_20) = 2.491$, $\delta(H) = +0.20$ and $\delta(O) = -0.40$ in good agreement with experimental values.

3. Hydrolysis of cations

The dissolution of inorganic salts in aqueous solutions leads to the solvation of positively charged metal cations by dipolar water molecules. $[M(OH_2)_n]^{z+}$ solute species are formed in which the oxygen atom of the water molecule is directed toward the metal cation. Water behaves as a Lewis base and the $M\text{-}OH_2$ "σ" bond involves some electron transfer from the bonding $3a_1$ molecular orbital of the water molecule toward empty atomic orbitals of the metal cation. This weakens the O-H bond within the water molecule which becomes more acidic. Spontaneous deprotonation then takes place as follows :

$$[M(OH_2)_n]^{z+} + h\ H_2O \longrightarrow [M(OH)_h(OH_2)_{n-h}]^{(z-h)+} + h\ H_3O^+ \qquad (7)$$

In dilute solutions this leads to a whole set of more or less deprotonated species ranging from aquo-ions $[M(OH_2)_n]^{z+}$ to hydroxy neutral species $[M(OH)_z]^0$ or oxy-anions $[MO_n]^{(2n-z)-}$ when all protons have been removed from the coordination sphere of the metal cation. As a general trend the coordination of metal cations converts from octahedral to tetrahedral as the pH increases *i.e.* when H_2O ligands are replaced by oxygen atoms. However very few structural information are available indicating where this transition occurs. The main point then is to know whether it is possible to predict the chemical nature of these species as a function of experimental conditions.

Following the electronegativity equalization principle it can be stated that deprotonation goes on until the mean electronegativity χ_h of the hydrolyzed species $[M(OH)_h(OH_2)_{n-h}]^{(z-h)+}$ becomes equal to the mean electronegativity of the surrounding aqueous solution χ_{aq}. According to thermodynamics the chemical potential of protons varies linearly with pH :

$$\mu(H^+) = \mu°(H^+) - \beta pH \qquad \text{where } \beta = 2.3RT \qquad (8)$$

Proton exchange reactions between H_3O^+ and H_2O species are very rapid in aqueous solutions so that $\chi(H^+)=\chi_{aq}$. As a consequence a linear relationship can also be established between the mean electronegativity and the pH of an aqueous solution.

$$\chi_{aq} = \chi_{aq}° - \lambda pH \qquad (9)$$

$\chi_{aq}°$ and λ depend on the reference state for proton and the electronegativity scale. Choosing $[H_5O_2]^+$ as a reference for proton at pH=0 and 2.49 as the mean electronegativity of water at pH=7 leads to :

$$\chi_{aq} = 2.732 - 0.035pH \qquad (10)$$

The mean electronegativity χ_h of hydrolyzed precursors can then be calculated as a function of the pH of the aqueous solution and the hydrolysis ratio h is deduced from the basic equations of the Partial Charge Model.

The charge of the hydrolyzed species $[M(OH)_h(OH_2)_{n-h}]^{(z-h)+}$ is :

$$z-h = \delta_{aq}(M) + n\delta_{aq}(O) + (2n-h)\delta_{aq}(H) \quad (11)$$

so that

$$h = [z-\delta_{aq}M(OH_2)_n]/[1-\delta_{aq}(H)] \quad (12)$$

Let us introduce the following formalism

$$\Delta(z,n,M) = z - \delta[M(OH_2)_n] \quad (13)$$

Relation (12) can then be written as :

$$h = \Delta_{aq}(z,n,M)/\Delta_{aq}(1,0,H) \quad (14)$$

Δ_{aq} means that the corresponding species, M^{z+} and H^+, are in equilibrium with the surrounding aqueous solution. Δ_{aq} therefore depends on pH.

Four different hydrolyzed Si(IV) precursors are found in aqueous solutions ranging from $[Si(OH)_3(OH_2)]^+$ (h=3) at pH=0 to $[SiO_2(OH)_2]^{2-}$ (h=6) at pH=14. In agreement with reported data $[Si(OH)_2(OH_2)_2]^{2+}$ (h=2) is too acid and $[SiO_3(OH)]^{3-}$ (h=7) too basic to be found in an aqueous medium [4]. These results already point out the complexity of the inorganic route compared to the metal organic one. A single $Si(OR)_4$ precursor has to be taken into account when silicon alkoxides are dissolved in their parent alcohol. Previous calculations can be applied to any metal cation. Highly charged cations such as Mn(VII) give oxo-ions. $HMnO_4$ is a very strong acid so that $[MnO_4]^-$ species cannot be protonated unless the pH is very low. The reverse occurs for low-valent cations such as Na^+. NaOH is a strong base and $[Na(OH_2)_n]^+$ aquo-ions cannot be deprotonated in usual conditions. Most cations exhibit several hydrolyzed monomeric species in the pH range 0-14.

The hydrolysis ratio h of any given precursor mainly depends on the pH of the solution and the oxidation state of the metal cation M^{z+}. A charge-pH diagram can then be drawn [6]. It shows three domains separated by two lines corresponding to h=1 and h=2n-1 respectively (Fig.1).

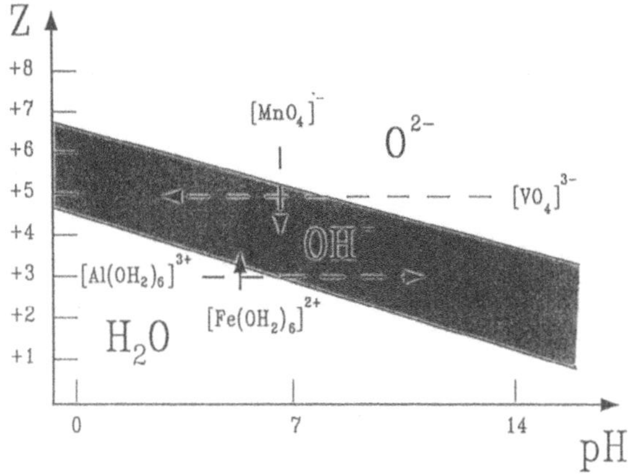

Figure 1. Charge-pH diagram showing the evolution of hydrolyzed species $[M(OH)_h(OH_2)_{n-h}]^{(z-h)+}$ as a function of pH and z^+.

4. Condensation of inorganic precursors

As for metal alkoxides, condensation becomes possible when at least one M-OH bond is formed i.e. in the intermediate domain of figure 1. Condensation is usually initiated via pH variations by adding a base to low-valent aquo-cations or an acid to high-valent oxo-anions. Redox reactions can also be used, MnO_2 gels for instance have been prepared via the reduction of permanganate salts $[MnO_4]^-$ by fumaric acid [10].

4.1. CONDENSATION OF LOW-VALENT CATIONS

Low-valent cations (z<4) give rise to aquo-ions at low pH. Condensation can be performed by increasing the pH. It follows an olation mechanism via the nucleophilic addition of negatively charged OH groups onto positively charged metal cations. This leads to the departure of coordinated water molecules and the formation of "ol" bridges as follows :

$$M\text{-}OH^{\delta-} + M^{\delta+}\text{-}OH_2 \longrightarrow M\text{-}OH\text{-}M^{\delta+}\text{-}OH_2{}^{\delta+} \longrightarrow M\text{-}OH\text{-}M + H_2O \qquad (15)$$

Olated polycations are formed during the first steps of condensation. Hydroxides or oxy-hydroxides are precipitated around the point of zero charge.

Trivalent cations (Al^{3+}, Cr^{3+}, Fe^{3+}) provide a good example of such a behavior [6]. They are rather similar but give different condensed species in aqueous solutions. Cr^{3+} gives stable polycations or gels. Al^{3+} gives the well known tridecamer "Al_{13}". Such species have never been evidenced for Fe^{3+} which gives amorphous colloidal particles in the same conditions. Hydrolyzed aluminium solutions under acidic conditions slowly convert to boehmite γ-AlOOH while goethite α-FeOOH is obtained with Fe^{3+}. Moreover hydroxides $Cr(OH)_3$ and $Al(OH)_3$ can be precipitated while only oxy-hydroxides have been obtained with iron. The instability of $Fe(OH)_3$ could be due to the larger electronegativity of Fe. A partial charge calculation performed on the following species

$$M(OH)_3 \longleftrightarrow MO(OH)(OH_2) \longleftrightarrow MOOH + H_2O \qquad (16)$$

gives a negative charge $\delta(H_2O)$=-0.01 for M=Al and positive one $\delta(H_2O)$= +0.07 for M= Fe. The positively charged water molecule cannot remain bonded to Fe^{3+} so that equilibrium (16) is displaced toward the formation of FeOOH rather than $Fe(OH)_3$.

Trivalent cations give hexa-coordinated aquo-ions $[M(OH_2)_6]^{3+}$ in acid solutions. Condensation is initiated by increasing the pH giving rise to positively charged polycations. Such species can be conveniently evidenced with Cr^{3+} which gives inert complexes because of crystal field stabilization effects arising from its electronic d^3 configuration. The first step begins with the h=1 $[M(OH)(OH_2)_5]^{2+}$ precursor. It gives rise to edge sharing dimers as follows :

$$2\,[M(OH)(OH_2)_5]^{2+} \longleftrightarrow [(H_2O)_4\text{-}M\begin{smallmatrix}OH\\ \diagup\ \diagdown\\ \diagdown\ \diagup\\ OH\end{smallmatrix}M\text{-}(OH_2)_4]^{4+} + 2H_2O \qquad (17)$$

Further hydrolysis leads to trimeric species formed by adding one hydrolyzed precursor (h=2) to the previous dimer. It could be described as follows:

$$[M_2(OH)_2(OH_2)_8]^{4+} + [M(OH)_2(OH_2)_4]^{+} \longrightarrow [M_3(OH)_4(OH_2)_9]^{5+} + H_2O \quad (18)$$

This trimeric polycation is often observed in aqueous solutions. It exhibits a compact cyclic structure in order to minimize electrostatic repulsions between M^{3+} ions in edge sharing adjacent $[MO_6]$ octahedra. Metal ions are bonded to a central μ_3-OH group with an sp^3 hybridized oxygen atom. Three of these sp^3 orbitals are engaged in M-O bonds while the fourth one gives the O-H bond. The acidity of this hydrogen atom therefore depends on electron transfers within the three M-O bonds.

Electron transfer toward the small and highly polarizing Al^{3+} ion with empty d orbitals should be rather strong making the μ_3-OH group highly acidic and therefore easily deprotonated.

$$[Al_3(OH)_4(OH_2)_9]^{5+} + H_2O \longrightarrow [Al_3O(OH)_3(OH_2)_9]^{4+} + H_3O^{+} \quad (19)$$

A nucleophilic oxygen atom is formed which can react with positive $[Al(OH_2)_6]^{3+}$ species. Such a reaction is limited only by difffusion rates. It should be rather fast. The nucleophilic addition of four deprotonated trimers onto a single aquo-ion leads to the formation of the well known Keggin like "Al_{13}" polycation in which one tetraedric $[AlO_4]$ is surrounded by twelve $[AlO_6]$ octahedra [11] .

Such trimeric species with a highly acidic μ_3-OH group could also account for the formation of boehmite γ-AlOOH. A neutral monomer $[Al(OH)_3(OH_2)_3]^0$ could be added to the trimer to form a critical $[Al_4(OH)_{12}(OH_2)_5]^0$ nuclei. This reaction can be described as the condensation of one monomeric aluminium species onto the compact trimer via two corner sharing $[AlO_6]$ octahedra. A third condensation reaction could occur either with the μ_3-OH group at the center of the trimer or with the μ_2-OH terminal groups of the adjacent $[AlO_6]$ sites. The μ_3-OH bridge is more acidic so that tilting occurs towards this nucleophilic oxygen atom giving rise to a skewed tetramer $[Al_4O(OH)_{10}(OH_2)_5]^0$. Further condensation via olation and oxolation then leads to the corrugated sheets typical of the boehmite phase (Fig.2).

A rather different behavior is observed with Fe^{3+} ions. The five 3d orbitals are singly occupied (electron configuration $3d^5$). Electron transfer toward Fe^{3+} are therefore not so important. The μ_3-OH bridge is no more acidic and large polycations such as "Al_{13}" cannot be formed. Amorphous gelatinous precipitates are currently obtained when a base is added to an aqueous solution of Fe^{3+}.

Crystalline oxy-hydroxides FeOOH are formed upon aging or gentle heating. However the goethite phase α-FeOOH isostructural with diaspore α-AlOOH is usually obtained rather than lepidocrocite γ-FeOOH which is isostructural with boehmite. This can be explained by the poorly acidic character of the μ_3-OH group. Following the same pathway as for Al^{3+}, condensation with the fourth $[FeO_6]$ octahedron would occur via olation with the μ_2-OH terminal groups leading to the formation of planar tetrameric species $[Fe_4(OH)_{12}(OH_2)_4]^0$. Polycondensation of these precursors via olation would give rise to double rutile-type chain polymers. Oxolation then leads to the formation of the three dimensional network of the goethite phase (Fig.2).

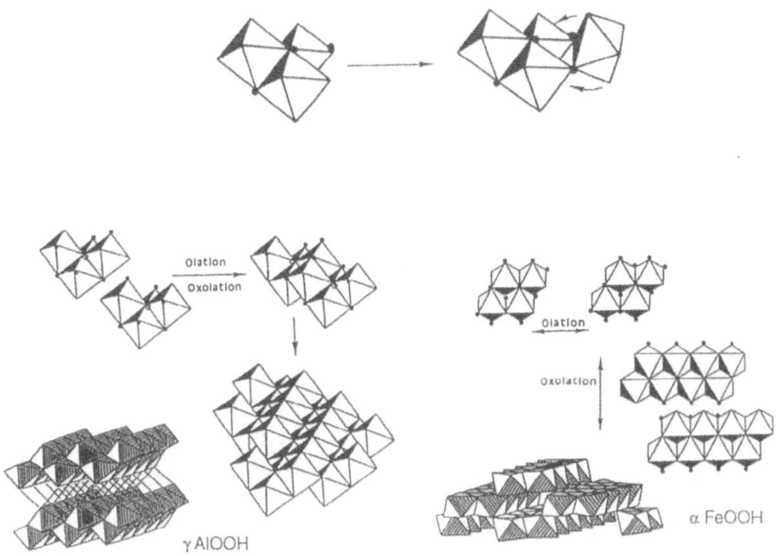

Figure 2. Possible condensation pathways leading to the precipitation of boehmite γ-AlOOH and goethite α-FeOOH from trimeric precursors.

4.2. CONDENSATION OF HIGH-VALENT CATIONS

High-valent cations (z>4) give oxy-anions $[MO_n]^{(2n-z)-}$ at high pH. Protonation leads to the formation of hydroxo species. Water molecules are not involved in the coordination sphere of the metal so that olation is not possible and condensation occurs via oxolation as follows :

$$\text{M-OH} + \text{HO-M} \longrightarrow \text{M-O}\overset{\text{H}}{\underset{\text{H}}{\rightleftarrows}}\text{O-M} \longrightarrow \text{M-O-M} + H_2O \qquad (20)$$

Such a reaction involves a proton transfer between two OH groups within the transition state (M-OH ...HO-M) in order to form the leaving water molecule. It should therefore be kinetically slower than olation. Negatively charged polyanions are formed during the first condensation steps. Hydrous oxides or hydroxy-oxides are precipitated around the point of zero charge.

V_2O_5,nH_2O gels exhibit a fibrous structure made of vanadium pentoxide double chains in which $[VO_6]$ octahedra share edges and corners. These gels can be conveniently synthesized via the acidification of metavanadate aqueous solutions by proton exchange resins [12]. Tetrahedral monomeric orthovanadates $[VO_4]^{3-}$ are the predominant species at high pH. Protonation occurs as the pH decreases leading to the formation of polyanionic species such as pyrovanadate $[V_2O_7]^{4-}$, metavanadate $[V_4O_{12}]^{4-}$ or decavanadate $[V_{10}O_{28}]^{6-}$[6]. The partial charge δ_V of vanadium atoms increases progressively and their coordination turns from 4 to 6 as the pH decreases. Metavanadates are made of $[VO_4]$ tetrahedra while decavanadates are based on $[VO_6]$ octahedra. Vanadium pentoxide gels are obtained around the point of zero charge (pH=2). Their formation from neutral (h=5) precursors can be described as follows :

i) Coordination expansion from 4 to 6 occurs via the nucleophilic addition of two water molecules. This gives hexacoordinated species $[VO(OH)_3(OH_2)_2]^0$ in which one water molecule lies along the z axis opposite to the short V=O double bond while the second one is in the equatorial plane opposite to an OH group.

ii) As a consequence one $V\text{-}OH_2$ and three V-OH bonds are formed in this plane so that x and y directions are not equivalent. Fast olation reactions along the $HO\text{-}V\text{-}OH_2$ direction give rise to olated corner sharing chain polymers (Fig.3).

iii) Slower oxolation reactions then lead to the formation of edge sharing double chains. This results into a layered structure made of water molecules intercalated between V_2O_5 ribbons [13].

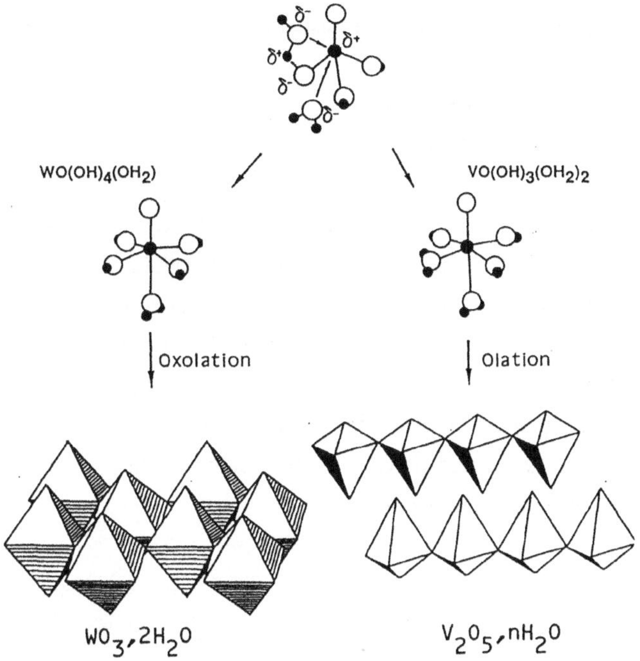

Figure 3. Formation of V_2O_5,nH_2O fibrous gels and WO_3,2H_2O layered precipitates from neutral molecular precursors.

A similar mechanism can be applied to the formation of WO_3,nH_2O gels upon acidification of tungstate solutions. Coordination expansion of neutral precursors (h=6) via the nucleophilic addition of two water molecules leads to $[WO_2(OH)_2(OH_2)_2]^0$ species (Fig.5). As previously one water molecule is bonded along the z axis opposite to a W=0 bond. The other one lies in the equatorial plane giving one W-OH_2 bond beside one W-O and two W-OH other bonds. According to literature W(VI) mono-oxo species are more stable than di-oxo ones [6]. A proton transfer would then occur in the equatorial plane from W-OH_2 to W-O leading to the formation of four equivalent W-OH bonds. Olation is no more possible and oxolation along equivalent x and y directions leads to the formation of layered WO_3,nH_2O gels.

4.3. THE CHARGE-ELECTRONEGATIVITY DIAGRAM

Condensation cannot occur unless stable M-OH bonds are formed. However it is well known that M-OH groups can undergo acid or basic dissociation in aqueous solutions :

$$M\text{-}OH + H_2O \longrightarrow M\text{-}O^- + H_3O^+ \tag{21}$$

$$M\text{-}OH + H_2O \longrightarrow M\text{-}OH_2^+ + OH^- \tag{22}$$

The acid or basic character of M-OH groups mainly depends on the electronegativity and the charge of the metal, i.e. on the electron transfer between M^{z+} and OH. Acid dissociation occurs with electronegative elements at the upper right of the periodic table (P(V), S(VI)) and highly charged transition elements (Mn(VII)). Basic dissociation occurs with large low-valent cations of low electronegativity (Na^+, Ba^{++}). A charge-electronegativity diagram (Fig.4) can then be drawn in the frame of the partial charge model evidencing a domain in which M-OH bonds remain stable toward acid and basic dissociation so that condensation can occur [6].

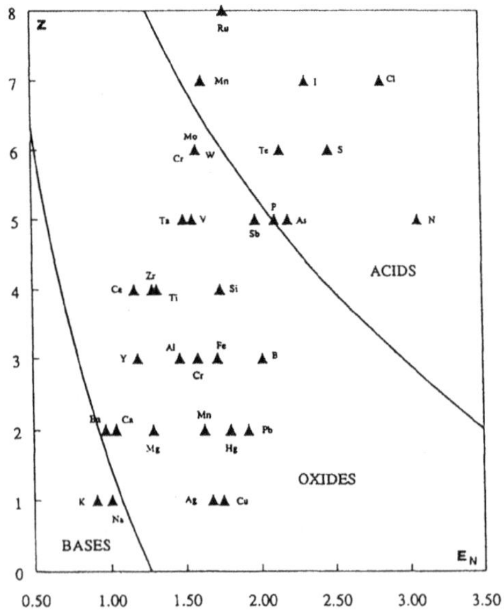

Figure 4. Charge-electronegativity diagram showing which cations are able to give condensed species in aqueous solutions.

Moreover both condensation processes, olation and oxolation, involve the nucleophilic addition of OH groups onto metal cations. Charge conditions have to be fullfilled for such reactions to be possible. OH groups must be negatively charged while the positive charge of metal cations should be large enough. Experimental data suggest that condensation does not occur unless $\delta_{OH}<0$ and $\delta_M>0.3$. A comparison of V(V) and P(V) aqueous chemistry illustrates this point. Both cations have a high charge and give monomeric tetrahedral species $[MO_4]^{3-}$ at high pH. Three protonated precursors $[H_nMO_4]^{(3-n)-}$ could lead to condensation. Partial charges δ_M and δ_{OH} show that vanadium precursors fullfill these conditions and can lead to condensed polyvanadate species. This is no more the case for phosphorus which is more electronegative than vanadium. Electron transfer from oxygen to phosphorus is more important. As a consequence the positive charge of P is not large enough in the first two precursors, $\delta_P=+0.03$ for $[HPO_4]^{2-}$ and $\delta_P=+0.2$ for $[H_2PO_4]^-$ while OH groups are no more negatively charged in H_3PO_4 which exhibits acid properties. Polyphosphates cannot be obtained in aqueous solutions at room temperature. They are all formed upon heating acid phosphates such as Na_2HPO_4 [14].

5. Complexation of metal cations by anions

The preceeding discussion shows that pH is the main parameter for the sol-gel synthesis of oxide powders from aqueous solutions. However foreign anions or cations are often added in order to improve the process. It is well known for instance that spherical hematite particles (α-Fe_2O_3) are obtained when aqueous solutions of $FeCl_3$ are heated under reflux while acicular particles are formed with $Fe(NO_3)_3$ in the presence of phosphate ions. Goethite (α-FeOOH) is precipitated from $Fe(NO_3)_3$ while the Akaganeite phase (β-FeOOH) is formed with $FeCl_3$. The thermo-hydrolysis of $Ti(SO_4)_2$ aqueous solutions leads to anatase TiO_2 while rutile is also formed in the presence of chloride or nitrate.

These examples show that anions can play an important role in the formation of condensed phase even if they are not chemically involved in these phases. Anions can lead to the complexation of aqueous cationic precursors and complexation could be considered as the inorganic analogue of the chemical modification of metal alkoxides by organic ligands [15].

Negatively charged anions X^{x-} can react with positively charged hydrolyzed species to give complexes as follows :

$$[M(OH)_h(OH_2)_{n-h}]^{(z-h)+} + X^{x-} <\!\!\longrightarrow [M(X)(OH)_h(OH_2)_{n-h-\alpha}]^{(z-h-n)+} + \alpha H_2O \quad (23)$$

α corresponds to the number of water molecules which have been replaced by X. α=1 for monodentate ligands such as Cl^- while α=2 for bidentate ligands such as $(SO_4)^{2-}$.

However "M-X" bonds have to remain stable in the presence of an excess of water which plays a double role :

i) Water is a solvent with a high dielectric constant (ε=80). It lead to the dissociation of ionic species : $(M^{\delta+}\text{-} X^{\delta-})_{aq} \longrightarrow M_{aq}^{z+} + X_{aq}^{x-}$

ii) Water is also a Lewis base (σ donor molecule) which can favor the hydrolysis of less nucleophilic ligands : $M\text{-}X + H_2O \longrightarrow M\text{-}OH + HX$

$[M(X)(OH)_h(OH_2)_{n-h-\alpha}]^{(z-h-n)+}$ complexes can only be formed when the following conditions are fullfilled :

$\delta(X) \leq x$ in order to avoid ionic dissociation in an aqueous medium

$\delta(HX) \geq 0$ in order to avoid hydrolysis by nucleophilic water molecules

Anionic ligands X^{x-} can be protonated when pH decreases.

$$X^{x-} + q\, H_3O+ <\!\!-\!\!> [H_qX]^{(x-q)} + q\, H_2O \quad (24)$$

This leads to (x+1) different species of general formula $[H_qX]^{(x-q)-}$ in which q ranges from 0 to x. The rough formula of the complexed precursor can be written as follows $[M(H_qX)(OH)_h(OH_2)_{n-h-a-q}]^{(z-h-n)+}$ in order to evidence $[H_qX]$ species. The total charge is given by :

$$z-h-n = \delta(H_qX) + \delta[M(OH_2)_{n-\alpha}] - (h+q)\delta(H) \quad (25)$$

The (x+1) values of q lead to (x+1) critical hydrolysis ratio defined by :

$\delta(H_qX) = q-x$ for ionic dissociation ($0 \leq q \leq x-1$)

$\delta(H_xX) = 0$ for hydrolysis (q=x)

Critical hydrolysis ratio h* can then be expressed as follows :

$$h_q^*+q = [z-\delta_q(M(OH_2)_{n-\alpha})]/[1-\delta_q(H)] = \Delta_q(z,n\text{-}a,M)/\Delta_q(1,0,H) \qquad (26)$$

In an aqueous solution the hydrolysis ratio depends on the pH as shown in relation (10). It is then possible to express h_q^* as critical pH_q^* values corresponding to each protonated form of anionic species. If we write h*=h , equations (10) and (14) lead to :

$$\Delta_{aq}(z,n,M)/\Delta_{aq}(1,0,H) + q = \Delta_q(z,n\text{-}\alpha,M)/\Delta_q(1,0,H) \qquad (27)$$

This equation provides a critical pH value for each protonated form $[H_qX]^{(x-q)-}$ of the anion.

According to the definition given in relation (13), we can write :

$$\Delta_i(z,n,M) = \Delta_j(z,n,M) + \sigma[M(OH_2)_n](\chi_j\text{-}\chi_i) \qquad (28)$$

where $\sigma[M(OH_2)_n]$ corresponds to the softness of the aquo precursor $M(OH_2)_n$.
All Δ_i values can then be expressed with respect to the same Δ_j. If we take the naked proton as a reference state, $\chi_p = \chi(H^+) = 4.071$, all χ_q values can be expressed as a function of χ_{aq} and therefore as a function of pH. Calculations lead to :

$$\chi_q = \chi_p - \frac{\Delta_p(z,n\text{-}\alpha,M)}{q\sigma(H) + \alpha\sigma(H_2O) + \Delta_p(z,n,M)/(\chi_p\text{-}\chi_{aq})} \qquad (29)$$

Since χ_{aq} varies linearly with pH, it is possible to express electronegativities as a function of pH, or critical pH_q^* values as a function of electronegativities. This leads to :

$$pH^*_q = \frac{28.6\ \Delta p(z,n,M)}{[\Delta p(z,n\text{-}\alpha,M)/\chi_p\text{-}\chi_q] - [q\sigma(H) + \alpha\sigma(H_2O)]} - 38.3 \qquad (30)$$

An "electronegativity-pH" diagram can then be drawn which could be very useful to describe complexation reactions in aqueous solutions (Fig.5).

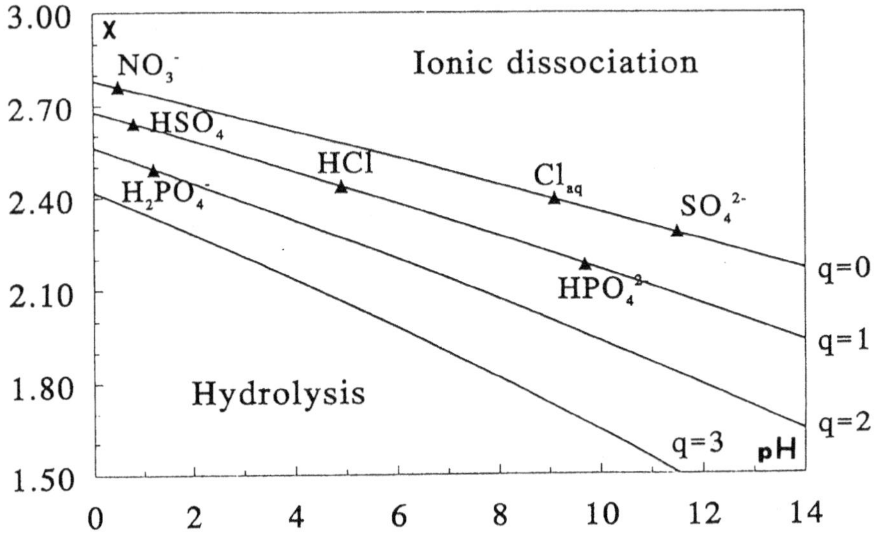

Figure 5. Electronegativity-pH diagram showing the pH range in which anions and their protonated form give complexes with Al^{3+} precursors. Outside this pH range, ionic dissociation or hydrolysis occur and complexes are no more stable.

According to eq.(30), the critical pH_q* value and therefore the complexing range, increases :

. for a given precursor, when the mean electronegativity of the anion decreases or when α increases.

. for a given anion, when the charge z+ decreases.

These calculations and the "electronegativity-pH" diagram provide a theoretical background to predict the complexing ability of anions toward cations. Three different cases can be observed.

i) Anionic species are not complexing. They just behave as counter anions and are not involved in the formation of condensed species. This is the case of Cl^- with zirconyl ions $[ZrO]^{2+}$. $ZrOCl_2,nH_2O$ crystals can be precipitated around pH 2. They are formed of cyclic tetrameric cationic species $[Zr_4(OH)_8(OH_2)_{16}]^{8+}$ in which zirconium is surrounded by four OH groups and four water molecules. Cl^- ions are not bonded zirconium [16]. At higher pH, amorphous hydrous zirconia ZrO_2,nH_2O is precipitated in which some Cl^- anions are only adsorbed at the surface of the gelatinous precipitate and can be removed by washing.

ii) Anions are strongly complexing. They remain bonded to the cation during the formation of condensed species giving rise to the precipitation of a salt rather than an oxide. This is typically the case of phosphates. Heating under reflux H_3PO_4 with an aqueous solution of $ZrOCl_2$ leads to the precipitation of the layered zirconium phosphate $Zr(HPO_4)_2,H_2O$ [17].

iii) The intermediate case corresponds to anions such as nitrates or sulfates. These anions are able to form first sphere or second sphere complexes but are only involved during the first steps of the hydrolysis-condensation process. They give rise to basic salts as long as the pH is low enough but are released at higher pH so that oxides or hydroxides are formed. The precipitation of an aqueous solution of zirconium nitrate around pH 2 leads to the formation of a basic salt $ZrO(NO_3)_2,5H_2O$ [18]. X-ray diffraction shows that these crystals are made of $[Zr(OH)_2(OH_2)_2(NO_3)]_n^+$ chains in which half of the nitrates are chelating zirconium atoms. The other nitrates behave as counter anions between chains [18]. Such anions are very important in the sol-gel synthesis of oxide materials. They prevent or at least slow down condensation reactions and can be used to control the size and morphology of solid particles. Their precise role is not easy to describe and would require an accurate characterization of all the intermediate species formed during the growth of oxide particles.

References.

1. Clearfield A., Stynes J.A., J. Inorg. Nucl. Chem. 26 (1964) 117.

2. Jones W., Chibwe M., in "Pillared layered structures" I.V. Mitchell Ed. Elsevier Science Pub. London, (1990) 67.

3. Pope M.T., in "Heteropoly and isopolyoxometallates", (1983) Springer Verlag Ed. Berlin .

4. Baes C.F., Mesmer R.E., in "Hydrolysis of cations" (1976) Wiley, New York.

5. Matijevic E., Ann. Rev. Mater. Sci., 15 (1985) 483.

6. Henry M., Jolivet J.P., Livage J., Structure and Bonding, 77 (1992) 153.

7. Sanderson R.T., Science, 114 (1985) 670.

8. Parr R.G., Donelly R.A., Levy M., Palke W.E., J. Chem. Phys., 68 (1978) 3801.

9. Allred A.L., Rochow E., J. Inorg. Nucl. Chem., 5 (1958) 264.

10. Bach S., Henry M., Baffier N., Livage J., J. Solid State Chem., 88 (1990) 325.

11. Johansson G., Lundgren G., Sillen L.G., Soderquist R., Acta Chem. Scand., 14 (1960) 769.

12. Livage J., Chem. Mater., 3 (1991) 578.

13. Legendre J.J., Livage J., J. Colloid Interface Sci., 94 (1983) 75.

14. Livage J., Barboux P., Vandenborre M.T., Schmutz C., Taulelle F., J. Non-Cryst. Solids (in press).

15. Livage J., Henry M., Sanchez C., Progress in Solid State Chem., 18 (1988) 259.

16. Clearfield A., Vaughan P.A., Acta Cryst., 9 (1956) 555.

17. Clearfield A., in "Inorganic exchange materials", (1981) CRC Press, Boca Raton.

18. Mc Whan D.B., Lundgren G., Acta Cryst., 16 (1963) A36.

POROUS STRUCTURES OF SOL-GEL AL_2O_3(-$3SiO_2$) GELS AND "GLASSES" INFILTRATED BY AQUEOUS SOLUTIONS.

V. VENDANGE and Ph. COLOMBAN
ONERA-OM, BP 72
92322 Chatillon cedex
France

ABSTRACT. Alumina and aluminosilicates are prepared by sol-gel process and then infiltrated with transition metal solutions. Modifications of the xerogel pore distribution after metal aggregate formation are studied by BET measurements. Possibilities of membrane-ionic solution reaction are discussed.

1. Introduction

Porous materials, optically clear or translucent, can be formed by the hydrolysis-polycondensation of alkoxides. These xerogels contain more or less water according to the heat treatment temperature (gel< 300°C, "glass"> 300°C) [1-2], the principal gel forming elements being Silicon, Aluminium, Germanium, Zirconium, Phosphorus and Titanium. With different hydrolysis conditions and compositions we obtain mesoporous (pore mean diameter 2-10 nm) or microporous (< 2 nm) materials [3]. For Alumina and Aluminosilicates the porous structure is preserved until 900°C (with a specific area upper than 200 m^2/g). A catastrophic densification results from the departure of the last protonic species which stabilised the porous polymeric network [3-5]. The crystallization simultaneously takes place. Such porous materials could find applications as scavengers and as catalysts.

In this work we show that the pore distribution, calculated from BET N_2 adsorption curves, is modified by metal aggregates produced after reduction of aqueous infiltrated solutions. Possibilities of membrane-ionic solution reaction are discussed.

2. Experimental

Alumina optically clear monolithic (ocm) gels (a few cm^3) have been prepared by slow hydrolysis of s-aluminium butoxide diluted in acetone (1/1), the atmospheric moisture reaching the surface of the solution though a plastic cover [5]. Alkoxide-solvent mixtures are prepared in water-free glove box using water-free solvents. A dense and sonorous monolith is obtain after about six months.

Gels with a Al/Si=1 ratio have been prepared using a mixing of water, propanol and $(OC_4H_9)_2Al\text{-}O\text{-}Si(OC_2H_5)_3$ "Al-O-Si" ester purchased from DYNA-FRANCE (formerly Dynamit-

C.A.C. Sequeira and M.J. Hudson (eds.), Multifunctional Mesoporous Inorganic Solids, 339–344.

Nobel, now Hulls-France). Water-propanol-ester volume ratios (wpe) are 1.3.6, 1.4.2, 4.12.3. Handling is possible in air, but the hydrolysis is performed in covered bottles or Petri dishes. Monoliths are extracted from the reaction vessel when they reach the sounding state. Drying is performed for ten hours at 40-60°C. The reaction time increases with the volume/surface ratio of the solution. However, batches prepared, simultaneously and with the same procedure, produced a gel after various lengths of time. The scattering of values reaches 30-50% and sometimes their optical properties differ. The water and hydroxyl amounts are calculated from TGA curves [3]. Immersion of Aluminosilicate gels or glasses in aqueous solutions containing transition metal ions leads to a diffusion of ions into the monoliths. After heat treatment under Hydrogen, nanoprecipitates are obtained . The electron microscopy of the samples was done on a JEOL 200CX. Diffuse Reflection IR spectra have been recorded using a NICOLET interferometer.

3. Porous Structures

3.1 HOST GELS AND GLASSES

The possibility to obtain highly porous fractal gels has been shown by recent studies on alkoxide hydrolysis [6]. But dense gels (apparent density >1.5) can also be formed by this method. Materials studied in this work belong to the last class. The scattered intensity profile shown in Figure 1a, corresponds to the xerogel 1.4.2 and a linear behaviour with a constant slope of about -4, in the Porod region is clearly observed, as in most samples [7]. There are only few cases like the 4.12.3 gel in figure 1b where a fractal domain could be noticed with a slope of -2.13. Gels are optically translucent. However, it must be noticed that (studying the reproductibility) the gelation parameters of ten batches prepared in the same initial conditions are not constant. For example Figure 2a shows the pore size adsorption distribution (PSAD) for white (w) and optically clear monoliths (ocm) from the same batch. In the following only normal case will be considered. It was observed that our alumina ocm gels have specific area higher than alumina gels prepared with acid and base catalyst [8].

3.2 HOST INFILTRATED SOLIDS

Few studies on membrane or porous infiltrated structures have been performed and mainly organic membranes are concerned [9]. Only Pope and Mackenzie [10] have been working on impregnation of a porous silica gel with a monomer followed by room temperature polymerization. However, main papers discuss the effect of solvent on stability, the ageing process and the porous structure of gels [11,12]. The preparation of ceramic-metal composites by alkoxides-hydrolysis was also studied [13-15]. In this work, the infiltration of the host solids takes place in the gel state (gelation is over and monoliths reached the sounding state) or in the glass state (i.e porous, amorphous and translucent solid).

Alumina. Figure 3 shows pore size adsorption distribution (PSAD) from BET curves for alumina in the glass state (heated at 600°C) and infiltrated with Iron (Fig 3a) and Cobalt (Fig 3b) aqueous solutions. Cobalt or Iron precipitates appeared after a reducing treatment at 600°C under Hydrogen. Typical examples of metal precipitates are shown on Figure 4. The specific area decreases after infiltration for the two samples but the distribution is slightly modified. It becomes broader with Iron infiltration and shift to smaller pore diameter with Cobalt.

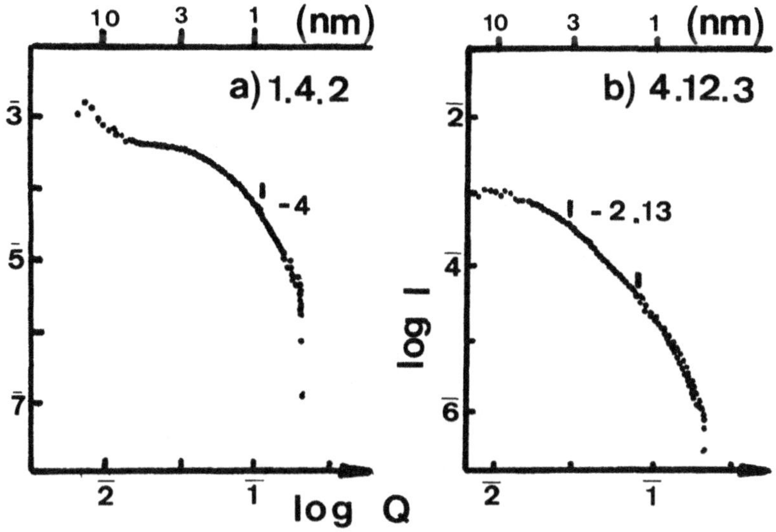

Figure 1: X Ray scattered intensity profiles from two xerogels: a) $Si_2Al_2O_{6.7}(OH)_{0.6}\cdot 8.4H_2O$ (wpe:1.4.2), b) $Si_2Al_2O_{6.5}(OH)\cdot 4.7H_2O$ (4.12.3).

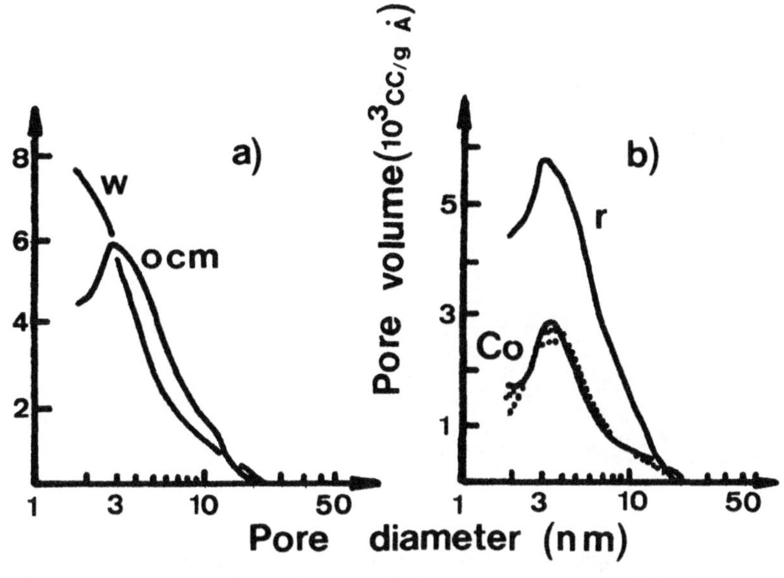

Figure 2: Pore size adsorption distribution (PSAD) for Aluminosilicate (4.12.3).
a) White (w) and optically clear monolithic gels (ocm) from the same batch.
b) Host solid in the gel state before infiltration (r) and solids infiltrated with different Cobalt concentrations from the gel state— and the glass state (heated at 600°C) ---.

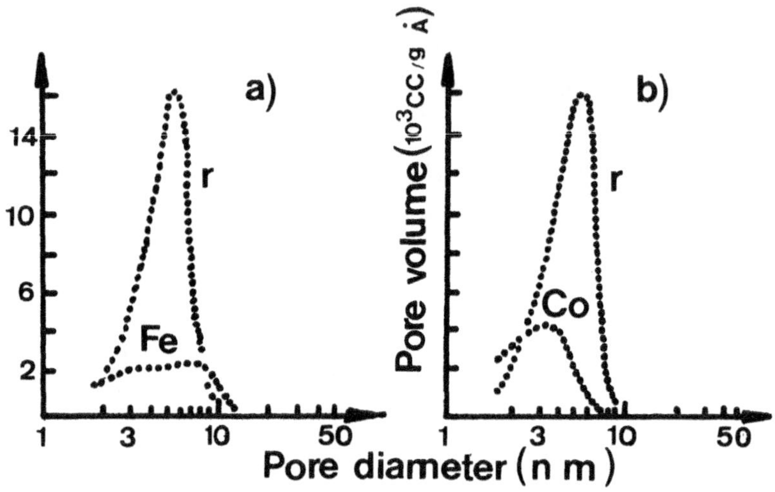

Figure 3: PSAD for Alumina glasses. Host solid in the initial glass state (r) and after infiltration with Iron (a) and Cobalt (b) solution and conversion into metal precipitates.

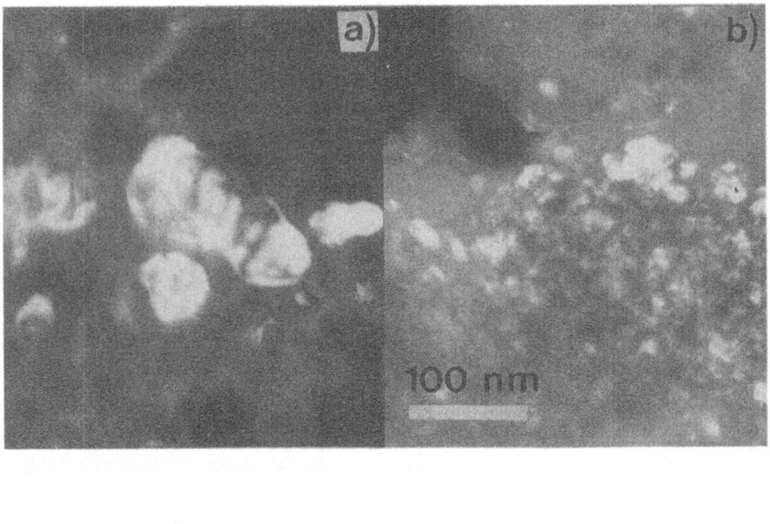

Figure 4: TEM micrographs (dark field) for alumina with Iron precipitates (a) and for $Al_2O_3 \cdot 2SiO_2$ (4.12.3) with Cobalt precipitates (b).

Aluminosilicates. Figure 2b and 5 give an illustration of the various kinds of PSAD encountered with the quasi fractal mesoporous and the microporous $Al_2O_3 \cdot 2SiO_2$ gels before and after Cobalt infiltration, respectively. These infiltrations with different Cobalt concentrations do not induced large modification in the pore distribution of the mesoporous solid 4.12.3. The initial state of the host solid (with --- or without—heat treatment at 600°C) does not seem to influence the porous stucture of the infiltrated solid. Conversely, the PSAD of the microporous solid is drastically modified. The pore distribution depends on the initial state of the host solid. Monoliths infiltrated with Cobalt solutions from the microporous gel state (Fig 5a) present a pore distribution with two broad peaks whereas monolith infiltrated from the glass state (Fig 5b) including a single broad distribution.

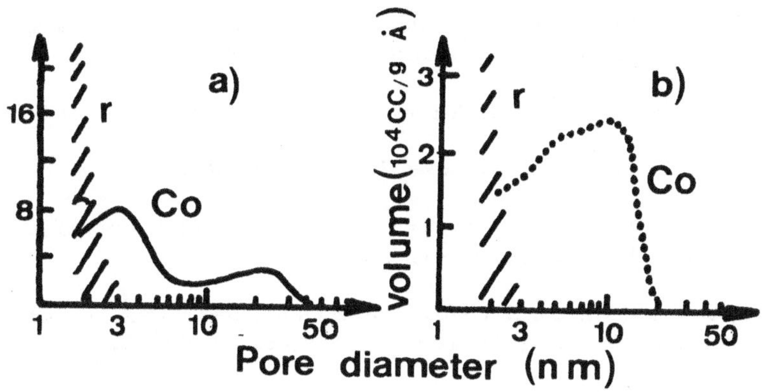

Figure 5: PSAD for Aluminosilicate (1.3.6): monoliths infiltrated with Cobalt from the gel state (a) and the glass state (b).The dashed area is a guide for the eyes of the host microporous distribution (r).

4. Conclusion

The porous structure defined by the polymeric network can be modified by infiltration of transition metal aqueous solutions. The evolution seems to depend on physical parameters (fractal structure or not) but also on chemical factors, like the composition of the host porous solids or elements of the aqueous solution. The shift of IR Si-O stretching mode (1040 cm^{-1} for the 1.3.6 reference gel, 1030 cm^{-1} after infiltration with Cobalt and 1080 cm^{-1} with Iron aqueous solutions) points out a partial reaction between metal ions and Si-O-Al polymeric network.

5. References

1 Klein, L.C. (1988) Sol-gel technologies, Noyes Publications, Park Ridge, New Jersey.

2 Colomban, Ph. (1989) 'Gel technology in ceramics, glass-ceramics and ceramic-ceramic composites', Ceramics. Int., 15, 23-50.

3 Colomban, Ph. and Vendange, V. (1992) 'Sintering of alumina and mullites prepared by slow hydrolysis of alkoxides: The role of the protonic species and of pore topology', J. Non-Cristalline Solids, to be published.

4 Colomban, Ph. (1991) 'Protonic defects and crystallization of sol-gel (Si,Ge) mullites and alumina, in P. Vincenzini (eds), Ceramics Today, To-morrow's Ceramics, (Proc. 7th CIMTEC, Montecatini-terme, 24-30 june 1990), Mater. Sci. Monographs 66B part B, 599-605, Elsevier, Amsterdam.

5 Colomban, Ph. (1989) 'Structure of oxides gels and glasses by IR and Raman scattering: I Alumina', J. Mater. Sci., 24, 3002-3020;'II Mullites', ibidem 3011-3020.

6 Schaeffer, D.W (1989) 'Polymers, fractals, and ceramics materials', Science, 243, 1023-1027.

7 Vendange,V. Colomban, Ph. and Larché, F. Unpublished results.

8 Rama Rao, G.V. Venkadesan, S. and Saraswati, V. (1989) 'Surface area and pore size studies of alumina gels', J. Non-Crystalline Solids, 111, 103-112.

9 Kurokawa, Y. and Haneda, K. (1987) 'Preparation of ultrafine magnetic particles by using finely porous membrane', J. Poly. Sci. Part C: Poly. Letts, 25, 369-372.

10 Pope, E.J.A. Asami, M. and Mackenzie, J.D. (1989) 'Transparent silica gel-PMMA composites', J. Mater. Res., 4, 1018-1023.

11 Sakka, S. and Adachi, T. (1990) 'Stability of sol-gel derived porous silica monolith to solvents', J. Mater. Sci., 25, 3408-3414.

12 Dumas, J. Quinson, J.F. and Serughetti, J. (1990) 'Hierarchy of pores and mechanical behaviour of wet silica gels', J. Non-Cristalline Solids, 125, 244-249.

13 Lopez, T. Bosch, P. Asomoza, M. and Gomez, R. (1992) 'Ru/SiO_2-impregnated and sol-gel -prepared catalysts: synthesis, characterization and catalytic properties', J. Catalysis, 133, 247-259.

14 Petrullat, J. Ray, S. Schubert, U. Guldner, G. Egger, Ch. and Breitscheidel, B. (1992) 'Preparation and processing of metal/ceramic composite materials', J. Non-Crystalline Solids, to be published.

15 Roy, R.A. and Roy, R. (1984) 'Diphasic xerogels: I. Ceramic-Metal composites', Mat. Res. Bull., 19, 169-177.

Acknowledgment: One of us (V.V) thanks the french "Ministère de la Recherche et de la Technologie" for a grant.

SECTION 5

OTHER METHODS OF CHARACTERISING INORGANIC MATERIALS

SOLID STATE NUCLEAR MAGNETIC RESONANCE SPECTROSCOPY

M.J. HUDSON and A.D. WORKMAN
Department of Chemistry,
University of Reading,
P. O. Box 224,
Whiteknights,
Reading,
Berkshire, U.K.
RG6 2AD

1. INTRODUCTION.

The area of mesoporous materials tends to suffer from a lack of detailed structural analysis, that is due to a dearth of well crystallised materials and ill-defined pore size. That is to say that the crystals produced are often too small to diffract X-rays coherently, let alone large and sufficiently free of defects for single crystal X-Ray analysis to be possible. This presents a problem to the structural chemist that must be solved by looking further afield than just conventional X-ray methods. One such technique that can cope with short range order is solid state nuclear magnetic resonance (nmr), used with magic angle spinning and also cross-polarisation techniques. This is not to say that other techniques (such as EXAFS, SAXS, Transmission Electron Microscopy, Neutron Diffraction, Nuclear Quadrupole Resonance and Mossbauer) can not cope with short-range order, or that they are not applicable, or that nmr is without problems, but that nmr is one technique that is readily available and with a wealth of previous studies.

Therefore, this paper will present a broad appreciation of nmr, and what might be possible with further study, that is applicable to the researcher working towards and on mesoporous materials.

2. EXPERIMENTAL CONSIDERATIONS.

2.1 THE THEORY OF MAGIC ANGLE SPINNIG NMR (1).

Line broadening of nuclear magnetic resonance spectra proves a problem in that the resolution of spectra are reduced, so diminishing the usefulness of the technique. This problem is more acute in the solid state than in solution, where the effects producing line broadening are overcome by the properties of a molecule or complex in solution. Line broadening falls mostly into two categories, namely chemical shift anisotropy and dipolar broadening, though spin-spin and spin-lattice relaxation and the extent of order in the solid are also involved.

C.A.C. Sequeira and M.J. Hudson (eds.),Multifunctional Mesoporous Inorganic Solids, 347–371.

2.1.1 *Chemical Shift Anisotropy.* This arises from non-spherical or non-cubic distributions of electron density around the nucleus under observation. The electron density surrounding the nucleus gives rise to the chemical shift recorded by shielding it from the applied magnetic field. Hence, in the case of an asymmetric electron density distribution, the chemical shift takes on a directional character, following a $(3\cos^2\theta-1)$ rule, where θ is the angle between the applied field and the principal axis about which the chemical shift is defined. This is seen most clearly in the case of a single crystal, where different spectra can be recorded along individual planes. In a powdered sample, individual crystallites occupy every possible orientation to the applied magnetic field and so a single broad line is seen, resulting from the coalescence of all the resonances.

Like the chemical shift itself, the chemical shift anisotropy is also subject to changes in the chemical environment of the nucleus. For example, in ^{13}C spectra, the chemical shift anisotropy of aliphatic carbon atoms is smaller than that of aromatic carbons, which is a reflection of the lower directionality of sp^3 hybrid orbitals. Chemical shift anisotropy is proportional to the applied magnetic field and so can be distinguished from other factors by recording spectra in fields of different strengths.

2.1.2 *Dipolar Broadening.* Dipolar broadening arises from the through space dipole-dipole interactions of atoms having spin of $I\neq0$. This induces local magnetic fields (B_{loc}) at the nuclei involved, and is given by:

$$B_{loc}=\pm\mu_S r_{IS}^{-3}(3\cos^2\theta_{IS}-1) \tag{1}$$

where I is the nucleus being observed
S is the interacting nucleus
μ_S is the magnetic moment of S
r_{IS} is the internuclear distance
θ_{IS} is the angle between the internuclear vector and the applied magnetic field, B_o

The plus or minus sign is included because the magnetic moment can be aligned with or against B_o. In powdered samples, the internuclear vectors reside in a number of different orientations to B_o, causing a distribution of different local field which affect the chemical shift of the nuclei being observed. As in the case of chemical shift anisotropy, the recorded spectra show the superimposition of all the perturbations caused to the chemical shift by the local fields, resulting in a broad peak. Both the line broadening arising from chemical shift anisotropy and dipole-dipole interactions can be overcome by modified experimental techniques.

2.1.3 *Magic Angle Spinning.* Many techniques can be used for narrowing lines in spectra, and all rely on being able to average the interactions causing broadening on a timescale that is rapid compared to the linewidths. Most commonly used is magic angle spinning, in addition to which other techniques may be used. In solution, rapid isotropic tumbling averages dipolar splittings and chemical shift anisotropy to zero and hence sharp lines are observed. As seen above, both of the two dominant line broadening phenomena contain an angular dependence factor, $(3\cos^2\theta-1)$. At a certain value of θ, this factor falls to zero. This is called the "magic angle" and equals 54° 44'. If the sample is spun at the magic angle to the applied magnetic field, the average orientation of the crystallite axes will be equal to the magic angle and a sharp line may be observed. The rate at which the sample is spun also contributes to the resolution of the spectra. If the sample is spun at a rate lower than the original linewidth, the averaging of the electron density symmetry is incomplete and the chemical shift anisotropy gives rise to a comb-like spectrum. The most intense peak in this comb is not necessarily the isotropic chemical shift, which is given by the central peak of the comb. The separation of the resonances of the comb is related to the spin rate. At spin rates higher than the linewidth of the band, the resonance due to the observed nucleus is reduced to a single peak, though often small residual spinning side bands are seen centred on the isotropic chemical shift.

2.1.4 *Dipolar Decoupling of Spectra.* Magic angle spinning is more effective at narrowing lines broadened by chemical shift anisotropy than those broadened by dipolar interactions and so there is often residual breadth to the resonances thus limiting resolution. This can be removed by heteronuclear decoupling. This is analogous to the situation in solution where the nuclei causing splittings are irradiated at their resonant frequency. This has the effect of scrambling the interactions between the two coupled spin vectors by causing the irradiated nuclei to flip between their high and low spin states thus overriding the dipolar broadening effect.

2.1.5 *Miscellaneous Effects.* Spin-spin and spin-lattice relaxation are dependent upon fluctuations of the magnetic field at the nucleus under observation, which are mainly caused by molecular motions, such as the tumbling of a molecule in solution. When the tumbling frequency is greater than the resonant frequency, the fluctuating magnetic field contributes equally to both spin-spin and spin-lattice effects. However, as the frequency of fluctuations tends towards zero, spin-lattice relaxation becomes less efficient but spin-spin relaxation becomes more important. The effect of a long spin-lattice relaxation time is to lengthen the interval between pulses in a multipulse experiment, otherwise the population of the excited state grows and the signal diminishes, eventually reaching zero. The affect of a fast spin-spin relaxation time is to broaden the peak by increasing the uncertainty in the energy gap between the excited and ground states. This can be understood from the Uncertainty Principle and is referred to as the natural linewidth of a peak (2). The Uncertainty in an energy can be related to the lifetime of a state by equation 2:

$$\Delta E \Delta t \sim \hbar \qquad 2$$

where ΔE is the uncertainty in the energy gap of the states
Δt is the lifetime of the excited state
and $\hbar$ is $1/(2\pi h)$

This means that the natural breadth of a line is proportional to the relaxation rate of the system. As the residence time of the nucleus in the upper spin state decreases, so the uncertainty in the energy of that state increases, which is observed as a broadened line. In solid state nmr, the peak widths are significantly broadened by spin-spin relaxation effects and yet it is often the case that several seconds must pass between pulses because of spin-lattice relaxation rate is slow.

Another factor in the broadening of MAS nmr lines is the amount of disorder in the solid. If the sample is crystalline, nuclei are in identical environments, which leads to sharp peaks because of the uniformity of chemical shift from each nucleus. As the samples becomes amorphous, nuclei in slightly different environments, give rise to non-uniform chemical shifts and the peak broadens. As disorder grows, therefore, so does the width of the peaks.

2.1.6 *Conclusions.* The combination of magic angle spinning and dipolar decoupling may remove some of the effect of chemical shift anisotropy and that of dipole-dipole interactions. Oftenly however, these do not sufficiently narrow resonances, as in the case where homonuclear dipole-dipole interactions are broadening the lines. Therefore different pulse sequences (WAHUHA (3), Cross Polarization (4) or MREV-8 (5)) are often used in addition to magic angle spinning. The combined use of these conditions is sometimes known by the acronym CRAMPS (6) (Combined Rotational And Multiple Pulse Spectroscopy).

2.2.2 *Experimental Conditions.* The nmr spectra we present from our own work were carried out by the SERC nmr service of the Industrial Research Laboratories at the University of Durham. All peak widths given are quoted as full width half maximum height (FWHM) measurements. Table 2.2.1 summarises the operating conditions that the spectra were recorded under.

Table 2.2.1 Operating Conditions for MASnmr.

Nucleus, Observation Frequency (MHz), and Spin rate (Hz).	Pulse Conditions	Spectral Bandwidth(KHz)	Relaxation Delay (s)	Contact Time (ms)
^{29}Si, 59.584, 4350	Crosspolar with gated decoupling	30.3	1.0	9.6
^{27}Al,78.152 11850	Crosspolar with gated decoupling	100	0.5	4.8
^{31}P, 121.4 5000	crosspolar with gated decoupling	100	60	1.0
^{119}Sn, 118.7 5000	crosspolar with flip-back	200	0.5	5.0
^{15}N, 30.4 5000	crosspolar with gated decoupling	30.3	1	1.0

2.3 SPECTRAL PARAMETERS.

2.3.1 *Chemical Shift.* Not only can the isotropic chemical shift of a nucleus reflects its immediate chemical environment. Factors which effect the value of the chemical shift are :-

- Number and type of the attached groups
- Strength and nature of bonding to the attached groups
- Bond angles between the nuclei and the attached groups.

The isotropic chemical shift give structural information, but for single crystals, so can the chemical shift anisotropy. As different axes of a non totally symmetrical crystal are aligned with the applied magnetic field then it presents different shielding tensors to the field and so the anisotropy changes (7).

2.3.2 *Line Intensity.* The intensity of a resonance peak is proportional to the "sensitivity" of a nucleus in its environment and the difference in the populations of the upper and lower spin states. In certain cases, where the sensitivity of the nucleus is not altered by its chemical environment (see 8), and the populations of the spin states have not been inhomogenously altered by the pulse sequence used, then the area under the peak is directly proportional to the number of nuclei present in that environment, as is the case with solution state ^1H nmr. This fact has much use in quantitative characterisation by nmr, such as the calculation of the ratio of different structural units, the number and amounts of different phases present in a sample and even the elemental analysis of some samples. This will be considered below.

2.3.3 *Linewidth.* The linewidth obtained in a spectrum is dependent upon three main factors:-

- Dispersion of isotropic chemical shifts aring from structural disorder.
- Unaveraged dipolar interactions with other nmr-active nuclei.
- The presence of paramagnetic impurities in the sample.

Where the nucleus being observed has a quadrupole, then the linewidth is also dependent on the electric field gradient at the nucleus (e.g. pentacoordinate ^{27}Al (9,10)).

3. *STRUCTURAL ANALYSIS.*

3.1 ^{29}SI AND ^{27}AL MAS NMR.

3.1.1 *Background.* Much work has already been carried out using nmr, especially of these two nuclei because these two atoms are the major constituents of the zeolitic framework. In the future, they may well useful for the study of mesoporous solids because, for example it is possible to examine the nmr of silicate solutions, from which mesoporous solids may be prepared.

3.1.2 Chemical Shift.

3.1.2.1 Dependence on Connectivity. The chemical shifts of silicate and silicic acid species extend over a range of about 60 ppm, that is from -60 to -120 ppm (11) and are seen to be partitioned into sub-ranges corresponding to the different connectivities possible for silicate species. The connectivity (Q^n) of an SiO_4 group is the number of other SiO_4 units attached to the SiO_4 group under observation. Hence, for silicate species there are five possibilities, Q^0 to Q^4. However, the field also encompasses aluminosilicates which are incorporated into the nomenclature by the addition of a bracketed term denoting the number of oxygen bridges to Al. Hence a description of an SiO_4 group in an aluminosilicate might be Q^2(1Al). This would mean an SiO_4 group attached to one other SiO_4 group and one AlO_4. Examples of the connectivity system are given in Figure 3.1.1.

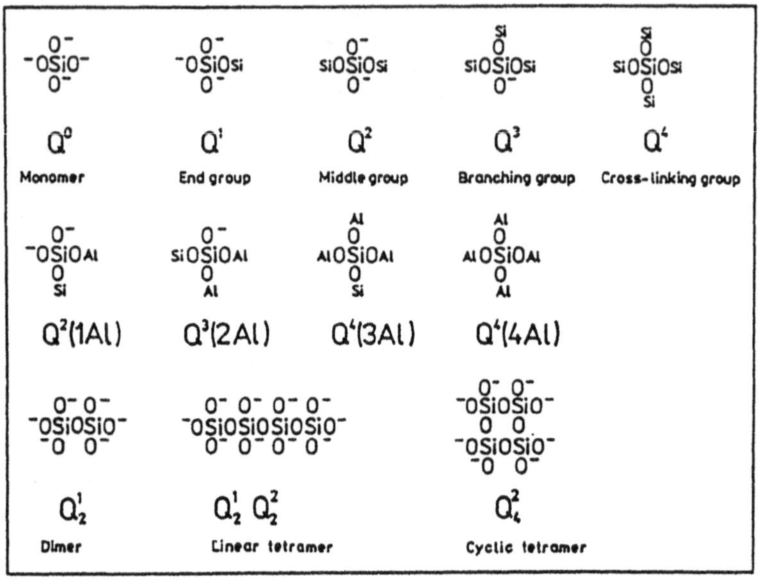

Figure 3.1.1; Connectivity Notation ($Q^n(mAl)$), for Condensed Silicates.

As previously stated, the connectivity of an SiO_4 group directly effects the chemical shift of the central silicon. This is depicted in Figure 3.1.2. It can be seen from this that, although where cyclical units are involved, there is little overlap between the ranges for each value of connectivity. As the connectivity increases, so generally does the linewidth. This is because of many factors, but is mostly due to the increase in the number of different distinct orientations of the SiO_4 group (12). These oreientations give rise to slightly different chemical environments which, when observed, overlap and result in a broad line. It is also noticeable that the range of chemical shifts is greater for Q^4 silicates than for any other. Similarly this is because, with a greater number of substituents, there are more possible combinations of and so the chemical environment of the silicon is open to more variation.

The intensity of a peak in ^{29}Si nmr is proportional to the number of Si units in a particular chemical environment and this has great use in structural characterisation when coupled with the connectivity system described above. The relative amounts of different constituent groups can be calculated directly from the ratios of the peak intensities An example is in the calculation of the Si/Al ratio of a tetrahedral framework. This makes use of Lowenstein's Rule, which states that in tetrahedral aluminosilicates two AlO_4 units will not be found bound to each other. From this, every Al atom in a $Q^n(mAl)$ unit is linked to n silicon atoms via oxygen bridges. Each Si-O-Al linkage of the $Q^n(mAL)$ therefore incorporates $^1/_n$ Al atoms, and the whole unit $^m/_n$ Al atoms. Since the individual $Q^n(mAl)$, peak intensity $I_{n,m}$ is directly proportional to the number of these units, the total amount of Al is equal

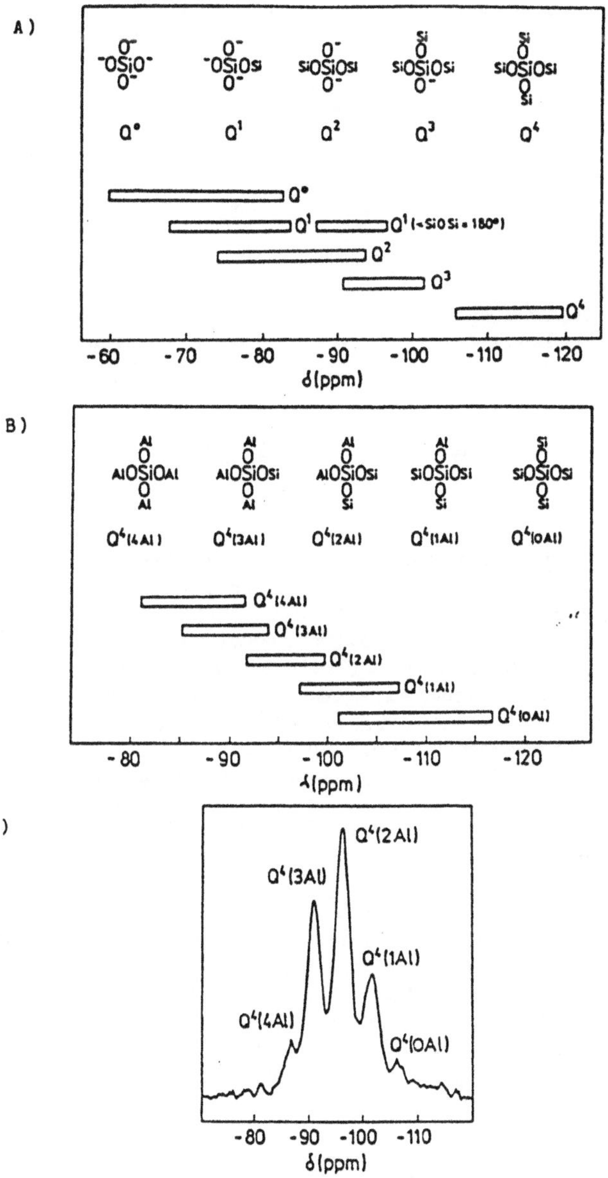

Figure 3.1.2;Solid State MAS nmr Chemical Shift Ranges of Silicon in Silicates and Alumino silicates. A)Variation of with n of Q^n;B)Variation with m of (mAl); C) Typical Spectrum (Zeolite)

to $\Sigma^{n}_{m=0}\,(^{n}/_{m})I_{n,m}$ and the total Si content to $\Sigma^{n}_{m=0}\,I_{m,n}$. Therefore the $^{Si}/_{Al}$ ratio is given by (13):-

$$\frac{Si}{Al}=\frac{\sum_{m=0}^{n} I_{n,m}}{\sum_{m=0}^{n} (\frac{m}{n}) I_{n,m}} \qquad 3$$

3.1.2.2 Dependence on Charge. The factor that defines the position of the chemical shift of a nucleus A has been termed the shielding constant and given the symbol σ^A. This can be broken down into three parts so that

$$\sigma^{A}=\sigma^{A}_{dia}+\sigma^{A}_{para}+\sum_{A\neq B}\sigma^{AB} \qquad 4$$

where "dia" denoted the diamagnetic screening term, "para" the paramagnetic screening term and the last term is the contribution from surrounding atoms (14). The relative contributions of each of these terms has yet to be settled. However, for the ^{29}Si nucleus it is generally agreed that the dominant term is the paramagnetic one. A semi-empirical quantum chemical approach (15,16), that uses Slater-type orbitals and Slater rules incorporates the difference in Pauling electronegativities of the central Si atom and the groups to which it is attached has been followed in order to calculate the dependence of the paramagnetic screening term on the charge on the Si atom. This has been shown to be in good agreement with the experimental data (16). This is illustrated in Figure 3.1.3. The region enclosed in the dotted box shows the range of values of electronegativity of groups bonded to the silicon via an oxygen bridge. It can be seen that, over this range, the correlation is virtually linear, and this has been used in empirical calculations of the dependence of $\partial^{29}Si$ with electronegativity and in the calculation of group electronegativities (17).

3.1.2.3 Dependence on Bond Angle and Length. Table 3.1.1 is a summary of empirical linear relationships between $\partial\,^{29}Si$ and mean Si-O-Si bond angles. A theoretical model has been constructed using the same rules and assumptions as for the net charge on the silicon atom and applying the model to estimate the effective orbital electronegativity of the oxygen and adjusting for the angle by interpreting it in terms of the degree of hybridisation of the oxygen s and p orbitals (18). This gives rise to the following equation which, since it has a curve similar to other functions of the bond angle, can also be used to explain other empirical linear relationships;

$$\delta=-A\left(\frac{\cos\alpha}{\cos(\alpha-1)}\right)+B+Cm \qquad 5$$

Where A, B and C are constants, α is the bond angle, and m is the number of Al tetrahedra attached

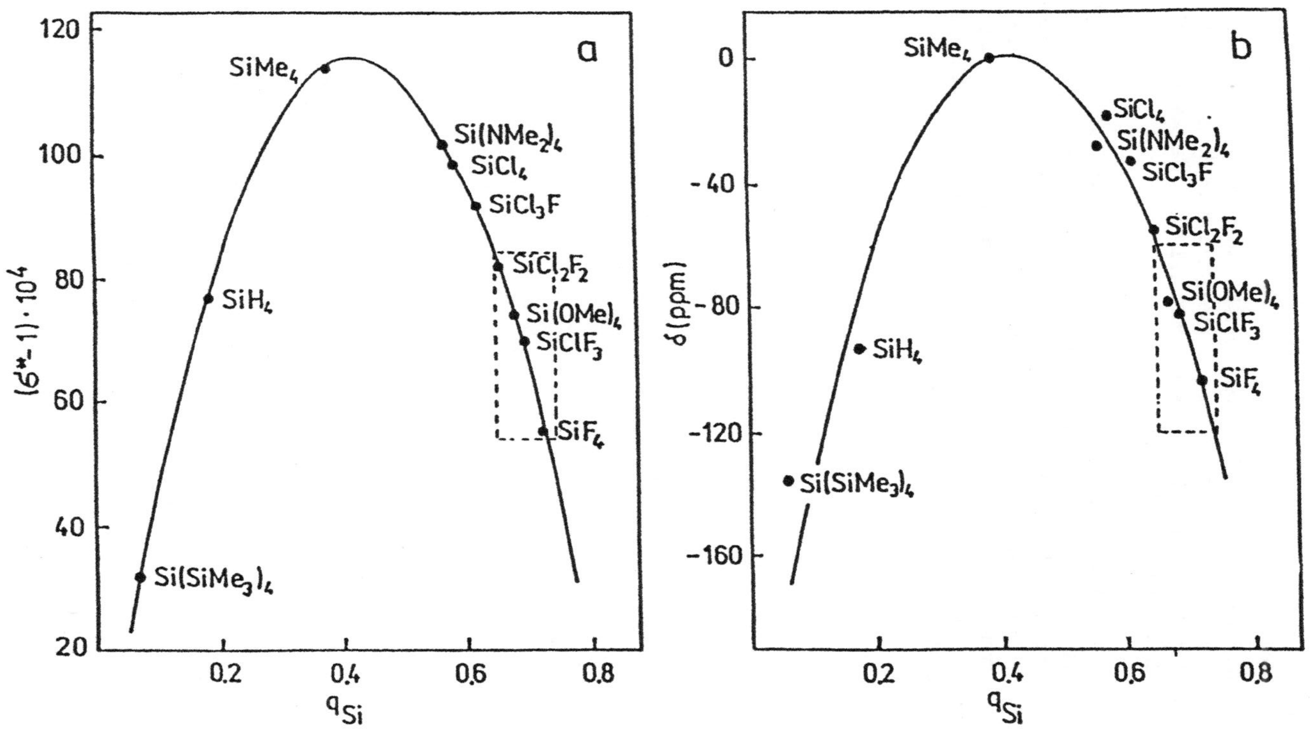

Figure 3.1.3; a) Variation of the Paramagnetic Screening Term with charge on the Silicon atom.
b) Variation of ChemicalShift with charge on the Silicon atom.

to the Q^n(mAL) unit.

The lack of accurate data for Si-O bond lengths makes a correlation between bond length and ∂ ^{29}Si difficult. However it has been concluded (19) that a change in bond length of 0.0001 nm corresponds to a change in the chemical shift of about 1 ppm, illustrating the sensitivity of the Si chemical shift towards the environment of the silicon atom.

3.1.2.4 Dependence on Cation. Whilst discussing ^{29}Si chemical shifts it is worthwhile noting that the chemical shifts observed for a particular zeolite are dependent on the cations themselves and in what proportions they are present. Table 3.1.2 shows some typical values of ∂^{29}Si for zeolites with different counter-ions. It is not possible to resolve bands arising from each particular type of cation present because of the rapid exchange of hydrated cations between different sites in the zeolite. On dehydration, the interaction of cation and exchange site is such so as to broaden the band so much to prevent resolution of bands arising from different cations.

Table 3.1.2. Change in ^{29}Si Chemical Shift for Zeolites X (20) and A (21) with change in Cation.

Zeolite[a]	-∂ (ppm)				
	Si(4Al)	Si(3Al)	Si(2Al)	Si(1Al)	Si(0Al)[b]
LiX (73)	82.5	87.8	93.2	97.6	
NaX	84.7	89.2	93.3	98.3	102.5
KX (100)	87.4	90.8	96.5	101.7	106
RbX (45)	84.5	89.2	94.0	98.8	102.6
CsX (48)	84.8	89.5	94.8	98.6	103.9
CsX (100)	85.6	90.0	94.8	100.6	
SrX (98)	88.1	93.2	97.7	103.1	108.3
BaX (94)	87.5	91.6	96.1	101.5	105.7
AgX (38)	84.3	88.9	93.8	98	
ZnX (59)	83.4	88.9	94.3	99.6	
CdX (96)	83.4	88.5	93.6	98.1	102.0
LiA	85.1				
NaA	88.9				
BaA	90.5				
AgA	87.5				
TlA	88.8				

[A] THE NUMBER IN PARENTHESES GIVES THE DEGREE OF CATION EXCHANGE (%) OF THE PARENT NAX SAMPLE.
[B] WEAK SIGNAL, NOT DETECTABLE IN ALL SAMPLES.

3.1.2.5 ^{27}Al MAS nmr. Further problems arise in the solid state nmr spectra of ^{27}Al, because the nuclear spin is greater than one half. This gives rise to problems of line-broadening from nuclear quadrupole interactions. It is possible to observe aluminium in octahedral and tetrahedral environments and the same rules for the relationship between connectivity and chemical shift apply, Figure 3.1.4. However, where the electric field gradient at the nucleus which is caused by the spatial arrangement of nearest neighbours is greater than zero, extreme line broadening can prevent a band from being observed (9,10).

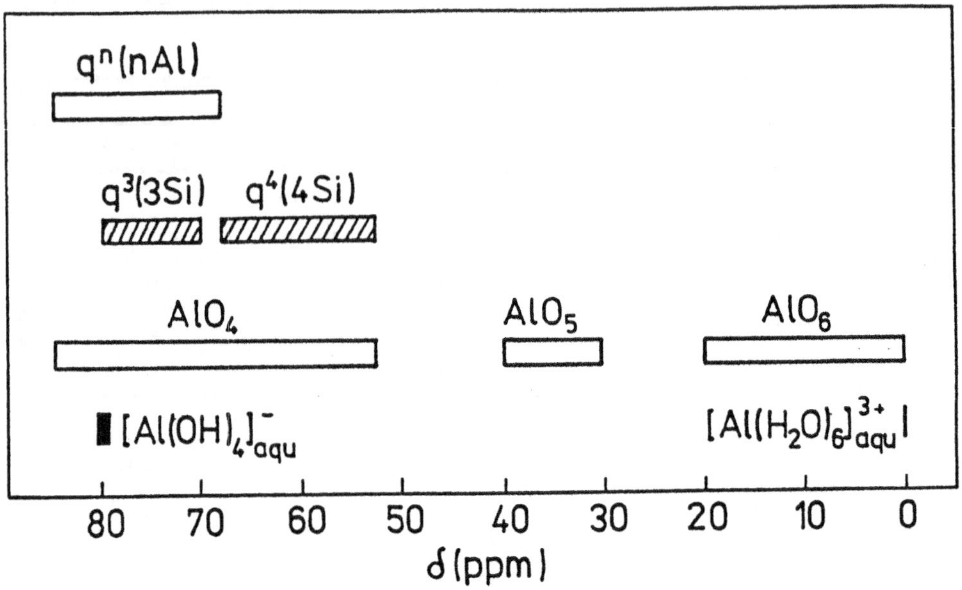

Figure 3.1.4; Variation of ^{27}Al Chemical Shift with Connectivity.

When interpreting ^{27}Al nmr spectra the line intensity is not necessarily proportional to the number of atoms in that environment. The intensities of the spectral lines depend on the length of the excitation pulse and the ratio of the quadrupole frequency to excitation frequency.

3.2 OTHER FRAMEWORK NUCLEI.

These rules have been shown over many years to be true for the above nuclei. However, they have not as yet been widely applied to other nuclei. This may be due to there being such a propensity of aluminosilicates, the range of porous materials dominated by elements other than silicon and aluminium is much more limited. However, with the recent synthesis of Cloverite, a gallium phosphate and the use of the layered metal(IV) hydrogenphosphates as starting points for the synthesis of mesoporous materials, other nuclei will become probes. As both of the above mentioned advances contain phosphorus, it is useful to use this as an example.

3.2.1 ^{31}P *nmr.*

3.2.1.1 Connectivity of Phosphate Groups. Figure 3.2.1 shows the relationship between the connectivity of the phosphorus (Q^n) and the isotropic chemical shift for zirconium and titanium hydrogenphosphates (ZrP and TiP respectively). The nomenclature is the same as that used to describe condensed silicates and is characterised by the number of bridging oxygens of the phosphate group, n. In the case of ZrP, each phosphorus is connected to three metal centres via bridging oxygens and so has a connectivity of three. In the case of γ-ZrP, there are equal amounts of two different phosphate groups, one connected to four Zr ions, Q^4, while the other is a dihydrogenphosphate connected to two Zr centres and so is Q^2 (22). It can be seen that there is a linear correlation between connectivity and the experimental ^{31}P chemical shift for the two compounds given. With this in mind, the suggestion that γ-TiP has the same layered structure was tested by producing a similar plot. As can be seen, a linear agreement was again obtained, from which it is inferred that γ-TiP has the same gamma structure as γ-ZrP. The direction of the slope is such that is indicative of increasing summed P-O bonds strength (23) with increasing connectivity.

3.2.1.2 Electronegativity of Metal Ion. The isotropic chemical shifts plotted as a function of Allred-Rochow electronegativity of the metal centre are shown in Figure 3.2.2. Primarily, the compounds of known structure, i.e. the α-hydrogenphosphates, were examined and then the principal peak from each of the amorphous hydrogen phosphates was added. In each of the former the spectrum is clear and well defined, though the width of the peak increases in the series titanium, zirconium and tin. It can be seen that there is a correlation between the electronegativity of the metal centre and isotropic chemical shift (∂_{iso}). The direction of the gradient shows increasing deshielding of the phosphorus atom, attributable to increased π electron density and hence the summed phosphate bond strength (23,24) with decreasing electronegativity. The phosphorus is increasingly able to remove electron density from the metal centre as electronegativity decreases. This extra electron density then contributes to dπ-pπ interactions and the chemical shift is moved upfield. The second line on the graph confirms this trend and simultaneously reveals information about the structure of the amorphous hydrogenphosphates.

3.2.1.3 Variation of Linewidth with Degree of Structural Disorder. The large differences in the chemical shifts of the phosphate groups with different connectivities has already been established in previous studies (24,25) and above. On this basis, the amorphous compounds consist of predominantly Q^3 phosphate groups, which are associated with the α-type structure. There are differences between the spectra, however, which can be divided into two categories, the width of the signal and the number of signals. To start with, the tin hydrogenphosphate spectrum (Figure 3.2.3) consists of a single peak, approximately 25 ppm wide. At this point it is worthwhile mentioning that only one resonance was observed in the ^{119}Sn MAS nmr spectrum, giving no evidence of hydrolysis to SnO_2. If any hydrolysed phosphate groups had been retained on or in the ion exchanger, these would have appeared at ∂_{iso} close to zero. The breadth is indicative of a low degree of crystallinity. The spectrum of zirconium hydrogenphosphate (Figure 3.2.4) consists of two peaks and a shoulder. The first (smaller) peak lies at -12.71 ppm, the second and principal peak at -20.35 ppm with a shoulder at -26 ppm. The estimated breadth of the of the central peak is 14 ppm. From previous work and the arguments above (24,25), each of these is immediately attributable to Q^2, Q^3 and Q^4 phosphate groups respectively. Again, the spectrum of amorphous titanium hydrogenphosphate (Figure 3.2.5) consists of three resonances, -16.02 ppm, -18.76 ppm and -30.22 ppm, though the band width is now much reduced, approximately 5ppm, indicating a higher degree of crystallinity. The peak at -16.02 ppm is unusual, as it does not appear low enough for a Q^2 phosphate. It may be that this is a doubly hydrated Q^3 phosphate, related to a known polyhydrated form. The X-ray powder diffraction pattern

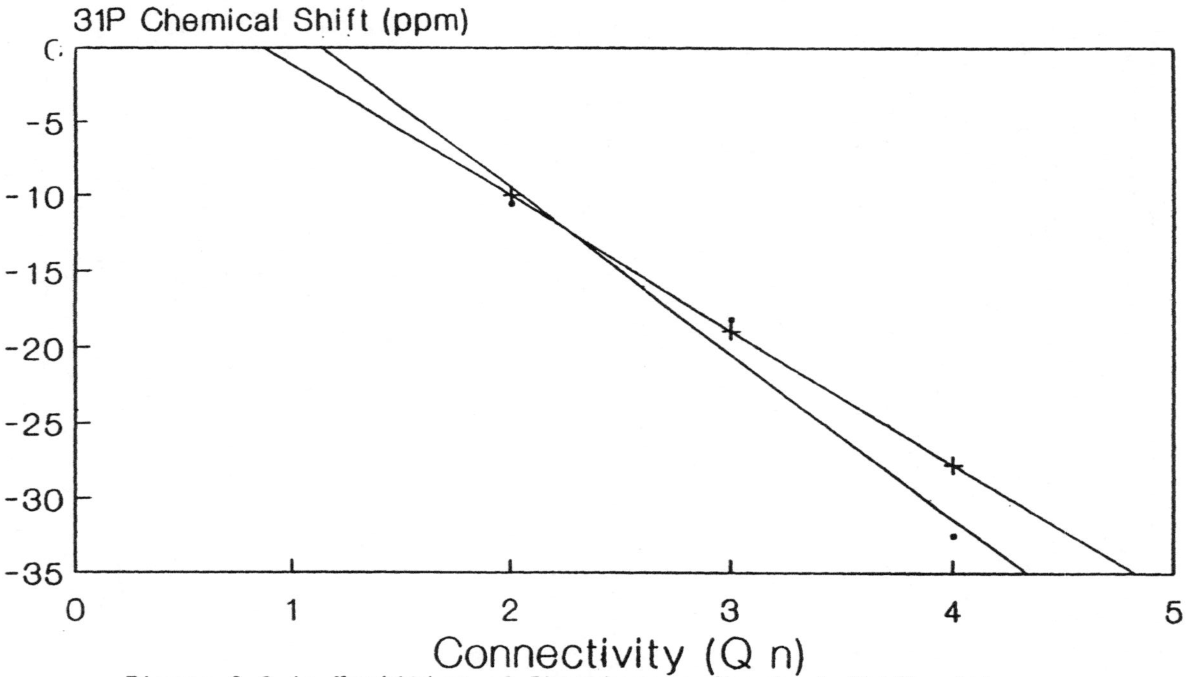

Figure 3.2.1; Variation of Phosphorous Chemical Shift with the connectivity of the phosphate group.

—•— Titanium Hydrogenphosphates —+— Zirconium Hydrogenphosphates.

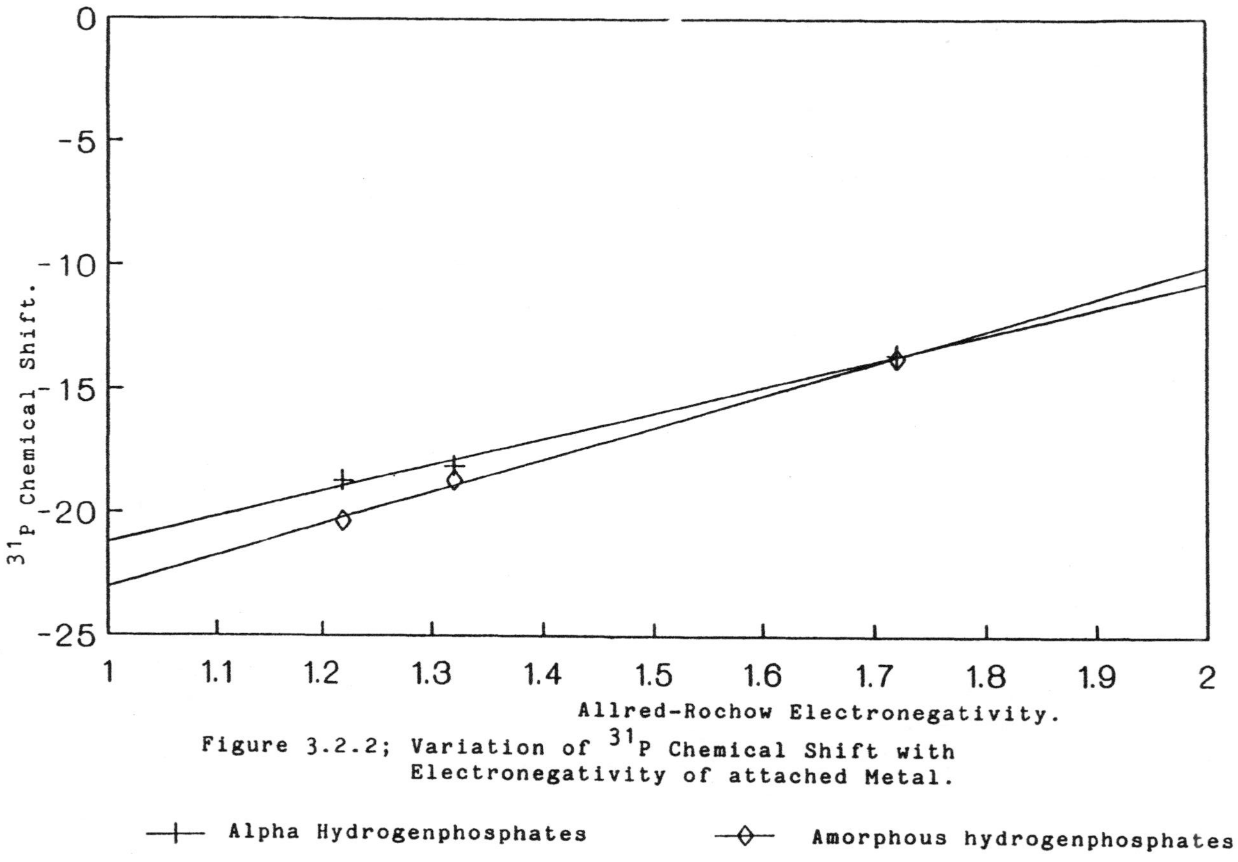

Figure 3.2.2; Variation of ^{31}P Chemical Shift with Electronegativity of attached Metal.

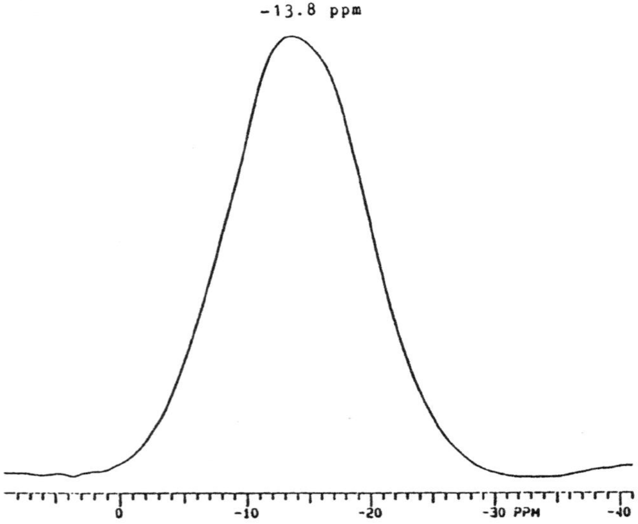

Figure 3.2.3; ^{31}P **MAS NMR Spectrum of** Amorphous Tin Hydrogenphosphate.

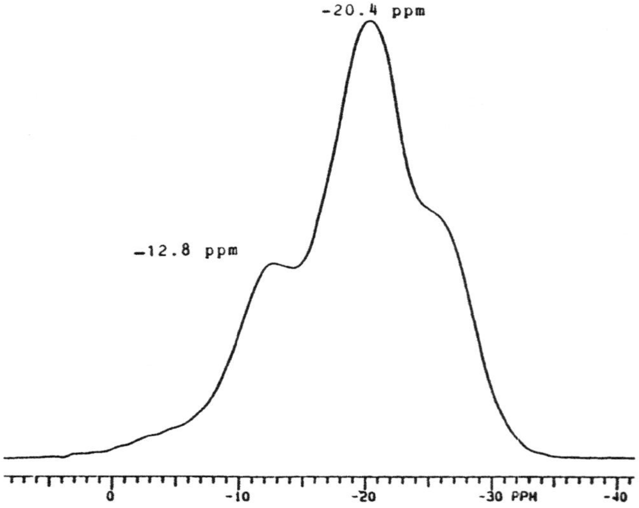

Figure 3.2.4; ^{31}P **MAS NMR Spectrum of Amorphous** Zirconium Hydrogenphosphate.

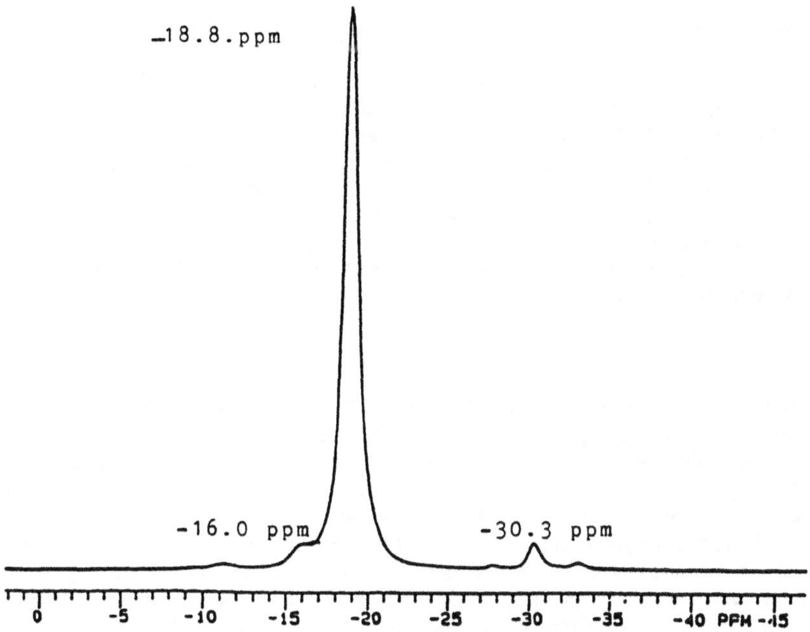

Figure 3.2.5; 31 P MAS NMR Spectrum of Amorphous Titanium Hydrogenphosphate.

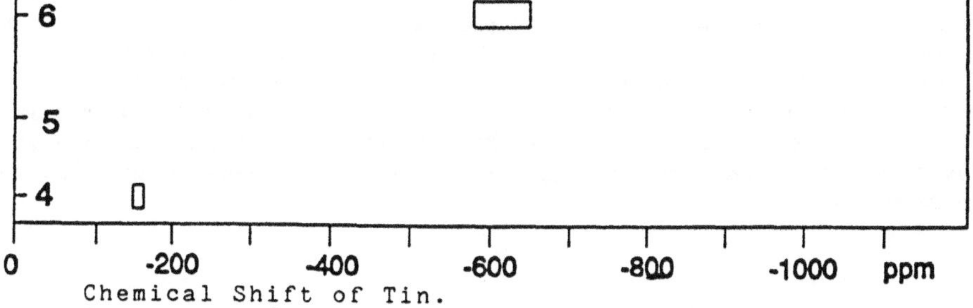

Figure 3.2.6; Variation of Chemical Shift with Coordination Number for Sn.

of this compound shows no structure at all (26). This can be rationalised on the basis of the compound having a high degree of order but only over a very short range, such that is undetectable by X-ray diffraction. This, therefore, is a is a truly microcrystalline compound. In the two spectra of multiple resonances, each peak can be assigned a phosphate connectivity, the closest to zero being Q^2, the next Q^3 and the last Q^4. Bearing in mind the preparation of the alpha and, where possible, the gamma hydrogenphosphates this is easily understood. Each was prepared as the amorphous compound that acted as a precursor which was then transformed into the required product. Hence, in the common precursor to these materials it is possible that there should be some component of each. The facts that γ-tin hydrogenphosphate has not been prepared and that the spectrum of the amorphous compound shows only a single peak attributable to α-SnP seems likely to imply that γ-SnP does not exist under these experimental conditions.

3.2.2 *^{119}Sn nmr.*

3.2.2.1 ^{119}Sn MAS nmr of SnP Intercalation Compounds. The ^{119}Sn MAS nmr spectra are all clear and well resolved, showing only a single resonance each. It can be seen from Figure 3.2.6 that all the ^{119}Sn peaks are around -600 ppm, typical of six-coordinate tin containing at least four directly bound oxygens (27). Figure 3.2.7 is a plot of ∂^{119}Sn as a function of ∂^{31}P. It can be seen that there is a linear correlation between these two terms. From the conclusions drawn above, it can be seen that, as the charge on the phosphate group increases ∂^{119}Sn falls. This movement in the isotropic chemical shift is attributable to increasing shielding of the tin nucleus by increasing electron density. A previous study (27) has shown that there are correlations between ∂^{119}Sn and the amount of electron density around the Sn nucleus, and confirms the direction of the change. This could be due to one of two reasons, either a transfer of charge density from the phosphate group to the tin, or contraction of the Sn d-orbitals due to electrostatic repulsion by the increasingly negatively charged phosphate groups, thus inducing an increase in the electron density around the nucleus.

In trying to resolve this dilemma it is useful to look further afield than the nmr spectra of these compounds. The first point of information is the mobility of the phosphate groups. In experiments performed by Clearfield and co-workers (28), metal hydrogenphosphates were found to rapidly exchange radioactively labelled phosphate rapidly from aqueous solution. There is also evidence to suggest that the pendant P-O bonds bend out to accommodate large guest species (28). Hence the phosphate group is mobile and relatively free to move suggesting that there are no strong covalent interactions within the layers. The tin Mossbauer spectrum of SnP also provides clues to the chemistry of the layer. The delta value is -0.361 0.005 mms^{-1} with a half height width of 1.05 mms^{-1}. These values resemble those for SnF_4 (29) which implies that the tin is highly ionic in character. Comparison with the Sn Mossbauer of SnO_2, ∂= -0.0015 mms^{-1}, with a large half height width of 2.52 mms^{-1} confirms the ionic nature of the tin centre in SnP. The nature of the interactions within the layers is, therefore, predominantly ionic suggesting that a transfer of charge causes the increase in electron density. If the increase in charge on the phosphate groups produced a contraction in the Sn 4d-orbitals, the increased ionic character of the tin would be seen as an opposite change in the isomer shift in the spectra.

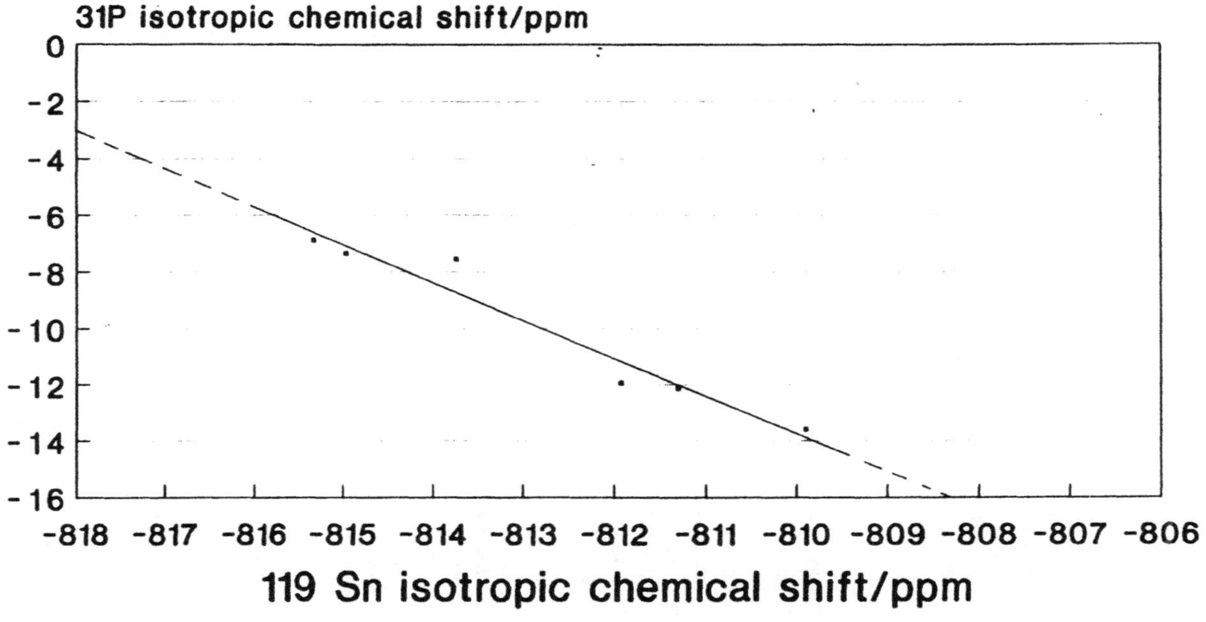

X= H, Ru(NH3)6, and protonated amines

Figure 3.2.7; Sn and P MAS NMR Correlations for Sn(IV)(PO4)2X2

4. Characterisation of Voids.

4.1 BACKGROUND.

The screening of a nucleus by its own electron density is susceptible to changes in the Van der Waals interactions of the atom with its environment. Hence, it is possible to use nmr to study adsorbed molecules entering the pores of a material or covering its surface. This has been done with ammonia (30), where there is a 15 ppm downfield shift in the ^{15}N chemical shift on going from gas to liquid and with various alkanes and alkenes (31), where the ^{13}C nmr peak can be shifted by up to 4 ppm on going from the gas to liquid. However, because of its vast sensitivity to density, ^{129}Xe nmr will here be discussed.

4.2 ^{129}XENON MAS NMR.

4.2.1 *Dependence of Chemical Shift on Density.* The chemical shift of the ^{129}Xe nucleus varies over a large range with density. With the chemical shift of the gas as a reference (i.e. $\partial Xe_{(g)} = 0$ ppm), the chemical shift of liquid xenon appears at 161 ppm downfield and solid xenon at 224 ppm (32). This large deshielding of the Xe nucleus is ascribed to the highly polarizable characterof the xenon atoms. The change in chemical shift of ammonia or the alkanes, it is 10 to 50 times less. The receptivity of the ^{129}Xe nucleus is around thirty times that of ^{13}C and is about twenty five times as abundant. A similar comparison to the ^{15}N nucleus, shows it to be some hundreds of time more receptive and about six times more abundant and like both of these nuclei, ^{129}Xe has no electric quadrupole moment. ^{129}Xe nmr has so far found three main applications. These are; 1) the physical characterisation of void spaces; 2) studying the location and electrostatic effects of ion exchanged cations (33) and 3) studying the nature of encaged particles and small chemisorbed molecules (34). It may also be used as a probe to ascertain the number of well formed supercages (35). As the first of these areas is most important to the field of mesoporous inorganic solids, this paper will mainly cover this topic but will also mention the effect of the valency of the metal counter-ion.

4.2.2 *Characterisation of Void Spaces.* The ^{129}Xe chemical shift has been found to consist of four components, expressed in the equation below;

$$\delta(Xe) = \delta_O + \delta_S + \delta_E + \delta(Xe-Xe)\rho_{Xe} \qquad 6$$

where ∂_O is the reference, ∂_E is a contribution from electrical fields generated by cations, $\partial(Xe\text{-}Xe)\rho_{Xe}$ is a contribution from Xe-Xe collisions and so is dependent on Xe density (ρ_{Xe}). The term ∂_S characterises the interactions of Xe with the walls of the cavity and takes the form $\partial(Xe\text{-wall})\rho_s$. The factor ∂(Xe-wall) is characteristic of the material under study and ρ_S is the apparent density of Xe in the cavity. Figure 4.2.1 shows the variation of chemical shift with the amount of Xe adsorbed. That the shifts for zeolites containing different amounts of ions of similar size show little variation is an indication that the contribution due to mean electric field is negligible (i.e. ∂_E 0). By extrapolating to zero Xe coverage, the term $\partial_S + \partial_E$ is determined but, as the second of these is negligible, the characteristic ∂_S is found. As the size of the counter ion increases, the mean free path of the Xe atom falls and the chemical shift increases (33). Studies have been carried out on microporous Y, A, L (36),

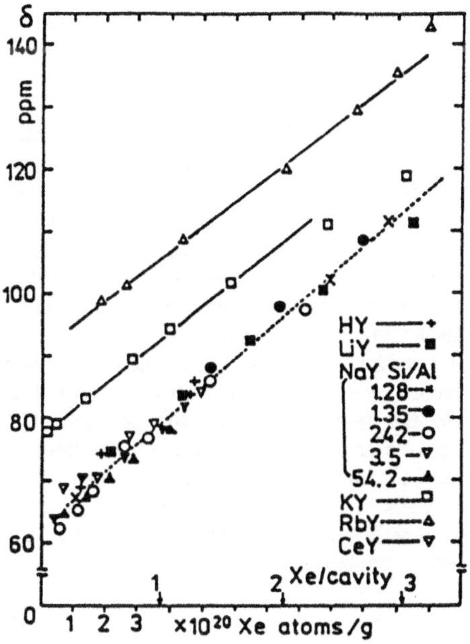

Figure 4.2.1;Variation of Xe Chemical Shift with Xe Pore Density for Zeolites containing Monovalent Ions.

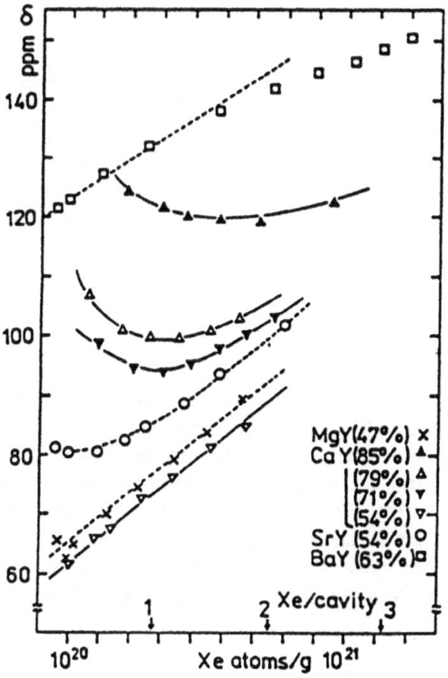

Figure 4.2.2; Variation of Xe Chemical Shift with Xe Pore Density for Zeolites containing Divalent Ions.

(36), ZSM-5, ZSM-11 erionite and offretite (37) zeolites and it has been found that the smaller the channels, the higher the chemical shift.

4.2.3 *Dependence of Chemical Shift on Cavity Environment.* Non-linear correlations of chemical shift and Xe uptake can occur when the material contains polyvalent (normally divalent) counter ions. This is illustrated in Figure 4.2.2. The deviations occur when the degree of exchange is high enough for the Xe-M^{n+} to dominate (50% exchange) and is due to the electric field effect of the divalent cations held in the supercages (36,38). The appearance of the nmr spectrum is deceptive in that the single peak is due to a coalescence of the peaks from Xe atoms interacting with the divalent ions in the lattice and with the parent structure. The nature of the variation and the heats of adsorption of Xe on divalent ion containing zeolites indicate that the Xe-M^{2+} interactions dominates. Therefore, at low Xe uptakes, each Xe atom must have a relatively long residence time on the M^{2+} adsorption site. The contribution of the divalent metal ion may be calculated by extrapolation to zero coverage of the plot of chemical shift and adsorbed Xe and then subtracting the term ($\partial_S + \partial_B$) for the material. As Xe coverage increases, ∂ falls because the residence time of each Xe atom on the cation decreases as rapid exchange occurs between atoms adsorbed onto the M^{2+} sites and the monovalent sites. As the Xe coverage becomes greater still, the Δ(Xe-Xe) term begins to dominate and so ∂ begins once again to increase. The effect of Mg^{2+} and Ca^{2+} on Xe chemical shifts has been calculated from their respective electric fields and has been shown to agree with experimental data (27).

4.3 ^{15}N MAS nmr

4.3.1 ^{15}N MAS nmr OF THE AMMONIUM ION WITH DIFFERENT STRENGTH CONJUGATE ACIDS. The isotropic chemical shift in the ^{15}N spectra ($\partial_{iso}{}^{15}N$) decreases with increasing net positive charge on the nitrogen centre (39) and so gives a convenient and accurate probe regarding the extent of protonation of the base. In the recorded spectra of the α-SnP-ammonium intercalation compound and that of $NH_4PO_4H_2$ there can be seen only one nitrogen environment. The reference for these spectra is the nitrate resonance of ammonium nitrate, the ammonium line of which comes at -353.3 ppm (40). The increase in $\partial_{iso}{}^{15}N$ (from -353.3 ppm to -351.6 ppm for $NH_4PO_4H_2$ and to 350.6 ppm for SnP) follows the increasing acidity of the counter-ion, nitric being a stronger acid than orthophosphoric which is in turn stronger than SnP.

5. Conclusions.

This paper has set out to show how modern solid state nmr techniques can throw light on the structure of poorly crystalline samples. From the wealth of work carried out on zeolites over the past ten years or so, much structural elucidation can now be successfully carried out. The future of solid state nmr continues to look promising as other nuclei become accessible and more frequently used (e.g. ^{71}Ga, ^{77}Se, and metal nuclei such as ^{59}Co, ^{95}Mo), as resolution increases and other methods of spectral interpretation (such as 2D nmr) become more prevalent. For instance, the two framework centres of Cloverite are both accessible to solid state nmr, as are all the framework metal nuclei of the series of materials recently patented by Mobil (41,42). This paper also has introduced the wide range of probe molecules, which will be of much interest to those who will be utilising mesoporous materials in the future.

References.

1. J.K.M. Saunders and B.K.Hunter, "Modern NMR Spectroscopy: A Guide for Chemists.", Oxford University Press, Oxford, 1989, and references therein.

2. R.S.Drago, Physical Methods in Chemistry, W.B. Saunders Company, London, 1977.

3. J.S. Waugh, L.M. Huber, U. Haeberlen, *Phys. Rev. Lett.*, 1968, ***20***, 180-2.

4. A. Pines, M.G. Gibby, J.S. Waugh, *J. Chem. Phys.*, 1973, ***59***, 569-90.

5. P. Mansfield, *J. Phys. C*, 1971, ***4***, 1444-52; W.K. Rhim, D.D. Elleman, R.W. Vaughn, *J. Chem. Phys.*, 1973, ***59***, 3740-59

6. L.M. Ryan, R.E. Taylor, A.J. Paff, B.C. Gerstein, *J. Chem. Phys.*, 1980, ***72***, 508-15.

7. N. Bloembergen and T.J. Rowland, *Phys. Rev.*, ***55*** (1955), 1679.

8. J.J. Papikse and T, Zoltai, *Am. Mineral.*, ***50*** (1965), 641; C.A. Fyfe, G.C. Gobi, J. Klinowski, A. Putnis, and J.M.Thomas, *J. Chem. Soc., Chem. Commun.*, 1983, 556.

9. N.C. Alma, G.R. Hays, A.V. Samoson and E.T. Lipmaa, *Anal. Chem.*, ***56*** (1984), 729.

10. G. Engelhardt and D. Michel, "High-Resolution Solid-State NMR of Silicates and Zeolites." Chp II, 2.1.3, pp 33-37, John Wiley and Sons, Chichester, 1987.

11. G. Engelhardt and D. Michel, "High-Resolution Solid-State NMR of Silicates and Zeolites.", John Wiley and Sons, Chichester, 1987, pp 75-83.

12. G. Engelhardt, D. Zeigan, H. Janke, D. Hoebbel and W. Wieker, *Z. Anorg. Allg. Chem.*, ***418*** (1975), 17.

13. G. Engelhardt, U. Lohse, E. Lipmaa, M. Tarmak and M. Magi, *Z. Anorg. Allg. Chem.*, ***482*** (1981), 49; J. Sanz and J.M. Serratosa, *J. Am. Chem. Soc.*, ***106*** (1984), 4790.13.

14. J.A. Pople, *Disscuss. Faraday Soc.*, ***34*** (1962), 7.

15. G. Engelhardt, R.Radeglia, H. Jancke, E. Lipmaa and M. Magi, *Org. Magn. Reson.*, ***5*** (1973), 561.

16. R. Wolf and R. Radeglia, *Z. Phys. Chem.*, ***257*** (1976).

17. N. Janes and E. Oldfield. *J. Am. Chem. Soc.*, ***107*** (1985), 6769.

18. G. Engelhardt and R. Radeglia, *Chem. Phys. Lett.*, ***108*** (1984), 271; R. Radeglia and G.Engelhardt, *Chem. Phys. Lett.*, ***114*** (1985), 28.

19. A.-R. Grimmer and G. Engelhardt, *Chem. Phys. Lett.*, ***106*** (1984), 263.

20. G. Engelhardt and U. Lohse, unpublished work.

21. J.M. Thomas, C.A. Fyfe, S. Ramdas, J. Klinowski, and G.C. Gobbi, *J. Phys. Chem.*, ***86*** (1982), 3061.

22. G. Alberti, U. Costantino, M.L. Giovagnotti, *J. Inorg. Nucl. Chem.*, ***41*** (1979), 643.

23. A.K. Cheetham, N.J. Clayden, C.M. Dobson, and R.J.B. Jakeman, *J. Chem. Soc., Chem. Commun.*, ***1986***, 195.

24. F. Taulelle, C. Sanchez, J. Livage, A. Lachagar and Y. Piffard, *J. Phys. Chem. Solids*, ***49*** (1988), 229.

25. N.J. Clayden, *J. Chem. Soc., Dalton Trans.*, ***1987***, 77

26. R.J.W. Adams, personal communication, 1990.

27. S.J. Blunden, D. Searle, and P.J. Smith, *Inorg. Chim. Acta*, ***116*** (1986), 131.

28. A. Clearfield, personal communication, 1990.

29. V.T. Goldanskii, V. Ya. Rochew, V.V. Khraphov, B.E. Dzevitskii, V.F. Sukherhoverov, *Isv. Sib. Otd. Akad. Nauk., Ser. Khim, Nauk*, ***1968***, 22

30. D. Michel, A. Germanus and H. Pfeifer, *J. Chem. Soc. Faraday Trans 1*, ***78*** (1982), 237.

31. D. Michel, W. Meiler, A. Gutsze and A. Wronkowski, *Z. Phys. Chem. (Liepzig)*, ***261*** (1980), 953.

32. T. Ito and J. Fraissard, *J. Chem. Phys.*, ***76*** (1982), 5225.

33. T. Ito and J. Fraissard, *J. Chem. Soc., Faraday Trans., 1*, ***83*** (1987), 451.

34. L. C. de Menorval, J. Fraissard, and T. Ito, *J. Chem. Soc., Faraday Trans., 1*, ***78*** (1982), 403.

35. M. Springuel-Huet, T. Ito and J. Fraissard, "*Structure and Reactivity of Modified Zeolites*", ed P.A. Jacobs, Elsevier, Amsterdam (1984), 13.

36. T. Ito, L.C. de Menorval, E. Geurrier and J. Fraissard, *Chem. Phys. Lett.*, ***111*** (1984), 271.

37. J. Fraissard, T. Ito, M. Springuel-Huet and J. Demarquay, "*New Developments in Zeolite Science and Technology, Proceedings of the 7th International Zeolite Conference.*", ed. Y.Muramaki A. Iijima and J.M. Ward, Elsevier, Amsterdam (1986), 393.

38. T. Ito and J. Fraissard, *J. Chim. Phys., Phys. Chim. Biol.*, ***83*** (1986), 441.

39. M. Comeau, M.-T. Beraldin, E.C. Vaulthier, and S. Fliszar, *Can. J. Chem.* ***63*** (1985). 3226.

40. D.C. Apperley, personal Communication, 1990.

41. International Patent: PCT # WO 91/11390 A (1991).

42. U.S. Patent: US# 50,98,684 (1992).

SOLID STATE ^{27}AL NMR STUDIES OF LDH-INTERCALATES

M. Vucelic and W. Jones
Chemistry Department, University of Cambridge, Lensfield Road, Cambridge CB2 1EW, United Kingdom

ABSTRACT ^{27}Al NMR - through conventional (e.g. magic angle spinning) and modified (e.g. proton decoupling) techniques - has been used in the study of organic-containing layered double hydroxides. The intercalates were synthesized via a two-stage process; preparation of the 'parent' material followed by intercalation via the calcined-precursor route. Three different organic anions have been incorporated: benzoate, terephthalate and 1-naphthoate. An aim of the NMR study was to determine the location of the lattice aluminium (i.e. whether in four or six co-ordinate sites) as well to monitor the arrangement of the anions around such sites. The following NMR techniques were used: ^{27}Al MAS, ^{27}Al ^{1}H decoupling, ^{27}Al CPMAS and ^{27}Al 2-D Nutation. In this paper 1-D NMR has been used primarily for the determination of aluminium coordination and the relationship between the aluminium and neighbouring protons. 2-D NMR techniques were used to detect the presence of octahedral aluminium in differing environments. Evidence for the existence of five-coordinate aluminium was also obtained.

1. INTRODUCTION

Layered solids are so-called because the intralayer forces binding atoms together are much stronger than the interlayer forces. As a result, the intercalation of guest species into the galleries between the host layers - which remain essentially undistorted - is relatively easy [1,2]. Within this family of solids (which includes graphite and cationic clay minerals) are the layered double hydroxides (LDHs) or the so-called anionic clays [3-5].

The structure of an LDH is similar to that of brucite, $Mg(OH)_2$, [6-9]. Magnesium is octahedrally coordinated to six oxygens in the form of hydroxide; the octahedral units then, through edge sharing, form infinite sheets. Individual sheets are then stacked via hydrogen bonding to yield the layered three-dimensional solid.

When some of the magnesium ions in brucite are isomorphously replaced by aluminium, the Mg^{2+}-Al^{3+}-OH layer carries a positive charge. If appropriate amounts of anions are intercalated, the structure becomes electrically neutral. In nature these anions are frequently carbonate. Species within the interlayer are relatively mobile and there is an interchange of water and carbonate through the formation of new hydrogen bonds both between themselves and with the hydroxyl groups of the Mg-Al-OH layer [6-9].

We discuss here the structural changes which occur when LDHs are intercalated with organic anions using the calcined precursor method reported by Chibwe and Jones [10]. We have used principally ^{27}Al NMR [11-14] as the structural probe.

C.A.C. Sequeira and M.J. Hudson (eds.),Multifunctional Mesoporous Inorganic Solids, 373–378.

2. EXPERIMENTAL

The preparation of the intercalates went through two stages - synthesis of the LDH followed by calcination and rehydration [10].

Preparation of Mg-Al-CO_3 LDH.
A solution containing 0.1 moles of $Mg(NO_3)_2{\cdot}6H_2O$ and 0.05 moles of $Al(NO_3)_3{\cdot}9H_2O$ in 70 ml distilled water was added with vigorous stirring to a solution of 0.35 moles NaOH and 0.09 moles of Na_2CO_3 (anhydrous) in 100 ml distilled water. The addition was over a period of 4 hours at 25 ^oC at a constant pH of 10. The heavy slurry which resulted was crystallized at 65 ^oC for 18 hours followed by cooling and washing several times with water. The sample was dried at 125 ^oC in an oven for 14 hours.

Intercalation of organic anions into the LDH.

The LDH was calcined in air at 450°C for 18 hours. A three necked round bottom flask equipped with a reflux condenser, thermometer, magnetic stirrer and electric heating mantle was charged with 100mls of distilled/deionized water. The system was heated up to 100°C and constantly bubbled with N_2 for an hour. 0.1M solutions of benzoic, terephthalic and naphthoic acids were independently added to water and left for another 40 minutes in N_2 atmosphere .1g of calcined material was then added to the organic solution and left under a N_2 atmosphere at 100 ^oC for 20 minutes.

The slurry was filtered and washed (under N_2) using boiled distilled water and the product left to dry in an oven for 18 hours at approximately 100 ^oC. This procedure follows closely that of Chibwe and Jones [10] except that instead of the acid salt, pure acids were used. The latter have a very low solubility in water.

The experiments were carried out on a 400 MSL Bruker machine, using standard 5-mm NMR tubes, a working frequency of 104.262 MHz and spectral width of 50 KHz. Chemical shifts are reported in ppm relative to $AlCl_3$ for the 1-D ^{27}Al measurements and $[Al(H_2O)_6]^{3+}$ for the 2-D measurements.

3. RESULTS AND DISCUSSION

3.1 1-D Aluminium NMR

The co-ordination number of the aluminium atom in AlO_n polyhedra dominates the value of the isotropic ^{27}Al chemical shift which is observed - tetrahedrally co-ordinated aluminium (AlO_4) between +50 and +80 ppm and octahedrally in the region -10 to +20 ppm. Resonances between +30 to +40 ppm have more recently been ascribed to five-coordinated aluminium (AlO_5) units [11-14]. These well separated shifts permit the co-ordination number of the AlO_n polyhedra present in layered double hydroxides materials to be readily determined from ^{27}Al NMR spectra (even under circumstances where relatively broad lines and uncorrected chemical shift data are available).

The ^{27}Al MAS spectra of the various intercalates are shown in Figure 1. For the benzoate-LDH the presence of octahedral aluminium is clearly indicated, resonating at 7.1 ppm. Although a rotation of 3 kHz was applied to the sample - a frequency usually sufficient for observing both four and six co-ordinated aluminium - only a slight indication of tetrahedral aluminium at ~80 ppm (a low-field shift characteristic of layered AlO_4 polyhedra [11-14]) is seen. For the terephthalate-LDH a similar spectrum is observed with a peak at 7.9 ppm. No signal for tetrahedral aluminium is observed (although as will be shown later such a species is present). There is a slightly improved signal-to-noise ratio for the terephthalate-LDH indicating a faster spin-spin relaxation, suggestive of a more rigid-type matrix. Finally, for the naphthoate-LDH a resonance at 7.5 ppm is observed with a significantly lower intensity than seen for the benzoate and terephthalate intercalates. There is no trace of four co-ordinated aluminium in this spectrum. From such spectra little further structural information can be obtained.

Heteronuclear high power proton-decoupling is a widely used technique for line narrowing in materials with pronounced dipolar coupling to protons. In this work we have attempted to eliminate the H-Al dipolar interaction using a modified proton decoupling technique, see Figure 2. All three LDH intercalates indicate the presence of octahedral aluminium at 8.8, 8.6 and 8.9 ppm, for the benzoate, terephthalate and naphthoate respectively. Significantly, in the proton decoupled spectra, there is stronger evidence for tetrahedrally co-ordinated aluminium, with resonances at 59.3, 58.9 and 63.2 ppm, respectively.

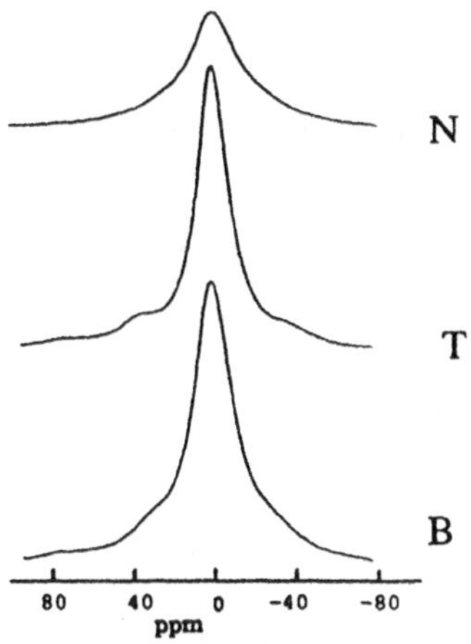

Figure 1. ^{27}Al MAS spectra of (B) LDH + benzoate, (T) LDH + terephthalate, (N) LDH + naphthoate.

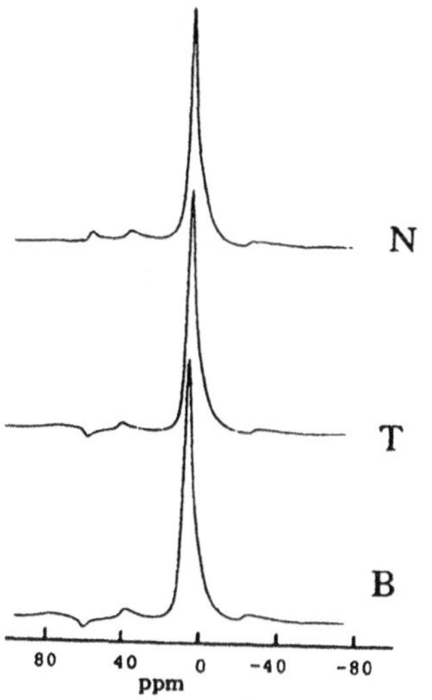

Figure 2. ^{27}Al ^{1}H decoupled spectra of (B), (T) and (N) - see Figure 1.

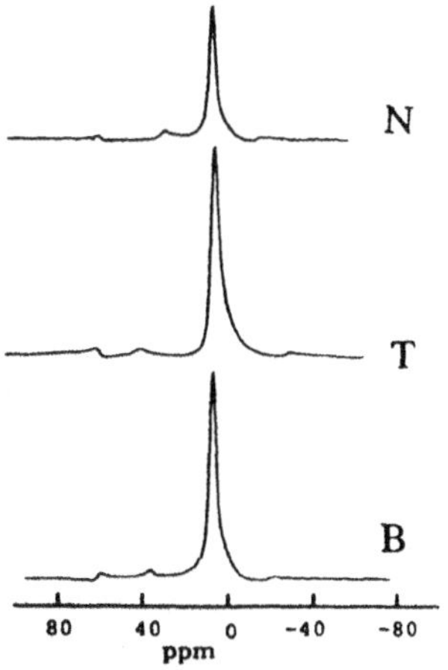

Figure 3. ^{27}Al CPMAS spectra of (B), (T) and (N).

Such differences in the ^{27}Al and the proton-decoupled ^{27}Al spectra may arise because firstly the amount of tetrahedral aluminium may be too small to be observed directly through conventional ^{27}Al NMR. Secondly, if the aluminium is in a five-fold environment, it may be only marginally visible by normal ^{27}Al NMR but much more visible when proton decoupled.

The proton decoupling experiments described above indicate an influence of the protons on the aluminium. In order to determine this influence an ideal method is cross polarisation magic-angle-spinning NMR (CPMAS). This technique is generally applied to dilute spin systems, (e.g. ^{13}C, and therefore not usually applied to the 100 % abundant ^{27}Al) and in line narrowing processes - and therefore appropriate for quadropolar nuclei such as ^{27}Al.

Figure 3 illustrates the observed spectra, with both tetra and octahedral aluminium present. Peaks at 9.2, 10.8 and 9.6 ppm, respectively, for LDH pillared with benzoic, terephthalic and naphthoic anions, are clearly associated with six co-ordinated aluminium. In the CPMAS spectra resonances for tetra co-ordinated aluminium are seen at 62.3, 59.7 and 60.4 ppm for the terephthalate, benzoate and naphthoate, respectively.

^{27}Al studies suggest that aluminium in non-framework species is preferably octahedrally co-ordinated by oxygen, although some evidence for the presence of tetrahedral has appeared [14]. This corresponds to the resonances at 70 to 80ppm, which we observed for our samples in ^{27}Al MAS spectra. The position of the AlO_4-signal at 55 to 60 ppm in the decoupled spectra and in the ^{27}Al CPMAS spectra of the organic-intercalated LDHs is similar to that seen for tetrahedral aluminium in framework aluminosilicates [12].

3.2 2D NMR spectra

Table 1 summarises the data obtained using two-dimensional (2D) Nutation experiments (after Samoson and Lippmaa [12]). The corresponding spectra are shown in Figure 4.

Table 1 Resonances observed in the 2D experiments for the various LDH-intercalates

Al Species	**Tetrahedral**		**Five-fold**		**Octahedral**	
	F1 kHz	F2 ppm	F1 kHz	F2 ppm	F1 kHz	F2 ppm
Benzoate-LDH	80	80	80	40	80	10
					130	10
Terephalate-LDH	140	80	140	40	140	10
					170	10
					220	10
Naphthoate-LDH	80	75	80	35	80	10
					130	10

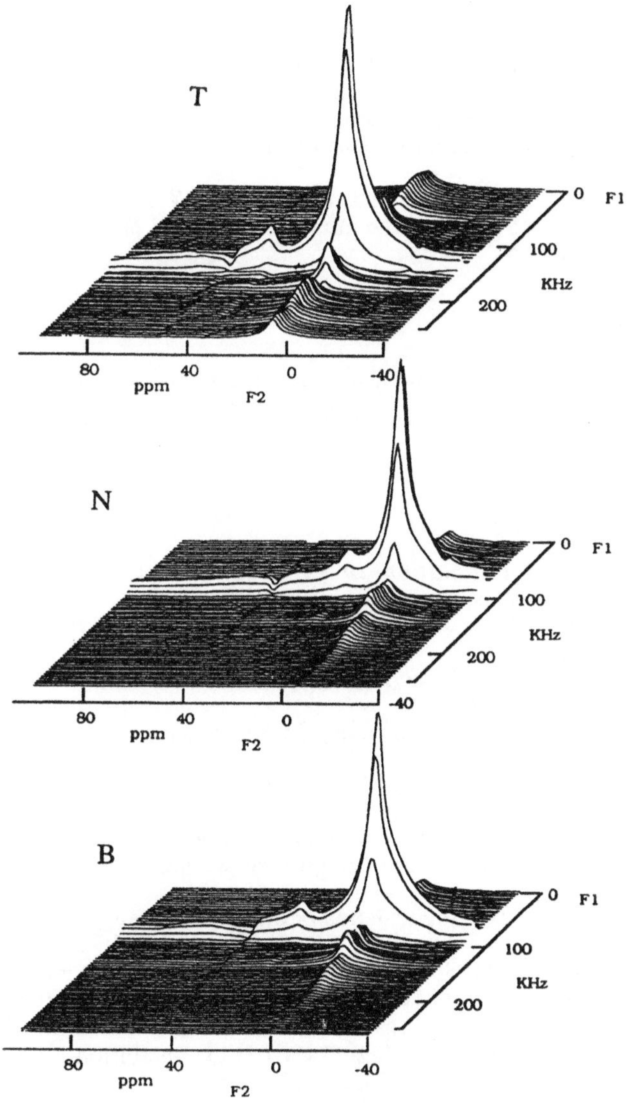

Figure 4 2-D data for the LDH containing benzoate, naphoate and terephthalate anions.

It emerges from these spectra that the local environments of the AlO_6 units are not the same - indicated by the different values of F1. Furthermore, there is strong evidence for the existence of AlO_5 units in all three samples. The behaviour of the terephalate-LDH, with three types of octahedral environments with quite different F1 values for the two environments seen for napthoate and benzoate, may result from the fact that it is incorporated into the interlayer space in form of the dianion.

We continue to investigate these materials using powder X-ray diffraction and IR to try and characterise further the structure of the intercalates and the processes which occur during calcination and rehydration.

ACKNOWLEDGEMENTS

We appreciate the provision of NMR facilities provided by Professor Dr D. Vucelic of the Institute of General and Physical Chemistry, University of Belgrade.

REFERENCES

1. S. A. Solin (1986) in *Intercalation in Layered Materials*, 145-154, M. S. Dresselhaus, Ed, Plenum Press, New York
2. M. S. Whittingham and A. J. Jacobson (1982) Eds, *Intercalation Chemistry*, Academic Press, London
3. W. T. Reichle (1985) Catalytical reactions by thermally activated, synthetic, anionic clay minerals, *J. Cat.*, **94**, 547-557.
4. W. T. Reichle (1986) Anionic clay minerals, *Chemtech*, 58-63.
5. W. T. Reichle (1986) *Solid State Ionics*, **22**, 135-142.
6. E. Manasse (1915) *Atti Soc. Foscana Sc. Nat., Proc.Verb.*, **24**, 92-103.
7. C. Frondel (1941) *Amer. Mineral.*, **26**, 295-315.
8. R. Allmann (1968) Crystal structure of pyroaurite, *Acta Cryst.*, **B24**, 972-977.
9. G. Brown and M. C. Gastuche (1967) Mixed magnesium-aluminium hydroxides, *Clay Miner*, **7**, 193-215
10. K. Chibwe and W. Jones (1989) Intercalation of organic and inorganic anions into layered double hydroxides, *J. Chem. Soc, Chem. Commun*, 926-927.
11. A. Samoson and E. Lippmaa (1983) Excitation phenomena and line intesities in high resolution NMR powder spectra of half-integer quadropolar nuclei, *Phys. Rev*, **B28**, 6567-6570.
12. A. Samoson, E. Lippmaa and M. Magi.(1986) High resolution ^{27}Al NMR of aluminosilicates, *J. Am. Chem. Soc.*, **108**, 1730-1735.
13. L. B. Alemany and G. W. Kirker (1986) First observation of 5-coordinated aluminium by MAS 27Al NMR in well characterized solids, *J. Am. Chem. Soc.*, **108**, 6158-6162.
14. M. C. Cruickshank, L. S. Dent-Glasser, S. A. Barri and J. F. Poplett (1986) Penta-co-ordinated aluminium: a solid-state ^{27}Al study, *J. Chem. Soc, Chem, Commun.*, 23-24.

CHARACTERISATION OF POROUS SOLIDS USING SMALL-ANGLE X-RAY AND NEUTRON SCATTERING

A.N.NORTH and J.C.DORE
Physics Laboratory
University of Kent at Canterbury
Canterbury Kent
CT2 7NR
United Kingdom

Abstract Small-angle scattering is introduced as a technique for characterising the structure of porous solids over the size range 1nm-2μm. Scattering theory and current methods of analysis are defined.Instrumentation and experimental procedures using synchrotron radiation and reactor neutrons are discussed. The methods are illustrated with reference to a broad range of materials of varying pore size, pore shape, surface texture and composition.

1 Introduction

Small-angle scattering of X-rays (SAXS) and Neutrons (SANS) is a powerful technique for studying structural features between 1 and 30nm[1,2]. The developing technique of Ultra-small-angle scattering (USAS) has extended the upper limit of the size range to 2μm[3]. Whilst the principle of SAS has been known for many years the increasing flux of primary beams (both X-rays and neutrons) at central facilities linked with an improvement in instrumentation, has contributed to a rapid increase in utilisation over the last decade. The major developments in use of SAS techniques have been in metallurgical, polymer and biological science but there have always been applications to other areas and there is a growing interest in the study of porous materials by this method in spite of the increased complexity and range of the micro-structural features.

Traditional methods of structure determination, based on optical and electron microscopy are often not applicable to porous solids due to the irregularities in the visual image; microscopy gives a localised two-dimensional image of a three-dimensional structure. In order to obtain an overall picture of a bulk sample many images must be obtained. For completeness the internal structure must also be observed, the sample under investigation being cleaved, a process which may in itself destroy the very structure of the material. Gas adsorption measures only pore volume and surface area. Scattering techniques on the other hand probe the 3-d structures directly. SAS may be used to interrogate non-destructively a large volume of material, providing statistically meaningful measurements of particle and/or pore size distributions and surface area/surface structure on a length scale up to a few microns.

The aim of this paper is to discuss the application of SAS to the study of porous materials. The theory of small-angle scattering will be presented starting with simple models and extending to complex fractal systems. Instrumentation for both X-ray and

C.A.C. Sequeira and M.J. Hudson (eds.),Multifunctional Mesoporous Inorganic Solids, 379–400.

neutron scattering will be discussed. Finally, examples of scattering studies of a variety of porous materials will be presented and discussed.

2. Theory

The theory of small-angle scattering has been widely presented in various review articles[1,2,4] and only a brief description will be given here. One of the most important parameters in small-angle scattering is the scattering length density $\rho(\underline{r})$. SAS is a low resolution technique, that is it can not be used to study structural features of less than about 10Å. Such that individual atoms within a sample can not be observed. The smallest volume that can be observed therefore contains a number of atoms. The scattering length density is the local average of the atomic factors in this volume element and as such represents its scattering strength. In the case of X-ray scattering, the X-rays are scattered by electrons and the form factor is proportional to the number of electrons within an atom. Therefore the electron density is a convenient measure of the scattering potential such that

$$\rho_x(\underline{r}) = \frac{\int n_e(\underline{r}) d^3\underline{r}}{\int d^3\underline{r}} \tag{1}$$

where n_e is the number of electrons in a volume dV.

In the case of neutron scattering, the neutrons are scattered by the atomic nuclei and the coherent scattering lengths b, vary as a function of atomic number (and even isotope) in an apparently irregular way.The integral can now be replaced by a summation over individual scattering centres such that the neutron scattering density is given by

$$\rho_n(\underline{r}) = \frac{\sum b_i}{V} \tag{2}$$

where the summation is over all nuclei in volume V.

The scattering amplitude $A(\underline{Q})$ for a scattering vector $\underline{Q}$ may be written as

$$A(\underline{Q}) = \int_V \rho(\underline{r}) \exp(i\underline{Q}.\underline{r}) \, d^3\underline{r} \tag{3}$$

where the magnitude of the scattering vector Q is defined by the experimental conditions as

$$Q = \frac{4\pi}{\lambda} \sin\frac{\theta}{2} \tag{4}$$

λ is the wavelength of the incident beam and θ is the angle of scatter. $\rho(r)$ is the distribution of scattering length densities across the sample and is therefore a measure of structural changes across the sample. The intensity profile I(Q) measured in a scattering experiment is simply given by

$$I(Q) = K|A(Q)|^2 \qquad (5)$$

where K is a normalisation constant, depenedent on experimental geometry and sample thickness.

The SAS intensity therefore contains information about $\rho(\underline{r})$ and hence the inhomogeneities within the material. It is however rarely possible to use inverse transform methods to determine $\rho(\underline{r})$ due to limitations in the Q-range of the experimental measurements[5]. The interpretation of the SAS intensity is therefore often based on the use of specified models. Outlined below are some of the models that have proved particularly useful in interpreting scattering data from porous solids.

In the case of isotropic materials containing well defined particles equation 2 may often be written as

$$I(Q) = N\, P(Q)\, S(Q) \qquad (6)$$

where N is the number of particles involved in the scattering process. P(Q), the particle form factor, represents the scattering from an individual particle and S(Q), known as the structure factor results from interference in the coherent scattering from individual particles.

Isolated particles

Consider a particle of arbitrary shape and scattering density $\rho(r)$ embedded in a homogeneous medium of scattering density ρ_m (figure 1) then the small angle scattering intensity for an isolated particle may be written as

$$I(Q) = \left| \int_V \left[\rho(\underline{r}) - \rho_m\right] \exp(i\underline{Q}.\underline{r})\, d^3\underline{r} \right|^2 \qquad (7)$$

where V is the volume of the particle. This equation is readily simplified if the particle is homogeneous, that is $\rho(r)$ is constant over all the particle and a well defined shape. For instance if the particle is a homogeneous spherical particle of radius R, such that $\rho(r)=\rho$ for $r<R$ then equation 7 becomes

$$I_P(Q) = 9\, (\Delta\rho)^2\, V^2 \left[\frac{\sin QR - QR\cos QR}{Q^3 R^3}\right]^2 \qquad (8)$$

where $\Delta\rho = (\rho_m - \rho_p)$ is known as the contrast. Similar expressions may be obtained for particles of other shapes i.e. cylinders, ellipsoids, sheets cubes etc.. Figure 2 shows the scattering that may be expected from a homogeneous sphere. The well defined minima are dependent on the particle radius. The minima moving to lower Q-values as the particle size is increased and to higher Q-values as the particle size is decreased.If there is no scattering contrast, between the particle and the surrounding medium ($\Delta\rho=0$) then no scattering will be observed. It should also be noted that there is no difference between the scattering due to a silica particle suspended in a vacuum ($\Delta\rho = (\rho_{Si} - \rho_v)$) and the scattering from an empty void of the same dimensions embedded in a solid block of silica ($\Delta\rho = (\rho_v - \rho_{Si})$). This arises because the scattering intensity depends only on size, shape and $(\Delta\rho)^2$. It can therefore be seen that the scattering from pores can be analysed using the formalism originally derived for particles and is often referred to as Babinet's principle

Several approximations may be made to the scattering profile from particles. In terms of porous solids the most important approximation is probably that due to Porod[1,2,4]. Porod's law as the approximation is known shows that for particles with well-defined

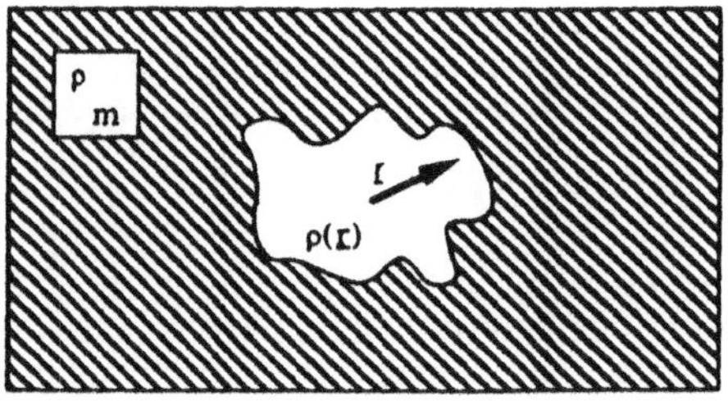

Figure 1 A particle of arbutrary shape and size with a scattering density ρ(r) embedded in a homogeneous medium of scattering density ρ_m

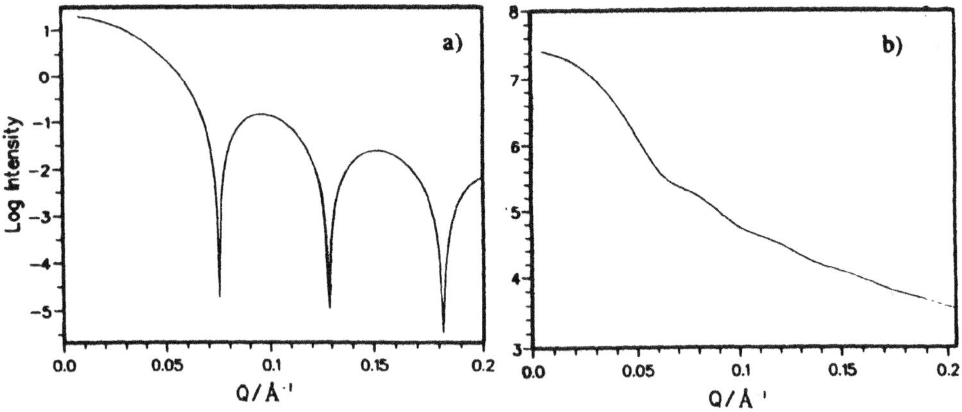

Figure 2. a)The theoretical scattering form a homogeneous spherical particle of radius 60Å. Form factors for different shaped homogeneous particles give a similar scattering profile.

b) The introduction of a spread of pore sizes results in a supression of the maxima and minima.

smooth boundaries that at high Q-values

$$I_{Q\to\infty} = (\Delta\rho)^2 \frac{2\pi S}{Q^4} \qquad (9)$$

where S is the surface area of the particle or pore.

Pore size distribution

If a material has a range of pore sizes then each pore will contribute to the measured scattering profile following the form of equation 7. So that the total observed scattering will be given by

$$I(Q) = \int_{r_1}^{r_2} w(r)\ P(Q.r)\ dr \qquad (10)$$

where w(r) is the number distribution of particles of size r, and P(Q,r) is the scattered intensity from a particle of size r at the Q-value of the measurement. r_1 and r_2, are respectively, the lower and upper limits of the pore size distribution. Figure 2 shows the effect of polydispersity on the scattering profile from spherical pores, the well defined maxima and minima are washed out and for a broad pore size distribution (PSD) the scattering profile will become a smooth curve. It is now the width of the scattering profile that gives an indication of the pore sizes, a broad scattering profile is indicative of small pores whilst a narrow profile suggests large pores.

Interference

When the number of particles/pores per unit volume is high, there is an additional contribution to the scattered intensity. This arises from the interference of the scattering from individual pores such that the observed scattering intensity may be written as

$$I(Q) = N\ P(Q)\ S(Q) \qquad (11)$$

where N is the number of pores/particles in the scattering volume and S(Q) is the inter-particle structure factor[4] given by the transform of the centre-centre pair correlation function g(r),

$$S(Q)-1 = \frac{4\pi}{\rho_o} \int r\left[g_c(r)-1\right] \sin Qr\ dr \qquad (12)$$

where ρ_o the pore number density. The pair correlation function, g(r), is a measure of the probability of finding another pore at a distance r from the centre of a chosen pore.

Fractal Structure Factors

Fractal-related properties have been found to exist in many porous systems[6,7] and the appropriate scattering regimes should be discussed. The subject of fractals is a complex topic beyond the scope of this paper however it is necessary to introduce some features in order to explain the SAS behaviour for different scattering regimes. In general, two different types of fractals are found in porous solids, surface fractals and mass fractals both giving rise to their own scattering characteristics[7,8]

A simple definition of a fractal system is that some of the spatial features do not scale according to Euclidean geometry[9]. For instance consider a length of coastline as shown in figure 3a. If the coastline is measured with a gauge of size L, then the coastline will have a length, l given by

$$l = L N_L \tag{13}$$

As the guage with which the coastline is measured comes down (i.e. L is decreased) it is to be expected that the length of the coastline would increase and eventually L would become sufficiently small that a limit would be reached for l.. However, Richardson[10] has shown that the length of the coastline approaches infinity as L approaches zero since there is structure on all length scales. The length of the coastline therefore diverges as

$$l = L^{D_s} N_L \tag{14}$$

where D_s is a non integral exponent greater than one, that is a fractal dimension. Is the interpretation of D_s as a dimension reasonable? Consider the length (l) of a rope for which it is clear that

$$l = L^1 N_L \tag{15}$$

or the area (A) of a floor for which

$$A = L^2 N_L \tag{16}$$

From these last two equations it can be seen that the measurement of interest depends on L raised to the topological dimension of the object. It is therefore reasonable to interpret D_s as a dimension [9].

Surface Fractals

A surface fractal may be considered to be the two dimensional version of the coastline above. The surface of a particle of radius r_o is given by

$$A = A_o r_o^{D_S} \tag{17}$$

where A_o is a constant and D_S is a dimension which, for smooth surfaces, would have a value of 2. However in a porous material large pores may have on their surfaces small-reentrant regions, which in turn have small irregularities etc. down to the atomic scale. Such pores can be visualised as fractal systems in which $2< D_s<3$.

Mass Fractals/Pore Fractals

Mass fractals consist of basic building blocks such that the mass of the object inside a sphere of radius r increases according to

$$M = M_o\, r^{D_m} \tag{18}$$

where M is the mass, M_o a constant and D_m the mass fractal dimension with $1<D_m<3$. A common example of a mass fractal is a system of aggregated particles such as colloidal silica gels. The network of pores running through a porous material can look somewhat similar to that of the aggregated system (figure 3b). By Babinet's principle the scattering obtained would be similar. Since by definition a pore has little or no mass, equation 18 may be re-written in terms of the volume and the fractals within porous materials referred to as pore fractals such that

$$V = V_o\, r^{D_p} \tag{19}$$

where V is volume, V_o is a constant and the pore fractal dimension is D_p.

Whilst the mathematical definition of a fractal requires the fractal to exist over all length scales[9] in a practical situation the size range over which a system exhibits fractal properties will be restricted. For instance in the case of a particle with a fractal surface, the

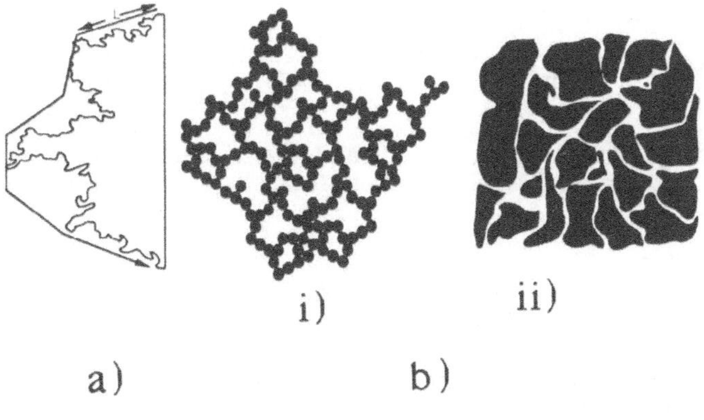

Figure3. a) The length of a coastline will vary depending on the guage used in measurement
b) An aggregated system of particles (i) will look similar to that of a network of pores running throughout a porous material (ii).

fractal surface could not exist on a length scale greater than that of the particle. It is possible that some porous systems exhibit both pore and surface fractal properties and these may be over different length scales[11]. The surface fractals would exist inside the pores whilst the mass fractal would occur due to an 'agglomeration' of pores. Both fractal regimes would be observed over different Q-ranges in the SAS profile.

Fractal structure factors

The theory of scattering from fractals has been covered in some depth elsewhere [12]. Bale and Schmidt give the structure factor for surface fractals to be [13]

$$S(Q) = \pi N \rho^2 \, \Gamma(5\text{-}d) \, \frac{\sin \frac{\pi(D_s-1)}{2}}{Q^{6-D_s}} \tag{20}$$

in d Euclidean dimensions, where N is a constant characteristic of the fractal boundaries [13], ρ is the scattering density of the sample, $\Gamma(x)$ is the gamma function and D_s is the fractal dimension of the boundary surface. If the limiting sizes of the surface fractal regime are well separated then it follows that

$$I(Q) \propto Q^{-k} \tag{21}$$

where $k=6-D_s$. If $D_s=2$ then there is a smooth interface and equation 21 is equivalent to Porod's law (equation 9).

For mass/pore fractals, Teixeira[14] has shown that the structure factor may be written as

$$S(Q)-1 = D_m \, \Gamma(D_m-1) \frac{\sin\left((D_m-1)\tan^{-1} Q\zeta\right)}{(QR)^{D_m} \, (1+(Q\zeta)^{-2})^{(D_m-1)/2}} \tag{22}$$

where $\Gamma(x)$ is the gamma function, ζ is a length that characterises the size of the mass fractal and D_m is the mass fractal dimension. In the limit as ζ approaches infinity then equation 22 becomes

$$I(Q) \propto Q^{-D_s} \tag{23}$$

over the range $\zeta^{1} << Q << R^{1}$. Whilst at high Q-values S(Q)=1 so that the scattering from the form factor dominates the scattered intensity at high Q-values.

Figure 4a shows the scattering that would be obtained from a pore fractal formed from an interconnected group of spherical pores of varying sizes. The major difference between the fractal and non-fractal case (figure 2b) is the increased scattering at low Q-values. Figure 4b shows the type of scattering profile that could be expected from a porous solid showing both pore and surface fractal properties.Measurements of the SAS intensity therefore reveals information about the $\rho(\underline{r})$ distribution which can be interpreted by the appropriate model description.

Instrumentation

The basic layout of a small angle scattering instrument is shown in figure 5. The monochromatic input beam is collimated to illuminate a finite beam spot at the sample position and the scattered radiation is measured by a one or two-dimensional multidetector. The sample thickness is usually optimised to maximise the scattering from the sample. The scattered intensity varies with the thickness t as

$$I \propto t\, e^{-\mu t}$$

where μ is the absorption coefficient of the sample, such that the maximum scattered intensity is achieved for an optimum thickness,t_{opt} [1]

$$t_{opt} = \frac{1}{\mu}$$

This restricts the samples to a typical thickness of 1-2mm for X-rays but larger values may occasionally be used for neutrons. The choice of wavelength to be used in small-angle scattering experiments is an important consideration in order to achieve the required Q-range, resolution and scattered intensity. The scattered intensity will be affected by changes in the absorption coefficient with wavelength as well as the incident spectrum. The longer wavelengths spread the scattering pattern over larger angles and so give a higher Q-resolution for a fixed sample to detector distance. In the case of neutrons typical wavelengths used are 4-15Å, whilst for X-rays the most commonly used is Cu-K_α radiation, from an X-ray tube, of wavelength 1.54Å. Larger wavelengths are rarely used in SAXS due to the sharp increase in absorption with increasing wavelength thus decreasing the optimum sample thickness and the maximum scattered intensity. Under favourable conditions the use of Chromium-K_α radiation (λ=2.29Å) is feasible.

Laboratory based SAXS

The discussion of scattering profiles assumes that the incident X-ray beam and detector system have point source and resolution characteristics. In the case of laboratory-based SAXS instruments which use X-ray tubes (usually with copper anodes) pinhole characteristics are rarely used due to the low incident beam intensity and hence very low count rates. The simplest laboratory based systems [1] use slit collimation in which two parallel slits produce beam collimation; the narrower the slits and the larger the space between them the lower the intensity but the higher the resolution. A third slit is frequently used to reduce the parasitic scattering from the first two slits. Many other specialised systems have been developed for specific experimental conditions [1], all of which attempt to find the best compromise between intensity and resolution for each specific experiment. The major problem encountered with laboratory-based SAXS is the smearing of the scattering profile due to the slit collimation and is often referred to as slit smearing. The measured SAXS profile is a convolution of the form of the aperture and the profile of the required scattering curve. This leads to a suppression of maxima and minima in the observed distribution and is of particular importance at

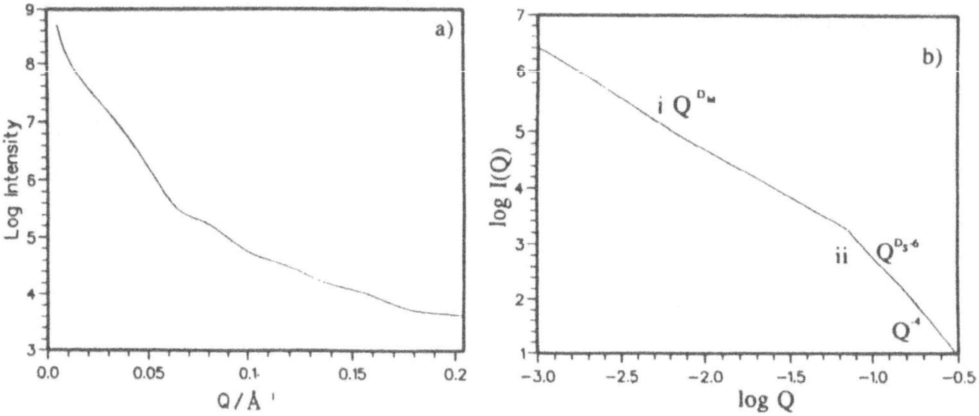

Figure 4. a) The theoretical SAS intensity from a pore fractal with a mean pore size of 60Å. b) If both pore and surface fractlas exist within a material the scattering profile may show two distinct regions due to i)the pore fractal and ii) the surface fractal structures.

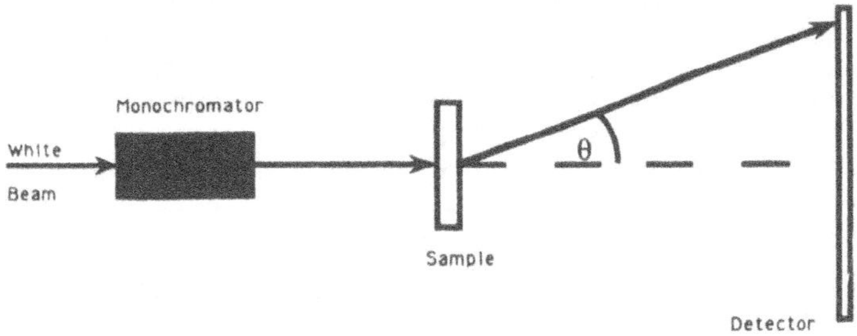

Figure 5. Schematic layout of a small-angle scattering instrument

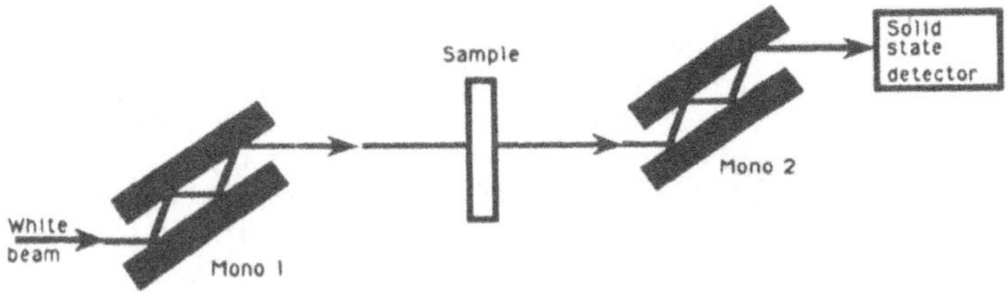

Figure 6. Schematic diagram of the USAXS instrument at the SERC Daresbury laboratory, Warrington UK.

the smallest-angles. Complex mathematical deconvolution routines have been developed to recover the required scattering curve from the smeared profile and many appear to give highly satisfactory results [15]. The effective Q-range available in laboratory based SAXS experiments depends critically on the short sample-to-detector distance and the resolution required but is in the region of $10^{-2} \leq Q \leq 0.3Å^{-1}$ depending on the experimental conditions. Such a Q-range allows studies of systems with structural features in the range of 10Å to 20nm.

Small-angle Neutron scattering

There are many small-angle neutron scattering instruments in operation and the performance details are well documented along with the general experimental requirements [16]. A mechanical velocity-selector is used to produce an incident wavelength at the sample of 5-15Å with a distribution of 5-10%. The instruments have a standard resolution imposed by the constructional features of the detector system. The detector system is typically a 64 by 64 cm^2 surface with $1cm^2$ cells with a variable sample-to-detector distance of 2-40m giving an overall-Q-range of 10^{-3} to $0.3Å^{-1}$. The incident beam being typically a few cm^2 in dimension the scattering angles are well defined except for regions very close to the beam stop. Problems may however arise from the wavelength spread and deconvolution techniques will often be required. SANS on a pulsed neutron source has different criteria which have been covered elsewhere[17]. The Q-ranges available in SANS allow structural features in the range of 10Å to 100nm to be investigated.

Synchrotron-based SAXS

Synchrotron-based small-angle X-ray scattering systems have undergone major development over the last 10 years [18]. Synchrotron radiation gives a flux at least two orders of magnitude greater than the best laboratory based X-ray generators. A silicon monochromator is used to select the required wavelength from the white beam with a bandpass $\delta\lambda/\lambda = 10^{-3}$. Relatively large sample-to-detector distances of 0.5-5m allow a Q-range of $5x10^{-3}$ to $0.3Å^{-1}$ giving real space measurements on systems with features in the range 1 to 50nm.The high incident flux and low angular divergence of the synchrotron beam give high count rates with good resolution. Linear- or area- detectors (of 0.3mm resolution) are usually available such that the convolution effects discussed under laboratory systems are negligible except at the smallest Q-values. Both SAXS and SANS instruments at central facilities are in great demand. The operation requirements of the systems are such that it is not feasible (unlike laboratory based SAXS) to change the operating parameters for specific experiments.

Ultra-small-angle X-ray scattering

The ultra-small-angle X-ray-scattering (USAXS) diffractometer uses two channel cut silicon monoliths, in an arrangement proposed by Bonse and Hart [19]. The first crystal of the USAXS instrument (Figure 6) collimates the incident beam and selects the wavelength and the second crystal (which can be rotated in the plane of the paper) is used to analyse the radiation scattered by the sample. Repeated Bragg diffraction using reflections from the **(111)** plane on opposite sides of the channel narrows the intrinsic reflection curves and suppresses the long-reaching tails of the Bragg-curve. The monochromatic beam emerges from the first crystal and is incident upon the sample situated at the centre of the rotation of the second crystal. Rotation of the second crystal from its position parallel to the first will result in only radiation scattered from the sample at the same angle passing through the second crystal to the detector.The angular collimation achieved with a channel cut crystal produces a primary beam with small angular divergence and large spatial dimensions (4mm by 10mm). Whilst the slit collimation system yields an increasing intensity for decreasing resolution the USAXS system produces a constant intensity and fixed resolution. Such that for small length scale systems SAXS techniques are preferred to USAXS [1]. The collimation in the USAXS camera is in one direction only and therefore corresponds to the 'slit geometry' used with a standard laboratory source. Complex desmearing routines developed for such laboratory instruments are therefore required.The higher resolution and lower Q-values ($<10^{-4}$)obtainable allow detailed studies of features on length scales up to several micrometres [20]

3 Examples

In the examples outlined below the SAXS data given were collected on station 8.2 of the SERC Daresbury Synchrotron Radiation Source (SRS), Warrington, U.K., the USAXS data were collected on station 2.2 of the SRS and the SANS data on the Orphee reactor at Saclay.

Isotropic scattering

Materials which give the same scattering profile regardless of the orientation of the material to the beam are said to be isotropic scatterers. Such scattering may arise from two sources, the first being a sample in which the scatterers are isotropic and hence present a similar view to the incident beam regardless of orientation i.e. spheres. The second case arises when anisotropic scatterers, such as rods, are present.Each rod will scatter differently depending on its orientation to the beam. If the rods are aligned then the small angle scattering profile will change as the sample is rotated with respect to the beam (Anisotropic scattering). However if the rods are randomly positioned within the sample isotropic scattering will observed because all orientations of the rods will be sampled by the scattered beam. In this section samples showing isotropic scattering will be presented and in the following section anisotropic scattering will be discussed in more detail.

Silica Gels

Several small angle scattering studies have been made on a variety of porous silica gels and in all cases they have been found to be isotropic scatterers. Figure 7 shows the SAXS scattering profile from two silica gels of nominal pore sizes 60 and 200Å. The general features to be noted are that the well defined maxima and minima predicted by theoretical models are not observed. This is an immediate indication of a range of pore sizes. Note also that the SAS pattern for the 60Å sample is much broader than that of the 200Å sample. This is in general agreement with the reciprocal relation between scattering vector and size. Figure 7 also shows a fit to the scattering profile (for the 60Å sample) using a pore size distribution based on spherical particles. Inset into figure 7 is the pore size distribution obtained. Also shown is the pore size distribution as calculated from nitrogen adsorption data along with a pore size distribution obtained from an NMR measurement using the suppression of the freezing point of cyclohexane [21]. In general the pore size distribution from NMR and Nitrogen adsorption are in agreement which is to be expected since the NMR technique was originally calibrated using nitrogen adsorption. What is

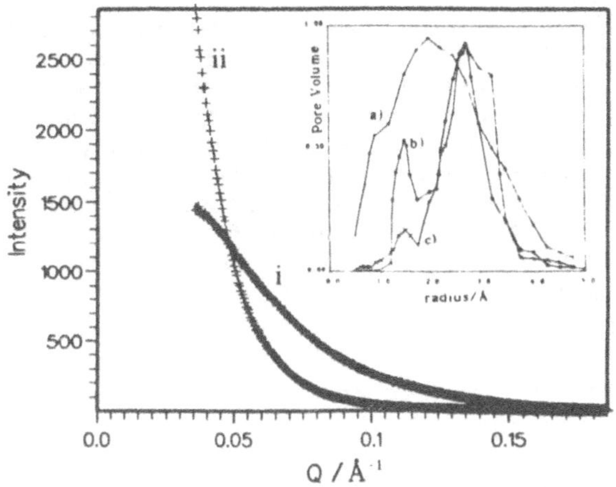

Figure 7. The scattering observed from two silica gels i) a nominal pore size of 60Å and ii)a nominal pore size of 200Å. The inset shows the pore size distribution obtained from a)SAXS, b)NMR and c)nitogen adsorption.

however clear is that such a distribution would **not** give the scattering profile measured. It appears that SAS is much more sensitive to the presence of smaller pores than is Nitrogen adsorption.

Another interesting feature of the small-angle scattering profile is only revealed in a log-log plot. At high Q-values a slope of -4.0+/-0.1 is found in agreement with Porod's law and implies a smooth surface. Ross and Li [22] have studied samples of Vycor,using SANS which have been dried at 200°C and have found slopes significantly less than 4 suggesting a fractally rough surface. However the addition of small amounts of water produced a smooth surface suggesting that the surface becomes fractured by intense drying.

Bakken shale

Bakken shale is a naturally-occuring oil-bearing shale. In order to understand the mechanism of oil migration from source rocks of relatively low porosity to reservoir rocks and to increase the mass of oil recovered an understanding of pore shape, size distribution and anisotropies are required. Mildner et al.[23] studied Bakken shale using small-angle scattering and found a broad pore size distribution with a peak around 50Å and a long tail extending to much higher values. A simple log I(Q) against log Q plot revealed a straight line of slope -3.39 with a surface fractal dimension D_s=2.51 for SAXS and a different slope for SANS data suggesting a surface fractal dimension of D_s=2.61. Figure 8 shows new SANS and SAXS data from Bakken shale in which the SAXS data is complemented by the addition of USAXS data. The fractal dimensions obtained are in agreement with those of Mildner et al. One surprising feature however is the apparent extent of the fractal surface as shown by the USAXS data. Since the largest pores are about 200Å and the average size is about 70Å the surface fractal would not be expected to extend much beyond the range measurable by conventional SAXS. However it is clear from the USAXS data that the fractal surface extends beyond 1000Å since there is no change in the slope of the log-log curve.

The different fractal dimensions obtained from the SAXS and SANS results have been explained by Mildner et al.[11] X-rays are sensitive to electron density differences whilst neutrons are sensitive to scattering length densities. If the shale was a simple two phase system, that is either solid or void then whilst the contrasts for SAXS and SANS would be different in magnitude they would be the same shape. Consequently the profile of the scattered intensities would have exactly the same form but of different magnitudes. However, the shale contains both hydrocarbons and water which is likely to be present at the pore interface resulting in different contrast profiles for X-rays and neutrons. The X-rays and neutrons are therefore likely to be scattered from different regions depending on molecular orientation, size and composition thus giving different fractal dimensions.

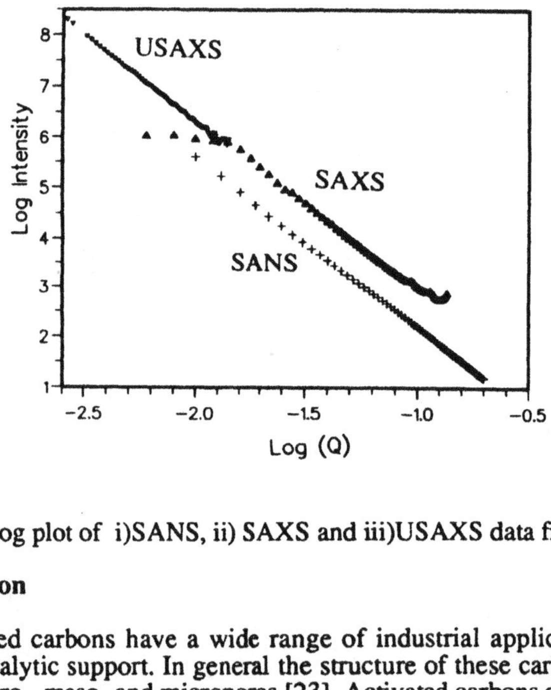

Figure 8. Log-log plot of i)SANS, ii) SAXS and iii)USAXS data from Bakken shale

Porous Carbon

Porous activated carbons have a wide range of industrial applications from molecular filtration to catalytic support. In general the structure of these carbons is highly complex containing macro-, meso- and micropores [23]. Activated carbons of varying porosity have been studied using SAXS [23]. One commercially available carbon, BP71 (Cabot Carbon Ltd.) shows many of the features generally seen in these materials. The SAXS profile from this carbon is shown in figure 9. Two comments should be made regarding this profile, firstly the rapid rise in the scattered intensity at low Q-values and secondly the very broad scattering out to high Q-values. The broad scattering suggest very small pores whilst the sharp rise at low-Q is suggestive of large pores or a pore fractal. In fact it is not possible to fit the scattering pattern with a bimodal distribution of large and small pores and a fractal structure factor must be introduced. Figure 9 shows the fitted function based on equations 8, 10, 11 and 22 using the pore volume distribution shown in figure 9. The pores have been assumed to be spherical though in practice they are likely to be slits[23]. Clearly from the fit the pores have a bimodal distribution of size ranging from 5Å to 150Å. This in in relatively good agreement with that obtained using the nitrogen adsorption method of Quirke [this ASI and 24] although there is a discrepancy in the volume of larger pores. However it is clear that the fractal structure factor may well account for the apparent loss of these pores in the scattering analysis. The pore fractal dimensionality was found to be 2.5. Electron microscope pictures of the BP71 show the carbon to have a highly complex surface, amplification of the image reveals very similar details, if this continues downwards in scale then the use of a fractal structure factor would be justified. The fractal dimension being a measure of the number of large pores within the system.

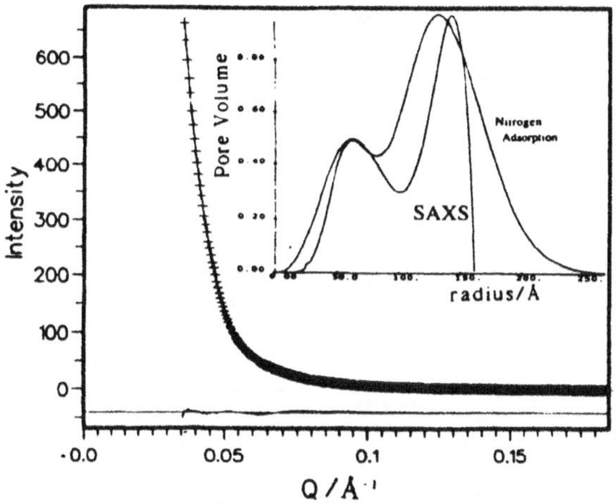

Figure9. The small-angle X-ray scattering from a porous carbon, BP71, the figure shows a fit to the data using a fractal structure factor. The fit gives the bi-modal pore volume distribution shown in the inset along with that due to Quirke [24]..

Anisotropic materials

Ceramic membranes

Synthetic membranes are used in a number of diverse applications such as filtration, bioreactors and supports for active materials. Narrow pore-size distribution, high pore density and thinness are often important attributes of the membrane[25].Anopore membranes produced by the anodic oxidation of aluminium have many of the required features. The anodizing voltage controls the pore size and pore density whilst the charge transfer determines the thickness of the membrane[26]. Pore sizes of between 10-250nm and film thicknesses of over 100μm are achievable[26]. Figure 10 shows the SAXS profiles obtained from one such membrane using the USAXS instrument at Daresbury. The membrane has two distinctive regions a layer, 300nm thick of cylindrical pores (of length 300nm) which are believed to have a circular cross-section of diameter 20nm. This layer sits upon a layer of 60μm thickness but having cylindrical pores of diameter 200nm. Figure 10a shows the scattering measured with the membrane perpendicular to the incident beam, that is with the cylinders parallel to the beam. The fit shown is for cylindrical pores of diameter 20nm and 300nm long. The larger pores of the second layer produce scattering at lower Q-values and therefore make little contribution to the measured profile. The scattering data in figure 10b were collected with the membrane at 45° to incident beam. The

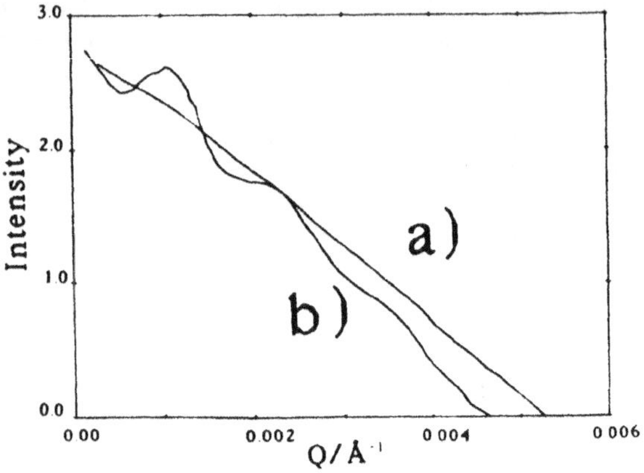

Figure 10. USAXS from an anopore membrane. a) The membrane is perpendicular to the beam. b) The membrane is at 45° to the beam, the oscillations suggest a well defined periodicity within the sample.

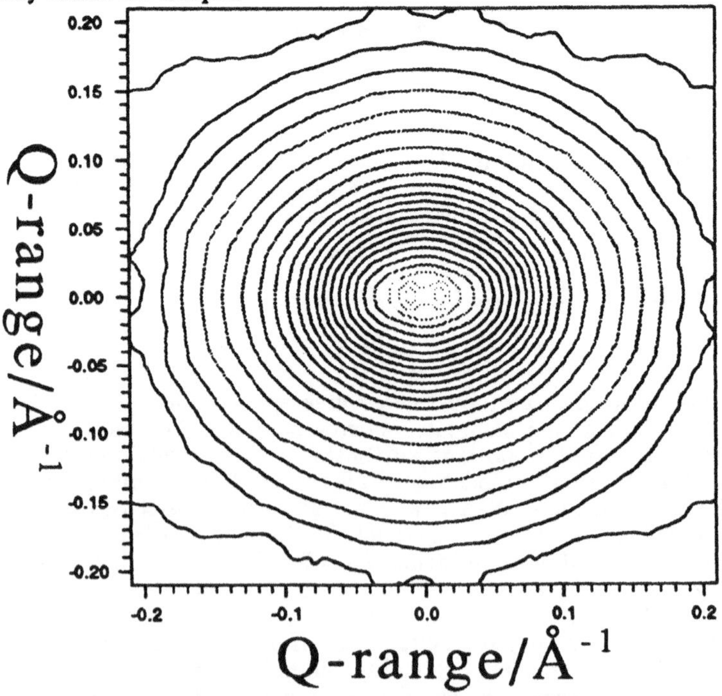

Figure 11. A two-dimensional scattering profile (SAXS) from a bunch of aligned alumina fibres.

data appear as an oscillatory form superimposed on a form factor suggesting that the oscillations result from a structure factor and hence some well defined periodicity in the structure. The repeat distance of the oscillation suggest a correlation length of 340nm which corresponds to features measured in the 200nm pore layer using electron microscopy. However these measurements require further examination and detailed analysis.

Alumina fibres

Ceramic oxide fibres derived from sol-gel methods show anisotropic scattering. The alumina fibre discussed in this section was made by ICI plc using a propriety sol-gel process and supplied supplied by M.H.Stacey. The fibres have a typical diameter of 0.4μm are meso-porous with a solid phase of η–alumina and an apparent crystallite size of 60Å [27]. The porosity is typically 30-40% and continuous in nature so that some kind of correlation effects (structure factors) are to be expected. The pores are believed to be cylindrical with a preferred ordering with the pore axes parallel to the fibre axis [27].For both X-rays and neutrons the contrast is large between the empty pores and the alumina and the material may be readily studied using both techniques. Stacey has used a standard Kratky camera for SAXS and a standard SANS instrument to study these materials in detail. Figure 11 shows the SAXS scattering from a bunch of aligned fibres these measurements were made at the SERC Daresbury laboratory, Warrington, UK. The resolution of this SAXS data being far greater than either the SAXS or SANS data of Stacey[27]. In agreement with Stacey a highly anisotropic scattering pattern is observed the main feature being an intense scattering below $Q=0.1Å^{-1}$ perpendicular to the direction of the fibre axes. There appear to be two independent pore populations. Half of the pores are randomly orientated generating isotropic scattering. The other half of the pores give anisotropic scattering which is characteristic of long cylindrical pores aligned parallel to the fibre axis.

4 Contrast matching

The section on Bakken shale has already discussed the differences that can be observed in the scattering profile when X-rays and neutrons are both used in SAS experiments. These arise from differences in the contrast due to the different ways in which X-rays and neutrons interact with matter. The aim of this section is to discuss how the contrast may changed in a controlled way and be used to give a greater insight into the structure of porous materials.

Equation 8 shows that the magnitude of the scattered intensity depends on the contrast between the disperse and continuous phases. Many systems will be far more complex than those already discussed in that the scattering length density distribution will have 3 or more distinct values.Table 1 shows the neutron scattering lengths of various elements, note the opposite sign of the scattering length for hydrogen and deuterium.These have been used to calculate the scattering length densities of several familiar materials which are given in table 2 along with the relevant X-ray scattering densities. It is immediately clear from this table that there exists a mixture of H_2O for which the scattering length density is the same as for silica such that the contrast between the two would be zero. A suspension of silica particles in such a liquid would give no small-angle scattering of neutrons. It is also clear that the magnitude of the scattering from particles suspended in H_2O/D_2O would be greater that from particles suspended in D_2O It is possible to coat the silica particles with polymers having a different scattering length. If the scattering length density of the polymer can be changed by H/D substitution a wide range of possibilities exist and these and the resulting scattering are shown in figure 12.

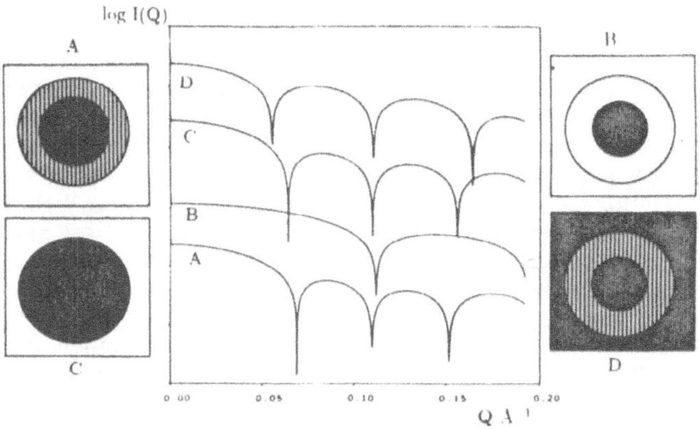

Figure 12. Scattering profiles from a coated particle for various H/D substitutions.

a)The water, silica and polymer may all have different scattering length densities and the SANS profile will contain information on the shape and size of the particle and polymer layer.
b)The water and polymer may have the same scattering length density which is different to that of the silica. In this case the SANS profile will depend on the size and shape of the silica particle only.
c)The polymer and silica may have the same scattering length density which is different to that of the water. The SANS profile will therefore contain information on the overall shape and size of the composite particle.
d)The water and silica particle may have the same scattering length density which is different to that of the polymer. In this case the scattering will be from a shell of polymer and will therefore yield information about the polymer thickness and shape.(Figure 2d)
Each of the above scenarios contains information about the system of interest some of which will be easy to extract and some which will not. The use of all four together will yield a great deal of interlinked (and cross-checked for consistency) information that would be difficult to extract from one scattering profile.
Unlike neutron scattering profiles the X-ray scattering profile depends on electron-density distributions it is therefore likely that the X-ray contrast for a given system will be different to that of neutrons. Such that SAXS will yield a different scattering profile and further useful information. Contrast matching by isotopic substitution is not applicable to X-rays (see table 2). However in certain cases it is possible to use anomalous X-ray scattering techniques to change the contrast.

Contrast matching can play an important role in investigating the structure of porous solids. One use is to investigate the structure and size of metal catalysts and contaminants within the pores. By contrast matching the pore matrix to the void by adding the correct mixture of H_2O/D_2O the scattering from the pores can be reduced to zero. Any scattering remaining will be due to the non-contrast matched catalyst particles and will consequently contain information on their size and shape. Outlined below are some preliminary measurements on the contrast matching of a silca gel.

Atom	Bound Scattering length
Hydrogen	-3.74
Deuterium	6.67
Carbon	6.65
Silicon	4.11
Nitrogen	9.37
Oxygen	5.80
Sodium	3.63
Aluminium	3.45
Chlorine	11.66

Table 1. The neutron scattering lengths of several atoms

Material	$\rho_{neutrons}$ 10^{10} cm^{-2}	$\rho_{X\text{-rays}}$ 10^{10} cm^{-2}
H_2O	-0.56	9.3
D_2O	6.34	9.3
C_7H_{16}	-0.48	7.9
C_7D_{16}	6.28	7.9
$Si\,O_2$	4.09	26.0
Na Cl	3.40	22.0
CCl_4	6.30	13.0

Table 2. The scattering length densities for a variety of materials. Note that the neutron densities change with isotopic substitution whilst the X-ray densities do not.

Contast-matching Silica gels

The silica gel has a nominal pore size of 60Å. The SANS profile from the dry silica gel is shown in figure 13. It is clear from the lack of minima and maxima in the data that the material has a large spread of pore sizes. Two other scattering profiles are shown in figure 13. The first is the scattering profile obtained when a mixture of H_2O/D_2O at the contrast match value (64% D_2O) completely fills all of the pores in the sample and the scattering can be seen to be negligable.This confirms that all of the pores are open to the general network. The second profile is from the sample completely filled with a mixture of H_2O/D_2O with 8% H_2O, this mixture is special because it has a scaterring density of zero. Such that when

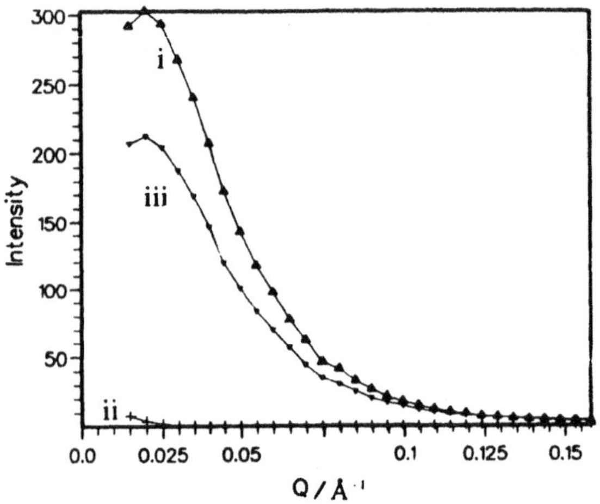

Figure 13. SANS from i) a dry silica gel of nominal pore size 60Å, ii)the same gel filled with a contrast matching liquid (the scattering falls to zero) and iii) the gel filled with a mixture of H_2O/D_2O that should result in the same scattering as the dry gel.

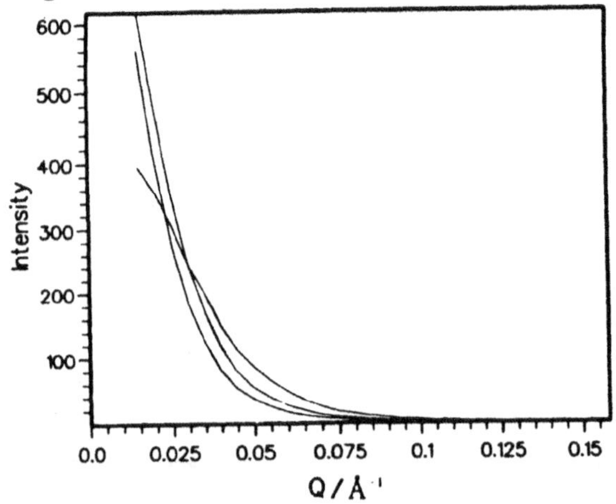

Figure 14. SANS from the 60Å silica gel with different partial fillings using the contrast match liquid. The scattering is indicative of 'pools' of liquid forming as opposed to a layering of liquid at the surface.

it fills the pores the contrast within the sample should be the same as for the dry sample. It is clear from the data that whilst the profile is the same the intensity has decreased. This is a clear indication that the density of the H_2O/D_2O mixture has changed on absorption into the pores.

If the filling of the pores with a contrast match liquid is restricted it is possible to investigate the effects of capillary condensation as shown by Li et al[22]. Figure 14 shows the scattering from the 60Å sol-gel porous solid when only partially filled with the contarst match liquid. There is a significant amount of scattering at low Q-values in agreement with measurements on Vycor reported by Li et al. A quantitative analysis based may be undertaken on the assumtion that the scattering at low Q-values arises from a fractal distribution. This fractal distribution would arise from the formation of pools of liquid within the porous material.

Summary

SAXS and SANS are powerful techniques for looking at the micro-structural features of porous materials. The high intensities of modern X-ray and neutron sources can be exploited in various ways, making use of a Q-range suitable for a particular problem. The high intensities of the sources make the techniques suitable for time-resolved studies. At the current time, various experimental groups are working on studies of the nucleation of zeolites, the formation of sol-gels, sintering and direct observation of capillary condensation and the spatial distribution of fluids

Acknowledgements

ANN would like to thank the SERC for funding in terms of an Advanced Fellowship that allowed this work to be undertaken. We also wish to thank the various people who have contributed to this work through the contribution of facilities, computational knowledge, experimental experise and specific samples; they include John Harries, Wim Bras, Jane Rigden, Michelle Gardner, Martin Stacey and the SERC (through SRFC and NBRC beam allocation).

References

1. O.Glatter and O.Kratky, *Smalll-angle X-ray scattering*, Acadenic Press, New York, 1984
2. A.Kostorz in *Treatise on materials science and technology*, p.227, volume 15, *Neutron Scattering*, Academic Press, New York, 1979
3. A.N.North, J.C.Dore, A.R.Mackie, A.M.Howe and J.Harries, *Nucl. Inst. Meths. Phys. Res.*, **B**47, 283, 1990
4. L.A.Feigin and D.I.Svergun, *Structure analysis by small-angle X-ray and neutron scattering*, Plenum press, New York, 1987
5. O.Glatter, *J.Appl. Cryst.*, **10**, 415, 1977
6. D.W.Schaeffer and K.D.Keefer, *Phys. Rev. Lett.*, **56**, 2199, 1986
7. D.Avnir, D.Farin and P.Pfeifer, *Nature*, **308**, 5956, 1984
8. D.F.R.Mildner and P.L.Hall, *J. Phys. D*, **20**, 61, 1987
9. B.B.Mandelbrot, *Fractals, Form and Chance*, Freeman, San Francisco, 1977
10. L.Richardson, *Gen. Syst. YearB.*, **6**, 139

11. D.F.R.Mildner, R.Rezvani, P.L.Hall and R.L.Borst, *Appl. Phys. Letts.*, **48**, 1314, 1986
12. J.E.Martin and A.J.Hurd, *J. Appl. Cryst.*, **20**, 61, 1987
13. H.Bale and P.W.Schmidt, *Phys. Rev. Lett.*, **53**, 596, 1984
14. J.Teixeira in *On growth and form*, eds. H.E.Stanley and N.Ostrowsky, Martinus Nijhoff publishers, Amsterdam, 1986
15. C.J.Vonk, *J. Appl. Cryst*, **9**, 433, 1976
16. J.Schelter in *Scattering techniques applied to supramolecular systems*, eds. S.H.Chen, B.Chu and R.Narsal, Plenum Press, London, 1981
17. R.K.Heenan, RAL report, **RAL-88-030**, 1988
18. E.Towns-Andrews, A.Berry, J.Bordas, G.R.Mant, P.K.Murray, K.Roberts, I.Sumner, J.S.Worgan and R.Lewis, *Rev. Sci. Inst.*, **60**, 2346, 1989
19. U.Bonse and M.Hart, *Appl. Phys. Letts.*, **6**, 155, 1965
20. A.N.North and J.S.Rigden Unpublished data
21. J.H.Strange. Private communication
22. J.C.Li, M.J.Benham, L.D.Howe and K.Ross in *Neutron and X-ray scattering: Complementary techniques* , eds. M.C.Fairbanks, A.N.North and R.J.Newport, Adam Hilger, 1989
23. M.A.Gardner, M.Sc. thesis, University of Kent at Canterbury, 1991
24. N.A.Seaton, J.P.R.B.Walton and N.Quirke, *Carbon*, **27**, 853, 1989
25. G.P.Crawford, L.M.Steele, R.Ondris-Crawford, G.S.Iannacchione, C.J.Yeager, J.W.Diane and D.Finotello, *J. Chem. Phys*, **96**,7788, 1992
26. R.C.Furneaux, W.R.Rigby and A.P.Davidson, *Nature*, **337**, 147, 1989
27. M.Stacey in *Neutron and X-ray scattering: Complementary techniques* , eds. M.C.Fairbanks, A.N.North and R.J.Newport, Adam Hilger, 1989

PULSED ESR APPLIED TO MICROPOROUS AND MESOPOROUS INORGANIC SOLIDS

D.R. Brown
Catalysis Research Unit
Department of Applied Science
Leeds Polytechnic
Leeds LS1 3HE
United Kingdom

ABSTRACT. Electron spin resonance (ESR) techniques, used with suitable paramagnetic probe ions, are applied to the structural characterisation of micro- and mesoporous inorganic solids. The structural information available is enhanced through the pulsed ESR technique, electron spin echo modulation (ESEM) spectroscopy, which can often resolve weak electron-nuclear hyperfine interactions unresolved in the continuous wave spectrum. Examples are described in which modulation associated with coupled ^2D nuclei is used to characterise interactions between paramagnetic probe ions and adsorbed solvents and organic reactant molecules on clays and zeolites. In clays and pillared clays spin echo modulation associated with weak electron-nuclear hyperfine coupling with ^{27}Al nuclei provides evidence for interactions between exchangeable cations and both framework and pillar aluminium atoms.

1. Introduction

Pulsed electron spin resonance spectroscopy, particularly in the form of electron spin echo modulation (ESEM) spectroscopy, has been successfully applied to the characterisation of zeolites [1-5], clays [6-10], silicas [11-18] and other inorganic solids. Using suitable paramagnetic probes, ESEM, combined with continuous wave electron spin resonance (ESR), has provided information on the structures of these solids, their interactions with adsorbates and their catalytic properties, complementary to that available from other techniques such as X-ray diffraction and solid state NMR.

Continuous wave ESR spectroscopy has been used by many workers to study micro- and mesoporous inorganic solids. Probably the most widely used paramagnetic probe has been the Cu^{2+} ion, providing information on the location, coordination and mobility of divalent exchangeable cations in zeolites [1], clays [19-24] and silicas [25,26]. Paramagnetic probes for monovalent cations are more difficult to find but the Ag^+ ion, reduced radiolytically to Ag^0 at 77K, has been one of the more effective [9,10,27]. Other ionic probes used with success have included the VO^{2+} ion [28,29] and the Mn^{2+} ion [30]. In addition, spin labelled surfactant molecules, such as doxylstearic acid, have been used to study the packing of adsorbed surfactant on porous inorganic solids [31-37]. ESR has also been used to study various iron centres, particularly in clays [38-41].

C.A.C. Sequeira and M.J. Hudson (eds.),Multifunctional Mesoporous Inorganic Solids, 401–412.

However, the structural information available from the ESR spectra of porous inorganic solids is limited, largely because the poor spectral resolution generally associated with this type of sample precludes the observation of weak hyperfine interactions. The usefulness of ESEM spectroscopy is that it can often resolve weak hyperfine interactions which are unresolved in the continuous wave ESR spectrum.

2. Pulsed ESR and ESEM

The comparatively large widths of ESR spectra which, in most cases, prevent the excitation of the entire spectrum with a single microwave pulse, mean that pulsed ESR does not show the same sensitivity advantage over the continuous wave technique that is exhibited by pulsed NMR over CWNMR. Applications of pulsed ESR are, therefore, restricted to areas where measurements in the time domain are of value. Furthermore, direct measurement of the electron spin free induction decay (FID) is difficult and, in general, it is instrumentally less demanding to generate and monitor spin echoes. This is because the rapid decay of the FID caused by inhomogeneous broadening is reversed in the spin echo experiment. As a result, emphasis in the development of pulsed ESR techniques has concentrated on electron spin echo (ESE) spectroscopy [42-44].

ESE spectroscopy is used successfully to measure spin relaxation times and other time dependent phenomena, but it is the electron spin echo modulation (ESEM) technique which has found application in structural studies. In the ESEM experiment a spin echo is generated following either a two or a three pulse sequence. The amplitude of the echo is monitored as the interpulse time is incremented. The echo intensity decreases as the interpulse time is increased, but the decay envelope may also exhibit modulation associated with weak dipolar hyperfine interactions between the unpaired electron and neighbouring magnetic nuclei. The modulation pattern can be interpreted, by simulation, in terms of the type of interacting nuclei, their distances from the paramagnetic centre and the number of such nuclei [43].

Two pulse and three pulse experiments are the most widely used (Figure 1). In the two pulse experiment a 90° microwave (MW) pulse is followed by a 180° pulse after a time interval τ. The echo is detected at time τ after the second pulse. The decay of the echo, as τ is incremented, is controlled by the phase memory time T_M which, in solids, is usually only a few microseconds, even at low temperature. The echo decay is modulated by weak anisotropic hyperfine interactions at approximately the Larmor frequency of the coupling nucleus and simultaneously at twice this frequency.

In the three pulse experiment three 90° pulses are separated by intervals of τ between the first and second and T between the second and third. The interval τ is fixed and T is incremented. The echo is detected at time τ after the third pulse and the echo amplitude recorded as a function of T. The three pulse technique offers the advantages that 1) echo decay is controlled by T_1, which is generally much longer than T_M, resulting in a longer modulation train, 2) the fixed value of τ can be chosen to enhance modulation associated with particular nuclei, while suppressing modulation from nuclei with different Larmor frequencies, and 3) modulation is at the Larmor frequency of the coupling nucleus only, with no contribution from twice this frequency.

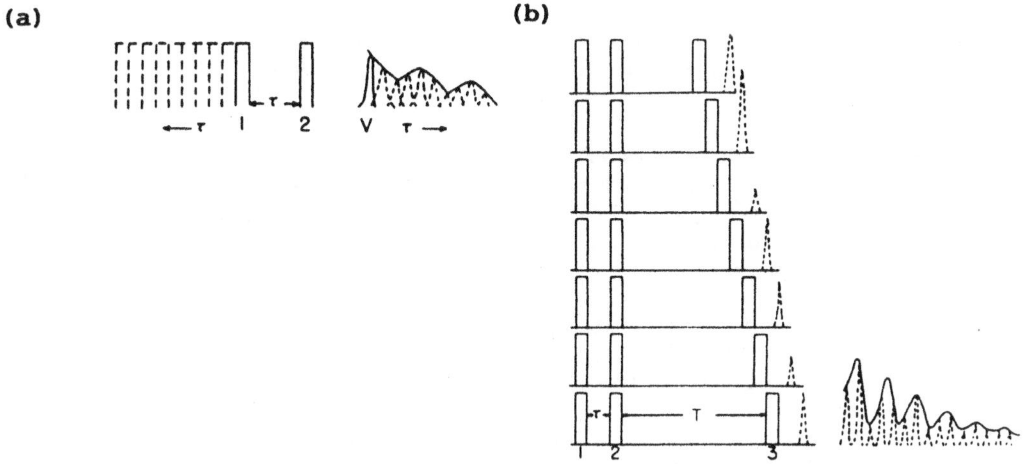

Figure 1. Illustration of modulation in (a) two- and (b) three-pulse electron spin echo decay. In both experiments, the echo (---) is detected at time τ after the last pulse. As τ (two pulse) or T (three pulse) is increased the echo amplitude changes and traces out an echo envelope which may be modulated.

Simulation of both two and three pulse modulation patterns can be readily generated [43], although assumptions that are usually made impose some restrictions on their validity, particularly at interaction distances of less than 0.25 - 0.30 nm. At distances greater than about 0.6 nm, depending on the number and type of coupling nuclei, interaction is weak and results in little detectable modulation. Quadrupole interactions are usually ignored; this is not significant for coupling to nuclei with small quadrupole moments such as ^{2}D, but is important for nuclei with larger quadrupole moments such as ^{27}Al and ^{14}N. Although it is possible to take quadrupole interactions into account with these nuclei [45,46], this is not routinely done and usually only semiquantitative data is obtained from the associated modulation patterns. For typical disordered systems where explicit geometry of coupling nuclei around the paramagnetic centre is not known, interactions are averaged over all orientations. This is a reasonable assumption and it has been shown that modulation patterns are not particularly sensitive to the angular distribution of coordinated ligands [2].

It is usual to record and to simulate ESEM decay envelopes in the time domain and all data reported here is in this form. Fourier transformation to the frequency domain is complicated by the occurrence of a dead-time at the beginning of the decay which, on transformation, distorts the frequency domain spectrum. However, for those who prefer to record data in the frequency domain, it is possible to minimise the effect of the dead-time by resorting to more complex pulse sequences or by reconstructing the distorted early part of the decay, allowing satisfactory transformation [44].

Instrumentally, it is generally easier to record modulation patterns for coupled nuclei with small, rather than large, magnetic moments. It is thus usual, when studying coupling of unpaired spins with hydrogen nuclei, to use deuterated materials and take advantage of the reduced magnetic moment of ^2D compared to ^1H.

3. Applications of ESEM to Porous Inorganic Solids

3.1. Cu^{2+} PROBES IN SMECTITE CLAYS

3.1.1. *Cu^{2+}-Exchanged Montmorillonite.* ESEM spectroscopy has provided information on the location and coordination of the exchangeable cations in Cu^{2+}-exchanged montmorillonite [6]. The ESR spectrum of Cu^{2+}-montmorillonite shows that there are three distinct sites occupied by the exchangeable cation as water is progressively removed from the clay interlayer.

When fully hydrated, ESR indicates a freely rotating Cu^{2+} ion in the clay interlayer. The ESEM spectrum (D_2O hydrated) shows modulation associated with coupled ^2D nuclei, which is simulated in terms of 12 equivalent ^2D nuclei at 0.29 nm, appropriate for directly coordinated D_2O molecules (Figure 2). Thus, a rotationally unhindered hexahydrate is identified, virtually identical to that found in bulk aqueous Cu^{2+} solution.

A different ESR spectrum is detected in partially hydrated Cu^{2+}-montmorillonite (d_{001} = ca. 1.5 nm), characteristic of a rotationally hindered ion. The orientation dependence of the spectrum indicates that the major symmetry axis of the ion is perpendicular to the plane of the clay lattice. ESEM is interpreted in terms of four equivalent, directly coordinated water molecules occupying equatorial coordination sites.

The third Cu^{2+} species is detected in the dehydrated state (evacuated at 150°C), in which the interlayer is essentially collapsed. ESEM indicates that only one water molecule remains coordinated to the cation under these conditions.

A similar study of the solvation of Cu^{2+}-montmorillonite by primary alcohols [7] shows that, under conditions of full solvation, methanol forms a four coordinate, freely rotating solvate in the interlayer (compared to a six coordinate solvate in bulk methanol solution), and the larger ethanol molecule forms a rotationally restricted four coordinate solvate under the same conditions. Figure 3 shows the ESR and ESEM spectra of ethanol solvated Cu^{2+}-montmorillonite.

Simulations of ^2D coupling on alcohols is made particularly reliable by the possibility of using alcohols selectively deuterated at more than one site. For instance, by simulating the modulation patterns associated with ethanol deuterated at the alpha carbon and then at the beta carbon, and also at the hydroxyl position, quite detailed geometries can be established.

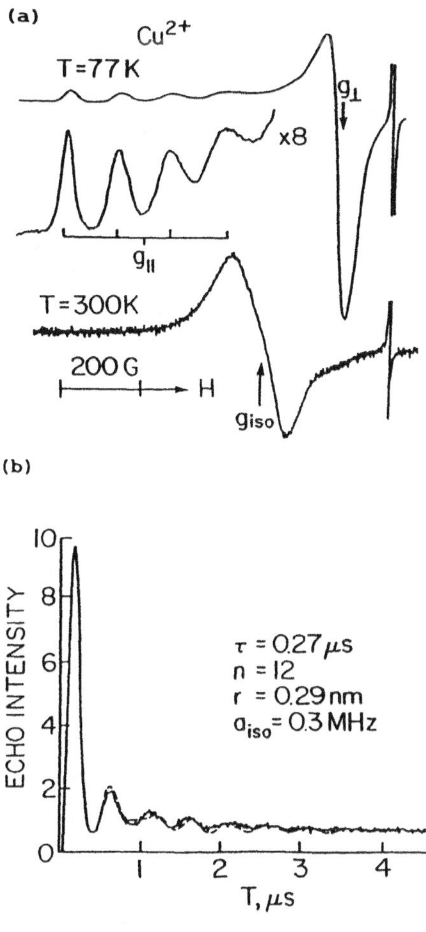

Figure 2. (a) ESR and (b) three pulse ESEM spectra of fully hydrated 5%Cu^{2+}/95%Mg^{2+} exchanged montmorillonite. The signal due to DPPH g marker (g = 2.0036) is shown at high field. The ESEM spectrum, measured on a sample hydrated with D_2O, is recorded at magnetic fields close to $g_\perp$ at 4K. τ is chosen to maximise ^2D modulation. The simulated ESEM spectrum (- - -) is shown superimposed on the experimental trace (—). The simulation parameters, n (number of interacting ^2D nuclei), r (their distance from the paramagnetic centre) and a_{iso} (estimated isotropic component of the hyperfine coupling constant), are shown.

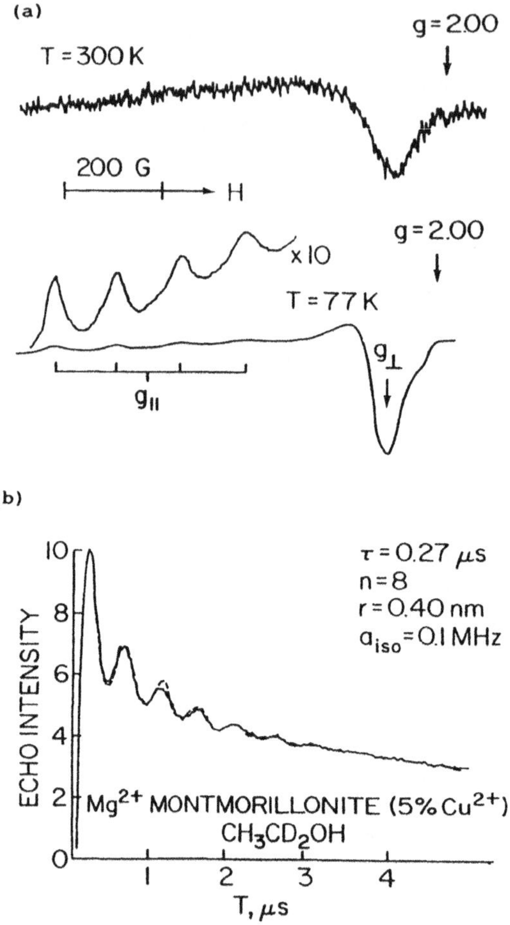

Figure 3. (a) ESR and (b) three pulse ESEM spectra of 5%Cu^{2+}/Mg^{2+} montmorillonite solvated with CH_3CD_2OH. Experimental (—) and simulated (- - -) ESEM decays are shown. τ is chosen to maximise ^2D modulation. Simulation is in terms of the number of interacting ^2D nuclei and their distance from the Cu^{2+} centre.

3.1.2. *Cu^{2+}-Exchanged Pillared Montmorillonite and Laponite.* The Cu^{2+} ion has also been used to probe the interlayer regions of pillared clays. In a study of the synthetic fluorohectorite Laponite pillared with Al_{13} alumina pillars and partially exchanged with Cu^{2+}, evidence for interaction between the Cu^{2+} ion and aluminium nuclei in the pillars is seen in the ESEM envelope, where weak modulation at the Larmor frequency of aluminium is detected [10]. Although not accurately simulated because of the complication of the quadrupolar interaction, the presence of modulation indicates that the cation is no more than 0.6 nm from the interacting Al nucleus. Since Laponite contains no framework Al this means that the cation must occupy a site within 0.6 nm of an Al atom in the Al_{13} pillar. This result is of significance in light of the recent suggestion that calcined alumina Al_{13} pillars in clay interlayers incorporate some intrinsic ion exchange capacity [47].

For these experiments, in which ^{27}Al modulation is studied, hydration of the clay is with H_2O, rather than D_2O. The much higher Larmor frequency of ^1H over ^2D avoids any risk of confusion with ^{27}Al modulation and, furthermore, because of the widely different nuclear Zeeman energies of ^{27}Al and ^1H, it is possible to set the power of the microwave pulses to excite only nuclear modulation from the lower frequency ^{27}Al nuclei.

3.2. Ag^+ PROBES IN SMECTITE CLAYS

The Ag^+ ion, radiolytically reduced at 77K to Ag^o, has been used as a probe of monovalent exchangeable cation behaviour in clay and pillared clay interlayers [7,8]. In order to use the ESR spectrum of Ag^o to probe the environment of the parent Ag^+ ion, it is necessary to assume that the coordination of the Ag^+ ion is retained following irradiation at 77K. This has been verified through parallel experiments at 4K, where molecular rearrangement on reduction is clearly prevented and in which similar ESR Ag^o species have been detected [48-50].

3.2.1. *Ag^+-Exchanged Montmorillonite.* The ESR spectrum of Ag^o generated in montmorillonite interlayers at 77K appears as an almost isotropic doublet (Figure 4). Unlike Cu^{2+}-exchanged montmorillonite, which shows three distinct Cu^{2+} ESR species depending on the degree of hydration of the clay, Ag^+-exchanged montmorillonite exhibits the same ESR spectrum (following irradiation) under all hydration/dehydration conditions. Furthermore, this spectrum is quite different from that associated with Ag^o generated in bulk aqueous solution, suggesting that the Ag^+ hydrate in the clay interlayer is, under all conditions, strongly influenced by, and presumably associated with, the clay surface. ESEM spectra of the Ag^o species provide further information on the coordination of the parent Ag^+ ion, showing four directly coordinated water molecules in the fully hydrated clay, and progressively fewer as the clay is dehydrated. It is significant that the ESR spectrum alone is not sensitive to this change in hydration.

In addition, ^{27}Al modulation, associated with interaction between the paramagnetic centre and lattice aluminium nuclei, is not observed, suggesting that the exchangeable cation is located at least 0.6 nm from atoms in the central layer of the clay. Thus, the cation, while associated with the clay surface, does not appear to be directly coordinated to the clay lattice. An NMR study of Cs^+ exchanged clays supports this view, and a similar model for the cation hydrate has been proposed, in which one coordinated water molecule acts as a bridging ligand between the cation and the lattice layer [51].

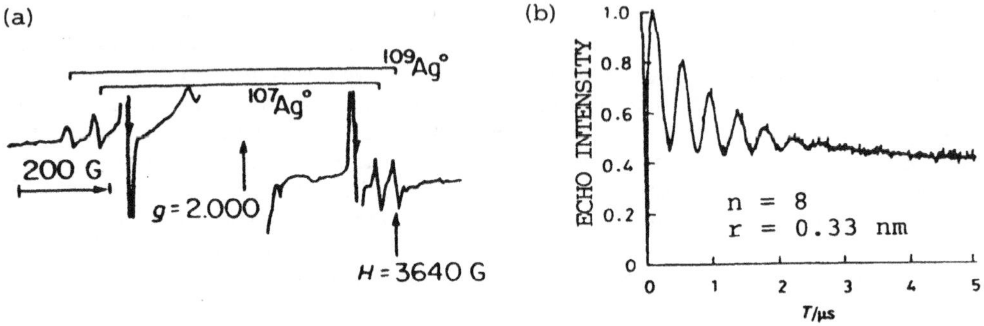

Figure 4. (a) ESR (77K) and (b) three pulse ESEM (4K) spectra of hydrated (D_2O) 10%Ag^+/Na^+ montmorillonite, following gamma irradiation at 77K. The intense doublet in the ESR spectrum inside the Ag^o features is due to trapped H atoms. The ESEM spectrum is recorded at a magnetic field of 3640 G. The experimental and simulated ESEM spectra are superimposed and simulation parameters (^2D coupling) shown. τ is chosen to maximise ^2D modulation.

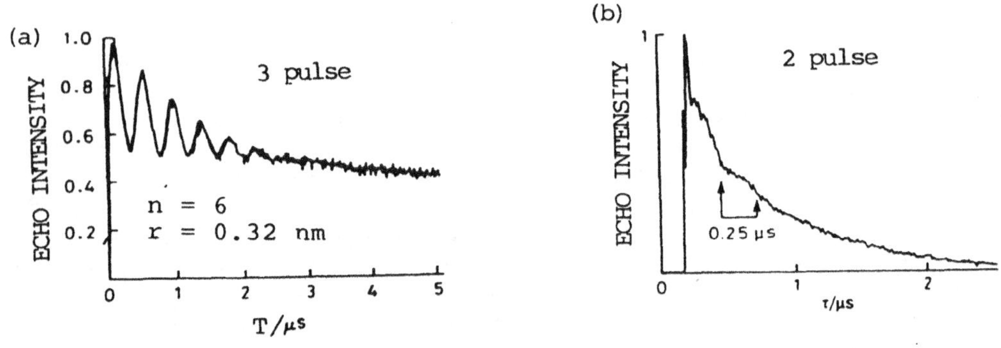

Figure 5. ESEM (4K) spectra of Ag^o formed on irradiation of hydrated Ag^+/Al_{13} montmorillonite. (a) Three pulse ESEM spectrum of sample hydrated with D_2O with τ chosen to maximise ^2D and minimise ^{27}Al modulation. The simulated spectrum, in terms of coupling ^2D nuclei, is superimposed on the experimental spectrum. (b) Two pulse echo decay, recorded on a sample hydrated with H_2O. The period associated with ^{27}Al modulation is shown.

3.2.2. *Ag^+-Exchanged Beidellite.* Similar ESEM experiments have been performed in the related mineral beidellite [9]. Beidellite differs from montmorillonite in that framework aluminium exhibits tetrahedral, rather than octahedral, geometry, and the lattice anionic charge is predominantly associated with substitutions in tetrahedral, rather than octahedral, lattice sites.

The ESR spectrum of Ag^o in beidellite interlayers is the same as in montmorillonite but ESEM shows somewhat different coordination. Under conditions of full hydration, for instance, ^2D modulation in the ESEM envelope indicates only two directly coordinated water molecules.

ESEM experiments in which H_2O is used show clear ^{27}Al modulation, suggesting direct coordination to the lattice. A model has been proposed in which the parent Ag^+ cation is directly coordinated to two lattice oxygen atoms and two water molecules.

3.2.3. *Ag^+-Exchanged Pillared Montmorillonite.* The Ag^+ ion has also been used to probe Al_{13} pillared clays [8,9]. The most significant result is found with hydrated Ag^+ exchanged Al_{13} pillared montmorillonite [8]. Again, the ESR spectrum of the resulting Ag^o centre is indistinguishable from that formed in unpillared montmorillonite, but the ESEM spectra, shown in Figure 5, are interpreted in terms of interaction with six, rather than eight, ^2D nuclei in the first solvation shell, suggesting three rather than four directly coordinated water molecules.

Furthermore, in contrast to the findings for Ag^+ in unpillared montmorillonite, quite intense ^{27}Al modulation is detected in the two pulse ESEM decay for samples hydrated with H_2O. Clearly, this must be due to direct interaction between Ag^+ and the Al_{13} pillar. It thus appears that only three of the coordination sites on the parent Ag^+ ion are occupied by water molecules and the fourth is occupied by a coordination centre on the Al_{13} pillar.

In summary, the ESEM results for both Ag^+ and Cu^{2+} exchanged Al_{13} pillared clays indicate direct interaction between exchangeable cation and pillar. Although no simulation of ^{27}Al modulation has been performed, it is possible in principle, and the resultant information on the distance of the cation from interacting Al nuclei would be particularly useful in the developing efforts to establish the state of hydrolysis, cationic charge and other structural features of Al_{13} pillars in clay interlayers, both before and after calcination.

3.3. Pd^+/X-ZEOLITES

There have been a large number of ESEM studies of paramagnetic probes in zeolite matrices [1-5]. The analysis of ^2D, ^{13}C, ^{27}Al and other nuclear modulation patterns, together with the ESR spectra of paramagnetic probe ions, have provided structural information on the location and coordination of exchangeable cations in these matrices. An example where ESEM has helped to elucidate a catalytic pathway is in the dimerisation of ethene on a Pd^+-containing faujasite type zeolite [52], in which the active valence state for the metal is not generally agreed upon.

On activation of a Pd^{2+}-exchanged zeolite and exposure to C_2D_4, ESR species are detected with parameters characteristic of Pd^+, shown in Figure 6. The first of these, species B, which forms at room temperature, shows no modulation in its ESEM spectrum, suggesting that it occupies a site inaccessible to ethene. On warming, the second two species C and D form, both of which show strong ^2D modulation, suggesting that Pd^+ becomes accessible to ethene at elevated temperature. Simultaneously with the detection of C and D the dimerisation product, butene, is detected by GC. ESEM results confirm that Pd^+(C) and Pd^+(D) involve direct complexation to ethene.

Simulation of the ESEM spectrum taken at the magnetic field corresponding to species C is in terms of four equivalent deuteriums at 0.43 nm, indicating coordination to a single ethene molecule. Simulation of the ESEM spectrum taken at a magnetic field corresponding to species D is in terms of a two shell model, involving four deuteriums at 0.40 nm and four at 0.47 nm, suggesting direct coordination of two ethene molecules in such a way that two deuteriums on each are closer to the Pd^+ than the other two.

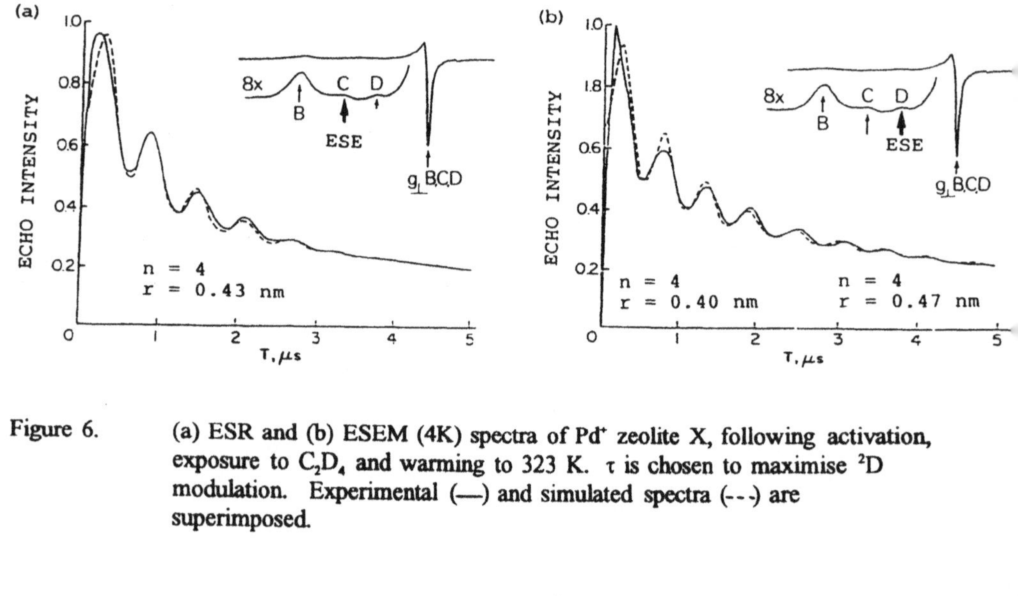

Figure 6. (a) ESR and (b) ESEM (4K) spectra of Pd^+ zeolite X, following activation, exposure to C_2D_4 and warming to 323 K. τ is chosen to maximise ^2D modulation. Experimental (—) and simulated spectra (- - -) are superimposed.

Thus, ESR and ESEM clearly show that Pd^+ is involved in the dimerisation process. It appears that species C is initially formed with one ethene coordinated to Pd^+, followed by coordination of a second reactant molecule to generate species D. Species D is then postulated to undergo a geometric isomerisation to bring the two molecules into correct orientation to form 1-butene, which then isomerises into an equilibrium mixture of *cis*- and *trans*-2-butene. This contrasts with another possible reaction mechanism involving the coordination of a single ethene molecule, followed by generation of an olefinic cationic species which then couples with a second molecule of ethene.

4. Conclusion

These examples of the application of ESEM show the types of structural information available on porous inorganic solids. As an adjunct to ESR the technique significantly increases the usefulness of paramagnetic probes in studying these systems. In the last few years commercial instrumentation for pulsed ESR has become widely available. As a result, ESEM/ESR is likely to develop as an important technique, complementary to many of the more widely used methods, in the characterisation of inorganic micro- and mesoporous solids.

5. Acknowledgement

The advice and cooperation in preparing this paper of Professor Larry Kevan, University of Houston, is gratefully acknowledged.

6. References

1. Kevan, L. (1987) Rev. Chem. Intermed., 8, 53.
2. Anderson, M.W. and Kevan, L. (1987) J. Chem. Soc. Faraday I, 83, 3505.
3. Sass, C.E. and Kevan, L. (1988) J. Phys. Chem., 92, 14.
4. Sass, C.E. and Kevan, L. (1988) J. Phys. Chem., 92, 5192.
5. Ghosh, A.K. and Kevan, L. (1988) J. Amer. Chem. Soc., 110, 8044.
6. Brown, D.R. and Kevan, L. (1988) J. Amer. Chem. Soc., 110, 2743.
7. Brown, D.R. and Kevan, L. (1988) J. Phys. Chem., 92, 1971.
8. Brown, D.R., Luca, V. and Kevan, L. (1991) J. Chem. Soc. Faraday, 87, 2749.
9. Luca, V., Brown, D.R. and Kevan, L. (1991) J. Phys. Chem., 95, 10065.
10. Kukkadapu, R.K., Luca, V. and Kevan, L. (1992) J. Phys. Chem., 96, 415.
11. Bogus, W. and Kevan, L. (1989) J. Phys. Chem., 93, 3223.
12. Ichikawa, T., Yoshida, H. and Kevan, L. (1981) J. Chem. Phys., 75, 2485.
13. Ichikawa, T., Yoshida, H. and Kevan, L. (1982) J. Phys. Chem., 86, 881.
14. Narayana, M., Zhan, R.Y. and Kevan, L. (1984) J. Phys. Chem., 88, 3990.
15. Narayana, M., Narasimhan, C.S. and Kevan, L. (1983) J. Catal., 79, 237.
16. Narayana, M., Narasimhan, C.S. and Kevan, L. (1985) J. Chem. Soc., Faraday I, 81, 137.
17. Narayana, M., Zhan, R.Y. and Kevan, L. (1985) J. Phys. Chem., 89, 636.
18. Zhan, R.Y., Narayana, M. and Kevan, L. (1985) J. Chem. Soc. Faraday I, 81, 2085.
19. McBride, M.B., Pinnavaia, T.J. and Mortland, M.M. (1975) J. Phys. Chem., 79, 2430.
20. Clementz, D.M., Pinnavaia, T.J. and Mortland, M.M. (1973) J. Phys. Chem., 77, 196.
21. McBride, M.B. (1982) Clays Clay Miner., 30, 200.
22. Kukkadapu, R.K. and Kevan, L. (1989) J.Phys. Chem., 93, 1654.
23. Kukkadapu, R.K. and Kevan, L. (1990) J. Chem. Soc. Faraday, 86, 691.
24. Dyrek, K., Klapyta, Z. and Sojka, Z. (1983) Clays Clay Miner. 31, 223.
25. Martini, G. and Burlamacchi, L. (1979) J. Phys. Chem., 83, 2505.
26. Martini, G. and Bassetti, V. (1979) J. Phys. Chem., 83, 2511.
27. Brown, D.R. and Kevan, L. (1986) J. Phys. Chem., 90, 1129.
28. Braddell, O., Barklie, R.C. and Doff, D.H. (1990) Clay Miner., 25, 15.
29. McBride, M.B. (1979) Clays Clay Miner., 27, 91.

30. McBride, M.B., Pinnavaia, T.J. and Mortland, M.M. (1975) Amer. Miner., 60, 66.
31. Waterman, K.C., Turro, N.J., Chandar, P. and Somarsundaran, P. (1986) J. Phys. Chem., 90, 6828.
32. Chandar, P., Somarsundaran, P., Waterman, K.C. and Turro, N.J. (1987) J. Phys. Chem., 91, 148.
33. McBride, M.B. (1976) J. Phys. Chem., 80, 186.
34. McBride, M.B. (1976) Amer. Chem. Soc. Symp. Ser., 34, 123.
35. McBride, M.B. (1977) Clay Miner., 12, 273.
36. McBride, M.B. (1977) Clays Clay Miner., 25, 6.
37. McBride, M.B. (1979) Clays Clay Miner., 27, 97.
38. Delvaux, B., Mestdagh, M.M., Vielvoye, L. and Herbillon, A.J. (1989) Clay Miner., 24, 617.
39. Luca, V. and Cardile, C.M. (1989) Clays Clay Miner., 37, 325.
40. McBride, M.B., Mortland, M.M. and Pinnavaia, T.J. (1975) Clays Clay Miner., 23, 162.
41. Goodman, B.A., Nadeau, P.H. and Chadwick, J. (1988) Clay Miner., 23, 147.
42. Schweiger, A. (1991) Angew. Chem. Int. Ed. Eng., 30, 265.
43. Kevan, L. and Schwartz, R.N. (1979) Time Domain Electron Spin Resonance, Wiley, New York.
44. Kevan, L. and Bowman, M.K. (1990) Modern Pulsed and Continuous Wave Electron Spin Resonance, Wiley, New York.
45. Romanelli, M., Goldfarb, D. and Kevan, L. (1988) Magn. Reson. Rev., 13, 179.
46. Goldfarb, D. and Kevan, L. (1987) J. Chem. Phys., 87, 6323.
47. Dyer, A. and Gallardo, T. (1990) in Recent Dev. Ion Exch. [Proc. Int. Conf. Ion. Exch. Processes], Eds. Williams, P.A. and Hudson, M.J., 2, 75.
48. Ichikawa, T., Kevan, L. and Narayana, P.A. (1979) J. Chem. Phys., 71, 3792.
49. Ichikawa, T., Kevan, L. and Narayana, P.A. (1979) J. Phys. Chem., 83, 3378.
50. Li, A.S.W. and Kevan, L. (1981) J. Phys. Chem., 85, 2557.
51. Grandjean, J. and Laszlo, P. (1989) Clays Clay Miner., 37, 403.
52. Ghosh, A.K. and Kevan, L. (1988) J. Amer. Chem. Soc., 110, 8044.

ELECTROCHEMICAL CHARACTERISATION OF POROUS MATERIALS: ELECTROKINETIC POTENTIAL MEASUREMENTS

P.S.D. BRITO, I.M.G.C.S. PAIVA and C.A.C. SEQUEIRA
Instituto Tecnológico para a Europa Comunitária
Av. Rovisco Pais
1096 Lisboa Codex
PORTUGAL

ABSTRACT. This paper underlines the importance of electrokinetic potential measurements for the characterisation of porous materials.

INTRODUCTION

The widespread interest in porous solids is well illustrated by the multifarious nature of the work reported in the open literature which was undertaken on materials of technological importance such as absorbents, catalysts and constructional materials and the solids studied include carbons, oxide, cements, clays, polymers, zeolites and metal films.

Among the porous solids, mesoporous materials (that is materials whose pore width is in the range 2-50 nm) are becoming remarkably importants due to their multifunctionality [1-5].

In view of this wide diversity of interest, it is clear that one outstanding problem includes the important requirement of predicting technological performance from the results of characterization measurements.

The measurement of adsorption at the gas/solid interface continues to be one of the most popular techniques for the study of microporous and mesoporous solids. It is not surprising therefore that many papers in the literature are concerned with the determination and interpretation of gas adsorption data. In this context, adsorption hysteresis deserves a note. It is well known that capillary condensation in mesopores is generally associated with hysteresis. Progress has been made in linking the characteristic shapes of certain hysteresis loops with the nature of the pore structure, and much is being done to explain the mechanisms of mesopore filling and emptying.

Mercury porosimetry is another popular method available for the characterization of a wide range of porous materials and the derived pore sizes are often quoted in the patent and technical literature. The method is based on the non-wetting nature of mercury and the application of the Washburn equation. The volume of mercury penetrating into a porous solid is determined as a function of the applied pressure,

C.A.C. Sequeira and M.J. Hudson (eds.), Multifunctional Mesoporous Inorganic Solids, 413–421.

which is assumed to be directly related to the pore width.

In spite of the growing popularity of mercury porosimetry and the ready availability of excellent automated equipment, the interpretation of the mercury intrusion-extrusion data is still far from clear. The values of surface tension and contact angle which must be inserted in the Washburn equation are still uncertain - as are the limits of applicability of the equation itself. Other problems include the reversible or irreversible deformation of the pore structure, which undoubtedly occurs with some corpuscular or weakly agglomerated systems.

Other techniques that play a vital role in the characterisation of many porous materials are microscopy (TEM, SEM), small angle neutron (SANS) and X-ray (SAXS) scattering, FTIR and NMR. FTIR and NMR may be applied in many different ways and for various reasons including study of the pore structure and the properties of adsorbed or occluded material. FTIR is especially useful for the characterization of surface species and the state of adsorbed molecules. NMR measurements have been carried out in the context of image analysis (NMRI) of pore structures, determination of diffusivities of adsorbed species and the pore structural analysis of wet materials.

The characterization of porous solids by electrochemical techniques is now attracting the attention of many distinguished scientists and technologists. In this paper, the main attributes of electrokinetic potential measurements for the characterization of fibers, films and mesoporous solids are discussed.

ELECTROKINETIC PHENOMENA

The electrokinetic surface properties of materials are important to many industrial processes [6] and also in numerous biological and medical applications. These properties result from the electrochemical double layer (EDL), which exists at the interface between a solid and the solution in which the solid is dispersed or suspended.

The net charge at the surface of a material in contact with a polar medium is usually the result of one or more of the following processes: ionization/dissociation of surface chemical groups, adsorption of ionic species from solution, and dissolution of ions from the solid into the liquid phase.

The surface charge is balanced by ions of opposite charge (counter ions) in solution which, according to the Stern model [7], form two distinct regions or layers. The layer closest to the surface, the Stern layer, is considered immobile, while the outer or diffuse layer allows diffusion of ions by thermal motion (Figure 1). The potential decreases linearly from ψ_o, the surface, or wall potential, to ψ_d, the Stern potential, and then decays exponentially to zero in the diffuse layer. Unfortunately, the fundamental interfacial property ψ_o (and, hence charge density σ_o) is experimentally inacessible [8]. What can be derived, however, is the electrokinetic potential, or zeta potential (ZP). This quantity is defined as the potential at the surface of shear - where any relative movement of the surface with respect to the solution will cause some of the counter ions to be sheared off-resulting

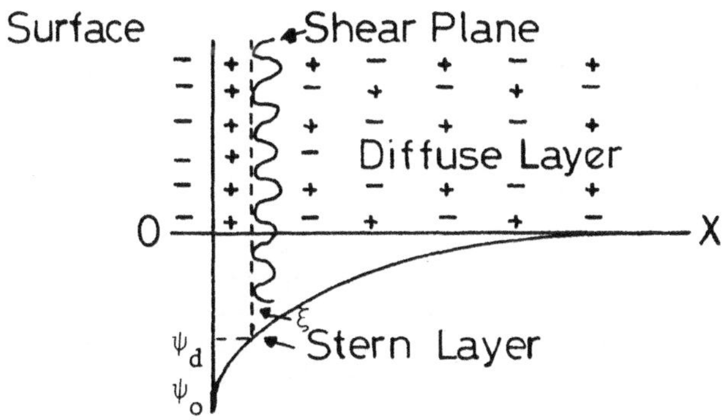

Figure 1. Stern model of the electrochemical double layer.

in only partial compensation of the surface charge.

THE CALCULATION OF ZETA POTENTIAL

Any relative motion between the rigid and mobile parts of the EDL will result in one of four related electrokinetic effects: electrophoresis, electroosmosis, streaming potential, or sedimentation potential. Measurements made using techniques based upon any of these effects should, in principle, result in the same calculated value for the ZP.

Among the most commonly used procedures is microelectrophoresis [9]. This requires that the material under investigation be in form of a stable dispersion of microscopically visible particles. Only extremely dilute sols (typically less than 0.001%) can be studied, and calculation of the ZP assumes a spherical particle. Most industrial dispersions are more concentrated, and dilution is seldom an innocuous process. Also, the particles under investigation must not be so large, nor so dense, that they sediment from the field of view before the measurement can be completed.

The new electroacoustic technique [10], which is an extension of the sedimentation potential method, will work with dispersion concentrations in the range of 1 to 50% by weight. However, the technique is in its infancy; various technical/practical complications arise as the particle concentration increases, making calibration difficult. Furthermore, the validity of calculations of ZP values from the raw (vibration) potential data has not been definitively established.

For many materials of interest, neither of the above approaches to ZP measurement is satisfactory. For samples such as hair or textile fibers, polymer films, metal foils, paper pulp, and coarse dispersions (and even solid specimens) of minerals, the streaming potential method is the most suitable - and often the only - choice.

As well as the simple instruments designed to monitor polyelectrolyte titrations via streaming current measurements, there is now available a more sophisticated system which uses the streaming potential /current (SPC) method to give actual values for ZP under controlled but variable conditions of pH, ion, and additive concentration. Its principles and application to ZP measurements on a variety of solid samples are described below.

THE STREAMING POTENTIAL/CURRENT METHOD

When an electrolyte solution is forced, by external pressure, through a porous plug of material or across a channel formed by two plates or down a capillary, a streaming potential developes between the ends. The potential difference developed for a given applied pressure difference can be measured by electrodes placed at either of the plug or capillary.

The ZP is calculated from:

$$\zeta = (E_s/\Delta p) \cdot (\eta/\varepsilon\varepsilon_o) \, (L/A) \, (1/R) \qquad (1)$$

Where ζ is the zeta potential, E_s is the streaming potential, Δp is the hydrodynamic pressure difference across the plug, η is the liquid viscosity, ε is the liquid permittivity, ε_o is the permittivity of free space, L is the length of the plug, A is the cross-sectional area of the plug, and R is the electrical resistance across the plug.

The various terms in Eq. 1 are either known or must be measured. The first term, $(E_s/\Delta p)$, is determined by direct measurement of the potential generated for a given constant, driving pressure. The precision and repeatability of detection can be significantly increased if the potential is monitored as a function of a continuously increasing pressure difference. The value of the quotient $(E_s/\Delta p)$ is then determined from the slope. For a given temperature, the values of η, ε, and ε_o are constant for the liquid used (usually water). The plug resistance, R, can be measured directly using an ac bridge at three different frequencies (500, 1000, and 2000 Hz). However, if both the streaming potential, E_s, and the streaming current, I_s, are measured simultaneously, then the dc resistance, R, can be determined indirectly from the ratio of the two quotients, $(E_s/\Delta p)/(I_s/\Delta p)$.

The final term, (L/A), consists of two parameters, neither of which can be easily measured. Various approaches have been suggested to address the problem.

In the Fairbrother and Mastin [11] approach, the term (L/A) is replaced by $(R_s.k_s)$, where R_s is the electrical resistance of the plug when the measurement cel is filled with an electrolyte whose specific conductance, k_s, is accurately known. A 0.1M KCl solution is used to avoid the problem of surface conductance.

Thus,

$$\zeta = (E_s/\Delta p) \, (\eta/\varepsilon\varepsilon_o) \, (R_s.k_s/R) \qquad (2)$$

This approach is limited to streaming potential measurements through single channels (membranes) or across surfaces with well-defined, fixed, open geometry (films).

For plugs of granular material and bundles of fibers, where the packing density is not well defined and cannot always be duplicated, the approach of Chang and Robertson [12] can be adopted. Using this, Eq. 1

can be written as:

$$\log_e(E_s/\Delta p)\ (L_m/R) = \log_e \zeta - Bd \qquad (3)$$

where L_m is the distance between the electrodes, B is a constant which depends upon the material under investigation, and d is the (fiber) concentration, or packing density, in the cell in g/cm^3. The ZP is obtained from the intercept of a plot of streaming potential versus packing density (Eq. 3).

The choice of cell design depends upon the nature of the sample under investigation. Cylindrical and rectangular cells are very commonly used. The cylindrical cell has two perforated electrodes (e.g. Ag/AgCl electrodes) whose position can be adjusted inside the glass body. A known weight of material in the form of fibers or granules is placed between the electrodes. The electrode spacing is measured precisely by means of micrometer screw drives. Either the FM or the CR method of ZP evaluation can be selected. In the latter method, the distance between the electrodes is decreased in a series of sequential steps to squeeze the plug and so change the packing density.

In the rectangular cell, a channel of well-defined dimensions is created using PTFE spacers. The sample is usually in the form of plates, film, or foil cut to size. Irregularly, shaped material can also be accommodated by mounting it in a suitable sealing material such as beeswax and using a modified version of the cell. Ag/AgCl electrodes are mounted at each end of the channel. Only the FM method can be used to calculate the ZP.

APPLICATIONS OF THE ZETA POTENTIAL

The following examples demonstrate the versatility and utility of the streaming potential technique for the investigation of the electrokinetic properties of surfaces of several materials used in important industrial processes.

Printing plates

Both the wetting and adhesive properties of solids depend on the dissociation constants (pK values) of the surface species [13, 14]. Since pK values can be estimated from ZP measurements, the latter provide a ready means of analyzing the effects of chemical and physical treatments of surfaces. For example, the quality of results obtained by offset printing is affected by the surface chemistry of the printing plates. The bright metal surface can be dimmed by corrosion and the various oxidative species produced on the surface reduce its hydrophobic character.

The magnitude of the ZP provides a sensitive measure of the dissociative nature of the surface groups; a ZP of zero indicates an isoelectric point (IEP), and a plateau in the plot of ZP versus pH is indicative of complete dissociation.

Figure 2 shows the pH dependency of the ZP determined for two

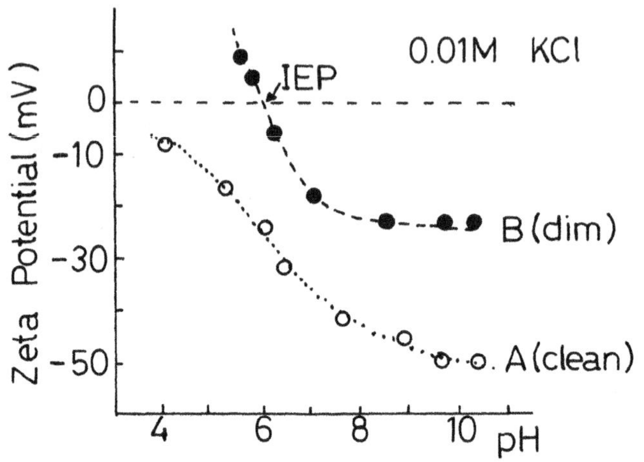

Figure 2. Zeta potential vs pH for two pieces of offset printing plate.

samples of printing plate that have received different surface treatments. Two pieces of each sample were cut to size to fit into the standard rectangular cell. Surface A is clean and has a bright metallic luster. Surface B has been exposed to corrosive elements and appears dim. The oxidized surface exhibits an IEP near pH 6, and total dissociation of surface groups can be attained at around pH 7. In contrast, dissociation is not complete for the clean surface until around pH 10, and there is no evidence of an IEP. The flowthrough nature of the streaming potential method also makes it ideal for the study of long--term corrosion or oxidation processes using different solution formulations.

Paper machine felts

The lifetime of paper machine felts is highly dependent upon the degree of contamination that occurs during paper manufacture [15]. The paper felt is a felt blanket that conveys the newly formed paper through press rollers (and steam-heated driers) where excess water is squeezed out of the paper and into the felt. The tendency of the felt to become polluted is controlled by the felt surface charge. The minimum contamination is found for very low, but negative, values of the ZP. Usually, cationic synthetic resins are added to felts during their manufacture to achieve this low ZP condition.

Table 1 lists results of ZP measurements made using samples of polyester fiber felt treated with different levels of resin. Pieces of felt were cut to fit the recesses of the modified rectangular cell and the ZP was determined using distilled water as the flow fluid. The results show that increasing amounts of resin lead to a decrease in the primary negative ZP. Overdosing with the resin produces a positive ZP -- an equally undesirable situation.

Thus, streaming potential measurements can help to optimize felt

TABLE 1. Polyester paper machine felt samples

Resin treatment	Zeta potential (mV)
Untreated	- 28
0.1 %	- 17
0.5 %	- 2
1.0 %	+ 10

manufacture and also improve the filtration process in paper production. In addition, it is possible to study variations in surface charge behavior caused by contamination that occurs during paper manufacture as well as the effectiveness of felt cleaning procedures.

Mining and minerals

In the mining industry, rock destruction or machining uses high-speed drills with diamond bits. The rate of diamond abrasion is a function of the machining speed. Drilling fluids that are used as coolants are formulated to contain auxiliary additives that critically affect both parameters, namely, drill abrasion and machine speed. Optimum conditions, i.e., minimum diamond abrasion and maximum machining speed, are achieved when the zeta potential of the rock surface is zero with respect to the liquid coolant [16]. To date, this enhanced drilling phenomenon is not well understood; the economic implications are, however, considerable. Also, no statistically reliable trend exists between rock property tests (such as hardness) and mechanisms proposed to explain this point of zero charge phenomenon.

For negatively charged rock surfaces, cationic surface active agents are used in the coolant formulation. Figure 3 shows the result of an experiment that used two platelike pieces of granite rock. The ZP was measured as a function of the concentration of dodecylpyridinium chloride (DoPyCl). For the material under investigation, only a relatively low concentration (less than 0.01 M DoPyBr) was needed to reverse the charge on the rock surface.

Thus, streaming potential measurements can be used directly to determine optimum additive concentrations for zero point of charge conditions of mineral surfaces.

For geological material, fracture of the surface can expose a completely different surface chemistry. Measurement of fines alone, using microelectrophoresis, can give incomplete, and possibly misleading, information. This may be one reason for the varied, often contradictory, ZP data for mineral surfaces. The streaming potential method uses material of macroscopic dimensions and is, thus, more appropriate for real, practical systems.

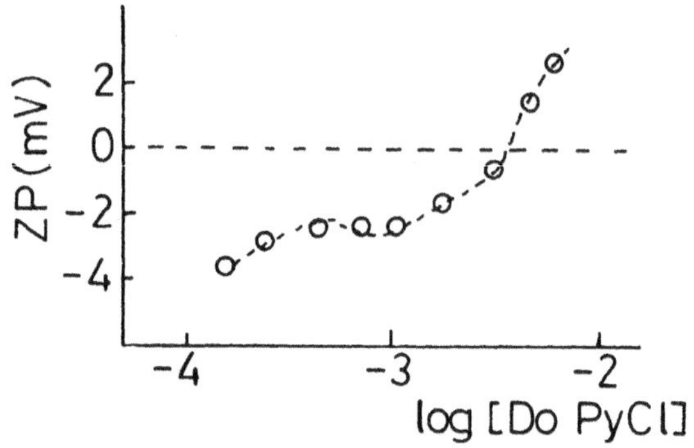

Figure 3. Reversal of charge of a granite surface using a quaternary ammonium salt.

CONCLUSION

The number of applications of the zeta potential measurements is limited at present. However, recent studies are giving striking demonstration of the potentialities of the use of the zeta potential, giving its strong correlation with many other relevant properties (namely electrochemical properties) of the solid materials. The mesoporous inorganic solids exhibit a broad spectrum of properties and given this fact, it is believed that the electrokinetic way can be a very fruitful approach for their investigation and characterisation.

References

[1] Cartlidge, S., Nissen, H.U. and Wessicken, R., (1989), Zeolites, 9, 346-349.

[2] Day, M. and Fletcher, R., (1987), Stud. Surf. Sci. Catal., 9, 491--499.

[3] Everett, D.H. and Rojas, F., (1988), J. Chem. Soc., Faraday Trans. 1, 84, 1455-1464.

[4] Gimblett, F.G.R., Rahman, A.A. and Sing, K.S.W., (1984), J. Colloid Interface Sci., 102, 483-490.

[5] Rupprecht, H., Kircher, W. and Unger, K., (1980), Pharm. Ztg., 125, 1482-1485.

[6] Hunter, R.J. (1981), Zeta Potential in Colloid Science, Academic Press, New York, NY.

[7] Kruyt, H.R. (ed.) (1949), Colloid Science, Elsevier, London, vol. 1.

[8] Sequeira, C.A.C. (1992) in C.A.C. Sequeira and A.K. Tiller (eds.), Microbial Corrosion, Institute of Materials, London, pp. 9-17.

[9] Shaw, D.J. (1969), Electrophoresis, Academic Press. London.

[10] Marlow, B.J., Fairhurst, D. and Pendse, H.P., (1988), Langmuir, 4, 611.

[11] Fairbrother, F. and Mastin, H., (1984), J. Chem. Soc., 75, 2318--2322.
[12] Chang, M. and Robertson, A., (1967), Can. J. Chem. Eng., 45, 66-69.
[13] Bolger, J.C. (1983) in M. Kittal (ed.), Adhesion Aspects of Polymeric Coatings, Plenum Press, New York, N.Y.
[14] Fowkes, F.M. (1983) in K. Mittal (ed.), Physicochemical Aspects of Polymer Surfaces, Plenum Press, New York, N.Y.
[15] Saltman, D. (1978), Paper Basics, Van Nostrand, New York, N.Y.
[16] Westwood, A.R.C., (1981), Colloids and Surfaces, 2, 1-4.

SECTION 6

SOME APPLICATIONS OF MULTIFUNCTIONAL MESOPOROUS INORGANIC MATERIALS

PILLARED ACID-ACTIVATED CLAYS : SYNTHESIS, CHARACTERISATION AND APPLICATION TO CHLOROPHYLL ADSORPTION

R. MOKAYA and W. JONES
Chemistry Department, University of Cambridge, Lensfield Road, Cambridge CB2 1EW, England

ABSTRACT. The treatment of montmorillonite clays with mineral acids is known to yield modified clays which exhibit enhanced adsorptive and catalytic properties. The structure, selectivity and thermal stability of such suitably acid-activated clays can be improved and developed by introducing polymeric hydroxy-Al cations into their interlayer region, resulting in pillared materials with improved thermal stability as well as well developed micro and mesoporosity. The pore size distributions, surface areas, pore volumes and surface acidities of these materials may be varied by changing preparation procedures. The materials show promise for the selective adsorption of chlorophyll from edible oils. Data summarising the results obtained which relate structure and acidity to chlorophyll adsorption capacity is presented.

1. INTRODUCTION

Montmorillonite clays may be modified in several ways so as to improve their adsorptive and catalytic properties. Modifications include cation-exchange, acid-activation and, more recently, the introduction of metal oxide pillars into the interlayer region of the clays to produce the so called pillared clays. Whereas cation-exchange affects only the acidity [1] and swelling properties [2] of the clays, acid-activation and pillaring both result in materials with altered chemical composition, surface acidity and structure. In particular, surface areas and pore volumes are significantly increased following acid treatment or pillaring. The porosity and stability of the products differ, however. Pillared clays exhibit a restricted pore size distribution (predominantly in the micropore range), show good thermal stability but have relatively low acidity [3,4,5]. Acid-activated clays, on the other hand, have a broad pore size distribution (in the mesopore and macropore range) and high acidity, but possess poor thermal stability [6,7,8].

There have been several reports concerned with the preparation of pillared clays with large pore diameters (>1nm) using larger pillaring species [9-13]. The rate and extent of exchange (into the interlayer region) of these pillaring species is influenced by their nature (e.g. charge/mass ratio) while their distribution in the interlayer region is dependent upon the density and distribution of charge on the clay layers. The nature of the pillaring species is determined by the concentration (of the metal ion), the basicity or degree of hydrolysis, the preparation temperature and the time of ageing of the pillaring solution [14-16].

Whilst previous attempts at extending the pore size range [9-11] and improving the surface acidity [17] of pillared clays have centred around varying the nature of the pillaring species, the approach outlined here has been to proceed via a modified host lattice by the introduction of pillars into an acid treated clay in the hope that the final pillared material would combine good thermal stability, high acidity and a pore size distribution intermediate between that of pillared clays and acid-activated clays. We describe the synthesis and characterisation of such pillared acid-activated clays and their use for edible oil improvement. Most of the discussion is on pillared acid-activated clays (PAM), although results for pillared non-activated clays (PM) will be presented for comparison.

C.A.C. Sequeira and M.J. Hudson (eds.), Multifunctional Mesoporous Inorganic Solids, 425–432.

2. EXPERIMENTAL

A natural montmorillonite clay of surface area 81 m^2/g (supplied by Laporte and designated M) was used as the starting material. The clay had an octahedral layer composition (expressed as wt% of clay) of 14.1% Al_2O_3, 8.7% Fe_2O_3 and 3.6% MgO. M was acid treated with concentrated sulphuric acid to yield the activated clay (designated AM) which had a surface area of 243 m^2/g. The surface area reduced to 203 m^2/g following calcination at 500 ^oC (sample designated CAM). The acid-activated clay had an octahedral layer composition (expressed as wt% of clay) of 10.9% Al_2O_3, 6.0% Fe_2O_3 and 2.5% MgO. The percentage of octahedral cations removed (~27%) was within the range in which the montmorillonite structure of the clay is preserved [18] and, therefore, ensuring its suitability for pillaring.

An aluminium chlorhydrate (ACH) solution (supplied by Albright and Wilson) of 6.4 M Al and with an OH/Al ratio of 2.5 was used as the source of the pillaring species. The precursor-pillared clays were prepared by adding the clay to a stirred pillaring solution at either 20 or 80 ^oC. The concentration of the pillaring solution was 0.06M and the clay was added to the solution at a solution/clay ratio of 50 ml/g. This corresponds to an Al/clay ratio of 6 mmole/g. The reaction was then carried out at for 1 hr with constant stirring. The resulting slurry was repeatedly centrifuged in fresh de-ionised water and reslurried until free of Cl^- ions (as tested by $AgNO_3$). The obtained solids were then re-dispersed in a minimum amount of de-ionised water and air-dried or freeze-dried to obtain the precursor-pillared clays designated as shown below :

M exchanged at 20 ^oC	PM20	M exchanged at 80 ^oC	PM80
AM exchanged at 20 ^oC	PAM20	AM exchanged at 80 ^oC	PAM80

The precursor materials were calcined in air at 500 ^oC for 4 hours to yield the calcined pillared clays (CPM20, CPM80, CPAM20, CPAM80).

Surface areas, pore volumes and pore size distributions were obtained using a Micromeritics ASAP 2000 analyser (at liquid nitrogen temperatures). Samples were degassed for 8-10 hours (at 50 ^oC for the precursor-pillared materials and 450 ^oC for the calcined materials) prior to analysis. Elemental analyses were obtained using a Phillips PW 1400 X-ray fluorescence spectrometer fitted with a rhodium target. Hammet indicators were used to determine the surface acidity in the pKa range -3 to 6.8. The bleaching test (to determine the chlorophyll adsorption capacity) was carried out by adding a known amount of clay sample to super degummed (SDG) rapeseed oil containing about 8000 ppb chlorophyll and stirring the mixture for 35 minutes at 100 ^oC. The adsorbent was filtered off and the amount of chlorophyll remaining in the oil measured using a CL500 Lovibond Colourscan.

3. RESULTS AND DISCUSSION

Nitrogen gas adsorption at 77K was used to characterise the (physical) surface properties of the materials. The sorption isotherms obtained for air-dried calcined samples of the acid-activated clay (CAM) and pillared materials (CPM80 and CPAM80) are shown in Figure 1. In the classification of Brunauer, Deming and Teller [19] CAM exhibited a type II isotherm (characteristic of non-porous powders or powders with pore diameters larger than micropores) while CPM80 had a type I isotherm (characteristic of microporous powders). CPAM80 had an isotherm intermediate between type I and type II. The hysteresis loops obtained were type H3 (for CAM) and type H4 (for CPM80) in the IUPAC classification [19]. The hysteresis loop observed for CPAM80 was somewhat intermediate between type H3 and H4. Among the materials showing this type of hysteresis are those which possess slit-shaped pores or plate-like particles with spaces between the parallel plates.

The sorption isotherms obtained for air-dried precursor PAM80 and PM80 (not shown) were similar and resembled that of CPM80 shown in Figure 1. However on calcination at 500 ^oC, a shift in the CPAM80 isotherm was observed (the CPM80 isotherm as indicated in Figure 1 remained largely unchanged). This difference in the isotherms of CPAM80 and CPM80 is attributed to differences in the pore size distribution of the materials. The shifting of the isotherm of CPAM80 towards that of CAM was due to the development of both meso and microporosity in the pillared acid-activated clay as opposed to CPM80 which developed pores predominantly in the micropore range. CPAM80 may thus be considered as being intermediate between CPM80 and CAM, representing a material with well developed micro- and mesoporosity.

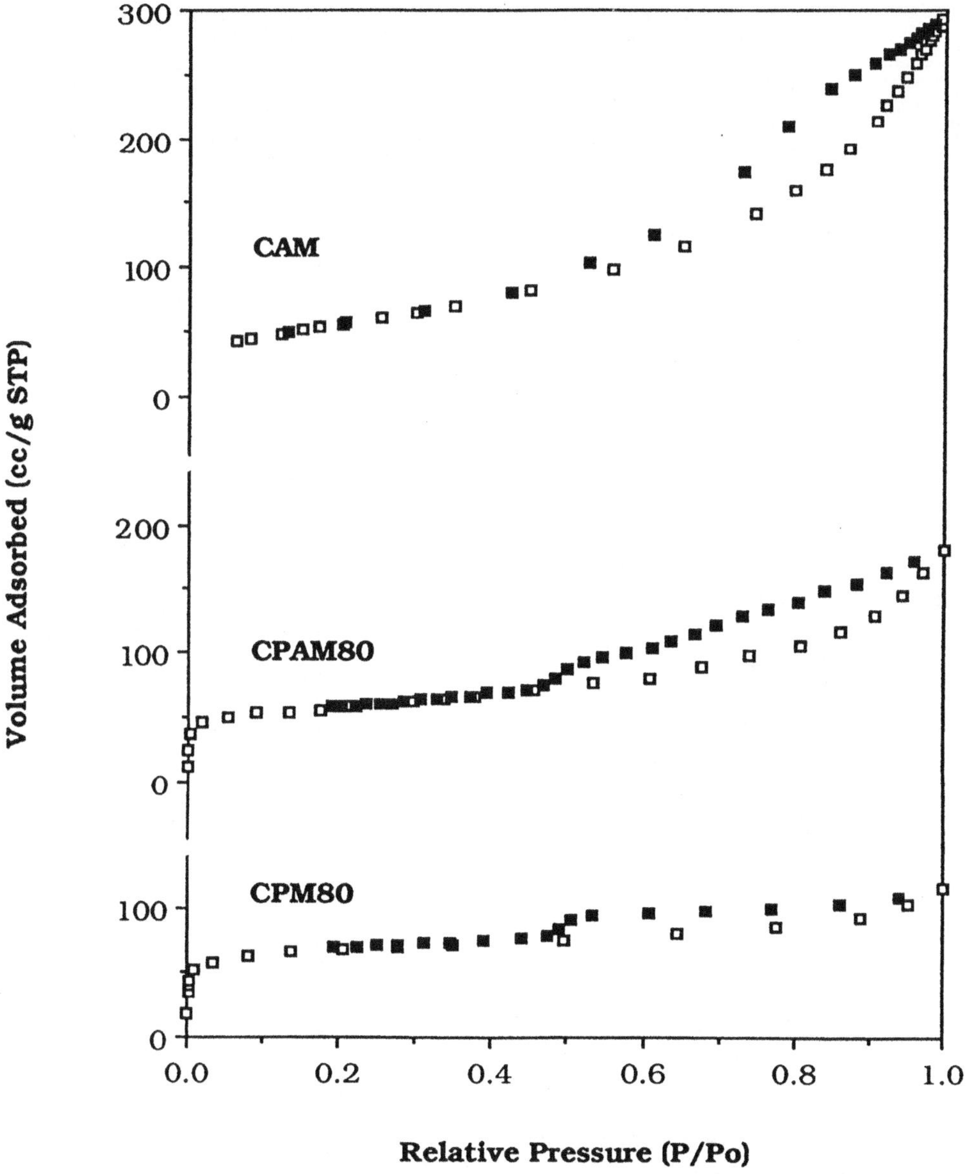

Figure 1. Adsorption-desorption isotherms for CAM, CPAM80 and CPM80.

TABLE 1. Surface areas, pore volumes and amount of alumina incorporated (Figures in parenthesis are for Freeze-dried samples).

Clay	BET surface Area (m^2/g)		Pore Volume (cc/g)		Al_2O_3 incorporated (mmole/g)
	Precursor	Calcined	Precursor	Calcined	
PAM20	191(223)	322(284)	0.27(0.31)	0.37(0.38)	2.04
PAM80	87(121)	257(259)	0.16(0.20)	0.29(0.30)	2.96
PM20	188(103)	172(200)	0.15(0.11)	0.17(0.18)	2.39
PM80	188(204)	270(293)	0.14(0.15)	0.19(0.21)	3.73

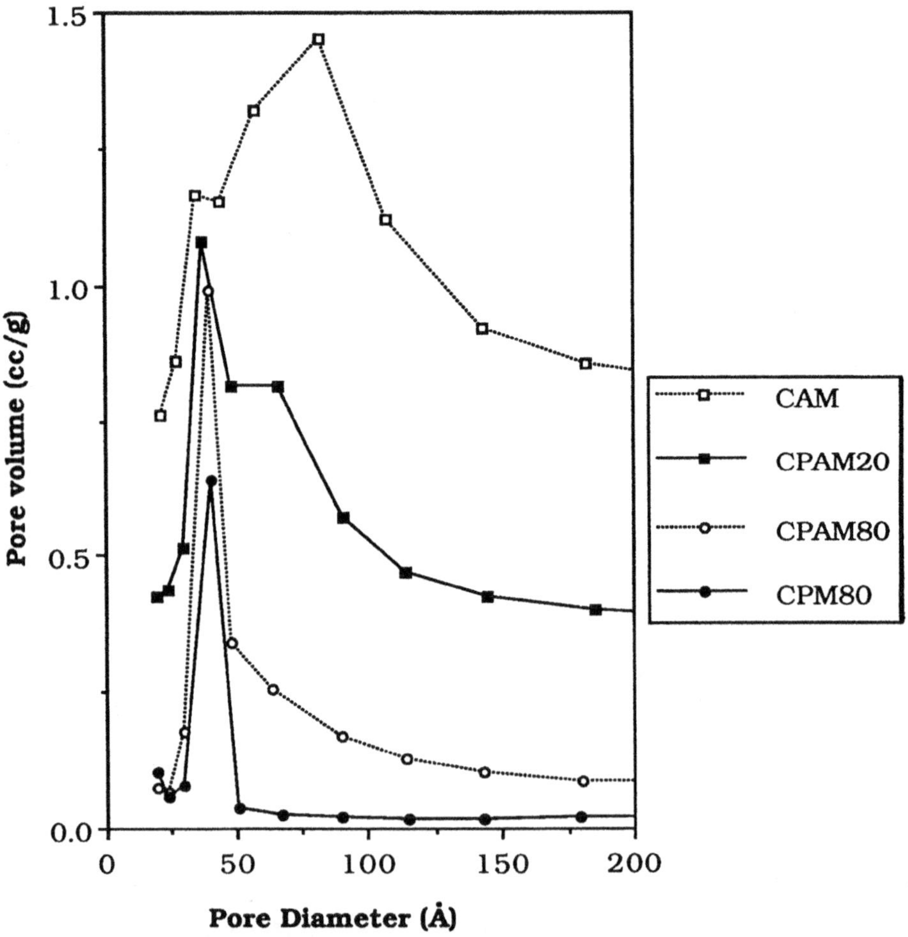

Figure 2. Pore size distribution curves for CAM, CPAM20, CPAM80 and CPM80. The curves for CPAM20 and CAM are offset (Y-axis) by 0.3 and 0.7 respectively.

The intermediate nature of the calcined pillared acid-activated clays was further highlighted by the pore size distribution curves (calculated from the desorption isotherm) for air-dried materials - see Figure 2. As expected CPM80 had a sharp pore size distribution characteristic of a microporous solid while CAM exhibited a broad pore size distribution in the mesopore range. The pore size distributions of the pillared acid-activated materials (CPAM80 and CPAM20) on the other hand indicated the existence of both micro- and mesoporosity. The relative amounts of micro- and mesoporosity were influenced by the temperature at which the exchange of the pillaring polycations was performed. Lower exchange temperature (CPAM20) led to the development of a higher proportion of pores in the mesopore range compared to those developed as a result of exchange at 80 ^{o}C. There was thus a gradual decrease in mesoporosity in the order CPM80 < CPAM80 < CPAM20 < CAM. This was reflected by the average pore diameters obtained for these materials i.e. 20.3, 32.5, 34.2, and 62.0 Å respectively. The curves shown in Figure 2 are for air-dried samples. A shift towards mesoporosity was observed for freeze-dried CPM80, CPAM80 and CPAM20 with their calculated average pore diameter increasing to 21.0, 33.8 and 38.6 Å respectively.

A summary of the surface areas, pore volumes and average pore diameters obtained for the precursor and calcined pillared materials is given in Table 1 along with the amounts of alumina incorporated. Freeze-dried materials generally had higher pore volumes compared to air-dried materials. In all cases calcination increased the pore volume while the surface area increased in all cases except for PM20. Non-acid-activated matrices incorporated more alumina (17% at 20 ^{o}C and 26% at 80 ^{o}C) than acid-activated matrices. In addition exchange at 80 ^{o}C favoured higher alumina incorporation (45% more for AM and 56% more for M) compared to exchange at 20 ^{o}C as a result of the higher mobility of the pillaring polycations at 80 ^{o}C.

It was further observed that for the pillared acid-activated materials, the temperature of exchange and method of drying affected the pore volume uniformly when either of the variables was kept constant. When the exchange temperature was increased from 20 ^{o}C to 80 ^{o}C a reduction in pore volume was observed. For precursor-pillared materials the pore volume of both the freeze-dried (FD) and air-dried (AD) samples reduced by 0.11 cc/g (a reduction of 35% for FD and 41% for AD) , while that of both their calcined analogues reduced by 0.08 cc/g (a reduction of ca. 21% for both FD and AD samples). On the other hand a change of the method of drying from air-drying to freeze-drying increased the pore volume of precursor-pillared materials by 0.04 cc/g while the pore volume of their calcined analogues increased by 0.01 cc/g.

These observations suggest that the temperature of exchange and the method of drying have quantifiable effects on the pore volume of the pillared acid-activated materials. The relative magnitude of the changes observed indicate that the temperature of exchange (and thus the amount of alumina incorporated) was more important in determining the pore volume than the method of drying. This was because the method of drying had a greater influence on the pore size distribution than the actual pore volume generated. This was due to the nature of aggregation during the drying process - whilst air-drying promoted face-to-face aggregation, freeze-drying promoted edge-to-face or edge-to-edge aggregation leading to relatively larger pores.

TABLE 2. Micropore volume distribution and chlorophyll adsorption capacity for calcined samples (AD = Air-dried, FD = Freeze-dried, Figures in parenthesis are for Precursor-pillared materials).

Clay	Micropore volume (cc/g)		% Micropore volume		Adsorption capacity % adsorbed/g clay	
	AD	FD	AD	FD	AD	FD
CPAM20	0.100	0.088	27	23	58.6(51.2)	59.3(55.6)
CPAM80	0.090	0.089	31	30	27.8(17.2)	36.0(22.3)
CPM20	0.075	0.077	44	43	2.7(2.6)	4.4(5.1)
CPM80	0.104	0.112	55	53	2.1(1.4)	5.8(5.3)

The porosity of the materials was further analysed by determining the micropore volume (Vm) from a_S plots [19]. [The micropore volume calculated for CAM was 0.055 cc/g representing 13% of its total pore volume. All the samples analysed (including CAM) had % macropore volume < 6%]. The micropore volumes obtained are given in Table 2 (with %Vm contribution to the total pore volume) along with the chlorophyll adsorption capacities expressed as %ads/g of the materials. The Vm values obtained for CPM20 and CPM80 are in agreement with those reported for conventional pillared montmorillonites [11,20]. Though the micropore volume does not vary significantly for the pillared materials, its contribution to the total pore volume does vary considerably. In all cases freeze-dried samples had a slightly lower % micropore volume compared to air-dried samples. What is more significant, however, is the increase in the % micropore volume as a result of exchange at 80 ^{o}C compared to exchange at 20 ^{o}C.

The exchange of polymeric cations into the acid-activated clay (containing an initial Vm of 0.055 cc/g) led to a significant increase in micropore volume, thus shifting the pore size distribution to lower pore sizes. It was observed that the pillared acid-activated clays had micropore volumes comparable to those of pillared non-activated clays. PAM's lower % micropore volume values are an indication of their higher total pore volume. The pillared acid-activated clays had mesopore volumes in the range 0.2 - 0.3 cc/g compared to ~ 0.1 cc/g for the pillared non-activated clays.

Figure 3 shows the variation of the number of acid sites with acid strength for calcined pillared materials. The method of drying had no effect on the acidity and thus the values given are for both air- and freeze-dried samples. The trends that emerge are :

1. Pillared acid-activated clays exhibit higher acidity compared to their pillared non-activated analogues.
2. Exchange at 80 ^{o}C yields materials with lower acidity compared to exchange at 20 ^{o}C.

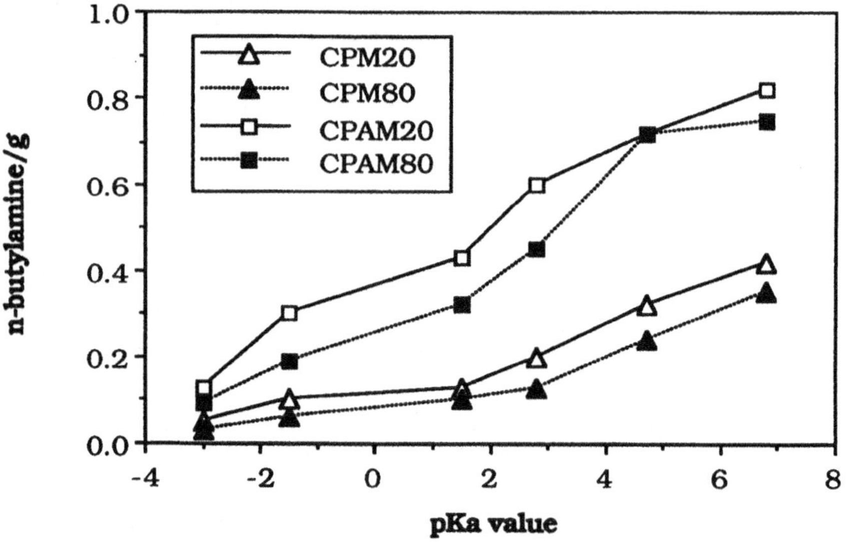

Figure 3. Variation of number of acid sites with acid strength (pKa value) for calcined-pillared materials.

Adsorption Properties.

The pillared acid-activated materials had significantly higher chlorophyll adsorption capacities compared to their non-activated pillared analogues. The trends that emerge from Table 2 are that improved adsorbents result from exchange at 20 ^{o}C, freeze-drying and calcination.

A comparison of the relative effects of exchange temperature and method of drying on the chlorophyll adsorption capacity indicated different trends for the pillared acid-activated clays (PAM) compared to the pillared non-activated clays (PM). For PAM the exchange temperature had a much greater effect on the adsorption capacity compared to the method of drying. For example, air-dried precursor PAM80 had an adsorption capacity of 17.1%/g as opposed to 51.2%/g for air-dried precursor PAM20. However on freeze-drying the precursor PAM80, its adsorption capacity increased only to 22.3%/g. Similarly treated samples of PM behave differently - increase from 1.4 to 2.6%/g was observed on changing the exchange temperature from 80 to 20 ^{o}C while a change to freeze-drying increased the adsorption from 1.4 to 5.3%/g.

For the pillared acid-activated clays these changes in chlorophyll adsorption capacity were consistent with the earlier observation that the temperature of exchange had a greater effect on pore volume than the method used for drying the clays suggesting that the adsorption capacity was responding to changes in pore volume. This hypothesis was put to test by relating the pore volume and the adsorption capacity (see Figure 4).

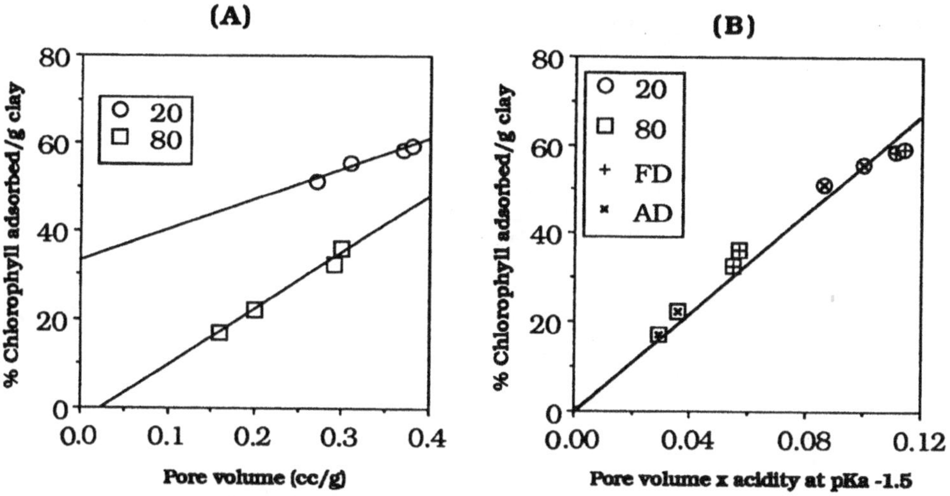

Figure 4. Relationship between chlorophyll adsorption capacity and (A) pore volume (B) a combination of pore volume and acidity at pKa -1.5.

Separate linear relationships were obtained for materials exchanged at 20 ^{o}C and 80 ^{o}C. However, when the pore volume was multiplied with the acidity at pKa -1.5, the obtained parameter had a linear relationship with the chlorophyll adsorption capacity. This was true for all the pillared acid-activated clays. The acidity at pKa -1.5 acted as a normalising factor, suggesting that the changes in adsorption capacity were as a result of the combined effect of changes in pore volume and acidity.

4. CONCLUDING REMARKS

By starting with an acid treated clay, we have been able to prepare pillared materials with improved thermal stability exhibiting characteristics intermediate between those of acid-activated clays and pillared clays. We have shown that the properties of these materials can be varied by changing the procedures used in their preparation. These pillared acid-activated clays possess micropore volumes similar to those of pillared clays but in addition they possess two to three times the mesopore volume of the latter thus making them superior in certain applications e.g. selective adsorption of chlorophyll from edible oils.

ACKNOWLEDGEMENTS

We are very grateful to Dr L Gladden and Mr P Grainge for performing the BET analyses and to Dr M E Davies and Mrs M Whittle for numerous discussions. The financial support of the Cambridge Commonwealth Trust and Laporte plc is appreciated.

REFERENCES

1. Frenkel, M. (1973) 'Surface acidity of montmorillonites', Clays and Clay Minerals 22, 435-441.
2. Mortland, M. M. and Raman, K. V. (1968) 'Surface acidity of smectites in relation to hydration, exchangeable cation and structure', Clays and Clay Minerals 16, 393-398.
3. Vaughan, D. E. W. and Lussier, R. (1980) 'Preparation of molecular sieves based on pillared interlayered clays (PILC)', in L. V. C. Rees (ed.), Proc. 5th Inter. Conf. Zeolites, Heyden, London, pp 94-101.
4. Vaughan, D. E. W., Lussier, R. and Magee, J. S. (1979) 'Pillared interlayer clay minerals useful as catalysts and sorbents', US Patent 4,176,090, 7pp.
5. Vaughan, D. E. W., Lussier, R. and Magee, J. S. (1981) 'Stabilised pillared clays', US Patent 4,248,739, 6pp.
6. Bodek, E., Ziotkiewicz, J. and Zyla M. (1983) 'Sorption properties and porosity of some cationic forms of montmorillonites and their acid-activated products', Mineral. Polonica 14, 45-54.
7. Zyla, M. and Olszar, M. (1984) 'Changes in the porous structure of montmorillonite as a function of its acid-activation time', Mineral. Polonica 15, 67-77.
8. Taylor, D. R. and Jenkins, D. B. (1988) 'Acid-activated clays', Soc. Min. Eng. AIME Trans. 282, 1901-1910.
9. Moini, A. and Pinnavaia, T. J. (1988) 'A new strategy for the synthesis of pillared clays with super galleries : Direct intercalation of metal oxide sols (DIMOS)', Solid State Ionics 26, 119-123.
10. Singh, S. S. and Kodama, H. (1988) 'Reactions of polynuclear hydroxyaluminium cations with montmorillonite and the formation of a 28-Å pillared complex', Clays and Clay Minerals 36, 397-402.
11. Sterte, J. (1991) 'Preparation and properties of large-pore La-Al-pillared montmorillonite', Clays and Clay Minerals 39, 167-173.
12. Yamanaka, S. and Brindley, G. W. (1979) 'High surface area solids obtained by reaction of montmorillonte with zirconyl chloride', Clays and Clay Minerals 27, 119 124.
13. Burch, R. and Warburton, C. I. (1986) 'Zr-containing pillared clays 1. Preparation and structural characterisation', J. Catal. 97, 503-510.
14. Bottero, J. Y., Cases, J. M., Fiessinger, F. and Poirier, J. E. (1980) 'Studies of hydrolysed aluminium chloride solutions I. Nature of aluminium species and composition of aqueous solutions', J. Phys. Chem. 84, 2933-2939.
15. Lahav, N., Shani, U. and Shabtai, J. (1978) 'Cross-linked smectites I. Synthesis and properties of hydroxy aluminium montmorillonite', Clays and Clay Minerals 26, 107-115.
16. Tokarz, M. and Shabtai, J. (1985) 'Cross-linked smectites IV. Preparation and properties of hydroxy aluminium-pillared Ce- and La-montmorillonites and flourinated NH_4^+-montmorillonites' Clays and Clay Minerals 33, 89-98
17. Tichit, D. and Figueras F. (1990) 'Preparation and modification of the thermal stabilities and acidities of Al, Zr and Si pillared smectites', in I. V. Mitchell (ed.), Pillared Layer Structures : Current Trends and Applications, Elsevier Science Publishers, London, pp. 149-158.
18. Heyding, R. D., Ironside, R., Norris, A. R. and Prysiazniuk, R. Y. (1960) 'Acid-activation of montmorillonite', Can. J. Chem. 38, 1003-1017.
19. Gregg, S. J. and Sing. K. S. W. (1982) 'Adsorption, Surface Area and Porosity, Academic Press, London.
20. Stacey, M. H. (1988) 'Alumina pillared clays and their adsorptive properties', Catalysis Today 2, 621-631.

STUDIES OF CLAY MODIFIED ELECTRODES

Dimitris Petridis
"Demokritos" National Center for Scientific Research
Institute of Materials Science
153 10 Ag. Paraskevi Attiki
Greece

ABSTRACT. The electrochemical properties of electrodes can be drastically changed by the deliberate chemical modification of their surfaces. Clays and to a less extent other inorganic solids have been employed as media for modification of electrode surfaces. The exchange properties of smectite clays allow a variety of redox couples to be incorporated readily on their intercrystalline surfaces. On the other hand, their high chemical and thermal stability as well as their unique and controllable structural characteristics offer additional advantages for their use as electrode surface modifiers. Understanding the mechanism for electron transfer and transport through the clay film, the movement of charge compensating ions and the structural features of the clay platelets responsible for electrochemical pathways to the electrode surface is very important for the design and potential value of clay modified electrodes in electrocatalytic reactions, sensor applications and solid state devices.

1. INTRODUCTION AND BRIEF OVERVIEW

During the past two decades the study of electrode surface modification has been an active area of research which has attracted increasing attention from the electrochemical, chemical and solid state communities. Innovative and fruitful work from various research groups has added a new dimension in understanding such fundamental aspects of electrochemistry as, for instance, the effects of adsorbed species on the electrical double layer and the effects of mono- or multimolecular coverage of electrode surfaces on the heterogeneous electron transfer and diffusion rates. At the same time the potential advantages of electrode surface modification for various technological applications were realized and stimulated constructive work in the fields of electrocatalysis, macromolecular electronics, semiconductors, organic synthesis, sensor electrodes, electrochromic displays and biological and medical chemistry [1-5]*.

*These and other related topics on chemically modified electrodes are thoroughly covered in the detailed reviews, references 1-5.

C.A.C. Sequeira and M.J. Hudson (eds.),Multifunctional Mesoporous Inorganic Solids, 433–450.

A modified electrode, in a general sense, is an ordinary electrode the surface of which is deliberately coated with a thin film of a chosen material for the purpose of immobilizing a specific redox reagent on its surface. The motivation for the modification process is to endow the electrode with the chemical, electrochemical, optical and other properties of the modifier so that the surface will display the reactivity and selectivity available in homogeneous systems [1-4]. In 1975 Murray and co-workers demonstrated how a designed chemical treatment could be employed for the deliberate modification of electrode surfaces. This initial work stimulated intensive studies in many research groups as is well documented in the thorough and authoritative review by Murray in 1984 [1]. Meanwhile the area of modified electrodes developed rapidly from organic monolayers to organic polymers and finally to inorganic crystalline materials.

A very attractive concept, originated from the work of Oyama and Anson [6], was the use of ion exchange polymers for casting coatings of high coverage on electrode surfaces. This potentially simple and very effective technique offers the possibility to incorporate into the polymeric matrix either redox active cations, including $M(bpy)_3^{2+}$ (M=Ru, Fe, Os, Co), $Co(terpy)_2^{2+}$, $Ru(NH_3)_6^{2+}$, viologen cations and others, or of $Fe(CN)_6^{4-}$, $IrCl_6^{4-}$ and $Co(C_2O_4)_3^{3-}$, for instance, as loading active anions [1,4]. Perfluorosulfonic acid membranes (Nafion), polyvinylsulfonic acid and polystyrene sulfonic acids have been used as cation exchange resins and protonated polyvinylpyridine or N-methylate polyvinylpyridine as anion exchangers (1,4,5).

In view of these developments the introduction of crystalline inorganic ion exchange materials for electrode surface modification was an event of no surprise. In 1983 the first example on the use of clays for electrode modification was reported by Ghosh and Bard [7]. However, the first inorganic materials to be used for this purpose were precipitated films of polynuclear transition metal cyanides of the formula $M_k[M'(CN))_6]$, where M and M' are transition metals with different oxidation numbers [8]. The deposition of Prussian blue (iron(III)hexacyanoferrate (II)) was first described by Neff in 1978 [9]. Layered double hydroxides (L.D.H.) have also been employed to modify electrode surfaces. Redox active anions, $Fe(CN)_6^{3-}$, $Mo(CN)_8^{4-}$ and $IrCl_6^{4-}$, have been intercalated into hydrotalcite films casted on tin oxide electrodes by immersing the electrodes into solutions of the desired complex [10]. Layered double hydroxides have also been used in composite electrodes [11]. The oxidation of catechol using L.D.H. composite electrodes showed improved charge-transfer kinetics and chemical reversibility which were attributed to the removal of protons from catechol aided by the basic properties of LDH. The use of zeolites for electrode modification was introduced by Rolison and co-workers [12]. Later Li and Mallouk used zeolites Y microparticles embedded into a polystyrene film for the construction of a novel modified electrode which, when loaded with metal complexes of appropriate size, allowed selective electron exchange reactions between surface and mobile species in the film [13]. In another interesting study Miller and Majda [14] prepared by anodic oxidation of aluminum surfaces in dilute phosphoric acid aluminum oxide films of controlled porosity (20 to 150 nm) according to the applied cell voltage. The inner walls of the pores were coated with a polyelectrolyte resin, for instance protonated polyvinyl pyridine, and then loaded with redox active $Fe(CN)_6^{4-}$ ions. Another attractive approach makes use of microporous glass coatings which are cast on conductive surfaces from glass sols by spin coating and thermal treatment. Experimental results show that redox active species diffuse freely through the glass films [15]. Additional inorganic supports tested for electrode modification include polyoxometalates [16], zirconium phosphate [17] and some oxides (CrO_2, RuO_2, MnO_2, VO_x) on titanium surfaces [18].

In the present article we will focus our attention to clay modified electrodes because most detailed and well understood work in the field of inorganic supports has been done with clays. A thorough review covering the literature on clay modified electrodes until 1990 has been published [19].

Smectite clays share many of the desirable properties of cationic exchange polymers for the modification of electrode surfaces. In addition to forming membrane-like films, clays are very effective cation exchangers and have certain other potential advantages compared to amorphous polymers. They present high chemical and thermal stability, known and controllable structural features, low cost, and can function as catalyst supports for dispersed metals or metal oxides [20] and semiconductors [21].

2. FABRICATION AND PROPERTIES OF CLAY MODIFIED ELECTRODES

The methods developed for the deposition of preformed polymer films on electrode surfaces can also apply for the preparation of clay modified electrodes. Droplet evaporation and spin coating are the methods most often used for casting clay films on conducting surfaces [7, 20-27]. In the first method a few microliters from a dilute dispersion of clay is spread and evaporated on the electrode surface. The method affords films of different thickness according to the dilution of the clay dispersion but suffers from nonuniformity of the film. In the spin coating method a clay dispersion is applied to a rapidly rotating disk electrode. Thick films can be produced by repeated application of very small volumes from a stock clay dispersion. The method gives more uniform and reproducible clay films. In their first publications on clay modified electrodes Bard and co-workers included polyvinyl alcohol (PVA) or colloidal platinum particles in the clay particles [7,20,22]. The addition of PVA or colloidal Pt yielded uniform films without or very few cracks and strongly adhered to the electrode surface. These films produced well developed cyclic voltammograms when transferred to solutions containing electroactive cations. It was concluded that the presence of PVA or colloidal Pt affected the morphology of the film which is important to the charge transfer process. Addition of PVA or colloidal Pt causes a high degree of particles edge-to-face association generating a house of cards structure that permits better communication among the redox species incorporated into the clay layers. However, later reports by Liu and Anson [26] and White and Bard [23] demonstrated that PVA or dispersed platinum particles were not necessary in the clay coatings for electroactivity of the intercalated redox reagents.

Anionic electroactive species, such as $Fe(CN)_6^{3-}$, $Mo(CN)_8^{4-}$, $C_2O_4^{2-}$, although repelled by the negative charge of the clay layers, penetrate the clay films readily but the current response is depressed because of electrostatic repulsions [22,26]. This result implies the presence of conductive pathways among the clay particles comprising the film. Lee and Fitch [28] made use of the electroactivity of $Fe(CN)_6^{3-}$ as a probe to study the morphology of the film for a variety preparations and electrolyte concentrations. A clay film prepared from spinning of the electrode gave, when immersed into 0.4 mM $Fe(CN)_6^{3-}$ and 0.02M NaCl solution a time invariant current in the reduction of $Fe(CN)_6^{3+}$. On the other hand, clay modified electrodes formed by air or oven drying generated a much smaller current and needed over 30 min time period in order to reach the value of current given by the spin-coated film. The authors suggested that the presence of edge-to-face cross-linking prevents or slows down the swelling of the air or oven dried films. The degree of swelling of the film is therefore vital for the electroactivity of redox ions freely penetrating the clay film. In addition they found that the current response for the reduction of 0.4 mM $Fe(CN)_6^{3-}$ depends upon the nature of the cation of the supporting electrolyte and its concentration. No current was observed in the presence of CsCl or KCl, whereas in the presence of LiCl or NaCl the peak current was dependent upon the concentration of the electrolyte. As the NaCl concentration increases the peak currents decrease and therefore the penetration of $Fe(CN)_6^{3-}$ becomes increasingly more difficult. In 1.75M NaCl the electrode becomes nearly insulating with respect to $Fe(CN)_6^{3-}$. This behavior was attributed to the change in the face-to-face interlayer distance arising from the known differences in hydration of the electrolyte cation. Similar results regarding the kind and concentration of the supporting electrolyte in the electrochemical response of $M(bpy)_3^{2+}$ and $Ru(NH_3)_6^{3+}$ complexes were first reported by Rudzinski and Bard [25].

Inoue et al investigated the effect of pH on the electrochemical response of $Ru(bpy)_3^{2+}$ and $Fe(CN)_6^{3-}$ ions using glassy carbon electrodes coated with montmorillonite films [29]. The motivation of the study was that the amount of the adsorbed ions on the clay surfaces and consequently the peak currents will be a function of the quantity of surface charge which in turn is expected to vary with the solution pH. The study showed that the peak current for the oxidation of $Ru(bpy)_3^{2+}$ was greatly depressed when the solution pH was below 4, whereas that of $Fe(CN)_6^{4-}$ was enhanced. These results were explained as a consequence of the uptake of protons by the clay platelets which in this way acquire a positive charge. Thus $Ru(bpy)_3^{2+}$ adsorption does not take place at low pH because of the positive charge of the particles, whereas positively charged voids enhance the amount of adsorbed $Fe(CN)_6^{3-}$ resulting in a higher peak current for the reduction of $Fe(CN)_6^{3-}$. The conclusion is that adsorption on edges and planar surfaces of clay platelets is fundamental for the electrochemistry of clay modified electrodes.

Methods for preparing pillared-clay layers on electrode surfaces have been reported [25]. The pillaring agents were solutions of polynuclear oxycations of aluminum or zirconium, aqueous solutions of trinuclear acetatohydroxoiron(III) cation, and an ethanol-dichloromethane solution of tetraethoxysilane. The pillared solution were prepared by soaking for several hours a Na^+-montmorillonite modified electrode in the pillaring solution, after which the sample was rinsed and finally heated to 350°C for 3 hours.

A novel pillared-clay modified electrode was conceived by Mallouk and co-workers [30]. The electrodes were made by binding $Al_{13}O_4(OH)_{28}^{3+}$-pillared montmorillonite to conductive SnO_2 or platinum surfaces via a thin coating of the cationic silane molecule

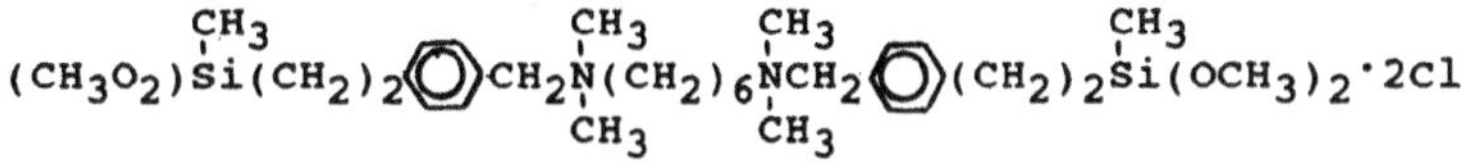

The preparation of this supramolecular functionalized electrode was accomplished by letting the reactive SnO_2 surface to react with the cationic silane in acetonitrile followed by immersing the electrode for 2 days in a stirred solution of the pillared clay in acetonitrile. Evidently this approach opens up new routes for the synthesis of a wide variety of modified electrodes comprised of functionalized organic and inorganic structures.

Clay modified electrodes possessing considerable organic function can be prepared by treating a Na^+-montmorillonite modified electrode with a tetraalkylammonium cation containing at least one long chain alkyl group [31,32]. In these organoclay modified electrodes a first layer results from the binding of the quaternary ammonium ion to the cation-exchange sites of the clay and a second layer is adsorbed by van der Waals interactions with the hydrophobic first layer surrounding the particles.

Finally electropolymerization methods have been used for the preparation of polypyrrole-clay modified electrodes [33,34].

3. ELECTROCHEMICAL METHODS FOR STUDIES OF CLAY MODIFIED ELECTRODES

For the elecrochemical characterization of clay modified electrodes cyclic voltammetry (CV) has been the principal method in use. The main advantage of the method lies in the variable time-scale of the experiment. In a CV experiment the potential of a small stationary electrode

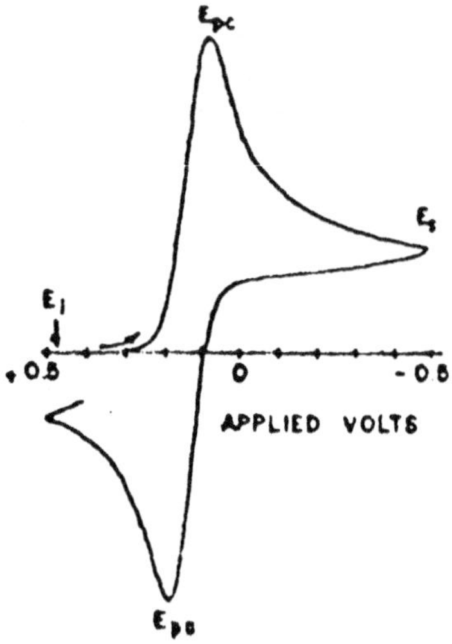

Figure 1. Cyclic voltammogram for a freely diffusing redox active species.

is swept linearly from an initial value, E_i, to a switching potential, E_s, and then back to the E_i, Figure 1.

During the cathodic scan the electroactive species near the electrode together with that which diffuses to the electrode surface in the time of the scan is reduced. The current response to such a linear sweep reaches a maximum value as the species in the immediate vicinity of the electrode is electrolysed and then decays in a characteristic "tailing" fashion. When the potential reaches the value of E_s the direction of the sweep is changed and the reverse sweep, involving the oxidation of the reduced material, is analogous to the forward sweep. The peak current is given by equation 1:

$$i_p = (2.69 \times 10^5) n^{3/2} A D^{1/2} C u^{1/2} \quad (1)$$

where n is the number of electrons transferred, A is the electrode area (cm_2), D is the diffusion coefficient ($cm^2.s^{-1}$), C is the bulk concentration (mol cm^{-3}) and u is the sweep rate (V s^{-1}). Thus for a freely diffusing species the peak current decays with a characteristic tail and its value is proportional to the square root of the sweep rate.

The response is quite different when the electrode surface is coated with a monolayer, approximately $1x10^{-10}$ mol cm^{-2}, of an electroactive reagent, Figure 2. The current again increases as we scan negatively and a peak current is again observed. However, in this case the current diminishes rapidly giving rise to symmetrical and mirror-image cathodic and anodic waves with identical peak potentials and current

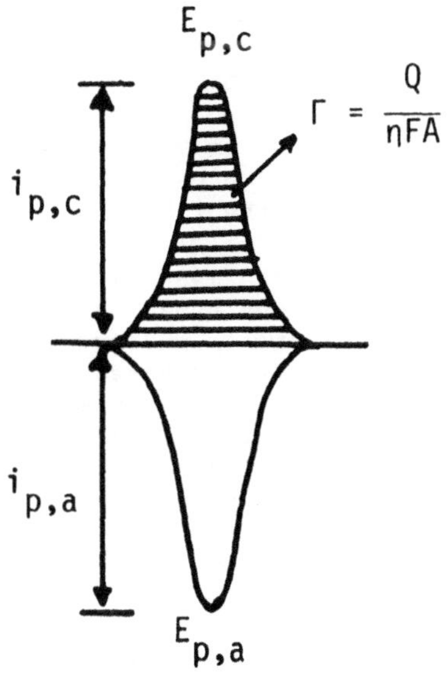

Figure 2. Cyclic voltammetric response for a surface immobilized redox couple at monolayer coverage.

levels. The different behavior arises from the fact that there is no diffusion of electroactive material towards or away from the electrode since it is now fixed at the electrode. Under these conditions the peak current is directly proportional to the rate of potential sweep (υ), and the surface coverage Γ can be obtained from the single relationship (2) since electrolysis of the entire film is completed during the voltammetric scan

$$\Gamma = \frac{Q}{nFA} \qquad (2)$$

The question of the kind of voltammetric response expected for a clay modified electrode can be answered by taking into consideration that in a swollen clay film charge transport takes place either via physical diffusion of the electroactive species to the electrode surface or via electron self-exchange where charge propagation occurs between oxidized and reduced neighbors within the film. Regarding the mechanism of self-exchange several studies have established that the site-to-site electron exchange can be treated as a diffusive transport process. The experimentally determined diffusion coefficient D_{exp} (often referred to as the apparent diffusion coefficient D_{app}) can be expressed as the sum $D_{exp} = D_o + D_{et}$ where D_o corresponds to physical diffusion and D_{et} to "diffusion" via electron hopping. The well known relationship (3) which describes the diffusion of ions in solution can be also applied to the motion of ions in a clay coating

$$\chi = (Dt)^{1/2} \qquad (3)$$

where x(cm) is the distance an ion travels in time t(sec) and D is the diffusion coefficient. This equation defines two limiting cases for charge transport in the film. If the film thickness, d, is smaller than the displacement (d<x) the entire film will be sampled during the experiment and a thin film response is anticipated. On the other hand, if d>x diffusional behavior will be observed. Since x depends on D and t these factors affect the voltammetric response. For voltammetric experiments this implies that as υ increases (t becomes shorter) there will be a shift from mirror-image in use symmetrical CV with i proportiona to υ to asymmetrical CV with i proportional to $\upsilon^{1/2}$. Both limits of diffusional (d>0.2 µm) and thin film behavior (d=30-200 nm) have been reported for clay modified electrodes [25].

Determination of Diffusion Coefficients

A straightforward method for determining D_{app} in multilayer clay modified electrodes is to use expression (1) giving peak currents for reversible waves and the corresponding plots of i_p vs $\upsilon^{1/2}$ [20]. Provided that n, C*(mole cm^{-1}) and A (cm^2) are known D_{app} can be estimated from the slope of the linear plots. The area of the electrode can be determined accurately by using ferrocyanide as a standard because of its well documented diffusion coefficient [35]. The concentration of electroactive species in the electrode film C*, can be determined by ultraviolet-visible absorption measurements at an optically transparent electrode [22,24] or by converting C* into the form Γ/d, where Γ (moles cm^{-2}) is the surface coverage of the modifying layer and d (cm) is the film thickness. The surface coverage Γ can be estimated by measuring the number of coulombs passed on a slow (<10 mV sec^{-1}) CV scan and using expression 3. Film thicknesses d are usually estimated using surface profilometry [22,24] or by other methods described in references [29,36]. D_{app} values can also be determined by chronocoulometry [25] by using the integrated form of the Cottrell equation

4. ELECTROCHEMISTRY AT CLAY MODIFIED ELECTRODES

In their initial studies Bard and co-workers established that a variety of redox active cations, $M(bpy)_3^{2+}$ (M=Fe, Os, Ru), $FcCH_2NMe_3^+$, $Ru(NH_3)_6^{3+}$ [7,22,25) and alkyl viologens [22,23] could be incorporated into smectite films and that these ions could participate in electron transfer with the underlying electrode. Intercalation of the redox cations takes place readily when the dried clay-electrodes were immersed into aqueous solutions of the ions for a period of time from 30 min to a few hours. Cyclic voltammograms could then be obtained either from these solutions or by transferring the electrode to pure electrolyte solution. Several important conclusions were drawn from these initial studies which are summarized below:

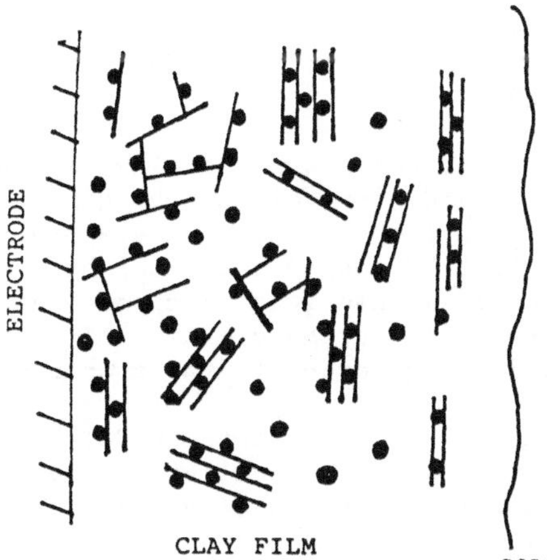

Figure 3. Schematic representation of the microstructure at a clay-modified electrode; o = $M(bpy)_3^{2+}$; ions of supporting electrolyte are omitted.

1. The clay framework acts as a preconcentrating surface. Enhanced peak currents, in comparison to those found at a bare electrode, were found in cyclic voltammograms of clay-adsorbed metal-bipyridyl cations and methyl viologen, despite their significantly lower diffusion coefficients in clay films with respect to those in solution. The local concentrations were so much larger, approximately 100-fold excess over the concentration of the contacting solution, that enhancement of current was found despite the fact that only a small fraction (~20%) of the absorbates were electroactive. Apparent diffusion coefficients for metal bipyridyl complexes were approximately $1x10^{-12}$ cm^2/sec, whereas that of $Ru(NH_3)_6^{3+}$ is a factor of 300 larger than the metal polypyridyl complexes. The $Ru(NH_3)_6^{3+}$, is expected to diffuse more rapidly within the film because of its ready access to narrow channels [25].

2. A key issue in the behavior of clay-modified electrodes is the fraction of the total absorbed species that can communicate via electron transfer with the electrode surface and the sites within the clay structure that the electroactive species occupy. Coulometric measurements show that only a fraction (15-30%) of the total complex confined in clay films is electrochemically active [22,25). The extent of electroactivity correlates in general with the external only surface of the clay (~15%) suggesting that an externally adsorbed complex is the primary contributor to charge transport in clay modified electrodes [25].

3. There is a leaching of the adsorbed cations from the film by exchange with ions from the supporting electrolyte. Leaching follows the order $Fe(bpy)_3^{2+}$ > $Os(bpy)^{32+}$ > $Ru(bpy)_3^{2+}$. Electrodes eventually

lose electroactive ions from the film by prolonged stay in aqueous solutions containing only electrolyte, yet the films retain the color of the inorganic ions [22].

4. Clay films are imperfect stacks of clay layers. They contain many defects such as holes and pores of various sizes. The stacking of the clay platelets in a film prepared by slow evaporation is expected to be more ordered than that of a spin-coated or heat evaporated film. In the last two cases the films had initial disordered states which afforded more defects and voids in the vicinity of the electroactive cation and thereby facilitating access to the electrode surface [28].

The above results have led to a general model for electroactivity shown in Figure 3. Ions reside in clay galleries, voids, and on clay surfaces. Most of the ions exchanged in clay films are electrochemically silent. Organophilic cations, located in clay galleries are electrochemically inactive due to their immobility, whereas those lying on the external surfaces are the main contributors to the observed electrochemical activity. Because ions reach the electrode by diffusing through channels around the clay particles, the swelling of the film and therefore the factors affecting swelling play an important role in the observed electroactivity.

The above proposed model though could explain many qualitative aspects of the observed electroactivity of clay modified electrodes, it could not offer a satisfactory answer regarding the extent to which the $M(bpy)_3^{2+}$ or MV^{2+} at the exchange sites of the clay contribute to the voltammetric response. The answer to this question was provided by the work of King et al. [27] who examined the behavior of films formed from pre-exchanged $Os(bpy)_3^{2+}$-montmorillonite. Montmorillonite electrode films were prepared in which 25, 50, 80 and 100% of the Na^+ ions had been replaced by $Os(bpy)_3^{2+}$. In this way, $Os(bpy)_3^{2+}$ was present on exchange sites both within the galleries and on the external surfaces. When these electrodes were transferred to solutions containing only pure electrolyte no evidence for any voltametric $Os(bpy)^{3+/2+}$ response was observed, Figure 4A. From these results it is clear that $M(bpy)_3^{2+}$ and MV^{2+} cations electrostatically bound by ion exchange to the galleries and external surfaces of the clay are rigorously electroinactive. However, electroactivity was restored by immersing the 80% $Os(bpy)_3^{2+}$-exchanged montmorillonite coated electrode into 0.2 mM solution of $Os(bpy)_3^{2+}$, Figure 4B.

The authors addressed also the question of transfer of electrons to the electrostatically bound surface cations by freely diffusing ions incorporated in clay films. The $Fe(bpy)_3^{2+}$ ions was used as a convenient probe of oxidation-reduction processes in 80% $Os(bpy)_3^{2+}$-exchanged electrodes. This electrode was placed into a solution of $Fe(bpy)_3^{2+}$, rinsed once with water and transferred to a cell containing pure electrolyte. The cyclic voltammogram, showed a well-defined $Fe(bpy)_3^{2+}$ wave with a peak current ratio of unity. Therefore $Fe(bpy)_3^{2+}$ did not shuttle charge to the $Os(bpy)_3^{2+}$ adsorbed at ion

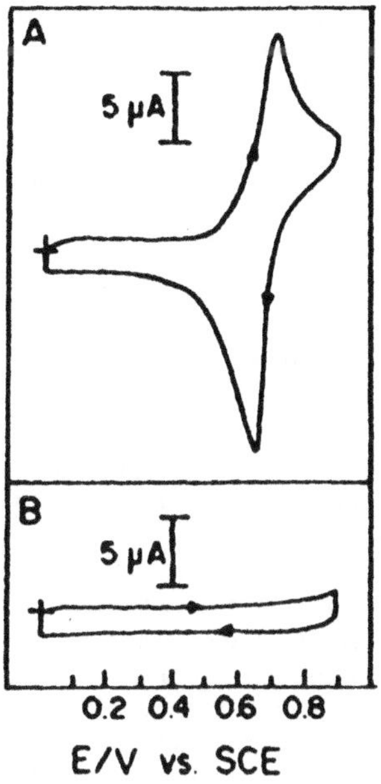

Figure 4. Cyclic voltammograms for (A) a sodium montmorillonite film in 0.2 mM $Os(bpy)_3^{2+}$ solution; (B) a pre-exchanged $Os(bpy)_3^{2+}$-montmorillonite film (80% exchange) in pure electrolyte. Scan rate: 50 mV s^1; 0.1M Na_2SO_4 (from ref. [27]).

exchange sites as might be postulated from the following reactions:

$$Fe(bpy)_3^{2+} \longrightarrow Fe(bpy)_3^{3+} + e^-$$

$$Fe(bpy)_3^{3+} + [Os(bpy)_3^{2+} - Mont]$$

$$Fe(bpy)_3^{2+} + [Os(bpy)_3^{3+} - Mont]$$

The above results established that $M(bpy)_3^{2+}$ and MV^{2+} ions bound electrostatically to the exchange sites of smectite clays do not undergo electron transfer with the electrode surface or with freely diffusing cations in the microchannels. Another important aspect of the electrochemistry on clay surfaces refers to the electroneutrality of the film during the electrolytic process. Keeping the clay layers neutral entails transfer of ions to or from the clay surfaces. Tris bipyryl metal complexes or methyl viologen are strongly adsorbed by clay surfaces and not likely to be dissorbed. Therefore the only possible mechanism for the reduction of electrostatically bound $M(bpy)_3^{2+}$ ions is to balance the charge change (2+ -> +3) by anion adsorption which, however, as the authors suggest is rather unlikely due to the negative charge of the clay layers. Therefore, they concluded that complexes strongly adsorbed by the clay layers are electroinactive when present in amounts less than the cation exchange capacity of the mineral. The electroactivity observed when the clay electrodes were dipped into a solution of $M(bpy)_3^{2+}$ or MV^{2+} was attributed to the presence of physically adsorbed $M(bpy)_3^{2+}/SO_4^{2+}$ or MV^{2+}/SO_4^{2-} ion pairs which exist in excess of the cation exchange capacity of the film. The ion pair mechanism is consistent with the electroneutrality of the clay film during the electrolytic event as depicted by equation

$$([2ML_3^{2+}, 2SO_4^{2-}], ML_3^{2+})_{clay} + [SO_4^{2-}] \xrightarrow{-2e^-} ([2ML_3^{3+}, 3SO_4^{2-}], ML_3^{2+})_{clay}$$

Support for ion pairing is provided by the dependence of the peak currents on the anion of the supporting electrolyte. For instance, the limiting current for an $Os(bpy)_3^{3+/2+}$ wave with Na_2SO_4 as electrolyte was 1.4 to 3.0 times as large as as the current observed with $C_2H_3O_2Na$ as the electrolyte over the ionic strength range 0.05-0.30M. The relative importance of ion pair formation on basal or edge surfaces was also addressed by examining the cycle voltammetric properties of clay films with different morphological properties. For this purpose synthetic laponite and fluorohectorite clays with particles sizes different than that of montmorillonite were chosen. The clay platelet sizes decrease along the series fluorohectorite (>1000 nm); montmorillonite (<200 nm); laponite (<50 nm). Decreasing the platelet size should increase the edge surface area greatly without altering considerably the basal surface area within the oriented film. The peak currents correlate with clay particle size: laponite > montmorillonite > fluorohectorite suggesting that the electrochemically active ion pairs may be associated with edge surfaces.

The ion pair mechanism has been manifested in other cyclovoltammetric studies related to the electroactivity of $Cr(bpy)_3^{3+/2+}$ pair [36], the organic dye thionine [37] and the surface active (ferrocenylmethyl)dodecyldimethylammonium ion [31] in smectite clay modified electrodes. These cations were electroactive only when the cation loading exceeding the C.E.C. of the mineral.

The ion pair mechanism offers the possibility to incorporate electroactive anions, such as $Fe(CN)_6^{4-}$ or $Mo(CN)_6^{4-}$, in clay films. This advantage has been exploited in an electrochemical study of the complex salts $[Fe(bpy)_3]_2Fe(CN)_6$ and $(MV)_3[Fe(CN)_6]_2$ assembled in montmorillonite films by ion exchange and ion pairing mechanisms [38]. The cyclic voltammograms showed, in addition to the $Fe(bpy)_3^{3+/2+}$ or $MV^{2+/+}$ response, waves due to the $Fe(CN)_6^{3-/4-}$ interconversion. The presence of electrochemically active $Fe(CN)_6^{4-}$ or $Fe(CN)_6^{3-}$ intercalated anions supports further the ion pair mechanism for the electroactivity of metal polypyridyl or other strongly adsorbed electroactive species on clay surfaces.

Smectite clays intercalate easily quaternary ammonium cations. When the ammonium cation is surface active, e.g. it has at least one long chain alkyl group, the clay retains the surfactant in amounts exceeding the CEC of the mineral [39,40]. The surface active molecules assemble spontaneously onto the clay surfaces and form bilayers which reverse the negative charge of the platelets and yield positively charged clay surfaces [40]. The counter anions of the bilayers can be ion exchanged with electroactive anions, such as $Fe(CN)_6^{3-}$, $Mo(CN)_8^{4-}$, $Fe(OX)_3^{3-}$, and consequently the clay films exhibit electrochemical response for their oxidation or reduction [31,32].

The uptake of 0.1 mM $Cr(bpy)_3^{3+}$ in 0.01M Na_2SO_4 by montmorillonite-modified Pt electrodes was studied by recording the increase in peak current with time [36]. Two peaks were observed, Figure 5. Initially peak II increases and then decreases as peak I increases. The species associated with peak I (more positive potential) was attributed to a product of some reaction involving the first appearing species in the film. To explain these results a model was advanced relating peak I to the docking of the tris(diimine) complex to the hexagonal holes on the face surface of the clay. The following steps were considered:

$$AX_n^* \longrightarrow AX_n^c$$

$$AX_n^c + n(\text{Na-clay}) \longrightarrow \text{A-clay} + n(NaX^c)$$

$$\text{A-clay} \longrightarrow \text{A-hole}$$

where AX_n^* is the bulk solution of ion-paired complex, AX_n^c and NaX^c are the ion-paired species in the vicinity of the face surface, Na-clay is the Na-exchanged clay, and A-clay is the complex-exchanged clay. The first equation expresses a diffusion step, the second an ion-exchange step and the third describes the docking of the complex. When the $Cr(bpy)_3^{3+}$ concentration is small (0.1 mM) the growth of only a single

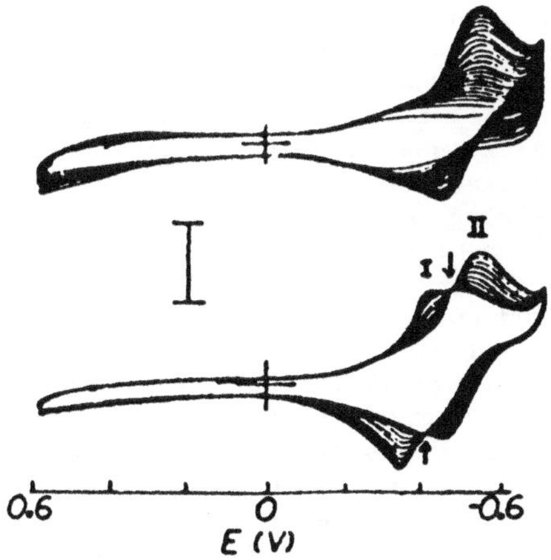

Figure 5. Uptake of 0.1 mM $Cr(bpy)_3^{3+}$ in 0.01 M Na_2SO_4 by SW_{y-1} montmorillonite-modified Pt electrode as measured by increases in current with time (bar marks 0.5 μA). Peak II first increases (upper) and then decreases as peak I increases (lower). Arrows mark location of isopotential points (from ref. [36]).

peak I (A-hole) is observed, indicating that the rate-determining step is the first reaction. When the concentration of $Cr(bpy)_3^{3+}$ was increased to 0.15 mM an isopotential point was developed indicating that the third reaction is rate limiting and that the total material was distributed between A-hole and C-clay. Since the equilibrium favors the docking of the complex to the hexagonal holes of the clay surfaces peak I becomes dominant.

An important question is whether all metal complexes must be co-adsorbed as ion pairs on clay surfaces in order to be electroactive. The answer is given in a recent publication describing in detail the electrochemical behavior of $Ru(NH_3)_6^{3+}$ and $Fe(bpy)_3^{2+}$ in clay-modified electrodes [41]. $Ru(NH_3)_6^{3+}$-montmorillonite films gave cyclic voltammograms with a well defined $Ru(NH_3)_6^{3+/2+}$ wave even when the complex ion loading was less than the CEC of the clay film. Chronocoulometry indicated that more than 80% of the $Ru(NH_3)_6^{3+}$ electrostatically bound at the cation exchange sites of the gallery surfaces was electroactive. The concentration and nature of the cation of the electrolyte greatly affected the peak currents of the homoionic $Ru(NH_3)_6^{3+}$ clay, whereas the electroactivity of the ion paired $Fe(bpy)_3^{2+}/SO_4^{2-}$ were not affected by the electrolyte cation. The peak current for the reduction of $Ru(NH_3)_6^{3+}$ was decreased with increasing electrolyte concentration because of the exchange of $Ru(NH_3)_6^{3+}$ by electrolyte cations.

Pillared clays have also been employed as media for electrode modification [24,25]. Tris(diimine) complexes intercalate into iron, aluminum, zirconium and silicon oxide pillared clays. The number of electroactive $Fe(bpy)_3^{2+}$ moles incorporated into an iron-pillared clay modified electrode corresponds to a 2-6% loading, probably because a large fraction of the ion-exchangeable sites are occupied by the pillaring reagent. Well defined cyclic voltammograms were obtained from pillared clay-modified electrodes in 0.1 $NaClO_4$/EtOH solutions, in contrast to the much attenuated peak current observed for sodium montmorillonite-modified electrodes under the same conditions. The question of whether an ion-pairing mechanism operates in pillared clay-modified electrodes remains to be answered.

Aluminum pillared montmorillonite was used to modify SnO_2 or Pt surfaces through the bridging action of a cationic silane molecule (see part II) [30]. The positive charges of the silane molecule bind electrostatically multiply charged anions, whereas absorption of tris(diimine) complexes occurs at the external surfaces of the clay. Because of spatial ordering of the electroactive anions and cations charge trapping behavior was observed in cyclic voltammetry. When the electrodes were loaded with both $Fe(CN)_6^{4-}$ and $Os(bpy)_3^{2+}$ the redox response of $Fe(CN)_6^{3-/4-}$ couple is facile since these ions are near the electrode surface, whereas direct oxidation or reduction of $Os(bpy)_3^{2+/3+}$ is very slow because these ions are far away for rapid electrode exchange with the electrode surface. The cyclic voltammogram shows only the peak arising from the $Fe(CN)_6^{3-/4-}$ interconversion. However, holding the electrode for prolonged periods of time at 1.0 V.

induces a pronounced change in the CV response. With very thin films ($Fe(CN)_6^{4-} \leq 10^{-10}$ mol/cm²) a cathodic current near the $Os(bpy)_3^{3+/2+}$ potential was observed and the cathodic peak for the reduction of $Fe(CN)_6^{3-}$ was significantly enhanced. The large cathodic wave at 0.15V was attributed to direct $Fe(CN)_6^{3-}$ reduction plus $Fe(CN)_6^{4-}$-mediated $Os(bpy)_3^{3+}$ reduction by the thermodynamically favorable reaction

$$Fe(CN)_6^{4-} + Os(bpy)_3^{2+} \longrightarrow Fe(CN)_6^{3-} + Os(bpy)_3^{2+}$$

5. CATALYTIC APPLICATIONS

The first catalytic use of clay modified electrode was to support colloidal Pt or colloidal RuO_2 near the electrode surface. A Na^+-montmorillonite/Pt/PVA modified tin oxide electrode was employed to study the voltammetric behavior of the uncharged propyl viologen sulfonate (PVS). A large catalytic wave was observed. The voltammograms exhibited a marked pH dependence with enhanced peak currents and positive shifts in the peak potential at lower pH values. The catalytic effct was attributed to Pt-mediated proton reduction and concomitant regeneration of PVS [22]. Similar catalytic effects were observed in the voltammetric behavior of the switterionic $Ru(bpy)_2[bpy(COO)_2]$ species on a clay/RuO_2/PVA-modified electrode. A catalytic current during the oxidation of $Ru(bpy)_2[bpy(COO)_2]$ was observed. In this case the oxidized Ru(III) complex shuttles electrons to RuO_2 immobilized near the electrode surface with regeneration of the Ru(II) complex. Concomitantly the oxidized RuO_2 mediates the oxidation of H_2O [20].

In a novel application Rusling and co-workers [42] have used a cetyltrimethylammonium-clay modified electrode to organize the electrocatalytic dehalogenation of 4,4'-dibromobiphenyl (4,4'-DBB) in aqueous solutions. The organoclay film provides a hydrophobic environment suitable to host the substrate. As charge shuttler the complex $Co(bpy)_3^{2+}$ was used which, on account of its positive change, could be incorporated into the clay layers. The cobalt complex can be reduced in two steps according to the equations:

$$Co(bpy)_3^{2+} + e \longrightarrow Co(bpy)_3^{+}$$

$$Co(bpy)_3^{+} + 2e^- \longrightarrow Co(bpy)_2^{-} + bpy$$

The $Co(bpy)_2^-$ is the reactive species involved in the reduction of 4,4'-DBB according to:

$$Co(bpy)_2^- + Ar_2Br_2 + H^+ + bpy \longrightarrow Co(bpy)_3^+ + Ar_2BrH + Br^-$$

It is important to stress here that the anionic $Co(bpy)_2^-$ remains in the clay film because it participates in ion-pair formation with excess $CTAB^+$ retained in the clay layers. The reduction of 4,4′-DBB in aqueous $CTAB^+$ at -1.5V represents a decrease in overpotential of at least 600 mV compared to the uncatalyzed reduction of 4,4′-DBB.

A very interesting catalytic effect in which the lattice iron of the clay plays a unique role was reported by Oyama and Anson [43]. The clay was found to catalyze the decomposition of H_2O_2 by $Ru(NH_3)_6^{2+}$ formed by electrochemical reduction of $Ru(NH_3)_6^{3+}$ at the clay electrode surface. The catalytic mechanism was attributed to the shuttling of electrons from $Ru(NH_3)_6^{2+}$ to Fe(III) centers in the clay to generate Fe(II) which in turn reduced the hydrogen peroxide as shown by the following equations:

$$Ru(NH_3)_6^{3+} + e \longrightarrow Ru(NH_3)_6^{2+}$$

$$Ru(NH_3)_6^{2+} + [Fe(III)]_{lat} \longrightarrow Ru(NH_3)_6^{3+} + [Fe(II)]_{lat}$$

$$[Fe(II)]_{lat} + H_2O_2 \longrightarrow [Fe(III)]_{lat} + H_2O + 1/2\ O_2$$

Further studies on this system have confirmed the electrocatalytic activity of lattice iron in smectite clay modified electrodes by using ferruginous clays, such as nontronite, which greatly enhances the reduction current of $Ru(NH_3)_6^{3+}$ in stirred solution voltammetry in the presence of hydrogen peroxide. On the other hand, the absence of any enhancement in the faradaic current in stirred solution voltammetry and the cathodic to anodic peak current ratio of unity for iron free clays, such as laponite and fluorohectorite, provide strong evidence that the catalytic activity truly originates from the lattice iron of the clay [44].

In conclusion this article describes fundamental principles of the electrochemistry of clay modified electrodes. The same principles should also apply to other inorganic ion exchange or porous materials. Research in the field has demonstrated the use of clay modified electrodes in catalytic applications. On the other hand, technical applications such as in electronic devices or the use of clays in batteries appear feasible and should encourage further research in the field.

ACKNOWLEDGEMENTS

Helpful discussions with Dr. P. Falaras are gratefully acknowledged. This work was partially supported by a NATO grant No. 890504.

REFERENCES

1. Murray, R.W. (1984), in Bard A.J. (ed.) "Electroanalytical Chemistry", 13, Marcel Dekker, New York, p.191.
2. Miller, J.S. (1982) "Chemically Modified Surfaces in Catalysis and Electrocatalysis", American Chemical Society, Washington D.C., p.1.
3. Snell, K.D. and Keenan, A.G. (1979) Chem. Soc. Rev., 8, 259.

4. Abruna, H.D. (1988) Coord. Chem. Rev., 86, 135.
5. Merz, A. (1990) in E. Steckham (ed.) "Electrochemistry IV", Topic in Current Chemistry, Springer-Verlag, p. 49.
6. Oyama, N., and Anson, F.C. (1980) J. Electrochem. Soc., 127, 247.
7. Ghosh, P.K. and Bard, A.J. (1983) J. Am. Chem. Soc., 105, 5691.
8. Itaya, K., Uchida, I. and Neff, V.D. (1986) Acc. Chem. Res., 19, 162.
9. Neff, V.D. (1978) J. Electrochem. Soc., 125, 886.
10. Itaya, K., Chang, H.-C. and Uchida, I. (1987) Inorg. Chem. 26, 624.
11. Shaw, B.R. and Creasy, K.E. (1988) J. Electroanal. Chem., 243, 209.
12. Murray, G.C., Nowark, R.J. and Rolison, D.R. (1984) J. Electroanal. Chem., 164, 205.
13. Li, Z., Mallouk, T.E. (1987) J. Phys. Chem., 91, 643.
14. Miller, C.J. and Majda, M. (1985) J. Am. Chem. Soc. 107, 1419.
15. Lundgren, C.A. and Murray, R.W. (1987) J. Electroanal. Chem., 227, 287.
16. Keita, B., Nadjo, L., and Haeussler, J.P. (1988) J. Elecroanal. Chem. 243, 481.
17. Lee, H., Kepley, L.J., Hong, H.-G., Akhter, S., Mallouk, T.E. (1988) J. Phys. Chem., 92, 2597.
18. Beck, F., and Gabriel, W. (1985) J. Electroanal. Chem., 182, 355.
19. Fitch, A. (1990), Clays and Clay Minerals, 38, 391.
20. Ghosh, P.K., Mau, A.W.-H. and Bard, A.J. (1984) J. Electroanal. Chem., 169, 315.
21. Enea, O., and Bard, A.J. (1986) J. Phys. Chem., 190, 301.
22. Ege, D., Ghosh, P.K., White, J.R., Equey, J.-F., and Bard, A.J. (1985) J. Am. Soc. 107, 5644.
23. White, J.R., and Bard, A.J. (1986) J. Electroanal. Chem., 197, 233.
24. Itaya, K. and Bard, A.J., (1985) J. Phys. Chem. 89, 5565.
25. Rudzinski, W.E. and Bard, A.J., (1986) J. Electroanal. Chem. 199, 323.
26. Liu, H.-Y. and Anson, F.C. (1985) J. Electroanal. Chem., 184, 411.
27. King, R.D., Nocera, D.G., and Pinnavaia, T.J. (1987) J. Electroanal. Chem., 236, 43.
28. Lee, S.A. and Fitch, A. (1990) J. Phys. Chem. 94, 4998.
29. Inoue, H., Haga, S., Iwakura, C. and Yoneyama, H. (1988) J. Electroanal. Chem., 249, 133.
30. Rong, D., Kim, Y. and Mallouk, T.E. (1990) Inorg. Chem. 29, 1531.
31. Brahimi, B., Labbe, P. and Reverdy, G. (1989) J. Electroanal. Chem., 267, 343.
32. Falaras, P. and Petridis, D. submitted to J. Electroanal. Chem.
33. Castro-Acuna, C.M., Fan, F.-R.F. and Bard, A.J. (1987) J. Electroanal. Chem., 234, 347.
34. Rudzinski, W.E., Figueroa, C. and Hoppe, C. (1988) J. Electroanal. Chem., 243, 347.
35. Adams, R.N. (1969) Electrochemistry at Solids Electrodes, M. Dekker, New York.
36. Fitch, A., Lavy-Feder, A., Lee, S.A. and Kirsh, M.T. (1988) J. Phys. Chem. 92, 6665.
37. Kamat, P.V. (1984) J. Electroanal. Chem., 163, 389.

38. Petridis, D., Falaras, P. and Pinnavaia, T.J. submitted to Inorg. Chem.
39. Greenland, D.J. and Quirk, J.P. (1962) Clays Clay Miner. 9, 499.
40. DellaGuardia, R.A. and Thomas, J.K. (1983) J. Phys. Chem. 87, 3550.
41. Kavirathna, P.D.S. and Pinnavaia, T.J., submitted to J. Electroanal. Chem.
42. Rusling, J.F., Shi, C.-N. and Suib, S.L. (1988) J. Electroanal. Chem. 245, 331.
43. Oyama, N., and Anson F.C. (1986) J. Electroanal. Chem. 199, 467.
44. Kaviratna, P.D.S., Petridis, D. and Pinnavaia, T.J. submitted to Clays and Clay Minerals.

MICROPOROUS-MESOPOROUS METAL-OXYGEN CLUSTER COMPOUNDS: ION EXCHANGE, STRUCTURE RETENTION AND THE OXIDATIVE DEHYDROGENATION OF ISOBUTYRIC ACID

G.B. McGARVEY AND J.B. MOFFAT
Department of Chemistry and Guelph-Waterloo
Centre for Graduate Work in Chemistry,
University of Waterloo, Waterloo, Ontario N2L 3G1 Canada

ABSTRACT. Metal-oxygen cluster compounds can be prepared with microporous-mesoporous structures by precipitation of certain of their monovalent cation salts from aqueous solution. Exchange of the cations of these salts can be achieved, although the exchange is not complete and is dependent upon the relative sizes of the entering and leaving cations. Measurements of surface areas, pore size distributions and lattice parameters, before and after cation exchange, show that the microporous-mesoporous structures are semiquantitatively retained, although not surprisingly shifted as a result of the change in size of the cation. The catalytic properties of the exchanged microporous-mesoporous metal-oxygen cluster compounds, as reflected in the conversion of isobutyric acid and selectivity to the various products, show a dependence on the nature and concentration of the cations as well as the expected relationship to the elemental composition of the anions.

Introduction

In 1974 the International Union of Pure and Applied Chemistry issued a classification and nomenclature for pore sizes in solids. Micropores have diameters less than 20 Å while mesopores fall in the range from 20 to 500 Å and macropores are larger. While the subdivisions are of course somewhat arbitrary they nevertheless serve a useful purpose in facilitating discussions of porous solids their properties and applications. Although, it is probable that the focus of attention during the last decade has fallen on microporous solids, however, as is evidenced from the 1992 NATO conference on Multifunctional Mesoporous Inorganic Solids, interest in the preparation, properties and application of solids with larger pores is now growing.

The potential applications for mesoporous solids are numerous. Physical separations of molecules larger than those currently accomplished with zeolitic molecular sieves will be possible. Shape selectivity, whose effects on reactants, products and transition states are well known in microporous systems may be extended to larger molecules with mesoporous solids. The trapping and thus separation, of a wide variety of materials in numerous processes can be envisaged with solids containing mesopores. Purification, preconcentration and recovery processes may be affected leading to competitive advantages not presently available with techniques currently employed.

In principle, methods of preparation of solids containing mesopores can be broken down into a number of approaches. Solids normally possessing no porous structure may be synthesized by

C.A.C. Sequeira and M.J. Hudson (eds.),Multifunctional Mesoporous Inorganic Solids, 451–472.

alternative methods which are capable of producing mesoporous structures. Ion exchange may be capable of introducing mesoporosity into solids where pores are not present or at least may enlarge pores already present. In the present work the former approach is employed to generate solids with porous structures which are altered through ion exchange.

The work to be described here is concerned with metal-oxygen cluster compounds containing four or more elements [1, 2]. These solids contain discrete cations and anions, in contrast to the network structures found for the latter in many zeolites. The Keggin anions (Figure 1) are high

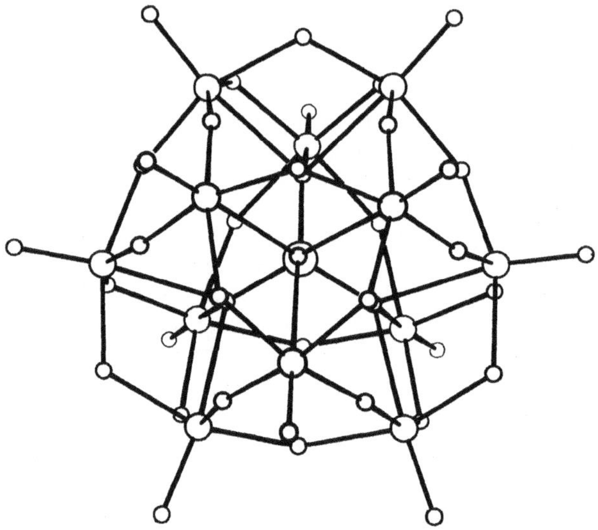

Figure 1. Metal-oxygen cluster anion of Keggin structure (KU). Small circles: central atom and peripheral metal atoms; large circles: oxygen atoms.

molecular weight cagelike structures containing a large number of oxygen atoms and at least two other elements. Each anion contains an atom such as phosphorus, silicon or arsenic at its centre with four oxygen atoms, arranged tetrahedrally, surrounding the central atom. Twelve octahedra with oxygen atoms at their vertices and a peripheral metal atom such as molybdenum or tungsten at their approximate centres surround the central tetrahedron of the anion and share oxygen atoms between themselves and the tetrahedron. Thus there are two types of bridging oxygen atoms found in the anion, those bridging two peripheral metal atoms and those connecting the latter with the central atom. In addition twelve terminal oxygen atoms protrude from the anion. It is important to note that, although the pictorial representation of these anions is reminiscent of the cages in certain zeolites, the metal-oxygen cluster anions are packed with atoms so that, unlike the cages in zeolites, even the smallest molecular species are unable to enter. The stoichiometry of the Keggin anion (sometimes referred to as the Keggin Unit, KU) may be written as

$$[(XO_4)\,(M_{12}O_{36})]^{-n}$$

where the value of n is dependent upon the oxidation numbers of the central atom X and of the peripheral metal atoms M.

Although a variety of cations may be employed to balance the charge of the anions, the simplest is the proton. With phosphorus as the central atom and tungsten as the peripheral metal element the proton is found to be located in the centre of and hydrogen-bonded to two of approximately coplanar water molecules (Figure 2) [3]. With the hexahydrate, each proton is hydrogen-bonded to two water molecules but, as a consequence of a two fold thermal disorder the positions of the water molecules may vary as shown. The water molecules are, in turn, hydrogen-bonded to the terminal oxygen atoms of the anions. As will be described in more detail later in this report the solid acid forms, that is where the proton is the cation, of the metal-oxygen cluster compounds are found to have low surface areas, generally less than 10 m^2/g.

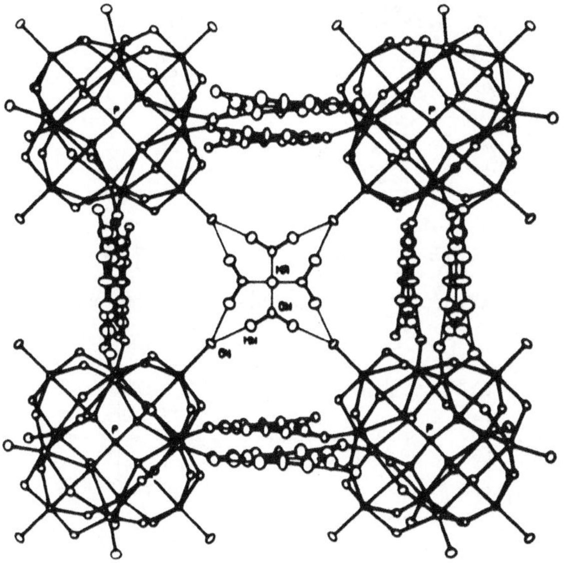

Figure 2. Arrangement of anions and diaquahydrogen cations in the hexahydrate of 12-tungstophosphoric acid [3].

While it is evident that a variety of these metal-oxygen cluster compounds can, at least in principle, be prepared, those of primary interest in this report contain phosphorus as the central atom and either molybdenum or tungsten as the peripheral metal atoms.

The thermal stability of the anions may be assessed by examining the photoacoustic FTIR spectra after heating 12-tungstophosphoric acid at various temperatures (Figure 3) [4]. A series of 5 or 6 bands between 800 and 1100 cm^{-1} is indicative of the anion structure. In particular two bands at approximately 1080 and 980 cm^{-1}, attributed to the triply degenerate asymmetric stretch of the central XO_4 tetrahedron and the stretch of the peripheral metal-terminal oxygen bond, respectively, may be employed as a measure of the presence of the anion. Although at 450°C the intensity of these bands is evidently reduced, nevertheless it is evident that the anion structure remains at that temperature.

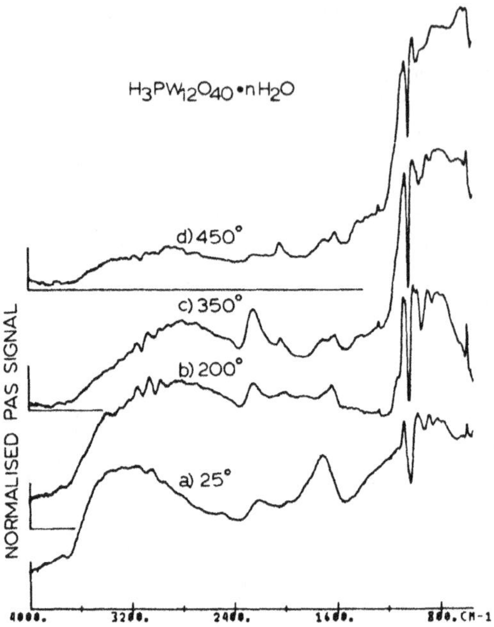

Figure 3. Photoacoustic FTIR spectra of $H_3PW_{12}O_{40} \cdot nH_2O$ after heating at various temperatures (°C) *in vacuo* [4].

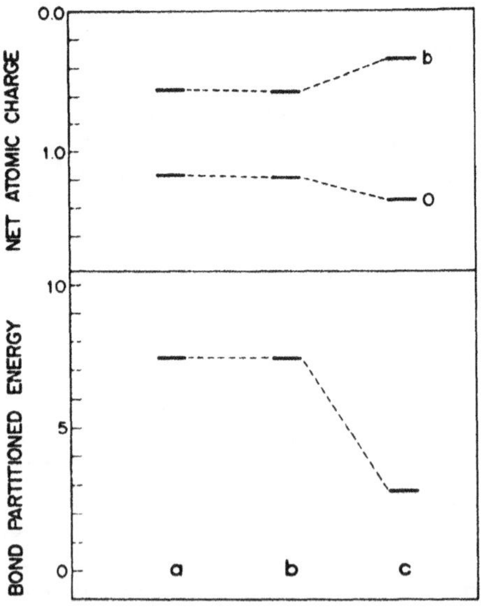

Figure 4. Net atomic charges on the bridging (b) and outer (o) oxygen atoms, and energies of the M-O (outer) bonds [5-7].

Semiempirical extended Hückel calculations on the anions $PW_{12}O_{40}^{-3}$, $PMo_{12}O_{40}^{-3}$ and $SiW_{12}O_{40}^{-4}$ predict the catalytic multifunctionality of the metal-oxygen cluster compounds (Figure 4) [5-7]. The bond energies of the peripheral metal-terminal oxygen bonds are higher for the anions containing tungsten than for those with molybdenum as the peripheral metal element while the charges on the terminal oxygen atoms are higher for the latter than the former. Thus the terminal oxygen atoms in the molybdenum-containing metal-oxygen cluster anions are more labile than those with tungsten while the protons in the compounds containing tungsten are more mobile and hence more acidic than where molybdenum is present. It is therefore expected that the tungsten-containing metal oxygen cluster compounds will be active in the catalysis of reactions where Brönsted acidity is required while the molybdenum-containing compounds should be effective in catalyzing oxidation processes. As will be shown later experimental results provide support for these predictions of catalytic multifunctionality.

One other important property of the metal-oxygen cluster compounds should be noted. Although the surface areas of the solid acid forms of these compounds, as obtained by applying the Brunauer-Emmett-Teller multilayer theory to the nitrogen adsorption isotherms measured at 78 K, are low, in the range from 5-10 m^2/g, nevertheless unexpectedly large quantities of polar molecules are sorbed. For example, gaseous ammonia is taken up in quantities equivalent to three molecules of ammonia per anion at room temperature, which corresponds to one molecule of ammonia for each proton present in the total structure, that is, on the surface and in the bulk of the solid [4]. The PAS FTIR spectra (Figure 5) show that the band at 1420 cm^{-1} which can be attributed to the ammonium cation, and which was not present prior to the introduction of the

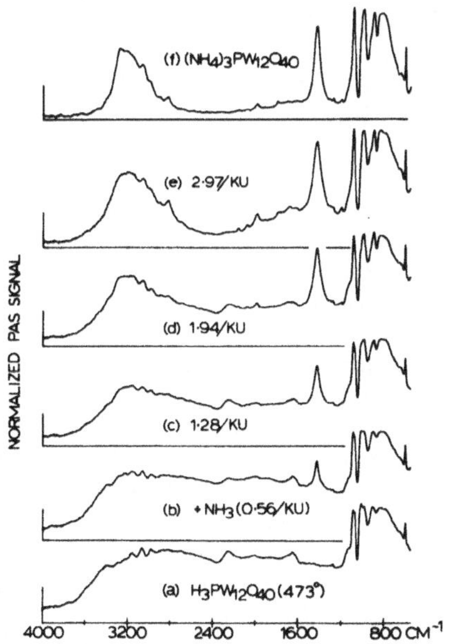

Figure 5. PAS FTIR spectra of HPW dosed stepwise with NH_3 (molecules sorbed/KU), (a) HPW in absence of NH_3; (b) - (e) after NH_3 at 423 K, pre-evacuated at 473 K; (f) NH_4 salt prepared by precipitation from the liquid phase [4].

gaseous ammonia, increases in intensity with the quantity of ammonia taken up by the solid. This observation demonstrates that the ammonia molecules are interacting with all the protons present both in the bulk and on the surface to form ammonium ions [4]. The PAS FTIR spectrum of the ammonium salt of 12-tungstophosphoric acid prepared by precipitation from aqueous solutions of the parent acid and of ammonium carbonate, while not quantitatively identical to that obtained for the parent acid containing three molecules of gaseous ammonia per anion, nevertheless is sufficiently similar to provide additional support for the contention that gaseous ammonia is capable of penetrating into the bulk structure.

Additional PAS FTIR studies have shown that gaseous pyridine [8] and methanol [9] are also capable of diffusing into the bulk structures of the metal-oxygen cluster compounds. Indeed the interaction between methanol and the protons forms protonated methanol at room temperature which is the first step in the process in which the metal-oxygen cluster compounds containing the 12-tungstophosphate anion convert methanol to hydrocarbons [9, 10].

The Synthesis of Metal-Oxygen Cluster Compounds With Microporous-Mesoporous Structures

As previously noted, the solid acidic forms of the metal-oxygen cluster compounds have low surface areas and no explicit pore structures, although polar compounds are capable of penetrating between the anions and cations into the bulk structures and thus could be considered as having implicit porosity. Earlier studies from this laboratory have shown, as noted above, that 12-tungstophosphoric acid ($H_3PW_{12}0_{40}$) and a number of its salts are effective catalysts in the conversion of methanol to hydrocarbons [11-15]. Surprisingly the ammonium salt of this acid was found to be more active and selective than the parent acid itself [16]. This led to further studies of the nature of ammonium 12-tungstophosphate (abbreviated to NH_4PW) which revealed that this solid possessed both a relatively high surface area and a microporous-mesoporous structure [17].

A variety of the solids prepared with monovalent cations as well as anions of selected elemental compositions have now been studied [17-26]. These were prepared by precipitation from aqueous solutions of appropriate compositions and concentrations. Parenthetically is should be noted that the preparative conditions have substantial effects on the subsequent morphologies of these solids [27].

Increased surface areas and microporous-mesoporous structures are not restricted to the ammonium salt of 12-tungstophosphoric acid. The adsorption-desorption isotherms of nitrogen at 78 K for the ammonium, cesium and potassium salts show that relatively large quantities of nitrogen are adsorbed at low relative pressures, indicative of the pressure of micropores or small mesopores (Figure 6) [17-26]. It should be noted that nitrogen molecules, being nonpolar, are unable to penetrate into the bulk structure of the metal-oxygen cluster compounds. The surface areas of the metal-oxygen cluster compounds with monovalent cations and various anion compositions are found to be appreciable for the potassium, ammonium and cesium salts but surprisingly, low for the sodium salts (Table 1). Values of the C parameter from the BET equation are also relatively large for those solids of high surface area (Table 2). Since, for a given adsorbate and adsorption temperature the sizes of these values for C are related to the energy of adsorption in the first layer, these observations again appear to be indicative of the presence of a porous structure. It is also evident (Tables 1 and 2 and Figure 7) that, while the surface areas are primarily dependent on the cation the anion effect is not negligible.

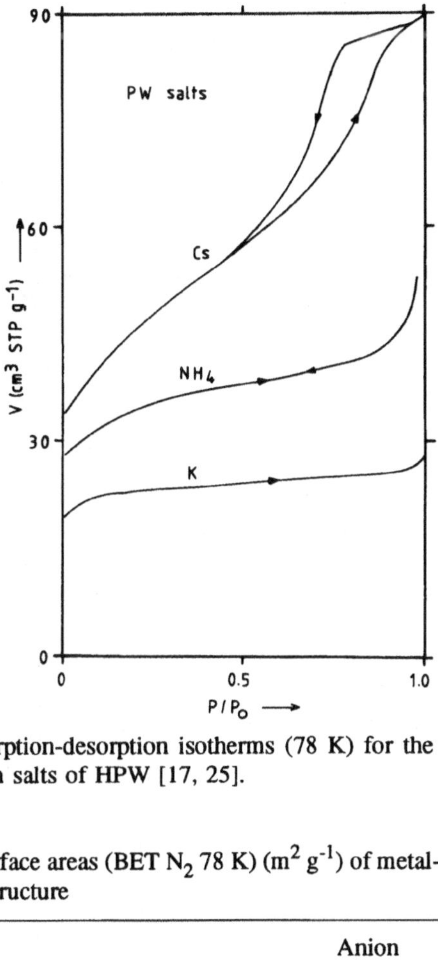

Figure 6. Nitrogen adsorption-desorption isotherms (78 K) for the high surface area cesium, ammonium and potassium salts of HPW [17, 25].

Table 1. Surface areas (BET N_2 78 K) ($m^2\ g^{-1}$) of metal-oxygen cluster compounds of Keggin structure

Cation	Anion			
	$SiW_{12}O_{40}^{-4}$	$PMo_{12}O_{40}^{-3}$	$PW_{12}O_{40}^{-3}$	$AsW_{12}O_{40}^{-3}$
H 3.2	(8)	(8)	6.5	
Na	1.3	3.5	3.7	-
Ag	-	-	3.0	-
K 3.3	39.9	90.0	46.0	
NH_4	116.9	193.4	128.2	82.1
Rb	116.3	-	-	101.4
Cs	153.4	145.5	162.9	65.4
$MeNH_3$	3.0	1.3	3.0	2.1
Me_4N	-	-	4.5	-

Table 2. BET *C* parameter for metal-oxygen cluster compounds of Keggin structure

Cation	Anion			
	$SiW_{12}O_{40}^{-4}$	$PMo_{12}O_{40}^{-3}$	$PW_{12}O_{40}^{-3}$	$AsW_{12}O_{40}^{-3}$
H	100	-	-	58
Na	473	142	18	-
Ag	-	-	101	-
K	127	1070	2430	22
NH_4	1919	877	760	1292
Rb	489	-	-	149
Cs	-	568	1720	1359
$MeNH_3$	-	-	261	292
Me_4N	-	-	113	-

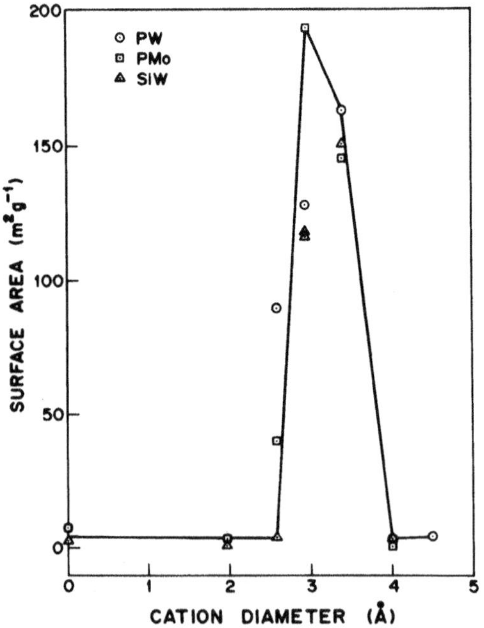

Figure 7. Surface areas (BET) and cation diameters for monovalent salts of 12-tungstophosphoric, 12-molybdophosphoric and 12-tungstosilicic acids [17, 25].

Pore size analyses employed the MP method [28] for micropores, a method based on capillary condensation and adsorbed layer thickening [29] for mesopores and standard isotherms for the nonporous analogues obtained from C values [30]. Average micropore radii are summarized in Table 3. Evidently values for $\bar{r}$ range from approximately 9 to 14 Å, and thus fall in the microporous-mesoporous range.

Table 3. Micropore average radii (Å) of high surface area heteropoly oxometalates

Cation	Anion			
	$SiW_{12}O_{40}^{-4}$	$PMo_{12}O_{40}^{-3}$	$PW_{12}O_{40}^{-3}$	$AsW_{12}O_{40}^{-3}$
K^+	-	9.3	8.8	11.2
NH_4^+	9.5	13.0	10.3	11.3
Rb^+	10.3	-	-	9.4
Cs^+	10.5	14.3	13.9	9.9

The derivation of the Brunauer-Emmett-Teller equation for multilayer physisorption involves summations over the number of adsorbed layers. In the usual derivation which generates the relation employed to calculate surface areas the summation is extended over an infinite number of adsorbed layers. If instead, the summation is restricted to a finite number of layers, n, the final relation becomes

$$V = \frac{CV_m x[1 - (n + 1) x^n + nx^{n + 1}]}{(1 - x)[1 + (c-1) x - cx^{n + 1}]} \tag{1}$$

where x is the relative pressure, P/Po, n represents the number of adsorbate monolayers and C and Vm are as defined with the infinite layer BET equation. This may be fitted to the measured adsorption isotherms, in the lower reactive pressure region, to provide values for the numbers of adsorbed layers. Where the surface area primarily results from small pores, in the microporous-small mesopores range, it is expected that the calculated values of n would be small, single digit numbers. That this is so in the present examples is evident from Figure 8 in which $\bar{r}$ is plotted versus the calculated values of n. As expected the correlation is linear.

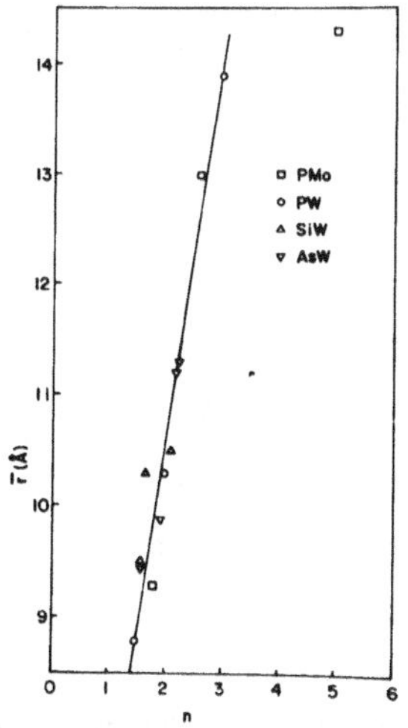

Figure 8. Average pore radius ($\bar{r}$) versus number of adsorbed layers (n) [25].

The pore size distributions calculated from the nitrogen adsorption-desorption isotherms show that, for a given sample, the pores are not uniform in size (see, for example Figure 9). Nevertheless a discernible maximum and reasonably symmetric distributions suggest that the preponderance of the pores are of a uniform size.

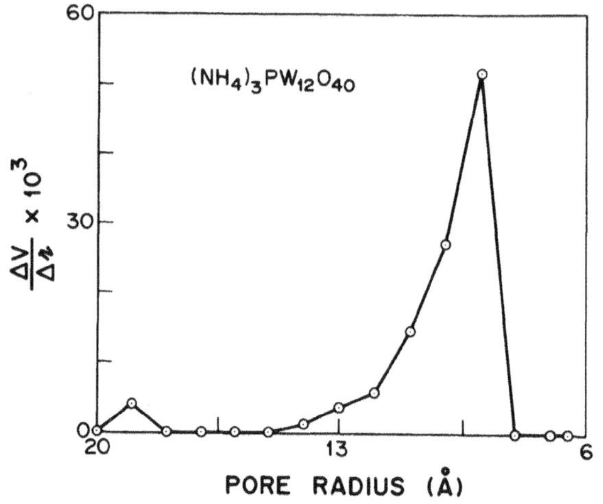

Figure 9. Pore size distribution for NH_4PW [25].

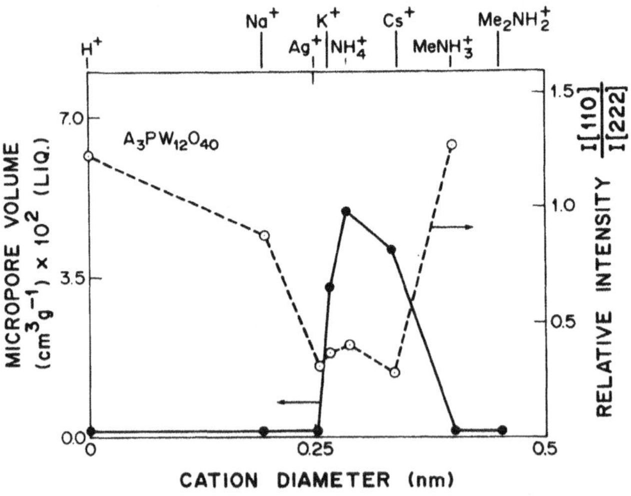

Figure 10. Micropore volumes and XRD [110]/[222] intensity ratio versus monovalent cation diameter for 12-tungstophosphates [17, 25].

The total micropore volumes for the parent acid and salts of 12-tungstophosphoric acid are shown in Figure 10. Semiquantitatively similar correlations have been found for the remaining anions shown in Table 1. Superimposed on the same figure are the X-ray diffraction intensities found for the [110] plane relative to the [222] plane. It is apparent that an approximately reciprocal relationship exists between the micropore volumes and the intensity ratios. Since the terminal oxygen atoms can be shown to lie in the [110] plane this data suggests that the pores are generated by a rotation and translation of the anions to remove (or reduce) the barriers between the interstitial voids found in the parent acid.

Cation Exchange in the Metal-Oxygen Cluster Compounds

It is well known, particularly with certain of the zeolites, that exchange of the cations can produce advantageous alterations in the pore sizes. Cation exchange has also been found to be possible with the microporous metal-oxygen cluster compounds [31]. However, since the precipitated salts contain residual amounts of protons, exchange with these and the predominant cation within the solid is possible. The numbers of cations (and hence by subtraction from three the numbers of protons in a number of the microporous metal-oxygen cluster solids are shown in Table 4. It may be noted that the tungsten-containing materials are more stoichiometric than their molybdenum-containing counterparts and those containing cesium contain fewer protons than the salts of ammonium and potassium.

Table 4. Molecular formulae for monovalent 12-heteropoly oxometalate salts as prepared from precipitation reactions with appropriate cations

Cation	Anion	
	$PW_{12}O_{40}^{-3}$	$PMo_{12}O_{40}^{-3}$
NH_4^+	2.69	2.08
	2.69	2.17
K^+	2.80	2.02
	2.84	2.73
Cs^+	2.91	2.78

Quoted values are number of cations per Keggin Unit

In general the cation exchange process reaches equilibrium prior to complete cation exchange (Table 5). However the exchange process is evidently rapid since equilibrium values were achieved, at least approximately, after one exchange (Figure 11). Interestingly, in the example shown in Figure 11 the quantity of the exchanging-in cation, in this case Cs^+, appears to be approximately independent of the nature of the cation being displaced from the solid. The maximum exchange capacity, that is, the quantity found at equilibrium is, however dependent on the relative sizes of the leaving and entering cations (Figure 12). The larger the cations in the solid the more difficult is their displacement by smaller cations, regardless of the composition of the anions.

Table 5. Concentration of cations in solid 12-tungstophosphate salts after ion exchange with monovalent cations

IN[a]	OUT[a]	Initial[b] IN[a]	Initial[b] OUT[a]	Final[b] IN[a]	Final[b] OUT[a]	Final[b] TOTAL
K^+	NH_4^+	0	2.69	1.35	1.45	2.80
Cs^+	NH_4^+	0	2.69	1.44	1.18	2.62
Cs^+	K^+	0	2.84	1.47	1.40	2.87
NH_4^+	K^+	0	2.84	2.01	0.67	2.68
NH_4^+	Cs^+	0	2.91	0.38	2.49	2.87
K^+	Cs^+	0	2.91	0.43	2.58	3.01

[a] In and out refer to the cation exchanging into and out of the solid.
[b] Initial and final refer to the concentrations of the cations (cations per anion) before the exchange and after six exchange reactions.

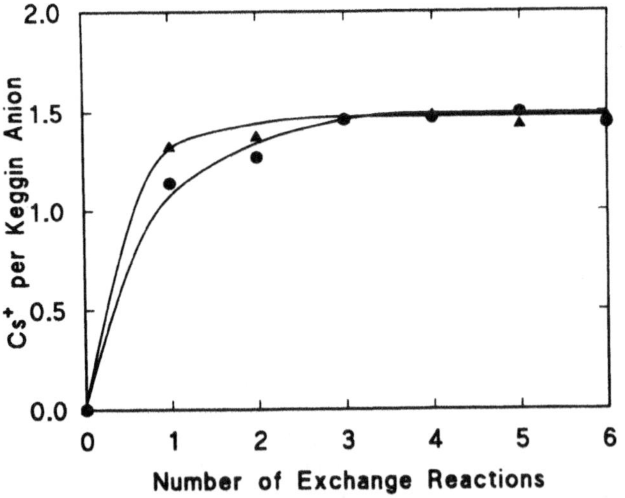

Figure 11. Cation exchange isotherm for Cs^+ exchanging into NH_4PW ● and $K_3PW_{12}O_{40}$ ▲ [31].

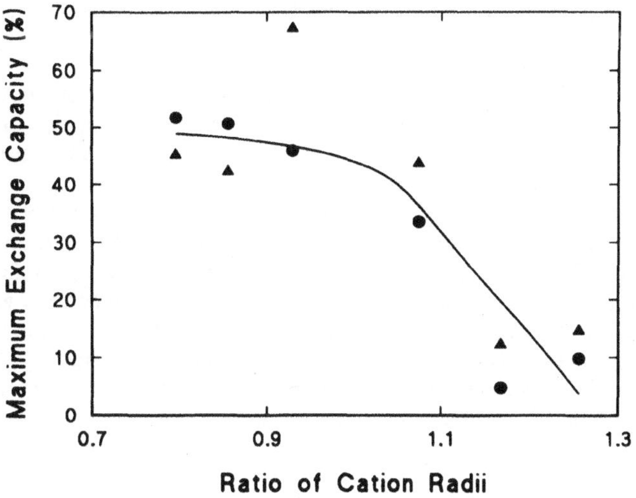

Figure 12. Maximum exchange capacities for various ratios of the radius of the cation exiting the solid to that of the cation entering the solid ● $PMo_{12}0_{40}^{-3}$ salts, ▲ $PW_{12}0_{40}^{-3}$ salts [31].

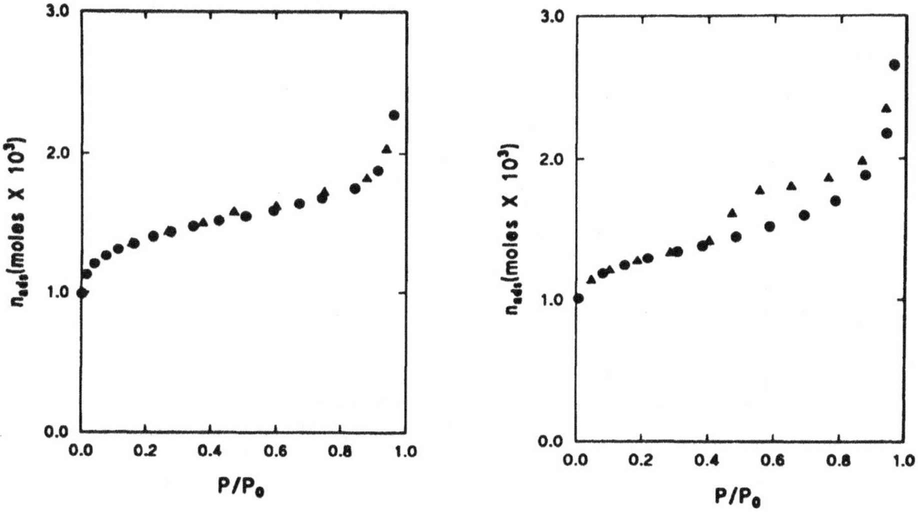

Figure 13. Representative nitrogen adsorption-desorption isotherms for (a) $KPWNH_42$ and (b) NH_4PWK3 [32].

Effect of Cation Exchange on the Pore Structure and Morphology of Metal-Oxygen Cluster Compounds

Equally important in the ion exchange process is the effect on the surface and bulk properties of the solid. With certain of the zeolites, for example zeolite A, it is well known that exchange of the sodium cations with calcium produces an enlargement of the pores, while the features of the structure are retained. With the microporous metal-oxygen cluster compounds it is thus natural to investigate the effect of the cation exchange. Is the structure destroyed or is there essential retention of the structural features? The nitrogen adsorption isotherms at 78 K for KPW exchanged twice with NH_4^+ and for NH_4PW exchanged three times with K^+ again show evidence for the presence of pores in the microporous region Figure 13 [32]. Further, the values of the BET surface areas for the partially exchanged samples fall between those obtained with the precipitated solids (Figure 14). Additionally the values of the lattice parameter behave in a similar manner (Figure 15). It is clear that the crystallographic structures, surface areas and pore structures are semiquantitatively retained but shift, not surprisingly, to reflect the changes in cation composition.

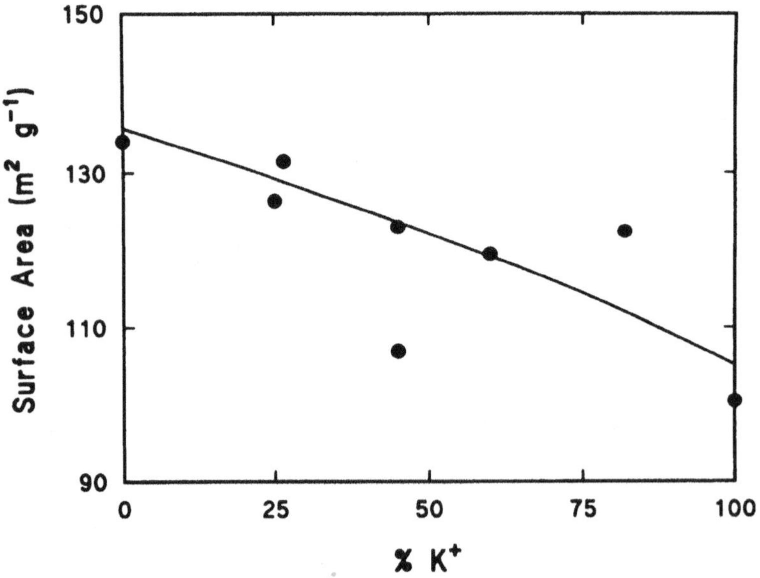

Figure 14. Surface area (BET) for various concentrations of K^+ in the ion-exchanged $K^+/NH_4^+/PW_{12}O_{40}^{-3}$ system [32].

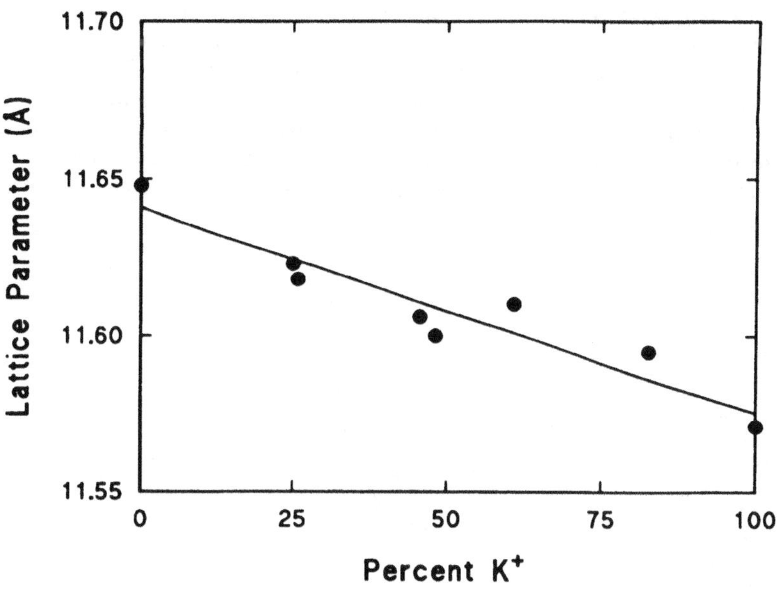

Figure 15. Lattice parameter for various concentrations of K^+ in the ion-exchanged $K^+/NH_4^+/PW_{12}O_{40}^{-3}$ system [32].

The Effect of Cation Exchange on the Heterogeneous Catalytic Properties of Metal-Oxygen Cluster Compounds

The availability of the microporous solids obtained with a variety of partially exchanged cation compositions provides an excellent opportunity to examine the effect of the nature and concentrations of the cations on catalytic processes [33, 34]. Isobutyric acid may, of course, be oxidized to carbon oxides. However, under suitable conditions and in the presence of an appropriate catalyst partial oxidation may occur to yield methacrylic acid, acetone and propene.

Table 6. Selectivities and conversions determined in preliminary catalytic tests for molybdenum and tungsten based 12-heteropoly compounds

Compound	Conversion (%)	Selectivity		
		MAA	Propene	Acetone
$(NH_4)_3PMo_{12}O_{40}$	23.5	55.8	10.4	32.1
$Cs_3PMo_{12}O_{40}$	71.0	40.1	9.9	49.0
$(NH_4)_3PW_{12}O_{40}$	29.6	0.9	99.1	0.0
$Cs_3PW_{12}O_{40}$	9.4	0.2	93.6	6.1

All catalytic runs at 300°C and W/F of ~ 1.4 mg min mL^{-1}

As noted earlier it is expected that tungsten-containing metal-oxygen cluster compounds would be less effective in oxidation processes than those containing molybdenum. This is borne out from a comparison of 12-tungstophosphoric and 12-molybdophosphoric acid (Table 6). With 12-molybdophosphoric acid containing various quantities of cesium cations introduced by ion exchange the selectivity is found to be markedly dependent on the number of the cesium cations present (Figure 16). Clearly, as the content of cesium cations approaches stoichiometric (3) the selectivity to methacrylic acid decreases to negligible values while concomitantly that to acetone increases to values approaching 90%.

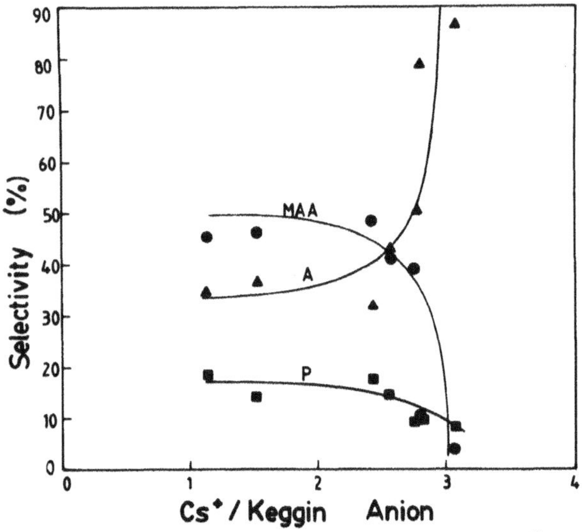

Figure 16. Selectivity of isobutyric acid conversion to methacrylic acid (●), propene (■), and acetone (▲) for concentrations of Cs^+ cations per anion ($PMo_{12}O_{40}^{-3}$) [33].

It is of interest to compare these results with those obtained for a second oxidation process. The partial oxidation of methane is currently of considerable interest both for its fundamental and practical implications. Deep oxidation of methane is, of course, thermodynamically preferred. However a suitable catalyst may direct the selectivity to formaldehyde and/or methanol. The relevant reactions may be represented stoichiometrically as

$$CH_4 + \tfrac{3}{2}O_2 \rightarrow CO + 2H_2O \qquad (1)$$

$$CH_4 + 2O_2 \rightarrow CO_2 + 2H_2O \qquad (2)$$

$$CH_4 + O_2 \rightarrow H_2CO + H_2O \qquad (3)$$

$$CH_4 + \tfrac{1}{2}O_2 \rightarrow CH_3OH \qquad (4)$$

Since methane is a nonpolar molecule it is convenient to support the metal-oxygen cluster compound, 12-molybdophosphoric acid on a high area solid such as silica [35-49]. The ion exchange process may then be carried out directly with the supported solid. Stepwise replacement of the proton by the cesium ion reduces the turnover number (TN) until, at approximately the stoichiometric concentration of the cesium cations the TN reaches that expected for the silica support itself (Figure 17). Concomitantly the selectivities revert to those found with the support (Figure 18).

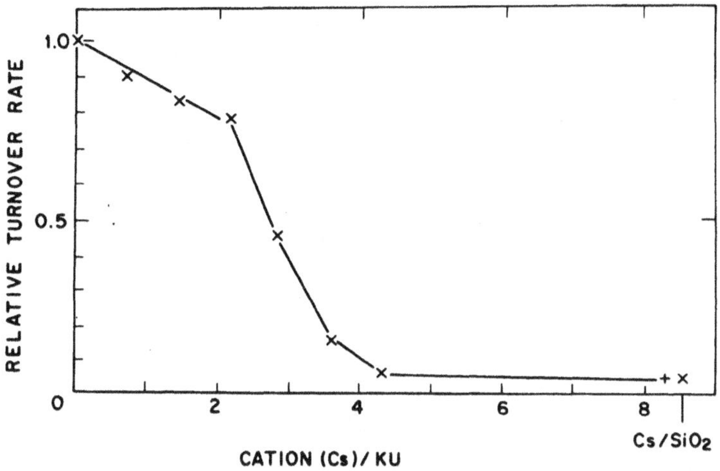

Figure 17. Relative turnover rate of the CH_4/N_20 reaction after addition of cesium carbonate (X) to silica-supported 12-molybdophosphoric acid (16 wt %). Reaction temperature 843 K, w = 0.5 g, F = 30 cm^3 min^{-1}, CH_4 (67 mol %), N_20 (33 mol %) [35].

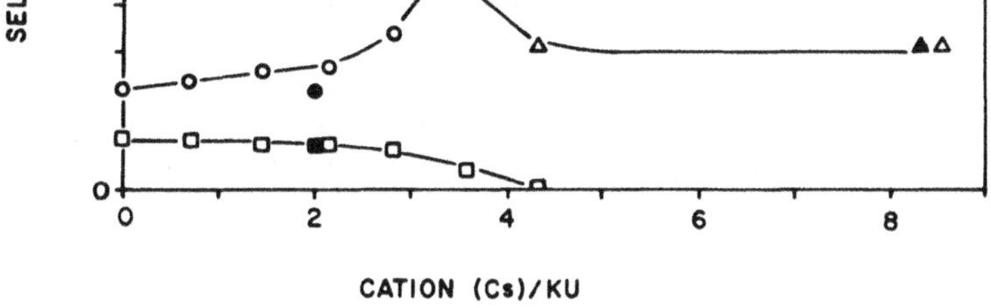

Figure 18. Selectivity of the CH_4 + N_20 reaction after addition of cesium carbonate to silica supported 12-molybdophosphoric acid (HPMo). Open symbols (16 wt % HPMo). Closed symbols (24 wt % HPMo). Open symbols not joined by line ($Cs/Si0_2$). Reaction conditions as in Figure 17. Δ; CO; 0 $C0_2$; ❑ H_2CO [35, 37].

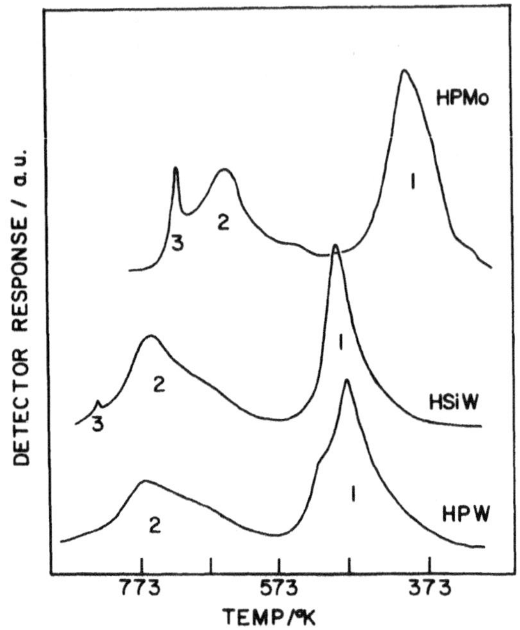

Figure 19. Temperature-programmed desorption spectra for 12-tungstophosphoric, 12-tungstosilicic and 12-molybdophosphoric acids after pretreatment at 298°K for 16 hours [50, 51].

Temperature programmed desorption studies of three of the acidic metal-oxygen cluster compounds show two major desorption peaks each due to the desorption of water (Figure 19) [50, 51]. The peaks labelled 1 which appear at temperatures in the range from 100-200°C and which are associated with low desorption energies are produced by the hydrogen-bonded water in the solids. The peaks labelled 2 and 3 have desorption energies of 120-400 kJ/mol and appear at temperatures between 400 and 600°C. Further the quantities of water desorbed at these higher temperatures are equivalent to the numbers of protons present in the solids. Evidently the protons are scavenging oxygen from the anions to produce the water which is then associatively desorbed. As noted earlier the structure of the anions remains largely preserved even at these elevated temperatures.

The loss of oxygen, stoichiometrically 3/2 for each anion, will generate a vacancy (or vacancies) in the structure. Since protons are apparently necessary for this process, the replacement of protons by cesium ions is believed to suppress the formation of oxygen vacancies. In view of the effect of cesium ions on the conversion of methane and of isobutyric acid it appears that the oxygen vacancies are necessary, but undoubtedly not sufficient for such conversion processes to occur.

Conclusions

1. Microporous-mesoporous metal-oxygen cluster compounds can be prepared, with appropriate monovalent cations by precipitation from aqueous solutions of the parent solid acids.

2. Cation exchange has been shown to be possible with the cation pairs NH_4^+/K^+, K^+/Cs^+ and NH_4^+/Cs^+ with the 12-tungstophosphate and 12-molybdophosphate anions.

3. The extent of the exchange depends on both the entering and leaving cations.

4. The extent of the exchange is strongly dependent on the size of the cation initially present in the solid relative to that being exchanged into the solid.

5. The residual protons in the solids are reduced in number, but not eliminated, during the ion exchange process.

6. The morphological properties of the solids are semiquantitatively retained after the ion exchange but the surface areas, pore structures and crystallographic parameters shift, reflecting the presence of the incoming cation.

7. The metal-oxygen cluster compounds display a catalytic multifunctionality. Those containing tungsten are active and selective in processes requiring Brönsted acid sites where those containing molybdenum are active and selective in oxidation processes.

Acknowledgements

The financial support of the Natural Sciences and Engineering Research Council of Canada and of Energy, Mines and Resources Canada is gratefully acknowledged.

References

1. Pope, M.T. and Müller, A. (1991) 'Polyoxometalate Chemistry: An Old Field with New Dimensions in Several Disciplines', Angew Chem. Intl. Ed. Eng., 30, 34.
2. Pope, M.T. (1983) 'Heteropoly and Isopoly Oxometalates', Springer-Verlag, Berlin.
3. Brown, C.M., Noe-Spirlet, M.R., Busing, W.R. and Levy, H.A. (1977) 'Dodecatungstophosphoric Acid Hexahydrate, $(H_5O_2^+)_3(PW_{12}O_{40}^{3-})$. The Time Structure of Keggin's 'Pentahydrate' from Single-Crystal X-ray and Neutron Diffraction Data', Acta. Cryst., B33, 1038.
4. Highfield, J.G. and Moffat, J.B. (1984) 'Characterization of 12-Tungstophosphoric Acid and Related Salts Using Photoacoustic Spectroscopy in the Infrared Region. 1. Thermal Stability and Interactions with Ammonia', J. Catalysis, 88, 177.
5. Moffat, J.B. (1984) 'A Comparison of the Catalytic and Structural Properties of Heteropoly Compounds: Semi-empirical Calculations', J. Molecular Catalysis, 26, 385.
6. Highfield, J.G., Hodnett, B.K., McMonagle, J.B. and Moffat, J.B. (1984) 'The Physical, Structural, and Surface Chemistry of Heteropoly Compounds and their Relationship to Catalytic Properties', Proceedings of the 8th International Congress on Catalysis, pp. 611, Dechema, Frankfurt am Main.
7. Moffat, J.B. (1984) 'The Electronic Structure and Catalytic Properties of Heteropoly Compounds', Proceedings of the 9th Iberoamerican Symposium on Catalysis, pp. 349, Lisbon.

8. Highfield, J.G and Moffat, J.B. (1984) 'Characteristics of 12-Tungstophosphoric Acid and Related Salts using Photoacoustic Spectroscopy in the Infrared Region. 2. Interactions with Pyridine', J. Catalysis, 89, 185.
9. Highfield, J.G. and Moffat, J.B. (1985) 'Elucidation of the Mechanism of Dehydration of Methanol over 12-Tungstophosphoric Acid using Infrared Photoacoustic Spectroscopy', J. Catalysis, 95, 108.
10. Highfield, J.G. and Moffat, J.B. (1986) 'Characterization of Sorbed Intermediates and Implications for the Mechanism of Chain Growth in the Conversion of Methanol and Ethanol to Hydrocarbons over 12-Tungstophosphoric Acid using Infrared Photoacoustic Spectroscopy', J. Catalysis, 98, 245.
11. Moffat, J.B. and Hayashi H. (1982) 'The Properties of Heteropoly Acids and the Conversion of Methanol to Hydrocarbons', J. Catalysis, 77, 473-484.
12. Moffat, J.B. (1984) 'Conversion of Methanol to Hydrocarbons on Heteropoly Compounds', invited paper in Catalytic Conversions of Synthesis Gas and Alcohols to Chemicals, R.G. Herman (Ed.), Plenum, N.Y.
13. McMonagle J.B. and Moffat, J.B. (1985) 'The Catalysis of Methanol and Ethanol Conversions on Molybdenum and Tungsten Heteropoly Compounds', J. Catalysis, 91, 132.
14. Moffat, J.B. (1988) 'The Conversion of Methanol to Hydrocarbons and the Oxidation of Methane on Heteropoly Oxometalates', in Methane Conversion, A Symposium on the Production of Fuels and Chemicals from Natural Gas, Auckland, April 1987, Elsevier, Amsterdam.
15. Hayashi, H. and Moffat, J.B. 'Methanol Conversion on Metal Salts of 12-Tungstophosphoric Acid', J. Catalysis, 81, 61.
16. Hayashi, H. and Moffat, J.B. (1983) 'Conversion of Methanol into Hydrocarbons over Ammonium 12-Tungstophosphate', J. Catalysis, 83, 1982.
17. McMonagle J.B. and Moffat, J.B. (1984) 'Pore Structures of the Monovalent Salts of the Heteropoly Compounds, 12-Tungstophosphoric and 12-Molybdophosphoric Acid', J. Colloid and Interface Science, 101, 479.
18. Moffat, J.B. (1986) 'Cation-Anion Effects in the Surface Chemistry and Catalytic Properties of Heteropoly Compounds', Polyhedron, 5, 261.
19. Moffat, J.B. (1987) 'Cation Effects in the Preparation of and Catalysis by Heteropoly Oxometalates', B. Delmon, P. Grange, P.A. Jacobs and G. Poncelet (eds.), Preparation of Catalysts IV, Studies in Surface Science and Catalysis, Vol. 30, Elsevier, Amsterdam.
20. (1988) 'Microporosity in Heteropoly Oxometalate Catalysts', in J. Ward (ed.), Surface Science and Catalysis, Vol. 38, Elsevier, Amsterdam.
21. Moffat, J.B., McMonagle, J.B. and Taylor, D. (1988) 'Microporous Heteropoly Oxometalate Heterogeneous Catalysts', Solid State Ionics 26, 101.
22. Taylor, D.B., McMonagle, J.B. and Moffat, J.B. (1985) 'Cation Effects on the Surface and Bulk Structure of the Salts of 12-Tungstosilicic Acid', J. Colloid and Interface Science, 108, 278.
23. McGarvey, G.B. and Moffat, J.B. (1988) 'The Microporous Structure of Arsenic Heteropoly Oxometalates', J. Colloid and Interface Science, 125, 51.
24. McMonagle, J.B., Nayak, V.S., Taylor, D. and Moffat, J.B. (1988) 'A New Class of Microporous Solids-Heteropoly Oxometalates', in M.J. Phillips and M. Ternan (eds.), Proceedings 9th International Congress on Catalysis, Calgary, 1988, Chemical Institute of Canada.

25. Moffat, J.B. (1989) 'Implicit and Explicit Microporosity in Heteropoly Oxometalates', J. Molecular Catalysis, 52, 169.
26. Moffat, J.B., McGarvey, G.B., McMonagle, J.B., Nayak, V. and Nishi H. (1990) 'Zeolitic Heteropoly Oxometalates', invited paper in D. Barthoment, E.G. Derouane and W. Hölderich (eds.), "Guidelines for Mastering the Properties of Molecular Sciences", NATO ASI Series Plenum Press, N.Y.
27. Lapham, D. and Moffat, J.B. (1991) 'Preparative Effects on the Surface Area and Pore Structure of Microporous Heteropoly Oxometalates', Langmuir, 7, 2273.
28. Mikhail, R.Sh., Brunauer, S. and Bodor, E.E. (1968) 'Investigations of a Complete Pore Strucutre Analysis. I. Analysis of Micropores, J. Colloid Interface Sci., 26, 45.
29. Pierce, C. (1953) 'Computation of Pore Sizes from Physical Adsorption Data', J. Phys. Chem., 57, 149.
30. Lecloux, A. and Pirard, J.P. (1979) 'The Importance of Standard Isotherms in the Analysis of Adsorption Isotherms for Determining the Porous Texture of Solids', J. Colloid Interface Sci., 70, 265.
31. McGarvey, G.B. and Moffat, J.B. (1991) 'Ion Exchange Properties of Microporous Monovalent Salts of 12-Tungstophosphoric Acid and 12-Molybdophosphoric Acid Catalysts', J. Catal., 128, 69-83.
32. McGarvey, G.B. and Moffat, J.B. (1991) 'Characterization of the Pore Structure and Morphology of Some Ion Exchanged 12-Heteropoly Oxometalates', J. Catal., 130, 483.
33. McGarvey, G.B. and Moffat, J.B. (1991) 'The Oxidative Dehydrogenation of Isobutyric Acid to Methacrylic Acid on Ion Exchange Modified 12-Heteropoly Oxometalates', J. Catal., 132, 100.
34. McGarvey, G.B. and Moffat, J.B. (1991) 'Observation of Nitriles Formed in the Oxidative Dehydrogenation of Carboxylic Acids Over Metal-Oxygen Cluster Catalysts', Catalysis Letters, 10, 41.
35. Kasztelan, S. and Moffat, J.B. (1987) 'The Oxidation of Methane on Heteropoly Oxometalates. I. Catalytic Properties of Silica-Supported Heteropoly Acids', J. Catalysis, 106, 512.
36. Kasztelan, S. and Moffat, J.B. (1987) 'Partial Oxidation of Methane by Oxygen over Silica', J. Chem. Soc., Chem. Commun. 1663.
37. Kasztelan, S. and Moffat, J.B. (1988) 'The Oxidation of Methane on Heteropoly Oxometalates II. Nature and Stability of the Supported Species', J. Catalysis, 109, 206.
38. Ahmed, S. and Moffat, J.B. (1988) 'Reactant Composition and Concentration Effects in the Oxidation of Methane on Supported 12-Molybdophosphoric Acid', Applied Catalysis, 40, 101.
39. Ahmed, S. and Moffat, J.B. (1988) 'The Conversion of Methane on Heteropoly Oxometalates in the Presence of Tetrachloromethane', Catalysis Letters, 1, 141.
40. Kasztelan, S. and Moffat, J.B. (1988) 'The Oxidation of Methane on Heteropoly Oxometalates. III. Effect of the Addition of Cesium on Silica-Supported 12-Molybdophosphoric Acid, Molybdena, Vanadia and Iron Oxide', J. Catalysis, 112, 54.
41. Kasztelan, S., Payen, E. and Moffat, J.B. (1988) 'The Formation of Molybdosilicic Acid on Mo/SiO_2 Catalysts and Its Relevance to Methane Oxidation', J. Catalysis, 112, 320.
42. Kasztelan, S. and Moffat, J.B. (1988) 'Partial Oxidation of Methane by Nitrous Oxide or Oxygen over Silica-Supported 12-Molybdophosphoric Acid', in M.J. Phillips and M. Ternan (eds.), Proceedings 9th International Congress on Catalysis, Calgary, Chemical Institute of Canada.

43. Kasztelan, S. and Moffat, J.B. (1989) 'The Oxidation of Methane on Heteropoly Oxometalates. IV. Properties of the Silica Supported Salts of 12-Molybdophosphoric Acid', J. Catalysis, 116, 82.
44. Ahmed, S. and Moffat, J.B. (1989) 'Methane Conversion on Silica-Supported 12-Molybdophosphoric Acid in the Presence of Dichloro- or Tetrachloromethane', J. Physical Chem., 93, 2542.
45. Ahmed, S. and Moffat, J.B. (1989) 'The Conversion of Methane on 12-Tungstophosphoric Acid in the Presence of Dichloro- or Tetrachloromethane', J. Catalysis, 118, 281.
46. Ahmed, S., Kasztelan S. and Moffat, J.B. (1989) 'Mechanistic Relationships in the Activation of Methane and the Conversion of Methanol on Heteropoly Oxometalates', invited paper in Faraday Discuss. Chem. Soc., 87, 23.
47. Kasztelan, S., Payen, E. and Moffat, J.B. (1990) 'The Existence and Stability of the Silica Supported 12-Molybdophosphoric Acid Keggin Unit as shown by Raman, XPS, and 31P NMR Spectroscopic Studies', J. Catal., 125, 45.
48. Kasztelan, S., Payen, E. and Moffat, J.B. (1991) 'The Interaction of Cesium Ions with Silica Supported 12-Molybdophosphoric Acid', J. Catal., 128, 479.
49. Payen, E., Kasztelan, S. and Moffat, J.B. (accepted, 1992) 'An In-Situ Laser Raman Study of the Effect of Water on Undoped and Cesium-Doped Silica-Supported 12-Molybdophosphoric Acid Catalysts', J. Chem. Soc., Faraday trans.
50. Hodnett, B.K. and Moffat, J.B. (1984) 'Application of Temperature Programmed Desorption to the Study of Heteropoly Compounds: Desorption of Water and Pyridine', J. Catalysis, 88, 253.
51. Hodnett, B.K. and Moffat, J.B. (1985) 'Study of Heteropoly Compounds with the Keggin Structure by Temperature Programmed Reduction and H-D Exchange', J. Catalysis, 91, 93.

CHEMICALLY MODIFIED ELECTRODES AND MESOPOROUS INORGANIC MATERIALS

L.F.F.T.T.G. Rodrigues, M.O.S. P.Caldeira and C.A.C. Sequeira
InstitutoTecnológico para a Europa Comunitária
Av. Rovisco Pais
1096 Lisboa Codex
Portugal

ABSTRACT. At the present time, the role of chemically modified electrodes (CMEs) is continuously increasing in nearly all of the branches of science and technology. As discussed in this course by Petridis and coworkers, mesoporous materials are potentially interesting for advanced CMEs.

In this paper, the design, preparation, characterisation and application of CMEs as sensors is discussed. Directions for the use of mesoporous materials as CMEs are envisaged.

Introduction

Following breakthroughs in electrochemical theory in the 1960s and in instrumentation in the 1970s, the future of electroanalytical chemistry now lies in the development of new useful working electrodes. Increasing the range of sensors available can expand the capabilities of voltammetry. Chemical sensing using CMEs can be improved by occluding interferences from the electrode surface using the principles of electrocatalysis to enhance signals or designing electrodes capable of preconcentrating species in a chemical environment which will enhance sensitivity. Ideally, chemical sensors, which perform optimally with no sample preparation, are required.

The ideal CME for chemical sensing should have the following properties.

a) Easily prepared.
b) Mechanically and chemically stable.
c) Electrochemically inert.
d) Ability to discriminate against matrix components.
e) Ability to resolve analyte signals.
f) High sensitivity.
g) High capacity.
h) Reusable.
i) Electrochemical processes well understood.

Several excellent reviews on CMEs already exist [1-3]. The purpose of this paper is to highlight the considerations necessary when these devices are to be used as chemical sensors and to give some useful examples. The design, preparation, characterization and application of CMEs as sensors will be discussed. Challenges, which need to be overcome to expand the

C.A.C. Sequeira and M.J. Hudson (eds.),Multifunctional Mesoporous Inorganic Solids, 473–483.

range of applications, will be highlighted. The potential ability of mesoporous inorganic materials for use as chemically modified electrodes will be considered.

Operation of a CME

Usually, CMEs which have been designed to act as chemical sensors employ chemical procedures used previously in solution analyses. CMEs which involve (i) ionic interaction (positive and negative); (ii) complexation; (iii) electrocatalysis including bioelectrocatalysis; or (iv) size exclusion have been previously employed.

The first two types may achieve extra sensitivity and selectivity by electrostatic attraction or complexation of the analyte onto the electrode surface, achieving a preconcentration effect similar to stripping voltammetry at a mercury-based electrode. Using charged modifiers to preconcentrate, on the basis of electrostatic attraction, will discriminate against species of similar charge to that of the modifier. Employing complexation to preconcentrate the analyte has additional benefits. The analyte can be trapped in a chemical environment which can be used to influence the nature of the electrode process. A selective complexing agent can be used to provide a selective sensor. The catalytic processes achieve extra sensitivity and selectivity by enhancing the electron transfer process for a particular analyte.

Choice of Substrate

The support matrix, whether organic or inorganic, may be classified according to electrical properties as shown in Table 1. All three material types may constitute the appropriate electrochemical interface for CMEs. A list of support materials, which have been used as electrodes or inert matrices to chemically linked reagents, appears in Table 2. Some less rigid polymers, such as gels and fibrous polymers, are used primarily for encapsulation or entrapment of reagents. The most frequently used materials are listed in Table 3.

TABLE 1. Electrical conductivity of materials.

Material	Conductivity σ (Ω^{-1} cm^{-1})
1. Conductor	$>10^{2}$
2. Semiconductor	10^{-10} - 10^{2}
3. Insulator	$<10^{-10}$

Derivatization of the Substrate

It is important at this stage to define first-order and second-order derivatization. First-order CMEs are prepared by adding the chemically active group directly to an electrode substrate.

TABLE 2. Commonly used support matrices.

1. Conductors
   - Pyrolytic graphite
   - Glassy carbon
   - Gold (Au)
   - Platinum (Pt)
   - Niobium (Nb)
   - Boron carbide

2. Semiconductors
   - Ruthenium oxide - RuO_2
   - Platinum oxide - Pt_2O_3
   - Titanium oxide - TiO_2
   - Tin oxide - SnO_2
   - Nickel oxide - NiO_2
   - Alumina - Al_2O_3
   - Hydroxyapatite
   - n-Ge
   - n-Si
   - Polyphosphides

3. Insulators

   A. Natural polymers
   - Carboxymethyl cellulose
   - Starch
   - Agarose
   - Sepharose
   - Sephadex

   B. Synthetic polymers

Poly(acrylamide)	Poly(urethane)
Poly(acrylic acid)	Poly(vinyl chloride)
Poly(glutamic acid)	Polyethene
Poly(styrene)	Glass
Nylon	
Cellulose	

There is a limited number of instances where this approach is appropriate. Second-order derivatization, which involves adding the chemically active group to a previously modified substrate, greatly expands the number of different CMEs which can be prepared:

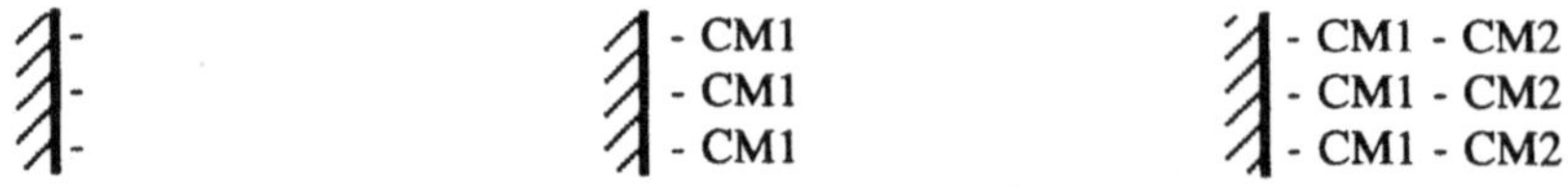

Bare substrate First order Second order

At present not many modified electrodes are derivatized beyond second order, although it is obvious that to do so would again expand the range of electrodes which could be prepared.

TABLE 3. Materials for encapsulation or entrapment.

1. Gels

   A. Natural

   Agarose
   Sepharose
   Sephadex
   Starch

   B. Synthetic

   Silica
   Poly(acrylamide)
   Poly(acrylic acid)
   Poly(glutamic acid)
   Cellulose

2. Fibrous polymers

   Cellulose acetate
   Ethyl cellulose
   Poly(acrylonitrile)
   Poly(butadiene)
   Poly(carbonate)
   Poly-Γ-methyl glutamate
   Polystyrene
   Polyvinyl chloride

Preparation of Composite Electrodes and Chemisorption

Composite CMEs have the modifier physically incorporated into the electrode material. The most common example involves carbon-paste-electrodes, which have a modifier of usually less than 5% (w/w), added to the graphite/Nujol mixture during electrode preparation [4-6]. Such electrodes are simple and inexpensive to prepare. They are easily regenerated by simply exposing a fresh electrode surface. The capacity of such electrodes is relatively high, in that surface coverages of 5×10^{-8} mol cm^{-2} have been reported. This is certainly greater than a monolayer coverage. G iven the porosity of carbon paste electrodes high capacities would be expected. The greatest problems with composite CMEs of this type are associated with leaching of the modifier from the electrode surface.

The chemisorption approach is most useful for modifiers which are insoluble in the analyte solvent. The modifier is usually dissolved in a volatile solvent. It is then pipetted onto the electrode surface and the solvent allowed to evaporate, thus leaving the modifier on the surface. It is imperative that the solvent/modifier completely covers the electrode surface to provide a homogeneous covering. To achieve this, an intermediary wetting solvent may be applied to the substrate before depositing the modifier.

The chemisorption approach has been widely used for coating electrode surfaces with polymers [7-9] as well as various other organic [10] and inorganic [11, 12] compounds. Uniform coatings varying in concentration from 10^{-10} — 10^{-7} mol cm^{-2} have been obtained using this procedure, which is extremely easy to implement. For stable, uniform coatings the electrode substrate must be meticulously prepared and often several hours are required for drying. Electrodes left to dry slowly appear to have better mechanical properties than those dried quickly. Electrodes prepared in this way suffer the same problems of instability as composite electrodes in that the modifier gradually leaches from the substrate [7].

Covalent Bonding

Using bonded-phase electrodes, the instability and difficulty to control and reproduce both composite electrodes and those prepared using chemisorption can be alleviated. Some of the more common methods used for bonding modifiers to electrode substrates are direct bonding [13], silanization [14-17], and the Williamson reaction [18].

Silanization may be represented as

$$\text{- OH} + XSiX_2R \longrightarrow \text{- O - }SiX_2R + HX, \qquad (X = Cl, OEt)$$

The silanizing agent ($XSiX_2R$) reacts with oxygen-containing groups on the electrode substrate, to produce a silanized surface. This surface can then be used as it is, or after derivatization.

To illustrate the process, consider the modification of a tin oxide electrode surface using a silane coupling agent:

Step 1 - Hydrolyze a silane coupling agent, e.g.,Γ-aminopropyltriethoxysilane,

$$(EtO)_3\text{-}Si\text{-}(CH_2)_2\text{-}NH_2 + H_2O \longrightarrow (HO)_3\text{-}Si\text{-}(CH_2)_2\text{-}NH_2$$

Step 2 - The hydrolyzed silane will then condense with the hydroxyls in the absence of water on the SnO_2 support surface;

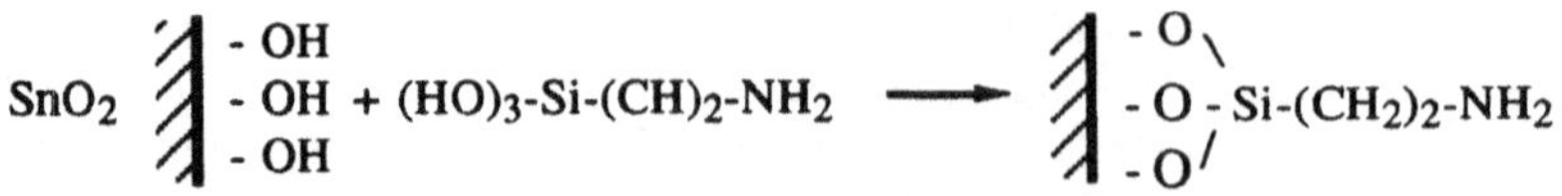

Step 3 - The amine functional group from the silane derivative can then react with glutaraldehyde through which a bridge is formed to a terminal reagent by the remaining carbonyl functional group at the chain end;

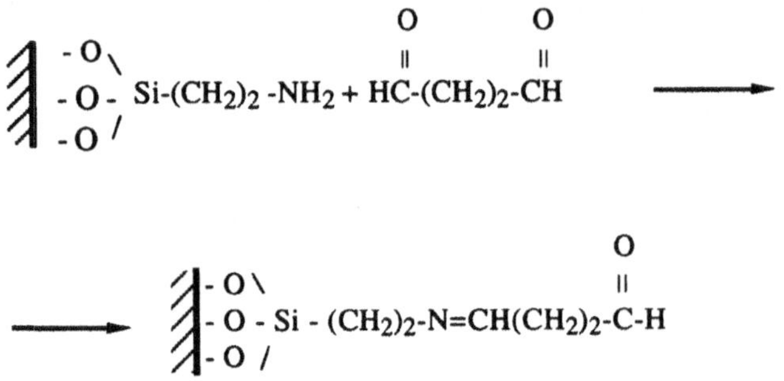

Step 4 - The terminal reagent, in this case an enzyme (H_2N-E_3), can now be coupled to the carbonyl intermediate through a Schiff base coupling;

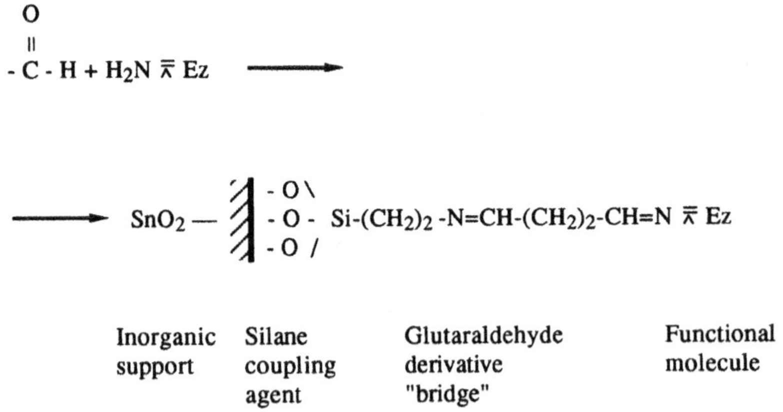

The intermediate step involving the formation of the glutaraldehyde bridge is essential to preserve enzyme activity.

Silane coupling agents have been successfully used as in the example just given and several such coupling agents are commercially available. Some representative silane coupling agents are listed in Table 4.

Silanizing agents are bifunctional and, once attached to the surface, can provide the functional group for terminal reagent attachment by, e.g., amide coupling reactions such as:

$$\phi \,\overline{\wedge}\, NH_2 + TR\text{-}COOH \longrightarrow \phi \,\overline{\wedge}\, NH\text{-}CO\text{-}TR$$

$$\phi \,\overline{\wedge}\, COOH + TR\text{-}NH_2 \longrightarrow \phi \,\overline{\wedge}\, CO\text{-}NH\text{-}TR$$

$$\phi \,\overline{\wedge}\, SH + TR\text{-}COOH \longrightarrow \phi \,\overline{\wedge}\, S\text{-}CO\text{-}TR$$

where ϕ = silanized solid support and TR = Terminal Reagent.

TABLE 4. Commercially available silane coupling agents.

Γ-aminopropyltriethoxysilane	$H_2N\text{-}(CH_2)_3\text{-}Si(OC_2H_5)_3$
N-β-(aminoethyl)-Γ-amino propyltrimethoxysilane	$H_2N\text{-}(CH_2)_2\text{-}NH\text{-}(CH_2)_3\text{-}$ $\text{-}Si(OCH_3)_3$
Γ-chloropropyltriethoxysilane	$Cl\text{-}(CH_2)_3\text{-}Si(OC_2H_5)_3$
β-trichlorosilyl-2-ethyl pyridine	$C_5H_4N\text{-}CH_2\text{-}CH_2\text{-}Si\text{-}Cl_3$
Γ-cyanopropyltrimethoxysilane	$NC\text{-}(CH_2)_3\text{-}Si(OC_2H_5)_3$
Γ-mercaptopropyltrimethoxysilane	$HS\text{-}(CH_2)_3\text{-}Si(OC_2H_5)_3$

In-Situ Polymerization

Another approach to electrode surface modification is to deposit a thin film of polymeric material comprising the electroactive molecule. The polymer is either modified with the appropriate terminal reagent or is itself electroactive/chemically reactive. Typical examples of highly conductive and redox active films consist of the organometallic polymers, poly(phthalocyanides), and poly(ferrocenes).

Electrodes prepared in this way have the functional reagent attached at the electrode-solution interface and, thus, the electrode serves a dual function of support and electron transfer agent. Surfaces modified with porphyrins, ferrocenes, etc., are more easily prepared than those with the more complex molecules, i.e., enzymes, co-factors, and immunoglobulins. Electrode surfaces derivatized by direct electrolytic reduction of terminal reagents such as viologen to form polymer films on the surface will retain both chemical reactivity as well as electroactivity. For example, heptyl viologen has been polymerized on Pt and SnO_2 [19], and methyl viologen on gold [20, 21] by anodic deposition.

The polyviologens retain the chemical reactivity so that they may mediate a chemical reaction such as methemoglobin or metmyoglobin reduction as follows:

$$Fe^{3+}(heme) + MVBr \longrightarrow Fe^{2+}(heme) + MV^{2+} + Br^{-} \quad (1)$$

$$MV^{2+} + Br^{-} + e^{-} \longrightarrow MVBr \quad (2)$$

The polyviologen acts as a mediator for the electron transfer process by presenting a lower activation energy for the transfer of electrons to hemoproteins, which would otherwise be unreactive, at the electrode surface. Hence, reaction (1) is a reduction/oxidation reaction between the viologen and the hemoprotein, and reaction (2) transfers the electrons between the metal-conductor and the electroactive molecule (viologen). In the case of heptyl viologen, the polymer also displayed piezoelectric properties. For selective electrode design, the electrode surface molecules should meet the following criteria:

(a) maintain electroactivity and reversibility; i.e., the molecule can be alternately oxidized and/or reduced without altering its reactivity;

(b) react selectively with the analyte sought.

Another promising procedure is that of coating of electrode surfaces with a polymer thin-film from monomers that have been derivatized with the terminal reactant. That is, the monomer is derivatized prior to casting the polymer film. For example, vinyl chloride can be modified with ferrocene, to form poly(vinyl-ferrocene). This material acts as a semiconductor and is susceptible to redox reactions, as expected from the ferrocene moiety [22]. Polystyrene may also be chemically modified to produce electroactive thin-film polymer-coated electrodes. Electropolymerization, as observed for polyviologens, has been demonstrated for a variety of vinyl polymerizations for olefin monomers and also ring-opening reactions of cyclic monomers [23].

Whereas polystyrene, nylon and poly(n-vinylcarbazole) are classified as non-conducting (or insulators) with specific conductivities of $10^{-14}\ \Omega^{-1}\ cm^{-1}$, their conductivities may be significantly increased by grafting or doping with redox active molecules [22]. Polymers may be rendered conductive by:

(a) increasing hyperconjugation of the polymer chain;

(b) branching with redox active molecules;

(c) doping with acceptor molecules or copolymers (charge transfer complexes and organometallic complexes).

Conductivities as high as $10^{3}\ \Omega^{-1}\ cm^{-1}$ are achievable, but more importantly, these polymer films are redox active and change their oxidation states and, thus, surface potentials.

Alternatively, substantial currents are produced for the measurement of Faradaic processes.

The above polymers, therefore, have all the attributes necessary for the fabrication of sensors. Polymers can be made into conductors, semiconductors, insulators, chemically reactive or inert, depending on the application at hand. To our knowledge, such redox-active/electroactive polymers have not yet been applied as sensors.

Electrode Characterization

Various surface analysis techniques have been used to characterize modified electrodes. Efforts should be made to characterize the CME in conditions similar to those it will be subjected to as a sensor. For example, if the CME is designed for solution analysis (the most common application), the electrochemical characterization methods, which can be performed in solution, are most relevant.

Electrochemical techniques are particularly amenable to solution analysis; however, the information obtained is usually limited and this necessitates the use of other analytical methods to gather more detailed, quantitative data. Both the physical properties and the chemical composition of polymers are important.

Cyclic and differential pulse voltammetry are the most commonly used electrochemical methods for characterizing electrode surfaces [24]. Cyclic voltammetry requires relatively simple and inexpensive instrumentation and the method is rapid.

Chronoamperometry [25] and rotating disk electrode [26] studies are primarily used to obtain information on the electrode process of interest, in particular the effective diffusion coefficient which is important in optimizing sensor performance. The effective diffusion coefficient accounts for the diffusion of charge which may occur in the modifier. This information is important as it is necessary to maximize the diffusion coefficient for the analyte while minimizing those for potential interferents.

As well as the above techniques, dry electrical conductivity measurements using the four-point probe technique may be useful. However, it is important to remember that this is likely to be much different from solution conductivity.

Apart from the electrochemical methods, several analytical techniques can be used to obtain useful information about new CMEs. These include x-ray photoelectron spectroscopy (XPS), Auger electron spectroscopy (AES), electron microprobe analysis (EPMA), fast atom bombardment mass spectrometry (FABMS), infrared spectroscopy (FTIR), etc.

Sensor Performance

To be employed as a chemical sensor, the CME must be capable of producing an electrical signal which is related to the concentration of the analyte. The signal may be due to a change in resistance of the CME, a change in the potential difference across the CME (potentiometric methods), or to generation of a current arising from oxidation or reduction of the analyte (voltammetric amperometric methods). Voltammetric methods give rise to the most sensitive techniques with the most rapid response times and, consequently, they have been used in the majority of applications to date.

Voltammetric signal generation at a conventional clean metal electrode surface depends on

(i) Transport of the species to the electron transfer boundary.

(ii) The rate of electron transfer at that boundary.

In general, optimum responses are obtained for species which are at a high concentration at the boundary and where electron transfer is rapid (reversible systems). This will result in well-defined responses with intensity proportional to analyte concentration. With CMEs similar processes occur, in that the analyte must be transported to the electron transfer boundary before the redox process occurs. However, the electron transfer boundary is not so clearly defined on CMEs as it on conventional metal electrodes. If so then an extra degree of complexity is added to the interpretation of the CME sensor response in that non-linear calibration curves as well as non-linear standard addition curves will be obtained. Furthermore, the electron transfer process is also dependent on the chemical environment. For example, the nature and concentration of the counter-ion in conductive polymers have a significant influence. In such cases, matrix matching between standards and samples must be carried out.

Chemical reactions, such as precipitation in the material subsequent to the electron transfer process as well as during preconcentration are also important. This re-emphasizes the fact that attention must be paid to the proper choice of a supporting electrolyte (counter-ion). The electroactive zone must be taken into account since it is necessary to electrolyse 100% of the trapped analyte for increased sensitivity, wich should be possible in mesoporous materials.

Application of CMEs

Examples of CMEs, which make use of the occluding power of modified electrodes, may be found in references [27-29]. Cellulose acetate is the most common coating used to date for this application and it has been used to occlude species from the substrate on the basis of molecular size. It has proved useful, for example, in occluding surfactants from a mercury film electrode [27] or proteins from a platinum surface [29].

Electrocatalytic CMEs are designed to increase the sensitivity for a particular element, either by mediating in the electron transfer process, by providing reactants or regenerating the analyte or by shifting oxidation reduction potentials to less extreme values. Examples can be found in references [30-32] as are examples of applications involving preconcentration [33-35]. Modified electrodes capable of preconcentrating the analyte onto the electrode surface have been employed to enhance the analytical sensitivity and selectivity. The most common applications appear to be in the area of metal analysis.

Complexation prior to the electron transfer step is particularly attractive, since species may be trapped in a chemical environment which is conductive to rapid electron transfer and facilitates more unusual electron transfer mechanisms (e.g. stabilization of unusual oxidation states). Conducting polymers have been employed for determination of electro-inactive species on the basis of counter-ion replacement [36].

Conclusion

The number of analytical applications of CMEs is limited at present. This is due to the difficulty of achieving proper design, understanding the mechanism of electron transfer, obtaining new clues from the methods of electrode characterization, and developing new materials for use as CMEs. The mesoporous inorganic solids exhibit a remarkably broad spectrum of structural, chemical, electronic, ionic, optical and magnetic properties. They can be used as absorbers, as charge carriers, as intercalation materials, as electrocatalysts and, given their remarkable properties and multifunctionality, it is envisaged that they will expand the number of applications of CMEs in the near future.

References

1. Bard, A.J., (1983), J. Chem. Educ., 60, 302-304.
2. Murray, R.W., (1980), Acc. Chem. Res., 13, 135-141.
3. Faulkner, L.R., (1984), Chem. Eng. News, Feb. 28-45.
4. Halbert, M.K. and Baldwin, R.P., (1985), Anal. Chem. 57, 591-595.
5. Baldwin, R.P., Christensen, J.K. and Kryger, L., (1986), Anal. Chem., 58, 1790-1798.
6. Geno, P.W., Ravichandran, K. and Baldwin, R.P., (1985), J. Electroanal. Chem., 183, 155-166.
7. Koval, C.A. and Anson, F.C., (1978), Anal. Chem., 50, 223-229.
8. Miller, L.L. and van de Mark, M.R., (1978), J. Amer. Chem. Soc., 100(2), 639-640.

9. Oyama, N. and Anson, F.C., (1979), J. Amer. Soc., 101(3), 739-741.
10. Gomathi, H. and Prabhakara Rao, G., (1985), J. Electroanal. Chem., 190, 85-94.
11. Taylor, L.R. and Johnson, D.C., (1974), Anal. Chem., (1974), 46, 262-266.
12. Larochelle, J.H. and Johnson, D.C., (1978), Anal. Chem., 50, 240-243.
13. Nowak, R., Schultz, F.A., Umana, M., Abruna, H. and Murray, R.W., (1978), J. Electroanal. Chem., 94, 219-225
14. Abruna, H.D., Meyer, T.J. and Murray, R.W., (1979), Inorg. Chem., 18, 3233-3240.
15. Moses, P.R. and Murray, R.W., (1977), J. Electroanal. Chem., 77, 393-399.
16. Elliott, L.M. and Murray, R.W., (1976), Anal. Chem., 48, 1247-1254.
17. Moses, P.R., Wier, L. and Murray, R.W., (1975), Anal. Chem., 47, 1882.
18. Jannakoudakis, N., Missaelidis, N. and Theodoridou, E. (1985), Synthet. Met., 11, 101-108.
19. Aldrich Chemical Co., 1979-80 Catalog.
20. Aldrich Chemical Co., 1979-80 Catalog.
21. Landrum, H.L., Salmon, R.T. and Hawkridge, (1977), J. Am. Chem. Soc., 99, 3154.
22. Pitman, C.V. and Surynarayanan, B., (1974), J. Am. Chem. Soc., 96, 7916.
23. Garg, B.K., Raff, R.A.V. and Subramanian, R.V., (1978), J. Appl. Polym. Sci., 22, 65.
24. Brown, A.P. and Anson, F.C., (1977), Anal. Chem., 49, 1589-1595.
25. Ikeda, T., Schmehl, R., Denisevich, P., Willman, K. and Murray, R.W., (1982), J. Amer. Chem. Soc., 104, 2683-2691.
26. Shigehara, K., Oyama, N. and Anson, F.C., (1981), J. Amer. Chem. Soc., 103, 2552-2558.
27. Wang, J. and Hutchins-Kumar, L.D., (1986), Anal. Chem. 58, 402-407.
28. Robinson, K.A., Gilbert, T.W. and Mark, H.B., (1980), Anal. Chem., 52, 1549-1551.
29. Gittampalam, G. and Wilson, G.S., (1983), Anal. Chem., 55, 1608-1610.
30. Cox, J.A. and Kulesza, P.J., (1984), J. Electroanal. Chem., 175, 105-118.
31. Johnson, D.C. and Resnick, E.W., (1977), Anal. Chem., 49, 1918-1924.
32. Austin, D.S., Polta, J.A., Polta, T.Z., Tang, A.D., Cabelka, T.D. and Johnson, D.C., (1984), J. Electroanal. Chem., 168, 227-248.
33. Wier, L.M., Guadalupa, A.R. and Abruna, H.D., (1985), Anal. Chem., 57, 2009-2011.
34. Guadalupa, A.R. and Abruna, H.D., (1985), Anal. Chem., 57, 142-149.
35. Baldwin, R.P., Christensen, J.K. and Kryger, L., (1986), Anal. Chem., 58, 1790-1798.
36. Honyama, Y. and Heineman, W.R., (1986), Anal. Chem., 58, 1803-1806.

SOL-GEL PROCESSED COATINGS

L. P. S. ARAÚJO, A. C. P. R. P. CARRASCO and C.A.C. SEQUEIRA
Instituto Tecnológico para a Europa Cumunitária
Av. Rovisco Pais
1096 Lisboa Codex
Portugal

ABSTRACT. The fundamental principles that govern the production of sol-gel coatings are briefly considered. Among sol-gel processed materials, some examples are given of thin films for abrasion-resistant, corrosion-resistant, optical, electrical and other commercial applications.

1. Introduction

Coatings are widely used in industry for many different purposes. They have permeated practically all areas of manufacturing, and it would be difficult to find a finished product that has escaped the application of a coating. Coatings are applied to protect bulk materials from corrosion and other detrimental effects of the ambient atmosphere. They are used to change surface properties and colour, gloss, and general appearance. Adhesive coatings are used in laminating and in the preparation of composites. Coatings are used as barriers for gases and liquids, and for many other reasons. Application techniques, coatings types, and their purposes make coating technology an extremely diverse field.

Although the literature on coatings can be traced back to the era of electrodeposition in general, and aluminising of iron in the early forties, much of the information on coatings relevant to achievement of optimum performance - production, control and replacement economics - has accumulated significantly in the seventies and to date.

The sol-gel process has been suggested as an alternative to conventional methods such as sputtering, chemical vapour deposition (CVD), and plasma spray for applying thin ceramic coatings. Sol-gel thin films are in a few notable cases technically sound alternatives to these methods as well as commercially viable alternatives [1-4].

The technology of sol-gel thin films has been around for more than 20 years. The process is surprisingly simple. A solution containing the desired alkoxide is prepared with a solvent and water. It is applied to a substrate by spinning, dipping, or draining. A tacky film forms that is typically 1 μm thick, uniform over large areas and adherent to the substrate. The sol-gel process is able to apply a coating simultaneously to the inside and outside of complex shapes. The film dries quickly to a microporous and mesoporous coating of oxide [5-9].

This paper describes the sol-gel coating technique and gives examples of sol-gel processed materials with prospects of successful commercial exploitation in many

C.A.C. Sequeira and M.J. Hudson (eds.),Multifunctional Mesoporous Inorganic Solids, 485–493.

technologies.

2. The Sol-Gel Method

There are four stages in the sol-gel coating process:
(i) preparation of the sol, (ii) spraying of the sol on the substrate, (iii) drying the sol to a gel, and finally, (iv) firing to obtain a consolidated and adherent coating [10-15]. Colloidal dispersions are not new. Their preparation is well characterized but not in the context of high temperature coatings. There can be aqua-sols or organo-sols, and these can be aggregated or non-aggregated in nature. Essentially, they are stable dispersions of particles in a fluid of colloidal units of hydrous oxides or hydroxides with particle sizes ranging from 20 Angstroms to 1 micron. Non-aggregated colloidal units are less than 100 Angstroms and such oxide-sols can form deposits of almost theoretical density even with low temperature firing (less than 850 °C). The same oxide in aggregated colloidal units of 200-300 Angstroms can form a porous layer of a high effective surface area. The degree of aggregation is controlled by the properties of the sol.

Sol-preparation can take two routes. Route 1 is hydrolysis followed by polymerization. Route 2 adopts precipitation and then peptization. Although only oxides have been considered for coating so far, it is possible that other ceramics may also have been attempted, e.g. Si_3N_4, AlN etc. It seems possible that cermets can be tried if they can be prepared as very fine aggregates. Oxide-sols of ceria, silica, zirconia, titania, thoria and alumina and gel-glass have been reported by the Harwell group. Table 1 shows the formation of hydrous alumina sols. It is probable that many of these are mesoporous materials. The extent of aggregation of the sol and its stability depends on factors such as the nature of the precursor material and the ratio of the anion to metal oxide. Absorption of H^+ on the metal oxide hydrates causes positive charging of the colloidal units. Anions in the solution associate with these and thus establish dipoles in the sol thus repelling each colloidal unit from its neighbours. The degree of aggregation depends on this inter- colloidal unit separation, and the anion to metal oxide ratio is a critical factor. A mole ratio of 0.32 for HNO_3/CeO_2 results in a gel density of 2.4 while a ratio of 1 yields a gel with 4.0 g cm^{-3} density.

Sols can be applied by immersion of the substrate, or by spinning, electrophoresis or spraying. Any drying or thickening of the sol during spraying has to be compensated for, so that the aggregation optimized for the purpose remains so. The method of application of the sol depends on the shape of the substrate which must be thoroughly cleaned and abraded at the outset.

The sol-to-gel transformation is done by careful, gradual drying; an over-rapid drying can cause gel-crazing. At this stage the process is reversible. The gel can be "refloated" to become the sol by merely adding the vehicle, e.g. water for aquo-sols.

Consolidation to a coating occurs when the gel-coated substrate is fired to high temperatures (440-1200°C) where sintering can occur, but the heating times are short, usually 15 minutes. A single sol-application gives a 0.1-2 micron thick coating which can be modified by controlling the rheology of the sol or by additives.

Unlike in slurry coats, a sol-gel material on a substrate may not have a distinct diffusion zone at the substrate-coat interface. But the coating oxide and the growing oxide do get interdispersed. Also, there may be mixed phases of the solid oxide. For instance, sol-gel silica

TABLE 1. Formation of hydrous alumina sols

Route 1: Hydrolysis

$$Al^{3+} + 3NO_3^- + H_2O = Al(OH)^{2+} + 2NO_3^- + HNO_3$$

$$Al(OH)^{2+} + 2NO_3^- + H_2O = Al(OH)_2^+ + NO_3^- + HNO_3$$

Polymerization

$$13Al(OH)^{2+} + 26NO_3^- + 27H_2O=$$
$$= [Al_{13}O_4(OH)_{24}(H_2O)_{12}]^{7+} + 7NO_3^- + 19HNO_3$$

Route 2: Precipitation

$$Al(NO_3)_3 + 3NH_4OH = Al(OH)_3 + 3NH_4NO_3$$

Peptization

$$nAl(OH)_3 + xHNO_3 = [Al(OH)_3]_n^+ + xNO_3^-$$

results in a mixture of beta-cristobalite with some alpha-quartz. TiO_2 forms coatings of density 4.07 and 4.11 g cm^{-3} when dried at 800 and 1000°C respectively, while the theoretical densities of anatase and rutile are 3.85 and 4.26 g cm^{-3}. CeO_2, on the other hand forms a single phase structured coating. Table 2 lists the densification effects of four sol-gel coating products.

Except for their oxidation resistance data sol-gel coatings have not been characterized in the published literature. No information can be given here on micrographs of the cross section, adhesion and cyclic thermal effects, and stress studies on the substrate-sol-gel-coated system.

3. Sol-Gel Coatings

The range of substrates that can be coated by the sol-gel process is not limited to those capable of withstanding a high firing temperature because of the associated low temperature for processing the gel. Thus, it is possible to put an inorganic sol-gel coating on a plastic. The majority of coatings have been applied by this technique in order to achieve some specific optical quality. Much of the progress leading to commercial coatings has been pioneered by Schott in Germany. Their process for coating architectural glass is simple. Large panes (4x3m) are chemically cleaned then dipped into a bath containing a hydrolysable metal compound. The pane is slowly withdrawn into a moist atmosphere which hydrolyses the film forming a transparent metal oxide layer on the surface.

TABLE 2. Densification effects of four sol-gel coating materials.

Calcination Temp., °C	Density, ($g\ cm^{-3}$) Silica Non-agg	Silica Agg	Titania Non-Agg	Titania Agg	Ceria Non-Agg	Zirconia Non-Agg
Dry Gel	1.8	0.5	2.6	1.5	4.3	3.6
400	1.5		3.2	1.5	5.2	3.9
600	1.5	0.4	3.4	1.6	6.0	4.8
800	2.1	0.5	4.1	2.6	6.9	5.2
1000	2.0	0.5	4.1	3.7	6.9	5.2
1200	2.0	1.8				
Theoretical density $g\ cm^{-3}$	2.2		4.3*		7.1	5.6
Melting pt, °C	1600		1840		2600	2715

(* Rutile; Anatase 3.9 $g\ cm^{-3}$)

Agg - Aggregate; Non-Agg - Non Aggregate

A simple relationship shows that the thickness of the film is proportional to the drawing speed to a power of 2/3.

$$d = k\ V^{2/3}$$

d = thickness
V = drawing speed
k = constant

The process is such that the film flows down the plate after wetting the surface. Because it wets the surface, it partly adheres and when the film reacts with the moisture in the atmosphere it undergoes hydrolysis and condensation. As mentioned previously, this film can be densified by a thermal treatment.

One major problem that arises here is anisotropic shrinkage. If the gel film attempts to shrink, it is prevented from doing so in the lateral direction by adhesion to the substrate.

One distinct advantage of this technique over others is the homogeneity and uniformity of films that can be obtained. Schott produce window panes known as Calorex and these show an excellent uniformity of coating. The technique is versatile and allows multilayer coatings to be produced which enhances the range of optical properties that can be produced. Glasses produced from alkoxides can also be doped with transition metals such as Co, Ni, Cr and Cu. Coatings have been developed for:

- anti-reflective purposes such as coating the surface of silicon solar cells
- partially reflective coatings
- filters for ultraviolet and colour cut-off
- oxidation resistance on metals such as iron and brass

- protection of moisture sensitive optical glasses
- strengthening of normal glass items by application of an appropriate sol-gel coating.

The important parameters in film formation for optical purposes are the refractive index of the film and its adsorption characteristics. Some examples of metal oxide coatings that absorb in the U.V. and visible are listed in Table 3.

There are several inherent advantages and disadvantages of using a solution method for coating. Firstly, it is possible to coat all over, inside and outside the component. The solution can coat everywhere uniformly which is a distinct advantage over vacuum coating but inside and outside coating may not be desirable. The capital equipment needed is not very sophisticated although the withdrawal from the solution needs to be as smooth and vibration free as possible. But nowadays this is easily achieved using high resolution stepping motors to drive frictionless lead screws and mounting the dip coater on an antivibration table. Commercial products that are made by the dip coat process are:.

- Sun Shielding Glass (IROX) - The interference layers allow some parts of the incident

TABLE 3. Metal oxide coatings that absorb in the UV and visible regions.

Absorbing in UV range		
Coating / nm	Refractive Index	Upper Absorption Limit
Al_2O_3	1.62	250
CeO_2	2.11	400
In_2O_3	1.95	420
Sb_2O_4	1.90	340
TiO_2	2.30	380

Absorbing in Visible		
Coating / nm	Refractive Index	Transmitted Light
CoO	2.00	Brown
Cr_2O_3	2.38	Yellow-orange
Fe_2O_3		Yellow-red
U_2O	1.95	Yellow
V_2O_5	2.0	Greenish-yellow

light spectrum to be transmitted whilst others are reflected. The coating is a colloidal titania in which certain colloidal metal particles are incorporated. This gives a scratch resistant coating which does not change colour in terms of transmitted and reflected light.

- Heat Mirrors - High intensity lamps such as spot lights and projector lamps use halogen lamps which not only generate a light of high intensity but also a great deal of I.R. (infra-red) radiation and it is desirable to reduce the heat produced. Filters can be made using sol-gel coatings which will transmit most of the visible light but reflect away the I.R.

- Cold Mirrors - This is the opposite to the above in that it transmits the I.R. region and reflects the visible. It too can be used to remove heat.

- Anti-Reflective Coatings - As the name implies, these are coatings that reduce the amount of light reflected from a surface. The panel glasses on instruments in air-craft often reflect shafts of sunlight and this is where such coatings are helpful. This is also helpful in automotive rear view mirrors.

Anti-reflective coatings are of three types:

1) Single layer coatings where $n_c=n_g$; n_c=refractive index of coating and n_g=refractive index of glass.
2) Multi-layer made from n_c1 and n_c2 where the two indices are different
3) Graded index where there is progressive change of index from the glass substrate to the external surface, usually air. This is achieved using a phase separating glass on the coating then etching one phase out to give a material the porosity of which varies with depth.

Anti-reflective coatings on silicon have been of considerable interest in solar cell technology. One great advantage is that the coating can be densified at 400°C, a temperature low enough not to cause excessive damage to the solar cell.

Further work has been done in spraying a mixture of colloidal silica and a partially hydrolysed silicon alkoxide onto glass surfaces. The alkoxide hydrolyses and condenses and helps form a bond between the colloidal silica particles as shown below:

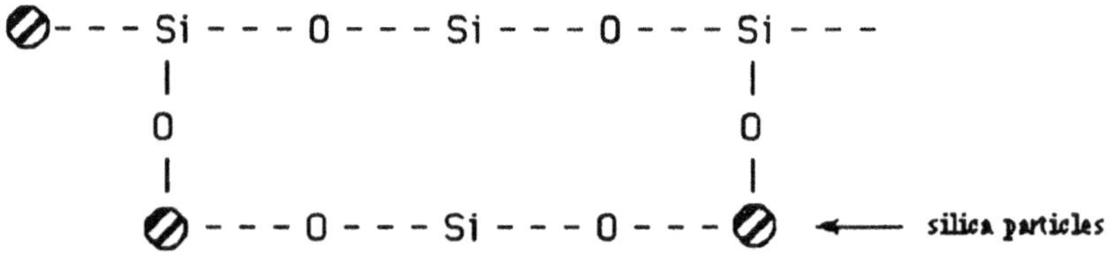

Colour conversion filters or interference layer coatings are another way of achieving colour filters on lamps. These coatings are applied to glasses with low thermal expansion coefficient which gives the filter thermal shock resistance.

With the advent of higher power lasers being used in many technologies, both lenses and mirrors require special characteristics if they are not to absorb light energy in such quantity as to cause irreversible damage. This requires materials of very high purity.

Many other devices can be produced for high energy laser optics including heat reflecting filters, dichroic beam splitters and polarizers. These coatings have been produced from colloidal titania and titania/silica films from colloidal gels.

Absorptive coatings are coatings that are absorptive but show no interference effects. This is achieved by incorporating finely divided platinum or palladium in a silica or alumina matrix. They give "grey" absorbtive coatings i.e. absorption is not really wavelength dependent. Less expensive alternatives have been provided using nickel oxide. They have found use in applications where glare is a problem.

Other coating methods can be employed for gels. Spin coating is a technology that has been well developed by the micro-electronics industry for the application of thin films of photoresist onto silicon wafers. This method can be applied here and consists of dropping a quantity of sol onto the substrate which is spinning at high speed, e.g. 3000 r.p.m. This gives uniform films and is ideal for a sol which hydrolyses very rapidly since it can be kept away from moisture prior to being dropped onto the surface. As it hits the surface and spreads, it can then react very quickly with water in the atmosphere and gel. Further advantages are that it is economical in usage of solution and only one surface need be coated. Disadvantages are edge effects on non axi- symmetric substrates and the mechanical problems associated with spinning large substrates. The thickness control is a little easier than dip coating and is governed by:

- spin speed
- viscosity of the solution

A final method is spray coating. In the past, the literature has criticized this technique as the least effective of the three but more recently advances in spraying and atomizing equipment have brought this technique into line with others. In particular, it is possible to spray onto a hot surface, e.g. glass below tg and obtain a very rapid reaction between the sol and the glass which gives coatings that develop excellent adherence and form in a very short time. The improvements to this technique have been brought about by modern techniques that enable very finely atomized jets to be produced by the use of ultrasonics or high pressure hydraulic atomisers. Very recently develoments have taken place in producing "molecular sprays".

Spraying has potential as a hot coating technique for applying coatings to glass surfaces in order to strengthen the object. Work by Schmidt at the Fraunhofer Institute and Uhlmann at M.I.T. has shown that conventional sol-gel coatings have a strengthening effect on glass. The mechanisms are far from understood but it is suggested that the coating penetrates existing flaws and that the gel forms bridges across the faces of the cracks. Uhlmann [16] has also reported nitriding the gel glass coating to increase its inherent strength. Ceramic Developments (Midlands) Ltd. has shown a strength increase on glass rods and containers coated by a hot process. A partially hydrolysed alkoxide is sprayed onto the hot glass surface between 400 and 600°C and this instantly hydrolyses and forms a coating which gives a significant strength increase. Further work is being carried out in this area. Apart from optical properties and strength improvement, it is also possible to apply electrically conducting layers to glass by sol-gel. Gonzalez-Oliver and Kato [17] have reported work on deposition of antimony-doped tin oxide on various glass panels.

Electroconductive layers are finding applications in display panel technology, and it is suggested that they will find other applications such as in transparent ovens or could be used to heat up the glass to prevent ice or moisture build up.

The techniques of dip coating and spray coating were contrated and in particular it was discovered that the spray coatings were more crystalline than dip coated films and that the spray coating showed preferred orientation.

There are many reported sol-gel coatings and some of them are listed in Table 4.

TABLE 4. Examples of sol-gel coatings

Single Oxide
SiO_2 TiO_2 V_2O_5 ZrO_2 Al_2O_3 In_2O_3
Various transition metal oxides
RuO_2 RhO_2 ThO_2
Binary Oxides
SiO_2-GeO_2 SiO_2-TiO_2 LiNb-O_3 SiO_2-Al_2O_3
Cd-SnO_4 SiO_2-ZrO_2 SnO_2-Sb_2O_3
Ba-TiO_3 CeO_2-TiO_2 Fe_2O_3-SiO_2
Multicomponent coatings
Pt-SiO_2 Pd-SiO_2 Pt-Al_2O_3 Pt-Al_2O_3
SiO_2-TiO_2-Al_2O_3 SiO_2-ZrO_2-TiO_2
SiO_2-TiO_2-(ZrO_2)-Al_2O_3
SiO_2-B_2O_3-Al_2O_3-Al_2O_3-Na_2O-K_2O
Fe_2O_3-P_2O_5-CaO Fe_2O_3-P_2O_5-SiO_2

4. Conclusions

Acceptance of the sol-gel process for thin mesoporous films is mostly a question of educating the user of thin films and coatings on how to apply these films. The number of commercial products available to the general public in this area is small. There are the spin-on coatings for doping silicon wafers (Accuspin from Allied-Signal Electronic Chemicals and Liquicoat from - Chemicals). There are the liquid oxides for various applications (Atolon from Nippon Soda). Many industries may have internal uses for what can be classified as sol-gel coatings or sputtered coatings.

To summarize sol-gel coatings, there are limitations. The main limitation is that only films around 1 μm thick can be applied with ease by a sol-gel process. Thinner films can be applied, but thicker films suffer from the same problems of drying, shrinkage, and cracking that plague bulk samples. However, for thin films there are the applications cited above and probably others that have been overlooked.

An important feature to emphasize is the mesoporous nature of the films in the early stages. It appears that mesoporosity has not been exploited for applications in which a high surface area coating is desired. Some applications to look for in sol-gel coatings are catalytic surfaces, membrane properties, and self-lubricating surfaces. In this context, it is our opinion that many opportunities exist for mesoporous inorganic structures used as advanced sol-gel coatings.

References

1. Gruninger, M., Wachtman, J.B., and Haber, R., (1986), MRS Symp. Proc., 54, 823.
2. Klein, L.C., (1984), Ceram. Eng. Sci. Proc., 5, 379.
3. Klein, L.C., (1985), Annu. Rev. Mater. Sci., 15, 227.
4. Klein, L.C., (1988), "Sol-Gel Technology for Thin Films, Fibers, Preforms, Electronics and Specialty Shapes", Park Ridge, NJ:Noyes Publications, 407pp.
5. Brinker, C.J., and Harrington, M.S., (1981), Solar Energy Mater., 5, 159.
6. Dislich, H., and Hinz, P., (1982), J. Non-Cryst. Solids, 48, 11.
7. Livage, J., and Lemerie, J., (1982), Annu. Rev. Mater. Sci., 12, 103.
8. Mukherjee, S.P., and Lowdermilk, W.H., (1982), J. Non-Cryst. Solids, 48, 177
9. Pantano, C., and P. Glaser, (1984), J. Non-Cryst. Solids, 63, 201.
10. Anon, (1980), Metallurgia, 47, 459.
11. Steffens, H.D., (1980), AGARD-LS-106 NATO, paper 5.
12. Nelson, R.L., et al., (1981), Thin Solid Films, 81, 329.
13. Spahn, H., (1981), Proc. 7th Intl. Conf. on Surface Technology, VDI Verlag, Berlin, p.427.
14. Bennett, M.J., (1983), High Temperature Corrosion, NACE, 6, 145.
15. Bennett, M.J., (1984), J. Vac. Sci. & Technol., B12, 800.
16. Fabes, B., Doyle, W., Silverman, L., Zelenski, B., and Uhlmann, D.R., Stronger Glass Via Sol-Gel Coating,
17. Gonzalez-Oliver, C.J.R., and Kato, I., (1986), J. Non-Cryst. Solids, 82, 165.

SUBJECT INDEX

C.A.C. Sequeira and M.J. Hudson (eds.), Multifunctional Mesoporous Inorganic Solids, 495–501.

GPSR Compliance

The European Union's (EU) General Product Safety Regulation (GPSR) is a set of rules that requires consumer products to be safe and our obligations to ensure this.

If you have any concerns about our products, you can contact us on ProductSafety@springernature.com

In case Publisher is established outside the EU, the EU authorized representative is:

Springer Nature Customer Service Center GmbH
Europaplatz 3
69115 Heidelberg, Germany

Batch number: 10809436

Printed by Printforce, the Netherlands